Peterson's

MASTER THE AP BIOLOGY EXAM

Glenn Croston

PETERSON'S
A nelnet COMPANY

PETERSON'S

A nelnet COMPANY

About Peterson's, a Nelnet company

Peterson's (www.petersons.com) is a leading provider of education information and advice, with books and online resources focusing on education search, test preparation, and financial aid. Its Web site offers searchable databases and interactive tools for contacting educational institutions, online practice tests and instruction, and planning tools for securing financial aid. Peterson's serves 110 million education consumers annually.

For more information, contact Peterson's, 2000 Lenox Drive, Lawrenceville, NJ 08648;
800-338-3282; or find us on the World Wide Web at: www.petersons.com/about.

Contributing editor: Doug Tarnopol

ISBN-13: 978-0-7689-2523-4
ISBN-10: 0-7689-2523-1

Printed in the United States of America

10 9 8 7 6 5 4 3 2 1 10 09 08

Petersons.com/publishing

Check out our Web site at www.petersons.com/publishing to see if there is any new information regarding the tests and any revisions or corrections to the content of this book. We've made sure the information in this book is accurate and up-to-date; however, the test format or content may have changed since the time of publication.

OTHER RECOMMENDED TITLES

Peterson's AP European History
Peterson's AP World History
Peterson's Master AP Calculus AB & BC
Peterson's Master AP Chemistry
Peterson's Master AP English Language & Composition
Peterson's Master AP English Literature & Composition
Peterson's Master AP U.S. Government & Politics
Peterson's Master AP U.S. History

Contents

Introduction

10 FACTS ABOUT THE AP BIOLOGY EXAM

1. The Advanced Placement Program Offers Students an Opportunity to Receive College Credit for Courses They Take in High School.

   The AP program is a collaborative effort of secondary schools, colleges and universities, and the College Board, through which students who are enrolled in AP or honors courses in any one of thirty-seven courses and exams across twenty-two subject areas may receive credit or advanced placement for college-level work completed in high school. While the College Board makes recommendations about course content, it does not prescribe content. As a result, the annual testing program ensures a degree of comparability among courses in the same subject.

2. More than 2,900 Colleges and Universities Participate in the AP Program.

   Neither the College Board nor your high school awards AP credit. You need to find out from the colleges to which you are planning to apply whether they grant credit and/or use AP scores for placement. It is IMPORTANT that you obtain each school's policy IN WRITING so that when you actually choose one college and register, you will have proof of what you were told.

3. The AP Biology Exam Measures Factual Knowledge and Analytical Skills.

   According to the College Board, the exam provides you with "the conceptual framework, factual knowledge, and analytical skills necessary to deal critically with the rapidly changing science of biology."

4. The AP Biology Exam Has Two Sections: Multiple Choice and Essays.

   Section I: The multiple-choice section of the exam consists of 100 questions that test your understanding of representative content and concepts drawn from across the entire AP biology course, to be answered in an 80-minute time period. This section counts for 60 percent of your total score.

Section II: For the free-response essays, you must analyze and interpret data or information drawn from laboratory experience and lecture material. These questions may require you to integrate material from different areas of the course. You will have four mandatory free-response essays to write; all four are weighted equally. This section counts for 40 percent of your exam grade.

5. The AP Biology Exam Covers Specific Areas of Knowledge.

   The AP Biology exam covers three areas of knowledge, with these percentage goals specified: Molecules and Cells, 25%; Heredity and Evolution, 25%, and Organisms and Populations, 50%. There are eight major themes that recur throughout AP Biology courses: Science as a Process; Evolution; Energy Transfer; Continuity and Change; Relationship of Structure to Function; Regulation; Interdependence in Nature; and Science, Technology, and Society.

6. There Is No Required Length for Your Free-Response Essays.

   It is the quality, not the quantity, that counts. Realistically, a one-paragraph essay is not going to garner you a high mark because you cannot develop a well-reasoned analysis and present it effectively in a single paragraph. An essay of five paragraphs is a good goal. By following this model, you can set out your ideas with an interesting introduction, develop a reasoned middle, and provide a solid ending.

7. You Will Get a Composite Score for Your Exam.

   The College Board reports a single score from 1–5 for the two-part exam, with 5 being the highest. By understanding how you can balance the number of questions you need to answer correctly against the essay score you need to receive to get at least a "3," you can relieve some of your anxiety about passing the exam.

8. Educated Guessing Can Help.

   No points are deducted for questions that you do not answer on the multiple-choice section, and don't expect to have time to answer them all. One-quarter of a point is deducted for wrong answers. The College Board suggests guessing IF you know something about a question and can eliminate a couple of answer choices. Let's call it "educated guessing."

9. The Exam Is Given in Mid-May.

   Most likely, the exam will be given at your school, so you do not have to worry about finding a strange building in a strange city or town. You will be in familiar surroundings, which should reduce your anxiety a bit. If the exam is given somewhere else, be sure to take identification with you.

10. Studying for the Exam Can Make a Difference.

    The first step is to familiarize yourself with the format and directions for each part of the exam. Then you will not waste time on the day of the exam trying to understand what you are supposed to do. The second step is to put those analytical skills you have been learning to work, dissecting and understanding the kinds of questions you will be asked. The third step is to practice "writing-on-demand" for the essays.

TOP 10 STRATEGIES TO RAISE YOUR SCORE

Preparing for the Exam

1. Read the *AP Biology Course Description* available from the College Board.
2. Review the Study Plan in this section.
3. Choose a place and time to study **every day.** Stick to your routine and your plan.
4. Complete the Diagnostic and other Practice Tests in this book. They will give you just what they promise: practice. Practice in reading and following directions, practice in pacing yourself, practice in understanding and answering multiple-choice questions, and practice in writing timed essays.
5. Complete all your assignments for your regular AP Biology class. Ask questions in class, talk about what you read and write, and enjoy what you are doing. The exam is supposed to measure your development as an educated and thinking reader.

The Night Before the Exam

6. Gather what you will need for the exam: your admission materials, four No. 2 pencils, two pens, a watch (without an alarm), and a healthy snack for the break. Put these items in a place where you will not forget them in the morning.
7. Don't cram. Relax. Go to a movie, visit a friend—but not one who is taking the exam with you. Get a good night's sleep.

The Day of the Exam

8. Wear comfortable clothes.
9. If you do not usually eat a big breakfast, this is not the morning to change your routine, but it is probably a good idea to eat something nutritious if you can.
10. If you feel yourself getting anxious, concentrate on taking a couple of deep breaths. Remember, you don't have to answer all the questions, you can use EDUCATED GUESSES, and you don't have to write four "5" essays.

STUDY PLAN TO PREPARE FOR THE AP BIOLOGY EXAM

The best study plan is one that continues through a full semester. The following plan should be followed for fourteen weeks. Then you have time to think about ideas and to talk with your teacher and other students about what you are learning, and you will not feel rushed. Staying relaxed about the test is important. A full-semester study plan also means that you can apply what you are learning here to classwork—your essay writing—and apply your classwork—everything that you are reading—to test preparation. The plan is worked out so that you should spend about 3 hours on each lesson.

Week 1

- Take *Practice Test 1: Diagnostic,* and check your answers. List the areas that you had difficulty with, such as timing and question types.
- Reread *10 Strategies for Acing the AP Biology Exam.*
- Reread *10 Facts About the AP Biology Exam* to remind yourself that a score of at least "3" is achievable.
- Review the list you made after you took *Practice Test 1: Diagnostic* to see what you need to learn about the multiple-choice section of the exam.
- Read Chapter 1, *All About the AP Biology Exam.*

Week 2

LESSON 1

- Read Chapter 3, *The Chemistry of Biology.*
- Answer the practice questions throughout Chapter 3 and the practice questions at the end of the chapter. Review your answers.

LESSON 2

- Review Chapter 3, *The Chemistry of Biology,* and review your answers to the practice questions at the end of the chapter.

Week 3

LESSON 1

- Reread *10 Strategies for Acing the AP Biology Exam.*
- Reread *10 Facts About the AP Biology Exam* to remind yourself that a score of at least "3" is achievable. It may seem boring by now, but it is important to remember that the test score does not ride on the essays.
- Review Chapter 1, *All About the AP Biology Exam.*
- Read Chapter 4, *Biological Macromolecules.*

- Answer the practice questions throughout Chapter 4 and the practice questions at the end of the chapter. Review your answers.

LESSON 2

- Review Chapter 4, *Biological Macromolecules,* and review your answers to the practice questions at the end of the chapter.

Week 4

LESSON 1

- Read Chapter 5, *The Cell.*
- Answer the practice questions throughout Chapter 5 and the practice questions at the end of the chapter. Review your answers.

LESSON 2

- Review Chapter 5, *The Cell,* and review your answers to the practice questions at the end of the chapter.

Week 5

LESSON 1

- Read Chapter 6, *Energy in Biological Systems.*
- Answer the practice questions throughout Chapter 6 and the practice questions at the end of the chapter. Review your answers.

LESSON 2

- Review Chapter 6, *Energy in Biological Systems,* and review your answers to the practice questions at the end of the chapter.

Week 6

LESSON 1

- Read Chapter 7, *Cellular ATP Production,* and Chapter 8, *Photosynthesis.*
- Answer the practice questions throughout Chapters 7 and 8 and the practice questions at the end of the chapters. Review your answers.

LESSON 2

- Review Chapter 7, *Cellular ATP Production,* and Chapter 8, *Photosynthesis,* and review your answers to the practice questions at the end of the chapters.

Week 7

LESSON 1

- Take *Practice Test 2* and check your answers.
- Compare this score to your score on *Practice Test 1: Diagnostic.* Which question types continue to be a concern?

LESSON 2

- Reread Chapters 3 through 8 as needed.

Week 8

LESSON 1

- Read Chapter 9, *Genetic Inheritance.*
- Answer the practice questions throughout Chapter 9 and the practice questions at the end of the chapter. Review your answers.

LESSON 2

- Review Chapter 9, *Genetic Inheritance,* and review your answers to the practice questions at the end of the chapter.

Week 9

LESSON 1

- Read Chapter 10, *The Genetic Code.*
- Answer the practice questions throughout Chapter 10 and the practice questions at the end of the chapter. Review your answers.

LESSON 2

- Review Chapter 10, *The Genetic Code,* and review your answers to the practice questions at the end of the chapter.

Week 10

LESSON 1

- Read Chapter 11, *Evolution and Diversity.*
- Answer the practice questions throughout Chapter 11 and the practice questions at the end of the chapter. Review your answers.

LESSON 2

- Review Chapter 11, *Evolution and Diversity,* and review your answers to the practice questions at the end of the chapter.

Week 11

LESSON 1

- Read Chapter 12, *Prokaryotes, Protists, Fungi, and Plants.*
- Answer the practice questions throughout Chapter 12 and the practice questions at the end of the chapter. Review your answers.

LESSON 2

- Review Chapter 12, *Prokaryotes, Protists, Fungi, and Plants,* and review your answers to the practice questions at the end of the chapter.

Week 12

LESSON 1

- Read Chapter 13, *Animal Structure and Function.*
- Answer the practice questions throughout Chapter 13 and the practice questions at the end of the chapter. Review your answers.

LESSON 2

- Review Chapter 13, *Animal Structure and Function,* and review your answers to the practice questions at the end of the chapter.

Week 13

LESSON 1

- Read Chapter 14, *Ecosystems.*
- Answer the practice questions throughout Chapter 14 and the practice questions at the end of the chapter. Review your answers.

LESSON 2

- Review Chapter 14, *Ecosystems,* and review your answers to the practice questions at the end of the chapter.

Week 14

LESSON 1

- Take *Practice Test 3* and check your answers.
- Check your results against the other two tests.

LESSON 2

- If you are still unsure about some areas of the exam, review the chapters relating to these areas, including the answers to the practice exercises.
- Reread Chapter 1, *All About the AP Biology Exam.*
- Reread *10 Strategies for Acing the AP Biology Exam* and *10 Facts About the AP Biology Exam* to remind yourself that a score of at least "3" is achievable.

YOU'RE WELL ON YOUR WAY TO SUCCESS

Remember that knowledge is power. You will be studying the most comprehensive guide available and you will become extremely knowledgeable about the exam. We look forward to helping you raise your score.

GIVE US YOUR FEEDBACK

Peterson's, a Nelnet company, publishes a full line of resources to help guide you through the college admission process. Peterson's publications can be found at your local bookstore, library, and high school guidance office, and you can access us online at www.petersons.com.

We welcome any comments or suggestions you may have about this publication and invite you to complete our online survey at www.petersons.com/booksurvey. Or you can fill out the survey at the back of this book, tear it out, and mail it to us at:

Publishing Department
Peterson's
2000 Lenox Drive
Lawrenceville, NJ 08648

Your feedback will help us to provide personalized solutions for your educational advancement.

PART I

AP BIOLOGY BASICS

All About the AP Biology Exam

chapter 1

OVERVIEW

- **Getting started**
- **Preparing for the AP Biology exam**
- **Making a study plan**
- **What the exam covers**
- **Getting to know the format of the exam**
- **How the AP Biology exam is scored**
- **The answer sheets**
- **Summing it up**

GETTING STARTED

There is no question that the AP Biology Exam is one of the most rigorous offered by the College Board. Chances are, however, that if you are planning to take the AP Biology Exam, you are the type of student who is ready to take on anything this test can dish out. You didn't take AP Biology by accident--you wanted to take it, and that's because you are the type of student who can handle challenges. So, although the test is difficult, don't be intimidated by it. You can beat this test!

PREPARING FOR THE AP BIOLOGY EXAM

Your first step is to understand that this test is not something to be taken lightly. Top athletes spend a great deal of time preparing their bodies for big competitions, and you, too, must prepare your mind for this test. But most athletes don't train by themselves. They work out with an experienced coach who knows the keys to being successful. In preparation for the AP challenge, you will need a coach to help you learn the best strategies for the test. **Your teacher should be your first coach.** Due to the time constraints that teachers face and the extensive amount of material covered in this course, you will also need an assistant coach to help you prepare. This book can be that assistant coach, but it is not designed to replace your teacher. He or she knows the course, the material, and you, but, unless your teacher can move in with you while you are preparing for the test, he or she can't provide you with **all** of the tools you will need to succeed. There aren't enough hours in the school day to teach you everything you need to know for the test. This book can be your personal tutor. The material between its covers will help to familiarize you with all of the material covered on the test, the structure of the test, and strategies to prepare for and to take the test and it can provide you with instant feedback about your performance. This book is also designed with the understanding that your time is valuable! You most likely have a very busy schedule before and after school, and the last thing you need is to add another time-consuming activity. Some of the information that you may have learned or that is in your textbook may go beyond the level of the AP exam. In other subjects, you may have learned less information than you need. This book attempts to get right to the point and to review only the material that you will need to know for the exam.

MAKING A STUDY PLAN

The fact that you bought this book is a step in the right direction for your success on the AP Biology Exam. And there are some strategies that will help you get the most out of it. The following are two key questions you need to answer before you proceed:

1. How much time do I have before the AP exam?
2. How much time can I realistically devote to test preparation?

Your answers to these questions will help you to set a pace for your review. If you have a long time before the test (two or three months), you can set a fairly relaxed pace. If you have a short time (one month or less), your pace will have to be more accelerated. Either way, this book is designed to accommodate a variety of situations. What follows is a brief description of the remainder of the book and how you can use it under different circumstances.

Read this section and then go on to Chapter 2 and take Practice Test 1: Diagnostic. The purpose of the test is to help you identify the strengths and potential weaknesses you'll need to know in order to design your personal study plan. Once you finish the diagnostic test, use these suggested guidelines to complete your test preparations.

The Complete Course

If you have plenty of time before the test (three months or more), it is recommended that you complete the entire course. Follow the Study Plan laid out in the Introduction. You will receive the maximum benefit if you complete the entire book. Even if you are planning to complete the entire book, it is recommended that you start with your most difficult areas first, especially those that you may not have covered much (or at all) in class. This way, if something happens and you don't have as much time as you thought, you've at least gone through the sections that will benefit you the most.

The Accelerated Course

If you are running out of time, you'll need to design an ambush approach to your studies. Don't attempt to complete the entire book. From the diagnostic test, select your weakest areas and plan to go through those first. Make a list of the topics you feel you can reasonably work through before the test. Be very careful during this process. Do not skip sections you know very little about, figuring, "Oh, that probably won't be on the test." If it is on the content outline, it will be on the test! And, if you are especially unfortunate, the section you skip could end up as one of the mandatory essay questions. Even if you don't have time for all of the practice problems in a section, get through what you can. If you know certain topics very well, skip them. It's not ideal, but if you are reading this section it is probably because you don't have time to do everything. Just remember, though, that every chapter you can work through is a bonus for you. Rather than be discouraged about not having enough time to finish everything, be encouraged about the sections you will finish. These are areas you wouldn't have known or would have done poorly on without your extra effort. Be positive!

WHAT THE EXAM COVERS

Each year, the College Board provides a content outline for the AP Biology course as well as a breakdown of the approximate percentage of the AP exam that will deal with certain topics. What follows is an outline of the content of the most recent exam.

Molecules and Cells (25 percent)

- Chemistry of Life
    - Water
    - Organic molecules in organisms
    - Free energy changes
    - Enzymes
- Cells
    - Prokaryotic and eukaryotic cells
    - Membranes
    - Subcellular organization
    - Cell cycle and regulation
- Cellular Energetics
    - Coupled reactions
    - Fermentation and cellular respiration
    - Photosynthesis

Heredity and Evolution (25 percent)

- Heredity
    - Meiosis and gametogenesis
    - Eukaryotic chromosomes
    - Inheritance patterns
- Molecular genetics
    - RNA and DNA structure and function
    - Gene regulation
    - Mutation
    - Viral structure and replication
    - Nucleic acid technology and applications

Organisms and Populations (50 percent)

- Diversity of organisms
    - Evolutionary patterns
    - Survey of the diversity of life
    - Phylogenetic classification
    - Evolutionary relationships
- Structure and function of plants and animals
    - Reproduction, growth, and development
    - Adaptations
    - Environment
- Ecology
    - Population dynamics
    - Global issues
    - Ecosystems

GETTING TO KNOW THE FORMAT OF THE EXAM

The College Board reserves the right to make changes to the test as it chooses; however, the structure of the test has undergone only very slight changes in the years it has been offered. To keep abreast of the specific details about the AP Biology Exam, refer to the College Board Web site at http://www.collegeboard.com. Click on "For Students" and then "AP" under "College Board Tests."

The test is 3 hours long and is divided into two sections. The first, Section I, consists of 100 multiple-choice questions. There is an 80-minute time limit on Section I, and the results make up 60 percent of the total grade. The second part, Section II, is the free-response section. It has four questions, is 90 minutes long, and makes up 40 percent of the total grade.

Section I: Multiple Choice. 100 questions. 80 minutes. 60 percent of total grade.

Section II: Free Response. 4 questions. 90 minutes. 40 percent of total grade.

HOW THE AP BIOLOGY EXAM IS SCORED

Multiple Choice

On the multiple-choice portion of the test, there are 100 questions. Scores are determined according to the following formula:

Score = Correct Score – 0.25(Incorrect Score)

The reason for the strange equation is to prevent uneducated guessing. There is no benefit to guessing if you are unable to reliably eliminate at least one answer choice.

Free Response

The free-response questions are graded by a group of more than 100 AP Biology teachers and college biology professors who gather at the beginning of the summer to grade all of the tests. Scoring rubrics are carefully designed for each question, and the graders spend a week grading the tests in a very thorough, unbiased manner. The key to your success on the free-response questions is your ability to write to these rubrics. While there is no way to predict in advance what topics will be covered in the free-response section (even though many people try), there are some strategies that will improve your chances. Hopefully, your teacher has had you practice writing AP-style essays, but even if not, you will still have an opportunity to try some in this book. You will also be able to evaluate your own performance using scoring rubrics that are similar to those used by the AP graders. This exercise will help you sharpen your writing skills and maximize your chances for a high score.

Your Composite Score

Once your multiple-choice and free-response scores have been calculated, each score will be used to calculate your composite score. This score ranges from 1 to 5, according to the scale shown below.

AP Grade	Qualification
5	Extremely well qualified
4	Well qualified
3	Qualified
2	Possibly qualified
1	No recommendation

One thing these numbers don't tell you is what it takes to get a 5 or a 4. The number of points to achieve these grades will differ slightly from year to year, but they remain relatively consistent over time. In most administrations, scores of 50–60 percent usually receive scores of 5.

THE ANSWER SHEETS

For Section I of the test (the multiple-choice questions), you will be provided a test booklet and an answer sheet. The answer sheet is a bubble sheet, and the test booklet contains all of the multiple-choice questions. Any extra writing you need to do can be done in the margins of the test booklet—not on the answer sheet! As far as the bubble sheet goes, first make sure you're using a #2 pencil—that way, the marks you make will be dark enough for the scoring machine to read. In addition, you must make sure that your answer ovals are neatly filled in.

For the free-response essays, you will be given a separate packet. The first part of the packet consists of your answer booklet. All answers are to be written here. You will also be given another packet that contains the questions and any reference material.

SUMMING IT UP

- Familiarize yourself with the structure of the AP Biology Exam. Knowing the test format will relieve test anxiety because you will know exactly what to expect on exam day.
- Make sure you fill in the bubble sheet neatly. Otherwise, the scoring machine won't give you credit for your answers.
- Pace yourself. You must work quickly and carefully throughout the test. You can still get a very high score without answering all of the questions. Answer as many as you can as quickly as you can, and then go back and try to fill in the others.
- Remember that uneducated guessing will have no effect on your score, but educated guessing will boost your score. So, if you've had time to read through a question and eliminate at least one choice, take a guess!
- In the free-response sections, be neat, thorough, and very clear. You do not want the graders having to guess what you wrote or what you meant.
- Remember--if you're working through this book, you're giving yourself the best preparation available for succeeding on the AP exam. Let your preparation give you the confidence you need to be calm and focused.

PART II

DIAGNOSING STRENGTHS AND WEAKNESSES

ANSWER SHEET PRACTICE TEST 1: DIAGNOSTIC

1. Ⓐ Ⓑ Ⓒ Ⓓ Ⓔ
2. Ⓐ Ⓑ Ⓒ Ⓓ Ⓔ
3. Ⓐ Ⓑ Ⓒ Ⓓ Ⓔ
4. Ⓐ Ⓑ Ⓒ Ⓓ Ⓔ
5. Ⓐ Ⓑ Ⓒ Ⓓ Ⓔ
6. Ⓐ Ⓑ Ⓒ Ⓓ Ⓔ
7. Ⓐ Ⓑ Ⓒ Ⓓ Ⓔ
8. Ⓐ Ⓑ Ⓒ Ⓓ Ⓔ
9. Ⓐ Ⓑ Ⓒ Ⓓ Ⓔ
10. Ⓐ Ⓑ Ⓒ Ⓓ Ⓔ
11. Ⓐ Ⓑ Ⓒ Ⓓ Ⓔ
12. Ⓐ Ⓑ Ⓒ Ⓓ Ⓔ
13. Ⓐ Ⓑ Ⓒ Ⓓ Ⓔ
14. Ⓐ Ⓑ Ⓒ Ⓓ Ⓔ
15. Ⓐ Ⓑ Ⓒ Ⓓ Ⓔ
16. Ⓐ Ⓑ Ⓒ Ⓓ Ⓔ
17. Ⓐ Ⓑ Ⓒ Ⓓ Ⓔ
18. Ⓐ Ⓑ Ⓒ Ⓓ Ⓔ
19. Ⓐ Ⓑ Ⓒ Ⓓ Ⓔ
20. Ⓐ Ⓑ Ⓒ Ⓓ Ⓔ
21. Ⓐ Ⓑ Ⓒ Ⓓ Ⓔ
22. Ⓐ Ⓑ Ⓒ Ⓓ Ⓔ
23. Ⓐ Ⓑ Ⓒ Ⓓ Ⓔ
24. Ⓐ Ⓑ Ⓒ Ⓓ Ⓔ
25. Ⓐ Ⓑ Ⓒ Ⓓ Ⓔ
26. Ⓐ Ⓑ Ⓒ Ⓓ Ⓔ
27. Ⓐ Ⓑ Ⓒ Ⓓ Ⓔ
28. Ⓐ Ⓑ Ⓒ Ⓓ Ⓔ
29. Ⓐ Ⓑ Ⓒ Ⓓ Ⓔ
30. Ⓐ Ⓑ Ⓒ Ⓓ Ⓔ
31. Ⓐ Ⓑ Ⓒ Ⓓ Ⓔ
32. Ⓐ Ⓑ Ⓒ Ⓓ Ⓔ
33. Ⓐ Ⓑ Ⓒ Ⓓ Ⓔ
34. Ⓐ Ⓑ Ⓒ Ⓓ Ⓔ
35. Ⓐ Ⓑ Ⓒ Ⓓ Ⓔ
36. Ⓐ Ⓑ Ⓒ Ⓓ Ⓔ
37. Ⓐ Ⓑ Ⓒ Ⓓ Ⓔ
38. Ⓐ Ⓑ Ⓒ Ⓓ Ⓔ
39. Ⓐ Ⓑ Ⓒ Ⓓ Ⓔ
40. Ⓐ Ⓑ Ⓒ Ⓓ Ⓔ
41. Ⓐ Ⓑ Ⓒ Ⓓ Ⓔ
42. Ⓐ Ⓑ Ⓒ Ⓓ Ⓔ
43. Ⓐ Ⓑ Ⓒ Ⓓ Ⓔ
44. Ⓐ Ⓑ Ⓒ Ⓓ Ⓔ
45. Ⓐ Ⓑ Ⓒ Ⓓ Ⓔ
46. Ⓐ Ⓑ Ⓒ Ⓓ Ⓔ
47. Ⓐ Ⓑ Ⓒ Ⓓ Ⓔ
48. Ⓐ Ⓑ Ⓒ Ⓓ Ⓔ
49. Ⓐ Ⓑ Ⓒ Ⓓ Ⓔ
50. Ⓐ Ⓑ Ⓒ Ⓓ Ⓔ
51. Ⓐ Ⓑ Ⓒ Ⓓ Ⓔ
52. Ⓐ Ⓑ Ⓒ Ⓓ Ⓔ
53. Ⓐ Ⓑ Ⓒ Ⓓ Ⓔ
54. Ⓐ Ⓑ Ⓒ Ⓓ Ⓔ
55. Ⓐ Ⓑ Ⓒ Ⓓ Ⓔ
56. Ⓐ Ⓑ Ⓒ Ⓓ Ⓔ
57. Ⓐ Ⓑ Ⓒ Ⓓ Ⓔ
58. Ⓐ Ⓑ Ⓒ Ⓓ Ⓔ
59. Ⓐ Ⓑ Ⓒ Ⓓ Ⓔ
60. Ⓐ Ⓑ Ⓒ Ⓓ Ⓔ

answer sheet

Practice Test 1: Diagnostic

60 QUESTIONS • 48 MINUTES

Directions: Each of the questions or incomplete statements below is followed by five suggested answers or completions. Select the one that is best in each case and mark the corresponding oval on the answer sheet.

1. Which of the following is *not* a property of water?

(A) Strong hydrogen bonding in liquid
(B) High molar heat of evaporation
(C) Strong cohesion
(D) Solid state is denser than liquid state
(E) High molar heat of melting

2. Which of the following contains phosphorous when it is first synthesized?

(A) DNA
(B) Triglyceride
(C) Protein
(D) Glycogen
(E) Chlorophyll

3. Which of the following is an example of active transport across membranes?

(A) Opening of voltage-gated sodium channels during an action potential
(B) The sodium-potassium pump that creates the resting potential
(C) Osmosis of water
(D) The role of aquaporins in creating concentrated urine
(E) The removal of carbon dioxide from cells

practice test 1

4. If intact cells are treated with a compound that allows hydrogen ions to flow freely across the inner mitochondrial membrane, which of the following will occur?

(A) ATP production will increase but oxygen consumption will decrease.
(B) ATP production will be unaffected but pyruvate consumption will increase.
(C) Oxygen consumption will increase but ATP production will decrease.
(D) Oxygen consumption will decrease, pyruvate consumption will increase, and ATP production will increase.
(E) The Krebs cycle and electron transport will cease.

5. Enzymes can do which of the following to a reaction?

(A) Change the equilibrium to favor products
(B) Block the reverse reaction
(C) Allow it to operate 100% efficiently
(D) Reduce the overall entropy of a system
(E) Stabilize the reaction state intermediate

6. Which of the following has a negative ΔG?

(A) Hydrolysis of phosphodiester bonds
(B) Peptide bond formation
(C) Transport of hydrogen ions from pH 5 to pH 3
(D) Transport of mitochondria on microtubules in cells
(E) Triglyceride synthesis

7. A complete digestive tract is found in which of the following?

(A) Cnidarians
(B) Flatworms
(C) Sponges
(D) Mollusks
(E) Paramecia

8. Which of these hormones is produced by the anterior pituitary?

(A) GnRH
(B) Thyroid-stimulating hormone
(C) Cortisol
(D) Oxytocin
(E) Glucagon

9. Protostomes have which of the following traits that differentiate them from deuterostomes?

(A) Notochord
(B) Radial cleavage
(C) Blastopore, which becomes the mouth in later development
(D) Lack mesoderm
(E) Centrioles

10. Which part of angiosperms is triploid?

(A) Endoderm
(B) Endosperm
(C) Cotyledons
(D) Female gametophyte
(E) Apical meristem

11. Which of the following mechanisms improves the accuracy of protein translation?

(A) After a protein is synthesized, proofreading machinery can replace errors in the mature protein with different amino acids.
(B) Aminoacyl tRNA synthetases check the accuracy of tRNA activation with amino acids.
(C) Errors in mRNA production are repaired by RNA polymerase.
(D) tRNA molecules have catalytic proofreading activity.
(E) Splicing removes protein errors.

12. Which of the following is not directly produced in the Krebs cycle?

(A) Pyruvate
(B) GTP
(C) NADH
(D) FADH2
(E) Carbon dioxide

13. Earthworms using hermaphroditic sexual reproduction would display which of the following traits?

(A) Reproduction with gametes produced mitotically
(B) No gamete production involved in reproduction
(C) Normal meiotic gamete production and production of a diploid zygote by fertilization
(D) A lack of genetic variation in offspring
(E) A lack of meiosis

14. At which stage of meiosis does meiotic recombination occur?

(A) Prophase I
(B) Prophase II
(C) Anaphase I
(D) Anaphase II
(E) Metaphase I

15. Which of the following does not take place in the eukaryotic nucleus?

(A) Removal of introns from mRNA
(B) PolyA tail addition
(C) Transcription
(D) Translation of RNA polymerase
(E) 5' cap addition

16. The growth of apical meristem in plants produces which of the following?

(A) Growth in width of tree trunks
(B) Elongation of roots
(C) Proliferation of cells in xylem
(D) Proliferation and elongation of sieve-tube members
(E) Maturation of fruits

17. In the kidney, the glomerulus is involved in

(A) Selective reabsorption of glucose from the urinary filtrate
(B) Secretion of urea into the urinary filtrate
(C) Establishing an osmotic gradient
(D) Response to vasopressin to concentrate urine
(E) Filtration of blood to create the initial urinary filtrate

18. The reflexive withdrawal of a hand from a hot stove involves which of the following?

(A) Sensitization
(B) Fixed action pattern
(C) Imprinting
(D) Perception of heat and pain after the motor response
(E) Integration of movement by the cerebral cortex after perception of pain

19. Of the choices below, the highest gross primary productivity would be found in:

(A) An oligotrophic mountain lake
(B) A pelagic ocean
(C) A benthic ocean
(D) Estuaries
(E) The Arctic Ocean

20. Hardy-Weinberg equilibrium cannot occur given which of the following?

(A) Increased migration
(B) Lack of mutation
(C) Large population size
(D) Random mating
(E) Lack of natural selection

21. Which of the following is involved in insect respiration?

(A) Malphigian tubules
(B) Tracheoles
(C) Alveoli
(D) Cells containing oxygen transporting proteins
(E) Circulation of air in one direction through air passages

diagnostic test

22. Which of the following is an example of a postzygotic barrier leading to reproductive isolation in conjunction with allopatric speciation?

(A) Two populations of birds display different courtship rituals that are not recognized by the other population.
(B) Lizards display rapid speciation after colonizing an island.
(C) Two populations can interbreed and produce offspring, but the offspring are sterile.
(D) Sperm from a population of sea urchins cannot bind to ova from another population of sea urchins.
(E) Physiological differences prevent mating between males and females in different populations.

23. Which of the following is true in all double-stranded DNA?

(A) The amount of A and G are the same.
(B) The amount of A and T are the same.
(C) A and T are 50% of the base content in DNA.
(D) G and C form two hydrogen bonds with one another.
(E) Phosphate groups point inward toward one another in the double helix.

24. In the binding of hemoglobin to oxygen, which of the following occurs?

(A) Oxygen induces the hemoglobin tetramer to dissociate into monomers.
(B) The first oxygen binds with high affinity to each hemoglobin tetramer, and additional oxygens bind with lower affinity.
(C) Oxygen causes hemoglobin monomers to associate together as a tetramer.
(D) The first oxygen binds with low affinity, causing a conformational shift that allows subsequent oxygens to bind with higher affinity.
(E) A more acidic pH causes hemoglobin to bind oxygen with higher affinity.

25. Maternal inheritance involves genes located in which of the following?

(A) Mitochondria
(B) Golgi body
(C) X chromosome
(D) Y chromosome
(E) Mitochondrial protein encoded in nuclear genome

26. At the DNA replication fork, what role does DNA ligase have?

(A) Brings together two regions of single-stranded DNA to make them double-stranded
(B) Removes RNA primers
(C) Degrades Okazaki fragments
(D) Covalently bonds free ends in double-stranded DNA to neighboring nucleotide
(E) Removes base mismatches in DNA repair

27. Which of the following is not a part of natural selection?

(A) A population must have genetic variation.
(B) A heritable variation is expressed in the phenotypes.
(C) Some phenotypes associated with heritable variation survive and reproduce differentially.
(D) Adaptations are the best possible responses to evolutionary pressure.
(E) Natural selection can occur indirectly through actions of genetically related social organisms.

28. Hemophilia is a recessive sex-linked trait. If a woman who has hemophilia marries a man who does not have it, what percentage of female children of the couple will develop hemophilia?

(A) 0%
(B) 25%
(C) 50%
(D) 75%
(E) 100%

29. Humans are related most closely to which of the following?

(A) Starfish
(B) Lobster
(C) Clam
(D) Earthworm
(E) Squid

30. The haploid gametophyte is dominant in which of the following?

(A) Mollusks
(B) Fungi
(C) Gymnosperms
(D) Mosses
(E) Protists

31. Which of the following is found in cnidarians?

(A) Muscle tissue
(B) Excretory system
(C) Circulatory system
(D) Respiratory tissue
(E) Neuronal cells

32. Which of the following occurs in the contraction of skeletal muscle?

(A) Action potential spreads across muscle cell plasma membrane.
(B) Neuron communicates with muscle cell through electrical synapse.
(C) Muscle cells release acetylcholine to carry action potentials to neighboring cells.
(D) Actin hydrolyzes ATP.
(E) Actin pulls myosin apart to lengthen muscle cells during the relaxation phase.

33. In the logistic model of population growth, which of the following occurs?

(A) Populations grow slowly to conserve resources for future use.
(B) Populations grow to exceed available resources until the increased rate of death causes the population size to crash.
(C) The population grows until the rate of births equals the rate of deaths.
(D) Populations fluctuate in size because of interactions between predators and prey.
(E) Competition for resources drives one species to become extinct.

34. The nitrogen cycle includes which of the following processes?

(A) Mycorrhizal fungi perform nitrogen fixation in mutualism with plants.
(B) Bacteria in the gut of ruminants perform nitrogen fixation from the atmosphere.
(C) Animals use nitrates to synthesize amino acids.
(D) Plants absorb nitrogen from the atmosphere for nitrogen fixation.
(E) Denitrifying bacteria in soil release N_2 into the atmosphere.

35. Disulfide bridges between two amino acids in the same polypeptide chain are part of which level of protein structure?

(A) Primary
(B) Secondary
(C) Tertiary
(D) Quaternary
(E) Unstable folding intermediates

36. Compared to other cell types, adipose cells contain larger quantities of which of the following?

(A) Protein
(B) Glycogen
(C) Triglyceride
(D) Phospholipids
(E) Cholesterol

37. In a tree, which force is most responsible for the movement of fluids in xylem?

(A) Water drawn upward to replace water lost by transpiration via leaves
(B) The osmotic potential of solutes pushing fluids down through the system
(C) Root pressure
(D) The pumping action of microvascular epithelium
(E) The repulsion of negative ions

38. Humans excrete nitrogenous wastes as urea because

(A) It is more soluble than ammonia.
(B) It optimizes water conservation better than uric acid does.
(C) It conserves more energy than releasing waste as ammonia.
(D) Urea is less toxic than ammonia.
(E) Urea does not contain carbon.

39. Fungi are

(A) Absorptive chemoautotrophs
(B) Ingestive chemoheterotrophs
(C) Obligate anaerobes
(D) Absorptive chemoheterotrophs
(E) Parasitic mutualists

40. Which of the following is a part of innate immunity?

(A) Cytotoxic T cells
(B) Macrophages
(C) Helper T cells
(D) Immunoglobulins
(E) Plasma B cells

41. Which of the following is *not* performed by transmembrane proteins in the eukaryotic cells?

(A) Establishing the resting potential
(B) Binding to peptide hormones
(C) Movement of glucose into cells
(D) Establishing the pH gradient for ATP production
(E) Movement of carbon dioxide in and out of cells

42. In a closed system, which of the following always occurs over time?

(A) Energy decreases
(B) Entropy decreases
(C) Entropy increases
(D) Energy increases
(E) Energy increases and entropy decreases

43. Cyclic AMP plays what role in eukaryotic cells?

(A) Stores energy
(B) Provides a building block for DNA
(C) Acts as a second messenger in signal transduction
(D) Is a high-energy electron carrier
(E) Protects against foreign pathogens

44. An organism is diploid and, during the G1 phase of the cell cycle, has eight chromosomes. How many different types of gametes can the organism form during meiosis through the mechanism of independent segregation of homologous chromosomes?

(A) 2
(B) 4
(C) 8
(D) 16
(E) 32

45. Ubiquitination of proteins has what effect on them?

(A) It activates enzyme activity.
(B) It causes protein dimerization.
(C) It targets proteins for proteolytic destruction.
(D) It stimulates receptor-mediated endocytosis.
(E) It causes secretion of the tagged protein from the cell.

46. Cloned mammals are produced by replacing the nucleus of an ovum with the nucleus from a differentiated somatic cell from outside the immune system that has been cultured to reverse its differentiated state. Although the mammalian clones that are generated are genetically identical to their parent, they are often not identical physiologically to the animal that served as the source of the donated genome. Most likely this is because

(A) Mammals have large complex genomes.
(B) Epigenetic changes in the genome that occur during development were retained and affected the development of the clone.
(C) Most differentiated mammalian cells lose sections of chromosomes that are not expressed in that tissue.
(D) Meiotic recombination is defective in the cloned animals.
(E) Differentiation does not alter patterns of gene expression.

47. Which of the following is displayed in MHC class II molecules?

(A) Peptide fragments released from viruses circulating in the plasma
(B) Secreted proteins released by cancer cells
(C) Immunoglobulins
(D) Peptides from antigens internalized by antigen-presenting cells
(E) Free fatty acids

48. In which of the following would one find permafrost below the surface, surface thawing in summer, and growth of low vegetation such as grasses?

(A) Antarctica
(B) The Arctic
(C) A tundra
(D) A taiga
(E) A boreal forest

49. How are sounds of different pitches distinguished by the human ear?

(A) Hair cells extended to different lengths have differing sensitivity to sounds of different pitch.
(B) Sounds of varying frequencies stimulate action potentials at different intervals.
(C) Sounds of different pitch stimulate vibration of the basilar membrane in varying sections.
(D) Different semicircular canals are stimulated by different sound amplitudes.
(E) Action potentials for higher-pitched sounds have greater depolarization.

50. A bell-shaped gradient distribution of a trait in a population would probably indicate which of the following influences on the trait?

(A) Environment and polygenic determination
(B) Co-dominance
(C) Noninvolvement of genetic factors
(D) Epigenetic influences
(E) Noninvolvement of environment

51. Cells in which of the following are most likely to be totipotent?

(A) Blastula
(B) Morula
(C) Gastrula
(D) Hematopoietic stem cell
(E) Intestinal stem cell

52. Vitamin C is primarily involved in which of the following processes?

(A) Calcium absorption
(B) Blood clotting
(C) As a precursor for a high-energy electron carrier
(D) Collagen synthesis
(E) Oxygen transport

53. In the angiosperm life cycle, the microspores give rise first to which of the following?

(A) Zygote
(B) Pollen grain
(C) Pollen tube
(D) Sperm
(E) Gametophyte

54. Bacterial genomes have which of the following traits?

(A) Circular chromosome
(B) Telomeres
(C) Introns
(D) Centromeres
(E) Histone proteins

55. The fluidity of the lipid bilayer membrane is increased by

(A) Increasing the length of fatty acid side chains
(B) Increasing the proline content of membrane lipids
(C) Increasing the number of double bonds in fatty acid side chains of phospholipids
(D) Decreasing temperature
(E) Association of transmembrane proteins with cytoskeleton

56. In carbon fixation in C3 plants, carbon from the atmosphere first appears in a stable form in which of the following?

(A) Glucose
(B) ATP
(C) Glyceraldehyde-3-phosphate
(D) Organic acids
(E) Rubisco

57. The first terrestrial plants had

(A) Seeds
(B) Flowers
(C) Vascular tissue
(D) Meristematic tissues
(E) Dominant sporophyte in alternation of generations

58. A significant layer of smooth muscle is found in the walls of all of the following *except*

(A) Uterus
(B) Pancreas
(C) Esophagus
(D) Stomach
(E) Small intestine

59. Which of these triggers an action potential in a motor neuron?

(A) Summation of excitatory post-synaptic potentials to reach the threshold
(B) Summation of inhibitory post-synaptic potentials to reach the threshold
(C) Spontaneous depolarization due to potassium leak channels
(D) The entry of chloride ions into the cell
(E) The exit of potassium ions from the cell

60. Which of the following statements about transcription is correct?

(A) Transcription occurs in groups of three nucleotides at a time.
(B) Transcription is highly accurate because of the proofreading activity of RNA polymerase.
(C) Transcription occurs from 3' to 5' of RNA.
(D) Transcription requires a primer to begin.
(E) Transcription is performed by three different RNA polymerases in eukaryotes.

ANSWER KEY AND EXPLANATIONS

1. D	13. C	25. A	37. A	49. C
2. A	14. A	26. D	38. D	50. A
3. B	15. D	27. D	39. D	51. B
4. C	16. B	28. A	40. B	52. D
5. E	17. E	29. A	41. E	53. B
6. A	18. D	30. D	42. C	54. A
7. D	19. D	31. E	43. C	55. C
8. B	20. A	32. A	44. D	56. C
9. C	21. B	33. C	45. C	57. D
10. B	22. C	34. E	46. B	58. B
11. B	23. B	35. C	47. D	59. A
12. A	24. D	36. C	48. C	60. E

1. **The correct answer is (D).** Water has unique physical properties stemming from strong hydrogen bonding between water molecules. This means that water molecules tend to stick together (cohesion is strong), and it takes a lot of energy to break hydrogen bonds to heat, melt, or evaporate water. Solid water (ice) also has the unusual property of being less dense than liquid water, which is why ice floats. Since (D) is *not* true, it is the correct answer.

2. **The correct answer is (A).** Both DNA and RNA contain phosphorous in the phosphodiester bonds between polymerized nucleotides. Triglycerides and most sugars contain only hydrogen, carbon, and oxygen, so (B) and (D) are not correct answer choices. Proteins also contain nitrogen and sulfur, but there are no amino acids containing phosphorous. The phosphorylation of proteins by protein kinases occurs after translation, so (C) is incorrect. Chlorophyll contains hydrogen, carbon, oxygen, nitrogen, and a complexed magnesium ion, but no phosphorous, so (E) is not the correct answer choice.

3. **The correct answer is (B).** Active transport involves pumping a substance against an electrochemical gradient. This takes energy, as with pumping water uphill. The sodium potassium pump uses the energy of ATP hydrolysis to pump sodium and potassium ions against electrochemical gradients and maintain the resting potential across the plasma membrane. Ion channels and aquaporins perform facilitated diffusion, with material flowing down a gradient with the involvement of a protein in the membrane. Carbon dioxide moves through the membrane by passive diffusion, down a gradient, without a need for a protein.

4. **The correct answer is (C).** ATP synthesis in mitochondria depends on a proton gradient across the inner mitochondrial membrane. Allowing protons to flow through the membrane diverts the energy of the proton gradient from ATP production, thereby decreasing or stopping it. The cells can continue the rest of oxidative phosphorylation and will actually increase the rate of glycolysis, the Krebs cycle, and electron transport

to restore the pH gradient and ATP production. This will increase oxygen consumption even as ATP production falls.

5. **The correct answer is (E).** Enzymes are biological catalysts that reduce the activation energy of reactions by stabilizing the reaction state intermediate. They do not change the equilibrium of reactions, and they catalyze both forward and backward reactions. Enzymes do not reduce the entropy of a system; the entropy of a closed system always increases.
6. **The correct answer is (A).** A negative ΔG means that a process is thermodynamically favorable and will occur spontaneously without additional energy. The processes named in answer choices (B) through (E) require energy to occur. Biosynthesis, including translation to form peptide bonds or the synthesis of triglycerides, requires energy, usually in the form of ATP hydrolysis. Pumping protons from a low concentration to a high concentration against a gradient takes energy. Moving mitochondria, or any other form of mechanical movement, also requires energy and would have a positive ΔG.
7. **The correct answer is (D).** A complete digestive tract has a separate mouth and anus with a tubelike structure connecting them. Cnidarians and flatworms have a simple gastrovascular cavity with a single opening, and paramecia and sponges do not have a digestive tract. Mollusks, however, do have a complete digestive tract.
8. **The correct answer is (B).** The anterior pituitary produces a variety of hormones in response to releasing hormones from the hypothalamus, including thyroid-stimulating hormone. GnRH is produced by the hypothalamus, cortisol is produced by the adrenal glands, oxytocin is secreted by the posterior pituitary, and glucagon comes from the pancreas.
9. **The correct answer is (C).** The mouth of a protostome, including annelids, arthropods, and mollusks, originates early in embryogenesis from the blastopore. In deuterostomes (echinoderms and chordates), the blastopore becomes the anus. Other traits of protostomes include spiral cleavage, in which cells rapidly become committed to a developmental fate, and what is called "schizocoelous" formation of the mesoderm and the coelom inside the mesoderm.
10. **The correct answer is (B).** The endosperm is formed from a double fertilization by two sperm. The portions of the plant that create the plant embryo are formed by a single fertilization event.
11. **The correct answer is (B).** The attachment of amino acids to each tRNA and the matching of tRNAs with each codon in mRNA is the key point at which accuracy of translation is achieved. No proofreading occurs after peptide bonds form during translation. Choice (C) describes transcription, not translation, and there is no proofreading in transcription anyway. Neither tRNAs nor splicing perform proofreading.
12. **The correct answer is (A).** Pyruvate is converted to acetyl-CoA to enter the Krebs cycle, but it is not produced by the Krebs cycle. The products in all of the other answer choices are produced as a result of this cyclic metabolic pathway.

13. **The correct answer is (C).** Hermaphroditic organisms such as earthworms produce male and female gametes, but gametes are still produced through meiosis, and zygotes form through union of haploid gametes.

14. **The correct answer is (A).** Meiotic recombination occurs during meiotic prophase I, with homologous chromosomes lining up and crossing over to exchange segments between chromosomes.

15. **The correct answer is (D).** RNA transcription and processing occur in the eukaryotic nucleus, including splicing, polyA tail addition, and 5' cap addition. Translation of all proteins takes place in the cytoplasm, even if the protein being translated is RNA polymerase.

16. **The correct answer is (B).** Apical meristem occurs at the tips of roots, shoots, and stems. Proliferation of cells at the apical meristem causes roots and shoots to grow longer. Lateral meristem makes branches and other parts of the plants grow thicker.

17. **The correct answer is (E).** The glomerulus is the ball of capillaries surrounding Bowman's capsule in each nephron in the kidney. Here, the initial filtration of blood occurs to create the urinary filtrate that is then processed in the rest of the nephron.

18. **The correct answer is (D).** In a reflex arc, a sensory neuron passes information in a simple circuit to a motor neuron, without waiting for information to be sent and processed in the central nervous system. The perception of heat and pain in the brain occurs after the reflex movement has already been initiated.

19. **The correct answer is (D).** A high level of productivity in aquatic ecosystems involves an abundance of sun and nutrients, both of which occur in estuaries. Oligotrophic lakes have very low nutrient levels, as do pelagic oceans. Benthic oceans consist of deep water with little or no light, which limits primary productivity. The Arctic is frigid and receives minimal light.

20. **The correct answer is (A).** For Hardy-Weinberg equilibrium to occur, a population must have no migration, no mutation, no natural selection, random mating, and a very large population size. Increased migration would mean that Hardy-Weinberg does not occur, so (A) is correct.

21. **The correct answer is (B).** Tracheoles are part of the tubules insects use to carry air between the tissues and the external atmosphere. Malphigian tubules are involved in insect excretion. Alveoli occur in mammalian lungs, not insects. Insects do not have special cells for carrying oxygen, and answer choice (E) describes bird lungs, not respiration in insects.

22. **The correct answer is (C).** A postzygotic barrier is a block to reproduction that occurs after fertilization. If offspring are produced when individuals breed but the offspring cannot themselves reproduce, it presents a postzygotic barrier that leads to reproductive isolation and speciation. Answer choices (A), (D), and (E) are prezygotic barriers—factors that would prevent mating or fertilization. Answer choice (B) says nothing about reproductive isolation mechanisms.

23. The correct answer is (B). In double-stranded DNA, the amount of A and T are always the same since A pairs with T, and the amount of C and G are also always equal since they pair with one another. Every A in one strand has a matching T in the other, but there is no reason why A and T must be 50% of the nucleotide content. G and C have three hydrogen bonds when they base-pair correctly, not two. Phosphates are on the outside of the double helix.

24. The correct answer is (D). Hemoglobin is a tetramer of subunits that displays cooperative binding of oxygen. The first oxygen binds to one subunit of the tetramer with relatively low affinity, but this binding changes the conformation of the other subunits so that they bind subsequent oxygen molecules with higher affinity. The tetramer does not dissociate normally, so (A) is incorrect. Also, it stays together whether or not oxygen is bound, so answer choice (C) is incorrect. (B) describes the opposite situation, which does not occur. More acidic conditions reduce the affinity for oxygen, making it diffuse into tissues that are actively metabolizing and where oxygen is needed the most, so (E) is incorrect.

25. The correct answer is (A). Mitochondria have their own small genome. When the egg is fertilized by a sperm, almost all of the cytoplasm, including mitochondria, comes from the egg cell and the mother. Any conditions that are inherited and are encoded in the mitochondrial genome will be passed from the mother to all children.

26. The correct answer is (D). DNA ligase completes DNA replication by sealing together the ends of DNA fragments that are side by side, covalently linking them with phosphodiester bonds.

27. The correct answer is (D). Natural selection is important in evolution, but it is not directed and does not necessarily result in the most optimal solution to the challenges faced by living things. It merely selects for the organisms with the highest fitness. It is, however, essential that a population has genetic variation if natural selection is to occur, so (A) is incorrect. It is also essential that this variation creates a variety of phenotypes that survive and reproduce with varying degrees of success, so answer choice (B) is incorrect. Sometimes kin selection occurs, in which an organism does not itself pass on its alleles to future generations, but it improves the fitness of related animals with which it shares alleles, so (E) is incorrect.

28. The correct answer is (A). A sex-linked trait is carried on the X chromosome. Women have two X chromosomes; men have one. Recessive alleles on the X chromosome are expressed in any male carrying the allele, since they do not have a dominant allele to mask them. Women, however, must be homozygous recessive to express a recessive allele on the X chromosome, so they express sex-linked traits much less frequently. The female hemophiliac must be homozygous recessive, since she expresses the trait, while the man who does not express the trait must carry one copy of the dominant allele. A female child will get one X chromosome from her father carrying the dominant wild-type allele, and one X from her mother carrying the recessive allele. All female children will be heterozygous and will have a normal phenotype, and 0% will have the hemophilia disease trait.

29. **The correct answer is (A).** Chordates and echinoderms may not look similar, but they are both deuterostomes and are more closely related than either one is to the protostomes such as mollusks, arthropods, and annelids.

30. **The correct answer is (D).** All plants have an alternation of generations between the haploid gametophyte and the diploid sporophyte. In the course of evolution, though, the sporophyte has become more dominant. In early plants—the nonvascular plants like mosses—the gametophyte is the larger, more obvious, dominant form. Gametophytes are found in plants, but not in animals, fungi, or protists.

31. **The correct answer is (E).** Cnidarians have few specialized tissues. They lack specialized excretory, circulatory, and respiratory systems, since all of their cells are in proximity to the external environment. They do not have muscle cells, but they do have a decentralized neuronal network that helps coordinate movement.

32. **The correct answer is (A).** Skeletal muscle contraction is triggered by an action potential that spreads throughout the membrane of muscle cells, triggering the release of intracellular calcium, and the sliding filament movement of actin and myosin past each other. Neurons have a chemical synapse with skeletal muscle at the neuromuscular junction, not an electrical synapse, so answer choice (B) is incorrect. Skeletal muscle cells do not convey action potentials to other muscle cells, so (C) is incorrect as well. Myosin, not actin, hydrolyzes ATP during contraction, which means that (D) is an incorrect answer choice, and because actin does not actively lengthen muscle cells, (E) is also incorrect.

33. **The correct answer is (C).** The logistic model describes early, rapid, exponential growth. This slows as density-dependent factors slow the birth rate and/or increase the death rate, and eventually it reaches a stable population size that no longer grows. At this stable population size, the rate of births and deaths is equal.

34. **The correct answer is (E).** While some bacteria fix nitrogen from the atmosphere into forms that other organisms such as plants can use, denitrifying bacteria reverse the process, returning molecular nitrogen to the atmosphere. Mycorrhizal fungi have an important symbiotic relationship with plants, but they do not fix nitrogen, so answer choice (A) is incorrect. Animals get nitrogen from eating plants, so (B) and (C) are incorrect, and because plants rely on microbes for nitrogen fixation, answer choice (D) is incorrect as well.

35. **The correct answer is (C).** Primary protein structure is the sequence of amino acids joined by peptide bonds in the linear protein sequence. Secondary structure is the folding into structures, such as an alpha helix or beta-pleated sheet, that relies on interactions between functional groups in the polypeptide backbone of amino acids near one another in the linear peptide sequence. Tertiary protein structure involves a variety of interactions within a polypeptide chain, including residues that are distant from each other in the linear sequence but are near one another in the folded structure. Quaternary structure occurs between different polypeptide chains. The formation of a cysteine bridge between residues in the same polypeptide chain falls within tertiary structure.

36. **The correct answer is (C).** Adipose cells are fat cells and carry large amounts of triglyceride as their dominant characteristic.

37. **The correct answer is (A).** The upward movement of water through xylem in a tree is driven by the transpiration of water via leaves and the cohesion of water in the xylem that draws water from the roots up through the xylem.

38. **The correct answer is (D).** All animals produce ammonia as a nitrogenous waste from protein metabolism, but ammonia is toxic and cannot be stored. Some aquatic animals excrete ammonia directly, but terrestrial animals generally convert it to some other less toxic form to avoid damage to their own tissues.

39. **The correct answer is (D).** Fungi are defined as absorptive chemoheterotrophs. They secrete digestive enzymes and absorb the digested material. As organisms that rely on organic compounds for carbon and for energy, they are chemoheterotrophs.

40. **The correct answer is (B).** Innate immunity does not respond to specific antigens, protecting generally against a broad variety of pathogens. Macrophages ingest and destroy a broad variety of potential infectious agents through phagocytosis. The other choices are all part of the adaptive immune response, which selects for cells that recognize and combat specific antigens.

41. **The correct answer is (E).** Carbon dioxide passively diffuses through the membrane without a need for a protein. All of the other functions mentioned require proteins in membranes.

42. **The correct answer is (C).** According to the laws of thermodynamics, in a closed system entropy always increases and energy is always conserved.

43. **The correct answer is (C).** Cyclic AMP (cAMP) is a second messenger in signal transduction. A hormone that binds to receptors in the plasma membrane can activate the receptor, which activates a G protein. This in turn activates the enzyme adenylate cyclase that produces cAMP. The cAMP can cause other subsequent events in signaling, such as activating protein kinases or transcription of specific genes.

44. **The correct answer is (D).** A diploid organism in G1 phase has its normal diploid complement of chromosomes, meaning that it has two copies of four different chromosomes—one from the mother and one from the father for each of the four pairs of homologous chromosomes. There are $2 \times 2 \times 2 \times 2$ (or 16) different ways these homologous chromosomes can be combined in gametes.

45. **The correct answer is (C).** Tagging a protein with ubiquitin tags it for destruction in the proteosome.

46. **The correct answer is (B).** If the clones have the same identical DNA sequence as their parent but are still not the same, there must be information affecting their development that is not contained in the DNA sequence of their genome. This information is epigenetic. It includes factors such as DNA methylation or inherited chromatin structures. Most mammalian cells still have most of their genome intact, even if they don't express it. Meiosis is not involved in the formation of the clone, because the genome is taken from a somatic cell, not a gamete.

47. **The correct answer is (D).** MHC class II is expressed in antigen-presenting cells such as macrophages. When these cells internalize and digest a foreign agent, they can present peptides from the foreign cell in MHC II. Material in the blood is not presented in the MHC.

48. **The correct answer is (C).** This is a description of a tundra. The Arctic and Antarctic have almost no vegetation. The taiga and a boreal forest are essentially the same and are dominated by large conifers, with no permafrost.

49. **The correct answer is (C).** The cochlea contains a basilar membrane along its length, with hair cells that detect and respond to vibration. High-pitched sounds, which have the shortest wavelengths, travel the shortest distance along the basilar membrane; low-pitched sounds have a longer wavelength and stimulate hair cells farther away.

50. **The correct answer is (A).** A bell-shaped gradient does not have discrete phenotypes but a gradient of phenotypes. The smoothing of the gradient may represent the effect of the environment in spreading out discrete phenotypes represented by genes. Having multiple genes involved in a trait may also smooth out the gradient because they provide a large number of closely related phenotypes.

51. **The correct answer is (B).** The earlier cells are found in development, the more likely they are to be totipotent—able to differentiate into all possible tissue types. The morula is a very early stage of development in which the embryo is a simple ball of undifferentiated cells. Stem cells are found in adults and have very broad developmental possibilities, but hematopoietic stem cells differentiate into blood cells, and intestinal stem cells probably differentiate only into intestinal cells.

52. **The correct answer is (D).** Vitamin C is required to properly form collagen. A lack of vitamin C weakens connective tissues in which collagen plays an important role, causing the nutrient deficiency known as scurvy.

53. **The correct answer is (B).** Microspores give rise to pollen grains, which give rise to the pollen tube and sperm cells. Sperm cells fertilize female gametes to create a zygote.

54. **The correct answer is (A).** Bacteria have circular DNA genomes. They lack the traits mentioned in the other answer choices, all of which are found in eukaryotes but not bacteria.

55. **The correct answer is (C).** Increasing the number of double bonds in fatty acids makes the side chain of phospholipids more bent, and it interacts less strongly with other membrane lipids. Reducing the interaction between neighboring lipids makes the membrane more fluid. Increasing the length of fatty acids makes the lipids interact more strongly with each other, reducing membrane fluidity. Membrane lipids have no proline. Decreasing the temperature reduces the kinetic energy of lipids and everything else, reducing fluidity. Anchoring membrane proteins in place will affect movement of proteins in the membrane, but it won't increase membrane fluidity.

56. **The correct answer is (C).** The carbon first appears stably in the three-carbon sugar glyceraldehydes-3-phosphate and subsequently in other sugars and other molecules. Organic acids refers to CAM plants and, although rubisco does involve carbon fixation, it is not itself the location of the fixed carbon.

57. The correct answer is (D). All plants have meristematic tissue, including the earliest terrestrial plants. Seeds, flowers, vascular tissue, and dominant sporophytes all evolved later.

58. The correct answer is (B). Smooth muscle has important functions in the reproductive system and the digestive tract. In the digestive tract, it helps to mix and propel food through peristalsis. The pancreas, however, does not contain smooth muscle.

59. The correct answer is (A). EPSPs can add up to reach the threshold depolarization and trigger an action potential in a neuron. IPSPs move in the wrong direction to reach threshold. Motor neurons do not have a great deal of spontaneous depolarization—this answer choice describes the situation in the cardiac pacemaker cells. The entry of chloride or the exit of potassium would make the membrane hyperpolarized and less likely to reach threshold.

60. The correct answer is (E). Eukaryotes have three different RNA polymerase enzymes that perform transcription. Protein coding genes are transcribed by RNA polymerase II, RNA polymerase I transcribes ribosomal RNA genes, and RNA polymerase III transcribes tRNAs. Transcription occurs one nucleotide at a time, although translation reads mRNA three nucleotides at a time in each codon. RNA polymerase does not proofread. RNA, like DNA, is synthesized from 5' to 3', and transcription does not require a primer.

PART III

AP BIOLOGY REVIEW

The Chemistry of Biology

chapter 3

OVERVIEW

- **The atom and radioactive elements**
- **Electrons in the periodic table and chemical bonds**
- **Water and pH**
- **The diversity of carbon bonds and functional groups**
- **Summing it up**

Biology studies life at all levels, all the way down to the atoms that all living things are made of. An appreciation for the unity of life on Earth must begin with an understanding of the building blocks at the atomic level. Much of what goes on in living things can be understood in terms of the molecules of life and the atoms and bonds that comprise biological molecules. From there, we can work our way up to molecules, cells, organisms, and populations, one step at a time. Let's get started.

THE ATOM AND RADIOACTIVE ELEMENTS

As the smallest unit of biologically relevent matter, the atom is a good place to start a discussion of chemistry. Each atom is built of subatomic particles, including **neutrons**, **protons,** and **electrons**. At the center of each atom is a small, dense nucleus, which contains most of the mass of each atom in the form of neutrons and protons. Protons each have one positive charge and neutrons have similar mass but no charge. Atoms of more than 100 different chemical elements are in the **periodic table.** The number of protons each atom has is its **atomic number**. Of these elements, the first ninety-two are naturally occurring on Earth; elements heavier than these have only been seen in man-made nuclear reactions.

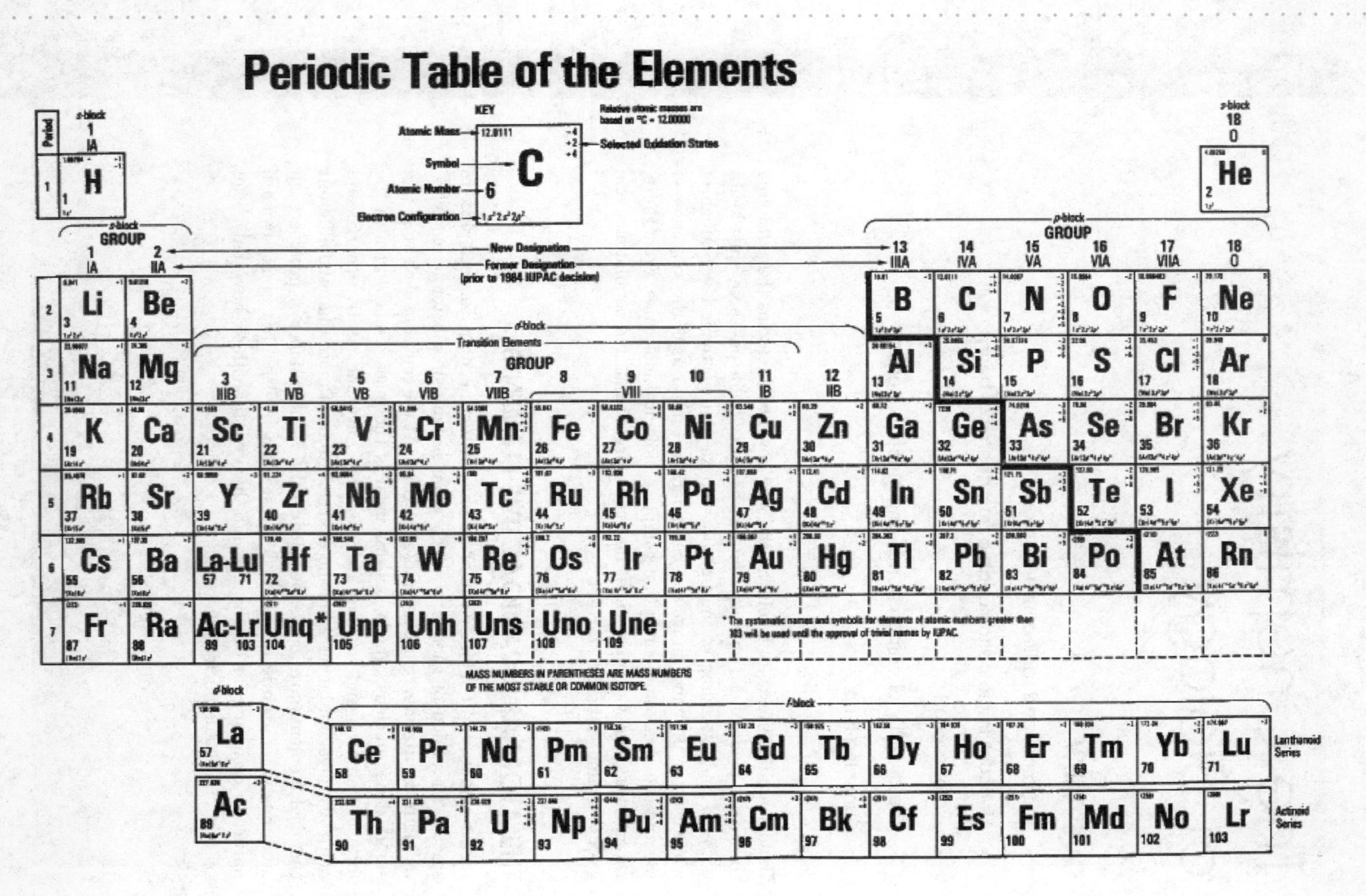
Periodic Table of the Elements
KEY
Atomic Mass
Symbol
Atomic Number
Electron Configuration
12.0111
C
6
Relative atomic masses are based on ¹²C = 12.00000
Selected Oxidation States
New Designation
Former Designation (prior to 1984 IUPAC decision)
s-block
p-block
d-block
f-block
Transition Elements
GROUP
Period
1 IA
2 IIA
3 IIIB
4 IVB
5 VB
6 VIB
7 VIIB
8
9 VIII
10
11 IB
12 IIB
13 IIIA
14 IVA
15 VA
16 VIA
17 VIIA
18 0
1 H
2 He
3 Li 4 Be 5 B 6 C 7 N 8 O 9 F 10 Ne
11 Na 12 Mg 13 Al 14 Si 15 P 16 S 17 Cl 18 Ar
19 K 20 Ca 21 Sc 22 Ti 23 V 24 Cr 25 Mn 26 Fe 27 Co 28 Ni 29 Cu 30 Zn 31 Ga 32 Ge 33 As 34 Se 35 Br 36 Kr
37 Rb 38 Sr 39 Y 40 Zr 41 Nb 42 Mo 43 Tc 44 Ru 45 Rh 46 Pd 47 Ag 48 Cd 49 In 50 Sn 51 Sb 52 Te 53 I 54 Xe
55 Cs 56 Ba La-Lu 57 71 72 Hf 73 Ta 74 W 75 Re 76 Os 77 Ir 78 Pt 79 Au 80 Hg 81 Tl 82 Pb 83 Bi 84 Po 85 At 86 Rn
87 Fr 88 Ra Ac-Lr 89 103 Unq* 104 Unp 105 Unh 106 Uns 107 Uno 108 Une 109
* The systematic names and symbols for elements of atomic numbers greater than 103 will be used until the approval of trivial names by IUPAC.
MASS NUMBERS IN PARENTHESES ARE MASS NUMBERS OF THE MOST STABLE OR COMMON ISOTOPE.
57 La
89 Ac
58 Ce 59 Pr 60 Nd 61 Pm 62 Sm 63 Eu 64 Gd 65 Tb 66 Dy 67 Ho 68 Er 69 Tm 70 Yb 71 Lu
Lanthanoid Series
90 Th 91 Pa 92 U 93 Np 94 Pu 95 Am 96 Cm 97 Bk 98 Cf 99 Es 100 Fm 101 Md 102 No 103 Lr
Actinoid Series

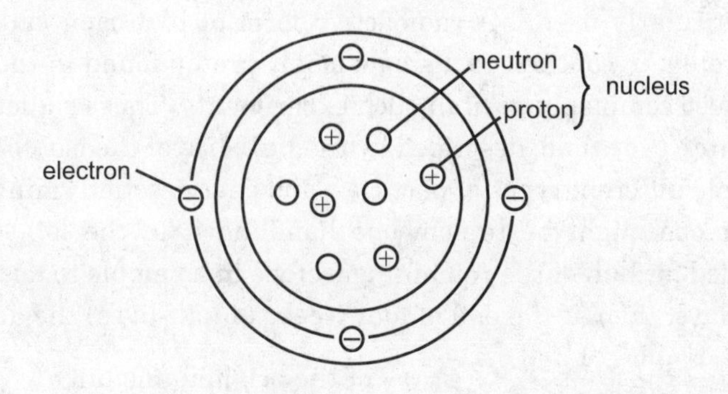

Typical atom (Beryllium)

Electrons surround the nucleus, each with one negative charge and a tiny mass compared to neutrons and protons (about 0.05% of the mass of a proton). An atom in its non-ionized state has the same number of electrons and protons, balancing positive and negative charges to have an overall charge of 0. The electrons are often depicted spinning around the nucleus like planets in a little solar system. The truth is a little more complicated. Electrons are both particles and waves, depending on how you look at them, and their position at a given moment is described not by a specific location but by a probability function. For the purposes of biology, it's a lot simpler to think of electrons as little balls spinning around the nucleus, even if this is not exactly correct. More about electrons later.

The number of neutrons and protons in an atom's nucleus is called its **atomic mass**. A carbon atom with 6 protons and 6 neutrons has an atomic number of 6 and an atomic mass of 12. Sometimes atoms that have the same number of protons (meaning they are all the same element) have varying numbers of neutrons. Atoms of an element that have different numbers of neutrons are called **isotopes.** For example, all carbon atoms have 6 protons, but they can also have 6, 7 or 8 neutrons, forming different isotopes of carbon with atomic masses of 12, 13, and 14. Carbon atoms with atomic masses of 12 or 13 (6 protons and 6 or 7 neutrons) are stable and will essentially last forever, as long as they are not involved in a nuclear reaction. Some isotopes are not stable. For example, carbon with atomic mass of 14 (also called ^{14}C) will spontaneously break down over time, emitting energy and particles. These isotopes are thus described as "radioactive." All elements with an atomic number greater than 83 are radioactive.

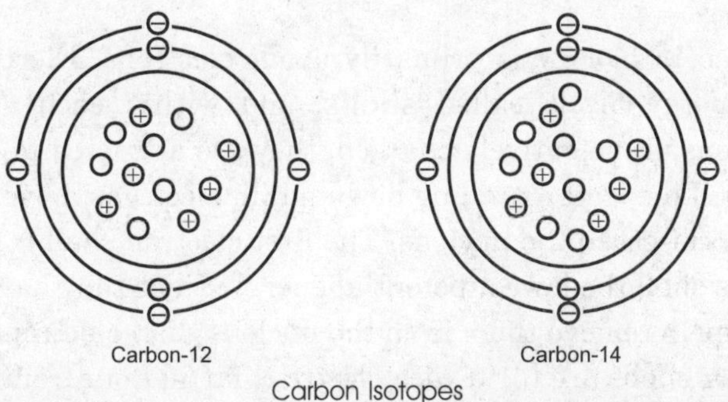

Carbon Isotopes

Biologists often use radioactive isotopes in experiments to label biological molecules and track them. Isotopes of elements found in biological molecules such as carbon, sulfur, and phosphorus are particularly useful. A radioactive form of hydrogen called tritium has two neutrons and one proton, in contrast to a single lonely proton found in the simplest isotope of hydrogen, and it is also commonly used in such experiments. These radioactive elements will decay at a constant rate over time, described by the **half-life** of the isotope. Different forms of radioactive decay yield different types of particles, and energies, determining how penetrating or hazardous an isotope might be to someone handling it in the lab. One half-life is the amount of time needed for half of the remaining isotope in a sample to decay. If the half-life of an isotope is seven days, then at the end of four weeks (4 half-lives) the amount of undecayed isotope remaining is $\frac{1}{2} \times \frac{1}{2} \times \frac{1}{2} \times \frac{1}{2} = \frac{1}{2^4} = \frac{1}{16}$ of the original amount.

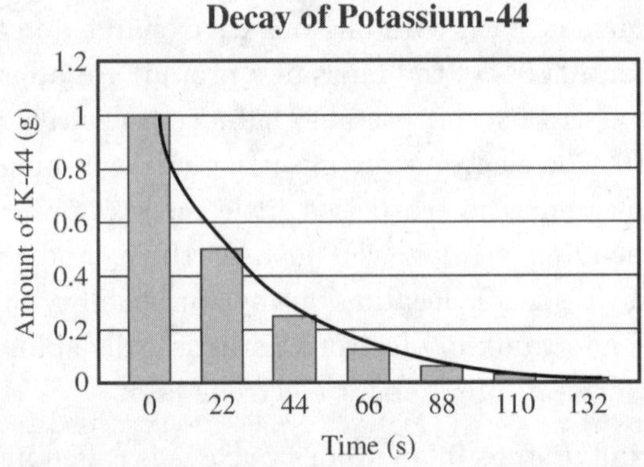

Q The half-life of tritium is about 12 years. If a sample of methane is prepared that contains 8 grams of tritium, how much tritium is left after 36 years?

A 36 years in this example is 3 half-lives for tritium. The amount of tritium left after 3 half-lives is 8 grams $\times \frac{1}{2} \times \frac{1}{2} \times \frac{1}{2} = 8 \times \frac{1}{2^3} = 8 \times \frac{1}{8} = 1$ gram.

ELECTRONS IN THE PERIODIC TABLE AND CHEMICAL BONDS

Electrons

The chemistry involved in biology is primarily about electrons. Electrons are organized in atoms at different energy levels called **shells**, and within each shell electrons occupy **orbitals.** Since electrons are negatively charged, they are attracted to the positively charged protons in the nucleus. Like water running down a hill, electrons move to the lowest possible potential energy level, closest to the nucleus. The first electron shell is closest to the nucleus, giving electrons in this shell the lowest potential energy. Since they have the lowest energy, it would take more energy to remove them from the nucleus than electrons located farther away from the nucleus. Other shells are filled with electrons farther out from the nucleus. When an electron receives energy—getting hit by a photon of light for example—it will move to an orbital in a higher shell. If the electron loses the energy again, emitting it back into its surroundings, it will move back down to a lower energy shell closer to the nucleus. Electrons cannot jump just any distance between higher and lower energy levels; they can only absorb or release discrete ("quantum") amounts of energy when they jump between specific energy levels. It is as if electrons must jump up and down a staircase, not slide up and down a sloping ramp. The quantum nature of energy given off by electrons is the reason that atoms have spectra with specific characteristic wavelengths of light that they absorb or release, or, "fluoresce."

QUANTUM NUMBERS AND ORBITAL CONFIGURATIONS FOR THE FIRST 3 PERIODS OF THE PERIODIC TABLE OF THE ELEMENTS

n	l	Sublevel	$m =$	Number of Orbitals	Electrons in Sublevel	Electrons in Energy Level
1	0	1*s*	0	1	2	2
2	0	2*s*	0	1	2	8
	1	2*p*	−1, 0, 1	3	6	
3	0	3*s*	0	1	2	
	1	3*p*	−1, 0, 1	3	6	18
	2	3*d*	−2, −1, 0, 1, 2	5	10	

Q If an atom absorbs a photon of light at a specific wavelength of 500 nm and then fluoresces to emit light at another wavelength, will the fluorescence wavelength be higher, lower, or equal to 500 nm?

A Higher wavelengths of light have lower energy. Although energy is conserved when atoms fluoresce, some of the energy is converted to heat. When an atom absorbs energy as light and then emits the energy again by emitting light, the light coming out has lower energy than the light that went in, and the additional energy is released as heat. The wavelength of fluorescence will be higher than the wavelength of absorbance.

Electrons sit in different shells and have different energy levels depending on what kind of atom the electrons are a part of. An atom of hydrogen has only one proton and one electron, so this electron will be in the first shell closest to the nucleus unless it becomes excited, absorbing energy to jump to a higher level. Helium has two protons and two electrons. As it happens, the first electron shell only has room for two electrons, so helium already has its first shell filled.

In the atoms of increasingly heavy elements—those with a larger number of protons—the accompanying electrons occupy more and more electron shells, filling one shell after another. The number of electrons in an atom's outer shell determines how the atom will behave chemically with other atoms. All atoms have the same goal in their atomic life: to have a filled outer shell of electrons. A lot of chemistry is about how atoms go about trying to accomplish this. Outer-shell electrons are often called **valence electrons**. Atoms with the same number of electrons in their outer shell usually behave in similar ways. The periodic table of the elements groups each element according to how many electrons it has in its outer shell. The column on the far left side of the table contains elements with a single electron in their outer shell. The column on the far right side of the table contains elements with a full outer shell. The first shell only has room for two electrons so there are only two elements on the first level of the periodic table, hydrogen and helium. The second and third electron shells have room for eight electrons, so there are eight elements on these levels in the periodic table.

Q In examining the periodic table, which of the following elements would be expected to behave with chemistry similar to sodium?

(A) Chlorine

(B) Helium

(C) Oxygen

(D) Lithium

(E) Carbon

A Chemical behavior is largely determined by the number of valence electrons, and the elements are grouped in columns in the periodic table that have the same number of valence electrons. Just by looking at the table and seeing that lithium and sodium are in the same column, it is reasonable to guess that they will have similar chemical properties.

An electron shell's electrons are distributed among specific orbitals in which electrons "live." The orbitals are not like the orbits of planets around the sun but rather are regions of space where an electron is statistically likely to be located. When filled, each orbital can hold two electrons. The two electrons have a different "spin," which we don't have to worry about.

Chemical Bonds

Up to this point we have only talked about atoms. Atoms are bonded together in various ways to form **molecules.** In a **covalent bond,** valence electrons are shared between two atoms, often completing the outer electron shell of the atoms involved in the bond. An atom of hydrogen has one electron in its outer shell and room for one more electron to fill its outer (and only) shell. One way for an atom of hydrogen to get this additional electron and fill its shell is by cozying up to another atom of hydrogen. If two atoms of hydrogen each share their single electrons with each other, they both have two electrons in their shells, and both are happy (happy for atoms, at least). This sharing of electrons is a chemical bond (a covalent bond in this case), between the atoms involved. When atoms are joined this way, they form a molecule. Sometimes atoms can share more than just two electrons. Atoms that share two pairs of electrons, each pair in different orbitals, form a double bond, which is much stronger than a single bond. Atoms can even share three pairs of electrons, forming a triple bond.

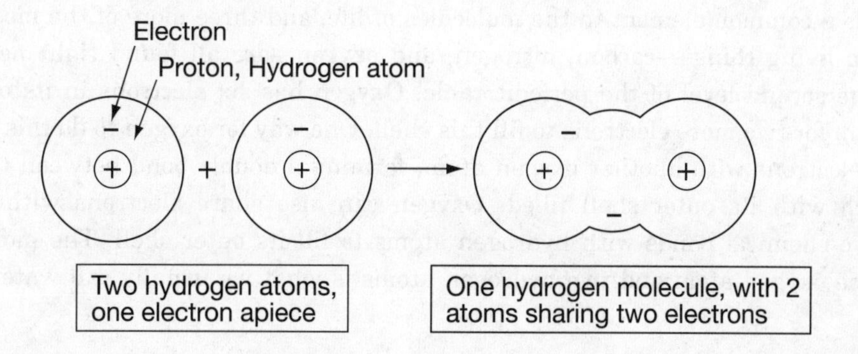

The bonds between atoms in molecules are represented in several different ways. Sometimes the individual electrons involved are sketched out, to better understand how the bonds form and the chemistry involved. This way of representing molecules is often called "Lewis Structures." At other times, the bond is represented by a line between atoms, in which the line represents the two shared electrons in a covalent chemical bond. A double bond with two pairs of electrons shared by two atoms is drawn as two lines between the atoms connecting them and a triple bond is drawn as three lines. With a little experience you can start to tell what sort of structures make sense (or not) by looking at structures.

The Lewis Structure for carbon dioxide, CO_2.

Ö::C::Ö or Ö=C=Ö

The Lewis Structure for the molecule PCl_3.

:Cl:P:Cl: or :Cl—P—Cl:
:Cl: | :Cl:

Generally, the rules for the most common elements in biological molecules are:

Element	Number of unpaired valence electrons	Number of single bonds an atom forms
Hydrogen	1	1
Carbon	4	4
Oxygen	2	2
Nitrogen	3	3

Element number 2 in the periodic table is helium, with two electrons in its outer shell. Since its valence electron shell is already filled, helium feels no great urge to share its electrons with anyone, and pretty much keeps to itself, chemistry-wise. This means that helium is not chemically reactive: it is resistant to forming any chemical bonds. This is true of all of the elements on the right-hand column of the periodic table, making these elements the "noble gases" (noble because too self-sufficient to mix with the more "needy" elements), including neon, argon, and xenon,

Hydrogen is a common element in the molecules of life, and three more of the most common elements in living things—carbon, nitrogen, and oxygen—are all found right next to each other on the second level of the periodic table. Oxygen has six electrons in its outer shell, leaving room for two more electrons to fill this shell. One way for oxygen to do this is to share two of its electrons with another oxygen atom, forming a double bond between two oxygen atoms, each with its outer shell filled. Oxygen can also share electrons with hydrogen, forming two chemical bonds with hydrogen atoms to fill its outer shell. The molecule that contains one oxygen atom and two hydrogen atoms is what we usually call water (more on this later).

Q Nitrogen has an atomic number of 7. How many single bonds can a nitrogen atom usually form?

A With an atomic number of 7, a nitrogen atom will have two electrons in the first shell and five electrons in the second shell. There are four orbitals in the second shell that contains five electrons. Each of the first four electrons has its own orbital, but the fifth electron must pair up with another electron. This leaves three unpaired electrons available to participate in a covalent bond, sharing electrons with another atom. The answer is three.

Q How would the chemistry of ^{13}C be different from the chemistry of ^{12}C?

A For the most part it wouldn't be any different. Nuclei don't play much of a role in chemistry—the electrons are the whole story. Whether a carbon nucleus has six neutrons or seven neutrons will change the weight of the atom, but it will have almost no effect on how the electrons are involved in forming and breaking chemical bonds.

When electrons are shared in a covalent bond, it is as if the nuclei of the two atoms are both pulling at the electrons in a game of tug of war. Sometimes the shared electrons in the bond are pulled equally by both atoms involved, and their charge is split equally between the atoms, making the chemical bond non-polar, meaning, with no apparent charge on either side. If the atoms involved in a bond are both the same element (as is the case in a molecule of oxygen, O_2), they will both pull equally at the shared electrons, leaving the electrons firmly in the middle and creating a non-polar bond. Some atoms, however, pull electrons harder than others in covalent bonds. Oxygen is very hungry for electrons, pulling harder on electrons in chemical bonds than other atoms. This tendency to strongly attract electrons is called **electronegativity.** It creates polar bonds in which the atoms in the bond have partial charges, one negative and one positive. Because oxygen is quite electronegative, in oxygen-hydrogen bonds (O-H) the oxygen will have a partial negative charge and the hydrogen will have a partial positive charge. Nitrogen is also fairly electronegative. Carbon and hydrogen draw electrons to almost the same degree, so carbon-hydrogen bonds (C-H) are only very slightly polar.

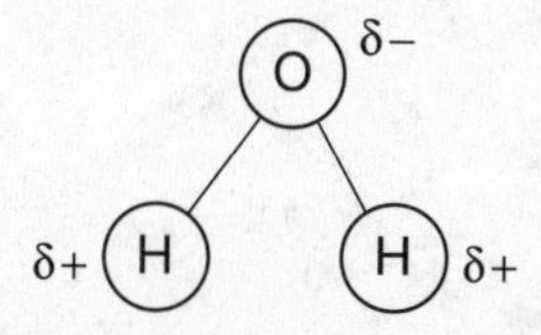

Electronegativity

In extreme cases, an atom will attract electrons so strongly from another atom that it strips the electron away from its partner. Sodium—and the rest of the elements sharing its column in the periodic table—has only a single valence electron, whereas chlorine has seven, needing only one more to have a full outer shell. Chlorine has such a strong attraction for that one electron it needs to complete its outer shell that it will, given the chance, steal an electron from sodium, leaving the sodium atom missing one electron and having a charge of +1. The chlorine, having picked up one extra electron, will have a charge of −1. The resulting, charge-based ("electrostatic") attraction between the +1 sodium atom and the −1 chlorine atom will strongly bind the two atoms after chlorine's theft of sodium's electron has occurred. This is called an **ionic bond.** An ionic bond does not form a molecule. In a salt crystal of sodium chloride, many atoms of sodium and chloride pair up in a regular lattice with no overall charge: each positively charged sodium atom pairs with one negatively charged chloride ion.

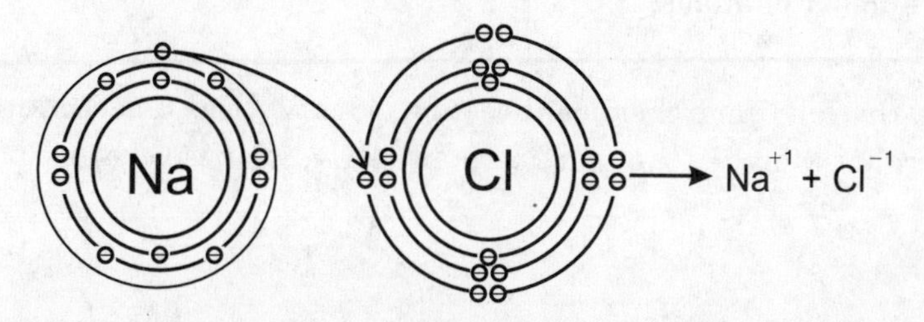

Formation of Ions

Other types of bonds are weaker but still essential for biology. When hydrogen forms a covalent bond with oxygen, oxygen draws electrons away from hydrogen, leaving a small positive charge on the hydrogen and a small negative charge on oxygen. Nitrogen does the same thing. These partial positive and negative charges can attract each other, linking two molecules together in **hydrogen bonds**. Hydrogen bonds are weaker than covalent bonds, holding less energy, and thus requiring less energy to break. They are quite common in biological molecules and thus they play a very important role in biology.

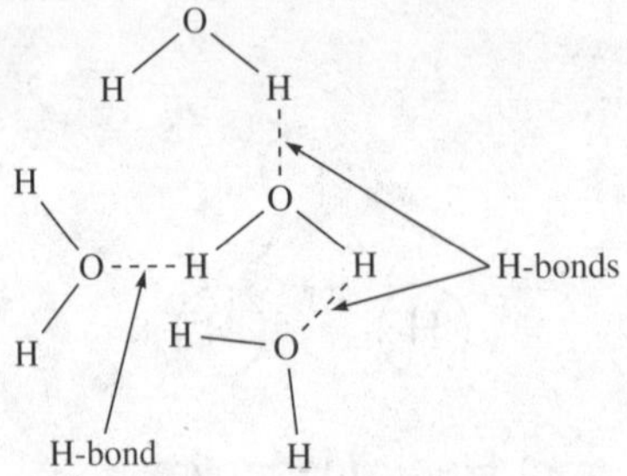

Another very weak interaction between molecules is called the **van der Waals attraction,** generally involving atoms in non-polar covalent bonds. Even in non-polar bonds, electrons are constantly moving around in shared orbitals, and this motion can create a momentary unequal distribution of charges in the molecule, which is known as a "transient dipole." Even though these charge differentials are small, and constantly changing, the net effect still creates a small attraction between non-polar molecules, such as the molecules called lipids in membranes, for example, as we shall see.

Chemical reactions occur when chemical bonds form or break. Chemical reactions are written in a form with molecules on two sides separated by arrows. The molecules on the left side are what the reaction starts with, the **reactants.** The molecules on the right side are the **products** formed as a result of the reaction. In a chemical reaction, matter is neither created nor destroyed. Therefore, the number of atoms must be the same on both sides of a chemical reaction. "Balancing" a chemical reaction entails making sure that both sides of the equation contain the same number of atoms.

Q Examine the following chemical reaction, identify the reactants and products, and determine whether the reaction is properly balanced:

$$C_6H_{12}O_6 + O_2 \rightleftarrows CO_2 + H_2O$$

A This reaction describes the breakdown of a sugar (like glucose) into carbon dioxide and water. The reactants, sugar and water, are on the left side, and the products, carbon dioxide and water, are on the right side. The equation is not properly balanced as written. There are six carbons on the left side and only one on the right side. Also, there are twelve hydrogens on the left side and only two on the right side. To balance the equation, there need to be six carbon dioxide molecules and six water molecules on the product side. This would then result in eighteen oxygen atoms on the right side, meaning that more oxygen is needed on the left side of the equation. Having six O_2 molecules, along with the six oxygens in the sugar, would equal eighteen oxygens on both sides of the equations. The balanced equation would then be:

$$C_6H_{12}O_6 + 6O_2 \rightleftarrows 6CO_2 + 6H_2O$$

Sometimes a chemical reaction can move forward as written, from left to right, or backward, from right to left. This can be written in a chemical formula by having arrows pointing in both directions as in the sugar example above. If you add more reactants, often the rate at which the products of the reaction form increases. If the forward reaction and the backward reaction are occurring at the same rate, then the reaction is said to be in **equilibrium.** Equilibrium in a reaction is not static, but dynamic: a balanced state in which reactions moving in both directions do so with equal speed. (We'll discuss equilibrium and related aspects of chemical reactions in the section about enzymes.)

WATER AND pH

Water

Water has many unique properties that make life on Earth possible. All of these properties stem from the structure of the water molecule, H_2O. Each water molecule contains an oxygen atom that forms covalent bonds with two hydrogen atoms. Oxygen has six valence electrons; in a water molecule it fills its outer electron shell by sharing an electron with each of the two hydrogen atoms. Since oxygen is highly electronegative, it draws the electrons in these bonds more tightly to itself, creating polar bonds with the two hydrogen atoms. The overall V shape of the water molecule comes from the angled arrangement of the electron orbitals in oxygen that form bonds with hydrogen. With the oxygen on one side of water drawing electrons toward it and the two positive hydrogen atoms sticking out on the other side, the water molecule as a whole is quite polar.

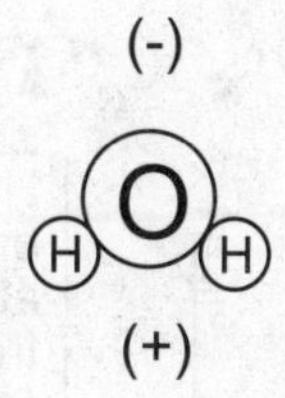

Water molecule

Polar water molecules form hydrogen bonds with other water molecules (or other polar molecules), each oxygen molecule attracting the hydrogen molecules in neighboring water molecules. At the same time, rapidly shifting hydrogen bonds connect networks of water molecules. The sticking of water molecules together with each other in this way is called **cohesion.** Water in a narrow tube displays the property of cohesion.

Surface tension is another key property of water molecules for living things. The water molecules found at the boundary with air bind to each other, forming a layer on the surface so strong that some insects (like water striders) can walk on it. Surface tension causes a drop of water to round up on waxed paper or when it is falling through the air. Surface tension is also an important factor in the lungs, at the interface of air with the surface of the lungs where gases diffuse.

The attraction between water molecules also affects their movement and heat in the liquid phase. The more molecules move, the greater their heat. The hydrogen bonds between water molecules make it harder for water molecules to move, meaning that more energy must be added to break the hydrogen bonds before the molecules will move and raise the heat. The relatively greater energy required to heat up water, as compared to other liquids, gives water a **high specific heat**. So, whereas it takes one calorie of heat energy to heat up 1 ml of water by 1°C, most liquids require less heat energy to achieve the same temperature increase. Living things generally need to have fairly constant temperatures both inside their bodies and in their immediate surroundings. Since most living things contain a large quantity of water, the high specific heat of water means that it takes a large amount of energy to change their internal temperature, protecting organisms against large or sudden changes in temperature. The high specific heat of water also means that bodies of water, like lakes and oceans, enjoy a fairly constant temperature. The relatively stable temperature of the oceans around the world not only helps the living things within those oceans but also helps keep the climate in neighboring landmasses more moderate.

The hydrogen bonds between molecules in water also hold molecules in the liquid phase, preventing them from escaping to form the gas water vapor. When water molecules escape and become a gas, they take a lot of heat energy with them, helping to cool the water left behind through **evaporative cooling**. Evaporative cooling helps land creatures keep their body temperature from growing too hot. On a hot day, you sweat, and as the sweat evaporates from your skin, it draws heat out of your body, helping to cool you. This is the only way you can maintain your body temperature below the external temperature on a very hot day, aside from immersing yourself in a convenient body of water!

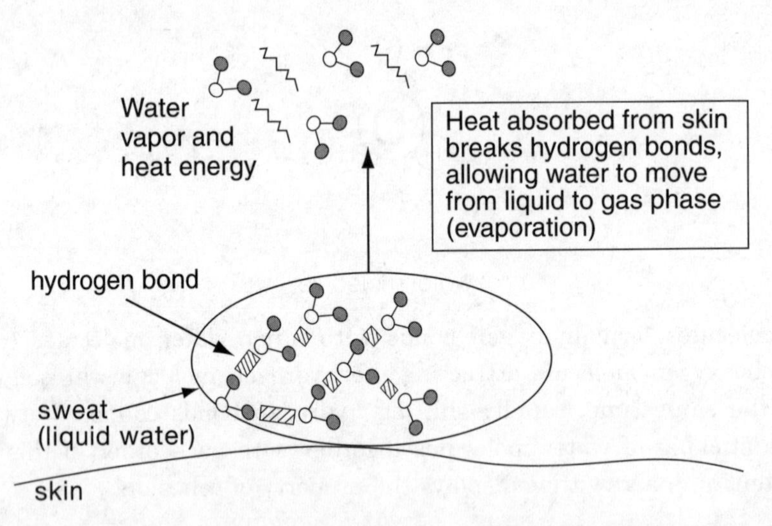

The bonds between water molecules are also found in solid water, more commonly known as ice. In ice, the hydrogen bonds become more permanent, frozen in place. Ice has the distinction of having lower density than water, an unusual property. Most solids are *more* dense than their corresponding liquid phase. As a result, ice floats, as you can see in your soda glass or in the arctic ice cap. If this were not true, if water sank when it solidified, this would affect a lot more than your soda. If ice were heavier than liquid water, then lakes and oceans would freeze solid *from the bottom up,* freezing all of the creatures living there as well. Not a good thing. Freezing from the top of the water in a lake or ocean down insulates the water beneath from the colder air temperatures above while also protecting creatures living beneath the ice.

Another distinct property of water is its ability to act as a solvent for a wide variety of substances, particularly charged or polar substances. This ability derives from the partially charged nature of the water molecule. When a water molecule encounters a positively charged substance like a sodium ion, the oxygen end of water gathers around, neutralizing the positive sodium charge with its own negative charge. Similarly, the hydrogen end of water gathers around a negatively charged chloride ion with its positive charge, neutralizing the charge and allowing the chloride ion to move into solution. With polar but non-charged molecules like sugars or ethanol, water molecules form hydrogen bonds that stabilize the molecule in solution. Our blood, our cells, and the oceans are full of dissolved material that is essential to life, all of which is dissolved in water in this fashion.

Water also dissolves gases that are important for life, including oxygen and carbon dioxide. For aquatic organisms like fish, the oxygen dissolved in the surrounding water allows them to extract the oxygen with their gills and use it to perform oxidative respiration (to be explained in a later chapter) underwater. Cold water holds more dissolved oxygen than warm water, so global warming may affect the ability of aquatic animals to extract the oxygen they need from their environment. In mammals, oxygen must diffuse through a thin layer of water lining the lungs to move into the blood, and from the blood into the tissues. Carbon dioxide also dissolves in water, forming carbonic acid, bicarbonate, and carbonate. These dissolved forms of carbon dioxide in blood transport carbon dioxide and prevent large changes in the acidity of blood, as we shall see.

While water is a great solvent, it does have its limits. Non-polar molecules have structures that do not favor interaction with water. These compounds are called "hydrophobic" and are composed mostly of carbon-carbon and carbon-hydrogen bonds. Examples are the fats and lipids found in animal fat or vegetable oils. Oil does not mix with water because the non-polar fatty compounds in oil favor interaction with each other rather than with water. The way that water interacts with these molecules is quite important in biology, affecting for example, both how membranes form around cells and organelles and how proteins fold.

One last property of water that is important for living things is the tendency of water molecules to occasionally break apart. Water is a very stable compound, but every so often a hydrogen from one water molecule will move over to another water molecule, leaving behind its single electron. The naked hydrogen ion is a proton, with a charge of +1, and the water molecule it leaves behind is called a hydroxide ion, OH^-. The proton and hydroxide ion can also rejoin each other to reform a new water molecule, reversing the reaction. Both processes happen at a low rate, leaving an average concentration in pure water of about 10^{-7} M for both hydrogen ions and hydroxide ions. "10^{-7} mol/L" is an example of the measurement of concentration of a solute in solution called **molarity.** A **mole** is defined as 6.022×10^{23} units of solute (either atoms or molecules); its symbol is "mol." The "L" part refers to "liters," a measurement of liquid volume.

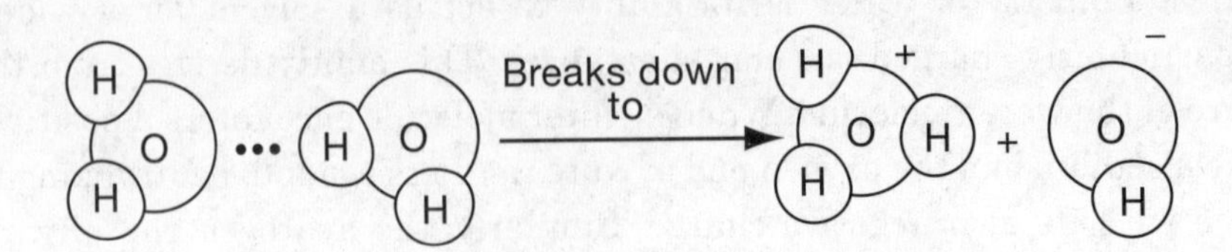

Spontaneous Breakdown of Water

Although water produces hydrogen ions infrequently, there are other substances that have a strong tendency to donate protons when they are dissolved in water, which increases the hydrogen ion concentration significantly. These substances are called **acids,** including strong acids like hydrochloric acid (HCl), and weaker organic acids like the acetic acid found in vinegar. When 0.1 M HCl is dissolved in water, virtually all of the HCl dissociates, creating a solution of 0.1 M hydrogen ions, as well as 0.1 M chloride ions. The additional hydrogen ions remove hydroxide ions from solution, so as the concentration of hydrogen ions increases, the concentration of hydroxide ions decreases. The relationship between the concentration of hydrogen ions and hydroxide ions is:

$$(H^+)(OH^-) = 10^{-14}$$

If the concentration of H^+ ions is 0.1 M, this is 10^{-1} M, which means that the concentration of hydroxide ions is 10^{-13} M, and the product of their concentrations is 10^{-14} ($= (10^{-1}) \times (10^{-13})$).

pH

Acid and base concentrations are conventionally represented as the negative log of their concentration, using a scale called pH. For example, the pH of a solution with 10^{-7} M hydrogen ions is:

$$pH = -\log(10^{-7})$$
$$pH = 7$$

A pH of 7 is called a neutral pH because the concentration of acidic H^+ ions and basic OH^- ions is equal. Because the pH scale is logarithmic, a change in pH that seems small, like going from pH 6 to pH 7, is really a ten-fold change in the concentration of hydrogen ions. A pH lower than 7 is acidic (more H^+ than OH^-) while a pH greater than 7 is basic (more OH^- than H^+).

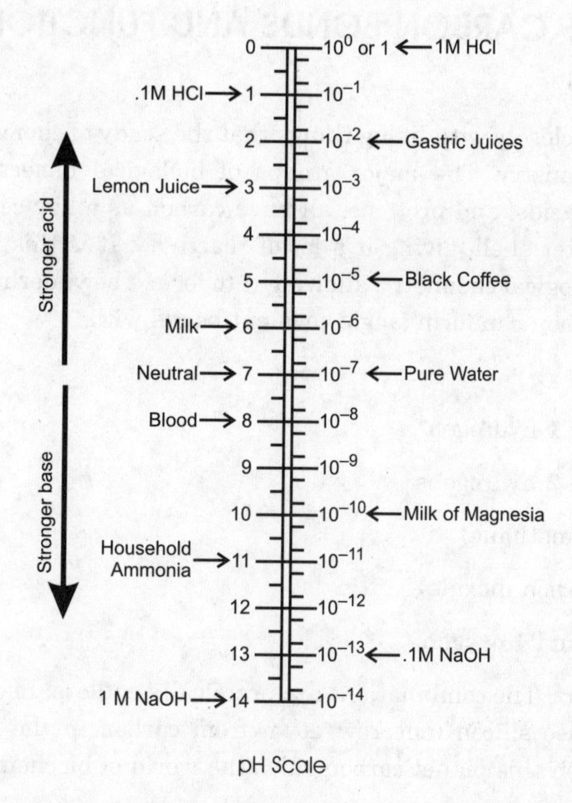

pH Scale

The pH of blood, the cytoplasm of cells, and most other fluids of the body is usually close to 7, and biochemical processes can be severely affected by even slight changes in pH. Molecules called "buffers" help to keep pH changes moderate, soaking up added hydrogen ions by acting as a proton acceptor. If hydrogen ions bind to buffer molecules, then they will not be available to change the pH of the water. Similarly, the buffer will donate protons back to the solution if a base is added, helping to keep the pH from becoming as basic as it would in the absence of the buffer. Biologists often use buffers to maintain a specific pH when they are doing experiments, and living systems use buffers in blood or other fluids to ensure a constant pH in the body.

Q If a solution of water is pH 10, what is the concentration of hydroxide ions?

A The pHs of hydrogen ions and hydroxide ions *always* add up to 14. At neutral pH, they are both at 10^{-7} M, or pH 7, and $7 + 7 = 14$. At pH 10, the concentration of hydroxide ions is $14 - 10 = 4$.

THE DIVERSITY OF CARBON BONDS AND FUNCTIONAL GROUPS

Carbon Bonds

Carbon is associated so closely with living things that the study of chemistry involving carbon is called "organic" chemistry. The major groups of biological molecules, including lipids, carbohydrates, nucleic acids, and proteins, all have carbon as a key component. Carbon has four electrons in its outer shell, giving it a lot of reactivity. It is this reactivity that makes carbon so central in biological chemistry, allowing it to form a bewildering diversity of possible structures. A single carbon can form single covalent bonds with:

- **4** carbons
- **3** carbons and **1** hydrogen
- **2** carbons and **2** hydrogens
- **4** hydrogens (methane)
- **2** oxygens (carbon dioxide)
- **3** hydrogens and **1** oxygen

And that is just the start! The combinations are practically endless. Life on some other planet in the universe may use silicon (one row down from carbon in the periodic table) as the foundation of life, but on this planet carbon rules the world of biochemistry.

Molecules built entirely of hydrogen and carbon are called **hydrocarbons**. Pure hydrocarbons are not usually found in living things. However, molecules like fatty acids, or lipids that contain fatty acids, do contain a hydrocarbon, making this part of the molecule very hydrophobic ("water-hating") or lipophilic ("fat-loving").

Functional Groups

Functional groups are parts of molecules that feature certain types of bonds. Functional groups help determine how molecules interact with each other. Here is a brief description of various functional groups:

Group	Structure	Properties
Hydroxyl	R*–O–H	Found in alcohols like ethanol, and in sugars, making them polar
Carbonyl	R=O	An oxygen with a double bond to a carbon
Carboxylic acid	R=O \OH	Acidic groups, found in amino acids, acetic acid (vinegar)
Amino	$R-NH_2$	Amino groups can often act as proton acceptors (a basic group) at neutral pH, and become $R-NH_3^+$, a charged group

*"R" in structures is often used to represent the rest of the molecule, whatever that may be.

Closely related chemical structures are called **isomers**. The molecular formula of a compound does not fully describe its structure. The molecule C_4H_{10} (butane), for example, can have either carbons all in a straight chain or three carbons all joined to a central carbon.

```
   H  H  H            H  H  H  H
   |  |  |            |  |  |  |
H—C—C—C—H        H—C—C—C—C—H
   |  |  |            |  |  |  |
   H  C  H            H  H  H  H
     /|\
    H H H
```

Compounds that have the same number of atoms connected differently are called **structural isomers**. Fructose and glucose are both $C_6H_{12}O_6$, but one has a carbonyl on the first carbon in the molecule and another has a carbonyl on the second carbon. It may not seem like much, but a difference like this in a chemical structure is a big deal for biological systems.

Stereoisomers are compounds that have the same number of atoms, and the same covalent bonds joining them all together, but with different three-dimensional structures. One special type of stereoisomer is called an **enantiomer,** which is one of two types of molecule that are mirror images of each other.

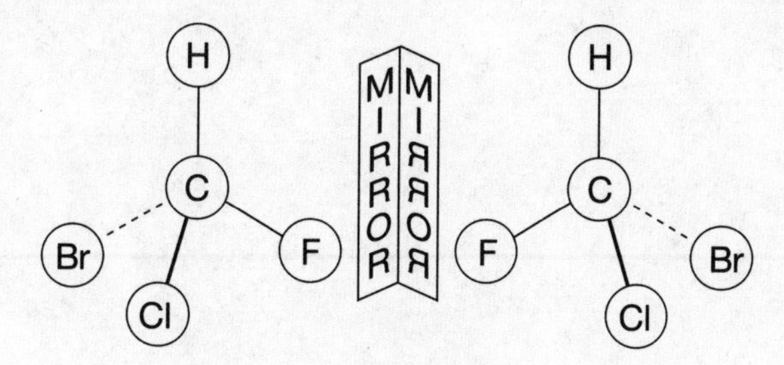

Enantiomers

Since carbon can form four covalent bonds, if the bonds connect the carbon to four different chemical groups, then there will be two different ways that the groups can be arranged around the carbon. Such a carbon is called a **chiral** carbon, and the arrangement of the four groups around the atom can result in two enantiomers, mirror images that cannot be superimposed no matter how much you twist or rotate them around. Many biological molecules contain chiral carbons, meaning that two different stereoisomers are possible, but only one is usually found in living things. For example, there are two possible stereoisomers of each amino acid, called the D and L forms, but only the L form is found in living things. This specific selection of stereoisomers in living things demonstrates the importance of stereochemistry.

Diastereomers are stereoisomers, but not mirror images. For example, sugars like glucose often have not just one chiral carbon, but several. This makes many different stereoisomers possible for compounds with the same reactivity, some of which are enantiomers of each other, but others of which are diastereomers of each other.

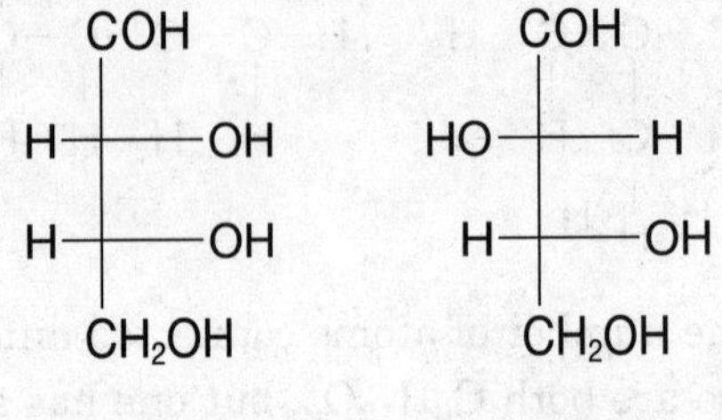

COH
H—OH
H—OH
CH_2OH
D-Erythrose

COH
HO—H
H—OH
CH_2OH
D-Threose

EXERCISES: THE CHEMISTRY OF BIOLOGY

1. If a nucleus has 6 protons and 7 neutrons, what element is it? (checking the periodic table is okay)

(A) Oxygen
(B) Carbon
(C) Aluminum
(D) Nitrogen
(E) Neon

2. For an isotope containing 6 protons and 8 neutrons, what is the atomic number?

(A) 6
(B) 8
(C) 14
(D) 16
(E) It is not possible to determine with the information given.

3. If ^{14}C has a half life of 5,700 years, and a gram of it was present in a sample of charcoal 11,400 years ago, how much remains today?

(A) 0.12 grams
(B) 0.25 grams
(C) 0.50 grams
(D) 0.62 grams
(E) 0.75 grams

4. How long does it take for exactly 0 grams of a 100 g sample of ^{14}C to be remaining?

(A) 0 half lives
(B) 20,000 years
(C) 100,000 years
(D) 10 half lives
(E) It is not possible to say.

5. Which of the following bonds does not involve sharing or transferring electrons, but attraction between transitory charges caused by uneven distribution of electrons in a non-polar molecule?

(A) Covalent single bond
(B) Ionic bond
(C) Hydrogen bond
(D) van der Waals interaction
(E) Covalent double bond

6. $\_\_H_2 + O_2 \rightarrow 2\ H_2O$

If hydrogen gas and oxygen gas are mixed and ignited, they burn, producing water and heat. How many molecules of hydrogen (H_2), indicated by the blank in the chemical reaction, are required to balance this equation?

(A) None
(B) 1
(C) 2
(D) 3
(E) 4

7. The pH of 10 mls of lemon juice is measured in the laboratory with a pH meter and found to be pH = 2.3. When 1 ml of a solution is added to the lemon juice and stirred, the new pH of the solution is found to be 4.3. How much did the concentration of OH^- ions change in this solution?

(A) It increased by 50%.
(B) It increased by 100%.
(C) It increased by 10,000%.
(D) It decreased by 50%.
(E) It decreased by 1,000%.

8. The molecular weight of glucose is 180 grams/mole. How much glucose does it take to produce 5 liters of a 0.2 molar solution of glucose?

(A) 1.8 grams
(B) 3.6 grams
(C) 18 grams
(D) 36 grams
(E) 180 grams

ANSWER KEY AND EXPLANATIONS

1. B	3. B	5. D	7. C	8. E
2. A	4. E	6. C		

1. **The correct answer is (B).** Any nucleus with 6 protons is carbon. A carbon atom with 7 neutrons is an isotope with an atomic number of 6 (the number of protons) and an atomic mass of 13.

2. **The correct answer is (A).** A nucleus with 6 protons and 8 neutrons is carbon with an atomic number of 6.

3. **The correct answer is (B).** If the half life is 5,700 years, then 11,400 years is two half lives. Starting with one gram, the amount remaining after two half lives is 1 gram $\times \frac{1}{2} \times \frac{1}{2}$, or 0.25 grams.

4. **The correct answer is (E).** This is a bit of a trick question. With each increasing half life, half of the remaining amount decays, but one half still remains. The amount remaining gets increasingly close to 0 with each half life, and will be exceedingly small as time passes, but can never be said to reach exactly 0 grams. Practically speaking, the amount of radiation remaining becomes so small at some point that it can no longer be detected above the background radiation, which for most of us is the same as 0. In this case, the answer would depend on the sensitivity of the instrument you use to detect the radiation, but this would be a different question.

5. **The correct answer is (D).** Since electrons are not shared or transferred, the bond is not covalent (A) or ionic (B). Since the interaction is based on transitory charges in a non-polar compound, not lasting partial charges in a polar molecule, the answer is not hydrogen bond (C).

6. **The correct answer is (C).** There must be the same number of atoms on both sides of a properly balanced chemical formula. In this formula, there are two atoms of oxygen on both sides of the formula already. On the right side, there are two molecules of water that each contain two hydrogens, for a total of four. To balance the equation, there must be two molecules of hydrogen on the left side, choice (C).

7. **The correct answer is (C).** The pH scale is logarithmic, and each 1 unit change in pH indicates a ten-fold change in the concentration of hydrogen ions. The concentration of hydrogen ions (H^+) and hydroxide ions (OH^-) are related to each other, so that the product of $(H^+)(OH^-) = 10^{-14}$, and a ten fold increase in one is always accompanied by a ten-fold decrease in the other. When the pH changes from 2.3 to 4.3, the concentration of H^+ ions goes down by 100 fold, and the concentration of hydroxide ions goes up by 100 fold. A 100-fold increase in hydroxide ions is the same as a 10,000% increase, choice (C).

8. **The correct choice is (E).** It will take the same amount of glucose to make 5 liters of 0.2 molar glucose as it would take to make 1 liter of 1 molar glucose. 1 liter of 1 molar glucose requires 1 mole of glucose (Molarity = moles/liter), which is 180 grams of glucose.

SUMMING IT UP

- Each element consists of atoms which contain a specific number of protons, known as the atomic number.
- Atoms of an element can have different atomic masses, depending on how many neutrons a given atom has. Atoms of the same element with different numbers of neutrons are called isotopes.
- Some isotopes are unstable and exhibit radioactive decay. An isotope decays according to a half-life that describes how long it takes for half of the remaining sample to decay.
- Elements have different numbers of valence electrons in their outer shells; these electrons are important in chemical reactions.
- In covalent bonds, atoms share electrons to fill their outer shells.
- The strong electronegativity of oxygen or nitrogen creates partial charges in some bonds.
- Hydrogen bonds between partial charges are weak but important bonds between water molecules and biological molecules.
- Ionic bonds consist of transfer of electrons from one atom to another, leaving each with a charge and strong electrostatic attraction.
- Van der Waals interactions are caused by transitory unequal distribution of charges within hydrophobic molecules.
- Water has many unique properties that are essential to life which are mostly due to strong hydrogen bonding between water molecules in liquid water and partial charges on each water molecule. These properties are: strong cohesion, strong surface tension, high specific heat, high molar heat of evaporation, low density of ice, and strong solvent properties.
- pH is a measure of the hydrogen ion concentration in water: $pH = -\log(H+)$

Biological Macromolecules

chapter 4

OVERVIEW

- **Carbohydrates**
- **Nucleic acids**
- **Lipids**
- **Proteins**
- **Summing it up**

All living things share four classes of biological molecules that carry out essential activities of living systems:

1. Carbohydrates
2. Nucleic acids
3. Lipids
4. Proteins

Many of the molecules of life are **macromolecules:** long chains of simple subunits that form larger molecules or, "polymers." The reactions that form macromolecules are generally reversible. **Dehydration** removes one water molecule when two subunits of a macromolecule are joined together.* Hydrolysis adds water to the two subunits of the macromolecule to break apart the bonds.**

*2 Subunits ⇄ 1 water + subunit-subunit

**1 water + subunit-subunit ⇄ 2 subunits

Generally, the assembly of biological polymers requires an input of energy as well, often in the form of the energy-carrying molecule. ATP (more on this crucial molecule later). The hydrolysis reaction is usually "energetically favorable"—meaning, not requiring the input of energy—and is part of the digestion process of food, as well as part of the recycling of biological molecules in the cell to construct new molecules.

CARBOHYDRATES

Carbohydrates are sugars. The simplest sugars are **monosaccharides**, consisting of a single relatively small sugar, like glucose. Carbohydrates are quite varied but have the general chemical formula $C_nH_{2n}O_n$. Simple sugars are often characterized by the number of carbons in each molecule. Sugars with five carbons are called **pentoses;** sugars with six carbons are called **hexoses**. Another feature used to distinguish sugars is whether they have a carbonyl group on the end of the carbon chain (a ketone), making them **ketoses**, or on a carbon inside the chain (an aldehyde), making them **aldoses**. One of the most important sugars for biology is **glucose**, which has six carbons, making it a hexose, and a ketone group, making it a ketose. Glucose is an important form of energy for almost all living things. It is the starting point for glycolysis and the Krebs cycle, which convert chemical energy into ATP (more on this in a later chapter). Another common monosaccharide is **fructose,** the sugar found in fruit. As a hexose, fructose is very similar to glucose, except that the carbonyl group is differently positioned as an aldehyde. Sugars are also used to build other macromolecules. Ribose, for example, is a pentose that provides one of the building blocks of nucleic acids.

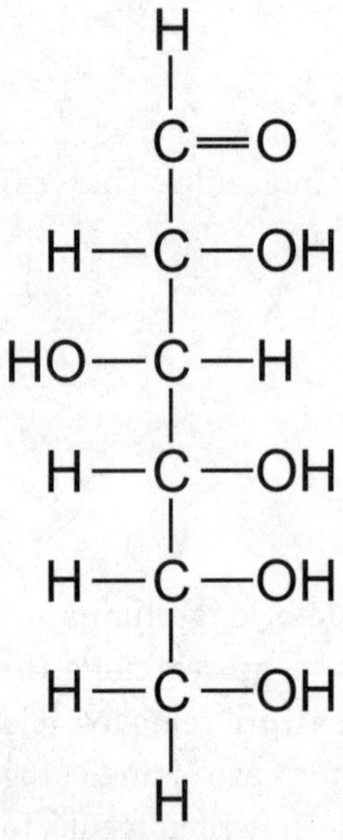

```
       H                  H
       |                  |
       C=O             H—C—OH
       |                  |
    H—C—OH               C=O
       |                  |
   HO—C—H             HO—C—H
       |                  |
    H—C—OH             H—C—OH
       |                  |
    H—C—OH             H—C—OH
       |                  |
    H—C—OH             H—C—OH
       |                  |
       H                  H
```

Glucose 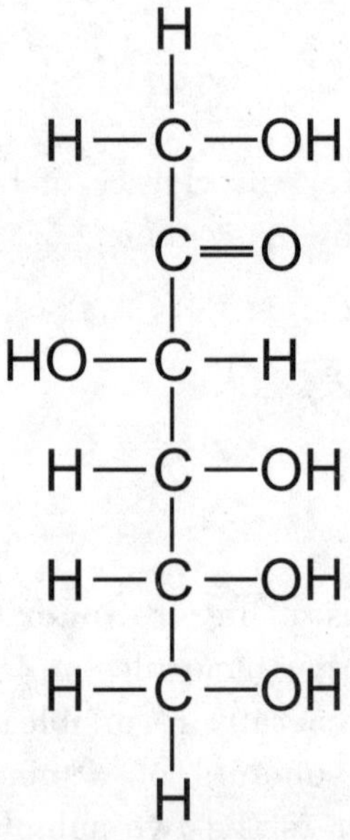 Fructose

The chemical structures of sugars are often drawn in the "stick format" shown above, with the carbons lined up in the middle. The stick format does not capture the entire story. Many sugars in water solution display a spontaneous intramolecular reaction: an OH group at one end binds with the carbonyl on the other end of the chain, forming a ring structure. The ring structure and chain structure can convert back and forth in solution, with one form dominating at equilibrium based on what is most energetically stable. There are two different versions of the ring structure that can form in solution; the particular ring structure formed depends on from which side of the chain the intramolecular reaction occurs.

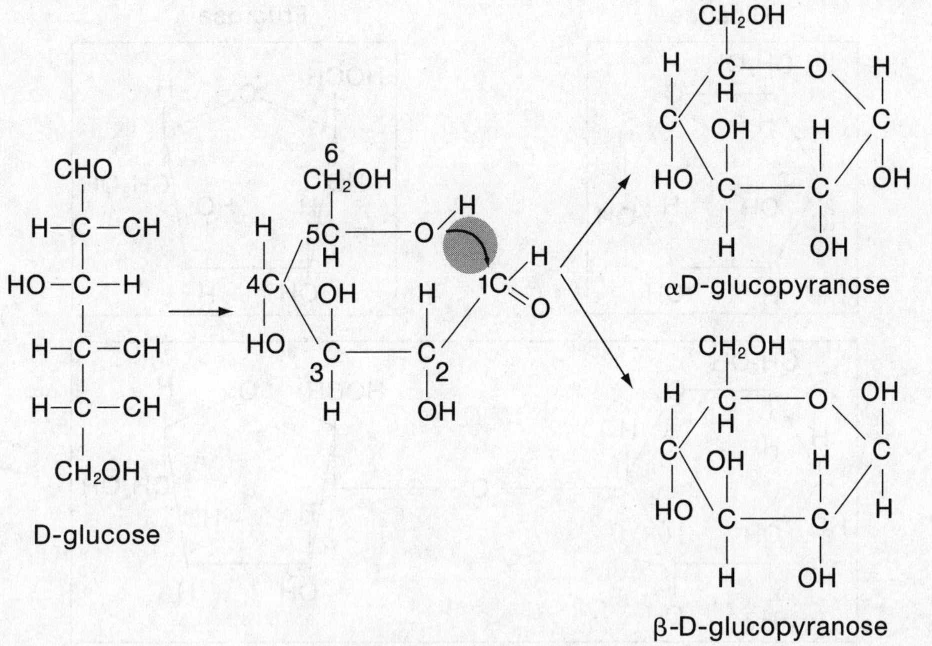

D-glucose

αD-glucopyranose

β-D-glucopyranose

Disaccharides consist of two sugar monomers joined together. One common example of a disaccharide is sucrose, "table sugar." Each molecule of sucrose contains one glucose unit that is linked to one fructose unit (see figure on next page). When a disaccharide is synthesized, a **glycosidic bond** is formed between the two sugar monomers. Different sugars contain different types of glycosidic linkages, varying the connecting bond of the sugars and the orientation of that bond. In sucrose, the glycosidic bond connects carbon #1 on glucose with carbon #2 on the fructose. Formation of a glycosidic bond is a dehydration reaction, removing water where the bond is formed. Reversing the reaction is a hydrolysis reaction in which a water molecule is added back into the disaccharide to break the glycosidic linkage and release the two sugar monomers again.

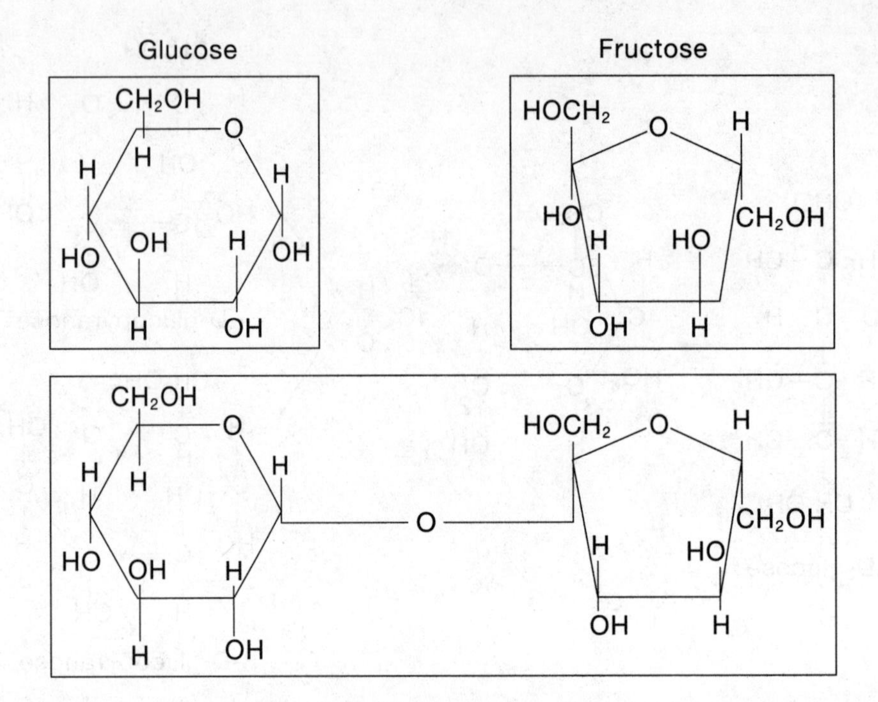

Sucrose and water

Synthesis of sucrose from glucose and fructose with removal of a water

Q How many water molecules are required to balance the reaction depicting the hydrolysis of sucrose into two monosaccharides:

Sucrose + __H_2O → glucose + fructose

A It takes one water molecule for every glycosidic bond that is broken. A disaccharide contains a single glycosidic linkage holding the two individual sugars together, so a single water molecule is needed per each sucrose molecule.

Polysaccharides

Glycosidic linkages can join sugar monomers in long chains to form polysaccharides, which sometimes contain thousands of sugar monomers linked together. Polysaccharides build support structures in the cell wall of plants, and can also be used to store energy as starch. The type of glycosidic linkages between sugar subunits distinguishes the structure and function of different polysaccharides. In animals, one of the major polysaccharides is **glycogen**, found in granules in liver and skeletal muscle cells. Glycogen is built of glucose molecules joined end-to-end, with many branches connected together to give it more of a tree-like structure than a simple chain. The many branches provide lots of "loose ends" where more glucose units can be quickly added or removed, as needed by the cell. Animals use glycogen as a short-term energy supply, releasing glucose from glycogen when energy is wanting and locking glucose back into glycogen when energy is ample. Although glycogen provides an important short-term energy supply that can be rapidly mobilized to release glucose when quick energy is needed, only a small percentage of the human body's overall energy is stored in glycogen. A much greater percentage is stored in fats.

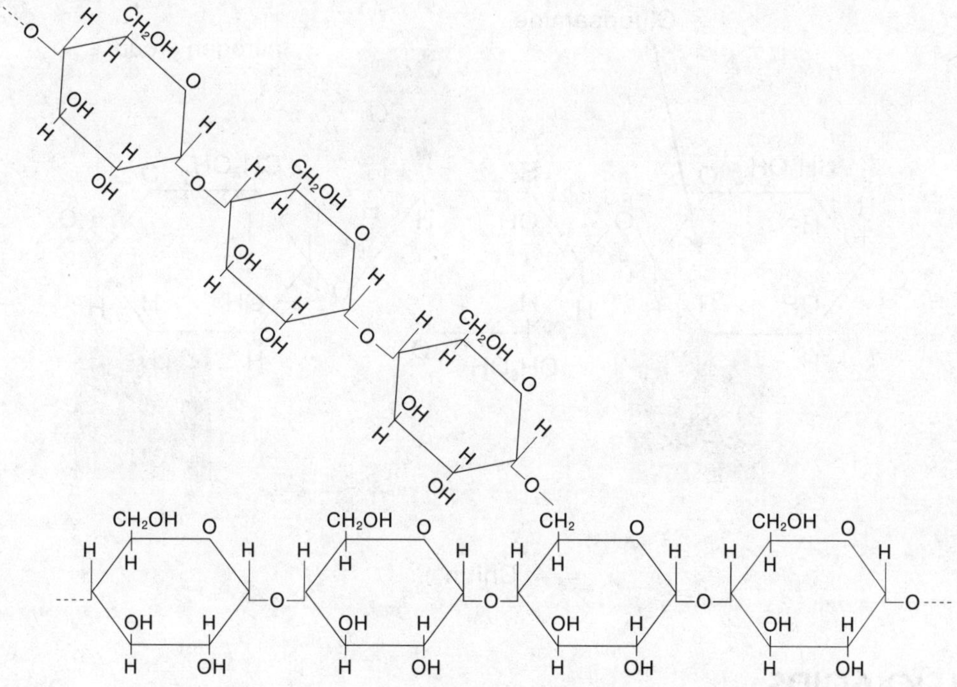

Part of a Glycogen "tree"

Plants do not make glycogen. But many plants store energy in a similar polysaccharide, **starch**. The starch polymer in potatoes is also composed of glucose, as in glycogen, but the starch is less branched. When plants have surplus energy, they can store some of their glucose in starch and then release the glucose as needed later. For animals that eat plants, the starch can also provide energy, starting with the hydrolysis of starch polymers to release glucose. The starch polymer is a long helix, with the glucose subunits spiraling.

Another important polysaccharide is **cellulose**, one of the most common biological molecules on the planet. Unlike glycogen and starch, cellulose does not store energy. Cellulose contains a different form of glycosidic linkage that locks the glucose subunits in a long, stiff chain, rather than the loose helix of starch. The stiff glucose chains in cellulose are cross-linked into tough, strong strands that are not easily broken. They are used to hold the plant up against gravity and other forces. Every plant cell is surrounded, supported, and protected by a cellulose cell wall, and the cell walls of many cells make up the stiff supporting structures we see in trees, grasses, flowers, and all other plants.

Many herbivores, like cows, deer, and elephants, eat only plants like grasses that contain a great deal of cellulose. But animals do not make the enzymes that break down cellulose into the glucose subunits that are needed for energy. These animals host microbes that do the job for them. Even termites depend on microbes to digest the wood they eat. These cellulose-digesting microbes live in a win-win relationship, called **symbiosis,** with their cow or termite hosts, a concept we will come back to in Chapter 14.

Another common polysaccharide is called **chitin**, which forms the cell wall of fungi and the exterior skeleton of arthropods, such as insects and crustaceans. Chitin is similar to cellulose but contains a modified form of glucose with a nitrogen group called glucosamine.

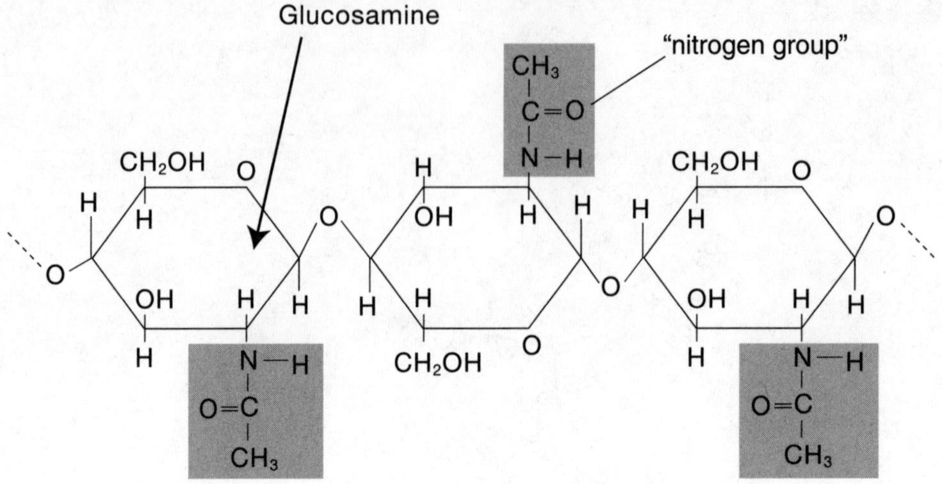

Glucosamine

"nitrogen group"

Chitin

NUCLEIC ACIDS

The two types of nucleic acids are DNA and RNA. Both are involved in storing hereditary information and converting that information into proteins that do the work of living things. Nucleic acids, like carbohydrates, are macromolecular polymers built from many small regular subunits strung together. But the types of subunits they contain, and how the subunits fit together, are quite distinct from the other biological macromolecules.

The DNA of an organism is known as its genome. It contains the hereditary information that defines an organism. The genome is contained in every cell, at least when the cell is first created, and is passed to new daughter cells in mitotic cell division, passing from generation to generation of organisms through reproduction. DNA is often described as the instruction book of the cell, or its operating system, containing all of the instructions for growth, metabolism, development, reproduction, and survival.

Each subunit of RNA or DNA is called a **nucleotide**, and the polymerized nucleotides are called **polynucleotides.** Each nucleotide subunit contains a nitrogenous base, a sugar (generally a form of ribose), and a phosphate (see figure on next page). There are two types of bases, purine and pyrimidine. DNA has two purines (adenine and guanine, or A and G) and two pyrimidines (cytosine and thymine, or C and T). In DNA the sugar group in each nucleotide is a form of the five-carbon sugar ribose called deoxyribose—hence: "**d**eoxyribo**n**ucleic **a**cid." The carbons in this sugar are numbered from 1 to 5 and, depending on which carbon is involved in bonding to other molecules, will be referred to as 3′ or 5′. When nucleotides are polymerized together in a chain, a covalent bond is formed between the phosphate of one nucleotide and the deoxyribose of the adjacent nucleotide. This bond is called a phosphodiester bond because the phosphate is joined by ester linkages to sugars on two sides. Each ribose is connected on the 5′ side to a phosphodiester bond to a neighboring nucleotide and on the 3′ carbon in deoxyribose to another neighboring unit. As the polynucleotide grows, the base group in each nucleotide subunit projects to the side away from the backbone. When synthesized in the cell, polynucleotides always grow from the 5′ end to the 3′ end.

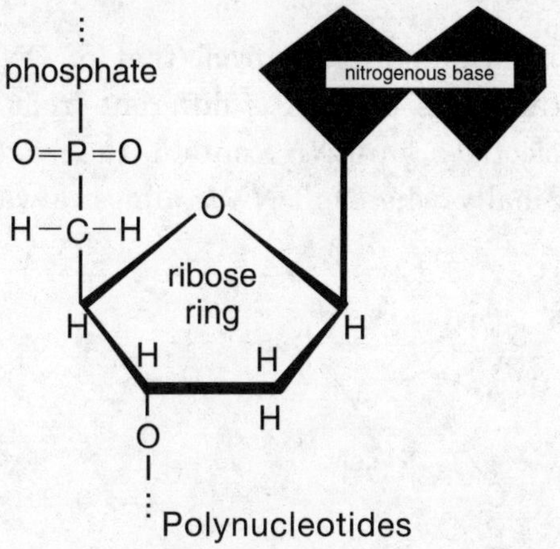

phosphate

nitrogenous base

O=P=O

H–C–H

O

ribose ring

H H H H

H

O

Polynucleotides

However, a single strand of a DNA polynucleotide does not make a complete DNA molecule. DNA usually occurs with two polynucleotide chains wound around each other in a shape called the **double helix**. One of the two strands in the double helix runs from 5′ to 3′; the other strand runs in the opposite direction or **antiparallel.**

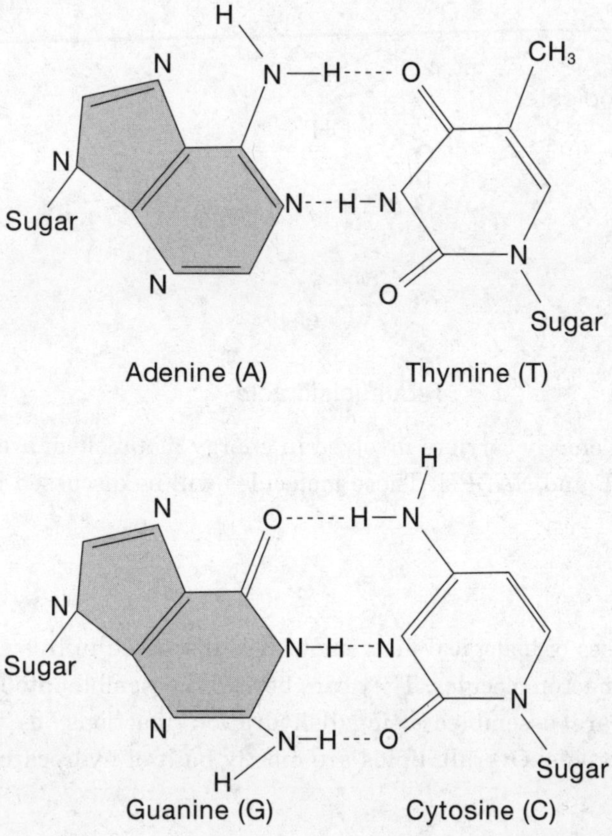

Adenine (A) Thymine (T)

Guanine (G) Cytosine (C)

In both cases, the phosphate-ribose backbones face outward toward the surrounding solvent and the base groups face inward. DNA always contains the same percentage of A and T, and of C and G. The reason for this is that the bases in each DNA strand pair with each other, and the pairing is highly selective. In DNA, A can only pair with T, and C can only pair with G. The pairing of bases is limited by the available space on the interior of the double helix, the orientation of the two strands as they wrap around each other and bring the bases together, and the hydrogen bonding of bases.

The structure of RNA is somewhat different from that of DNA. RNA also contains four different bases, but one of the bases in RNA is different from DNA—uracil (u) instead of thymine (T). Also, the nucleotides in RNA contain the sugar ribose rather than the deoxyribose found in DNA. Finally, whereas DNA is almost always double-stranded, RNA is mostly single-stranded.

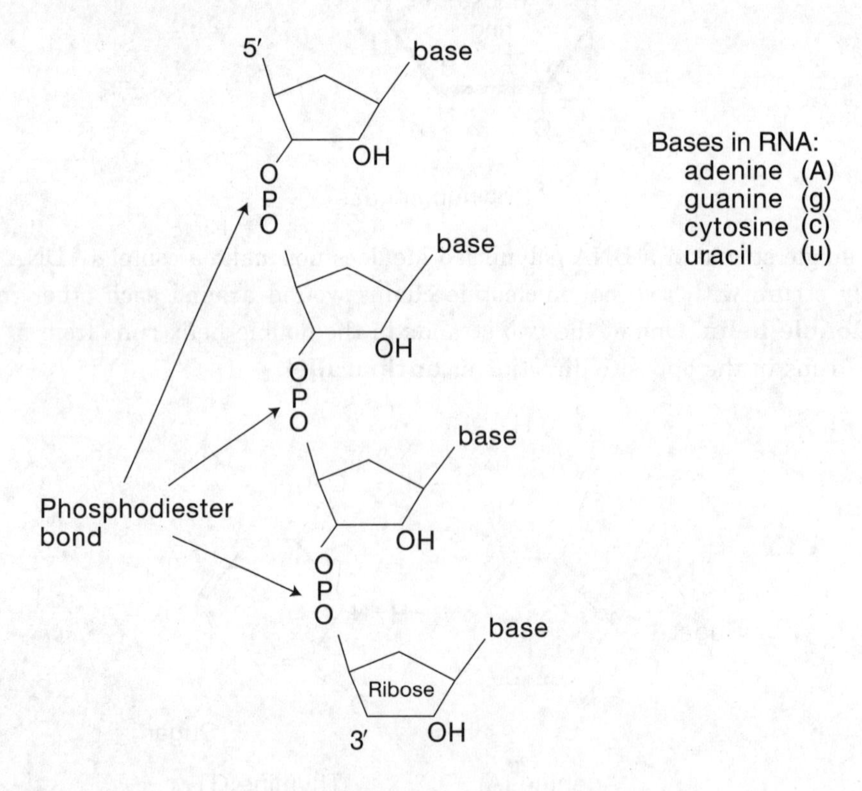

Ribonucleic acid

Some of the essential energy carriers involved in energy metabolism are nucleotides as well, including ATP, NADH, and NADPH. These molecules will be discussed in Chapter 7.

LIPIDS

Unlike the other classes of biological macromolecules discussed in this chapter, lipids or fats are not polymerized macromolecules. They can, however, assemble into large structures. The individual subunits are assembled using dehydration reactions, as with carbohydrates, nucleic acids, and proteins. Overall, lipids are mostly built of hydrocarbon chains that repel water. Lipids do not mix with water, tending to prefer to interact more with other non-polar lipids, but they will generally dissolve in more non-polar solvents like hexane.

Fatty acids are a common component in lipids. A fatty acid contains a long hydrocarbon chain that is very hydrophobic, with a carboxylic acid group at one end. Fatty acids can contain carbon chains of varying length, often as long as twenty carbons. If the fatty acid contains only single bonds between carbons, it will contain the greatest possible number of hydrogens, and the fatty acid is called **saturated**. **Unsaturated** fatty acids have a double bond somewhere along the length of the carbon chain, which causes them to bend in the middle. The long, straight fatty acids in saturated fats make the fatty acids tend to line up and stick to each other, solidifying at a relatively low temperature. Unsaturated fatty acids don't pack together as well due to the kink in the middle, so they tend to remain liquid at higher temperatures than saturated fats. Animal fat is high in saturated fat, while vegetable oils are high in unsaturated fats. This is why butter is a solid at room temperature and vegetable oils are liquids. The saturated fats we ingest when eating meat are more likely to lead to heart disease, in part because they accumulate and solidify more readily in the lining of the arteries.

```
       H   H   H   H   H   H   H
 O\\    |   |   |   |   |   |   |
    C–C–C–C–C–C–C–C–H
HO/     |   |   |   |   |   |   |
       H   H   H   H   H   H   H
```

Saturated

```
                                H
 O\\                             |
    C–C–C–C=C–C–C–C–H
HO/     |                   |   |
       H                   H   H
```

Unsaturated

Fats

One form of lipid used for energy storage is **triglyceride**, or **triacylglycerol**. A triglyceride contains a single glycerol, with three hydroxyl groups, each of which is bound to a fatty acid carboxyl group. There are three fatty acids bound to a single glycerol in a triglyceride. The synthesis of a single triglyceride involves the removal of three water molecules, and reversing the reaction to form free glycerol and fatty acids requires three waters to be added. When fats are eaten, their digestion involves the hydrolysis of triglycerides so that the fatty acids can be absorbed. Since fatty acids can have different lengths, triglycerides can also contain fatty acids of varying length, even in a single triglyceride molecule. Adipose cells in animals contain large quantities of triglycerides in storage.

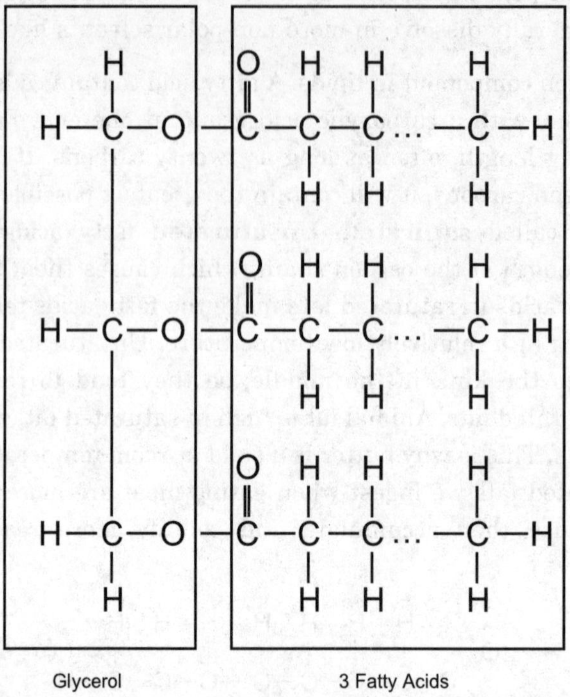

Glycerol 3 Fatty Acids

Triglyceride Structure

Triglycerides are a more concentrated form of energy storage than polysaccharides like glycogen. They are also more compact, meaning they are a more efficient way for mobile animals to store energy. Animals generally store the majority of their energy reserves as lipids and only a small fraction as glycogen or other carbohydrates.

Triglyceride deposits in animals also are very effective at insulation against a cold external environment. Animals living in cold climates, like seals, whales, walruses, and polar bears, tend to have a thick layer of fat under their skin that insulates them from the cold.

Phospholipids

Another type of lipid, the **phospholipids** are like a detergent, with a hydrophobic fatty acid tail on one side and a polar, hydrophilic end on the other. If phospholipids are mixed with water, they spontaneously arrange themselves to minimize contact of the fatty acid tails with water, arranging the tails in a layer facing each other, with the polar heads facing out toward the water. These spontaneous phospholipid structures can either be large sheets with two layers of phospholipids or rounded vesicles, in which the two sheets curl up into a ball. The arrangement of phospholipids in sheets forms the membranes that surround cells and the organelles inside eukaryotic cells, selectively forming a boundary to keep material in or out of the cell as needed.

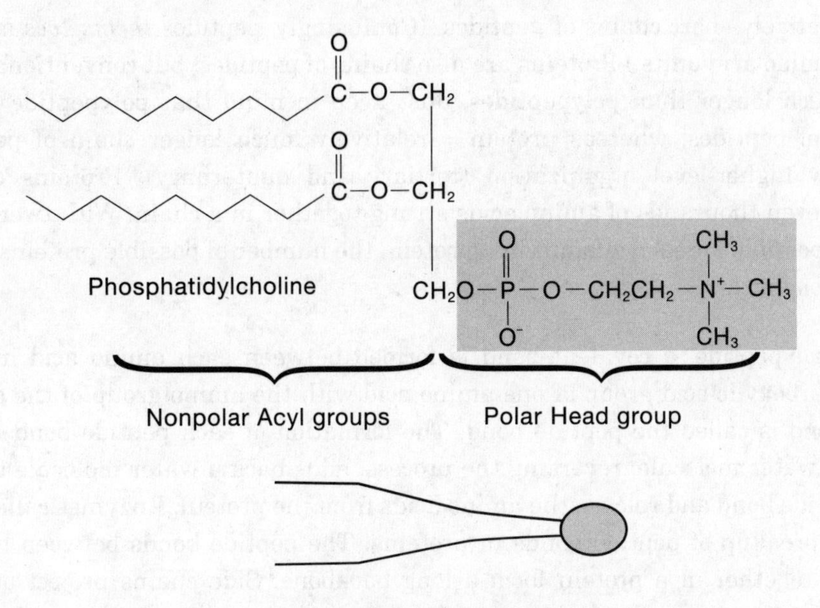

Phospholipids are Amphipathic

PROTEINS

While nucleic acids store and transfer information, the information they carry is used to build the real workers of living things, proteins. Proteins do a wide range of jobs based on the structure of each individual protein, and they have a huge variety of structures.

Proteins are macromolecular polymers built from subunits called amino acids. There are twenty different amino acids with the generalized structure having a carboxylic acid on one end, an amino group on the other end, and a carbon with a varying side chain (denoted here by the placeholder, "R") attached.

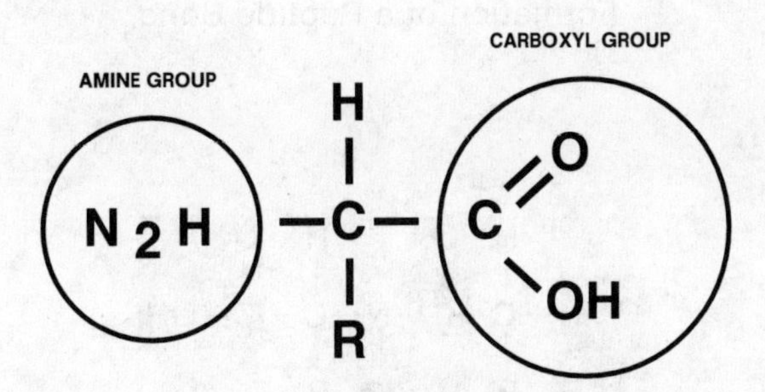

Typical Amino Acid Structure

The nature of the side chain is what makes the twenty amino acids different from each other. Some side chains are hydrophobic and non-polar, built of hydrocarbons (valine, leucine, isoleucine, phenylalanine, and alanine). Other side chains are polar, but not charged, (serine, threonine, etc.), and some will have a charge at the pH at which they are normally found (lysine, arginine, glutamate, aspartate). Even if a polypeptide contains only four amino acids strung together, there are $20 \times 20 \times 20 \times 20$ possible polypeptides that could be made, or 160,000 different polypeptides. A polypeptide is a chain of peptides that is conventionally used to denote relatively short chains of peptides. (Confusingly, peptides *themselves* are polymers made up of amino acid units.) Proteins are also chains of peptides, but conventionally taken to relatively much longer than polypeptides. Just keep in mind that polypeptide = relatively short chain of peptides, whereas protein = relatively much longer chain of peptides that usually show higher-level organization (tertiary and quaternary). Proteins can contain hundreds or even thousands of amino acids strung together in a chain. With twenty different amino acids possible for each subunit in a protein, the number of possible proteins that can be made is astronomical.

To make a polypeptide, a covalent bond is formed between each amino acid in the chain, joining the carboxylic acid group of one amino acid with the amino group of the other amino acid. This bond is called the peptide bond. The formation of each peptide bond in a protein removes one water molecule; reversing the process, adds back a water molecule to break the covalent peptide bond and release the amino acids from the protein. Enzymes called proteases catalyze the breakup of peptide bonds in proteins. The peptide bonds between many amino acids strung together in a protein form a long backbone. Side chains project out from the backbone at regular intervals.

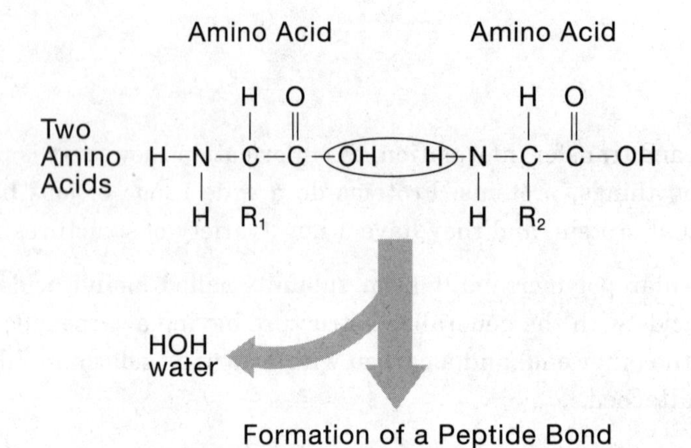

Formation of a Peptide Bond

Peptide Bond Linkage

N-terminus

Formation of many peptide bonds

C-terminus

Polypeptide H_2N--AA_1--AA_2--AA_3--AA_4--AA_{n-2}--AA_{n-1}--AA_n--COOH

Formation of a Polypeptide

The first amino acid in a protein always has an unbonded amino group remaining as more amino acids are added to the protein, so this end of the protein is often called the **amino terminus.** The last amino acid in the chain, on the other hand, has a residual carboxylic acid, and is called the **carboxy terminus**.

What makes every protein unique is the sequence of amino acids along its chain. The order of the amino acids makes each protein fold up into a different three-dimensional shape, and the job it does is based on the shape it folds into. There are four different terms that are used to describe protein structure: **primary**, **secondary**, **tertiary**, and **quaternary** protein structure. The primary structure is simply the order of the amino acids that are strung together in the protein's polypeptide chain. The order of the amino acids in a protein comes from the gene that encoded it, and for a given type of protein the primary sequence is always the same.

The secondary structure is related mainly to interactions along the polypeptide backbone itself, not the side chains that vary from one amino acid to another. Each amino acid along a polypeptide can rotate around the bonds that hold the backbone together. These rotations can bring together amino groups and oxygens that form hydrogen bonds, locking together certain shapes of the folded protein. There are two main types of secondary structure in proteins: **alpha helix** and **beta-pleated sheet**. In an alpha helix, the polypeptide backbone forms a spiral, with N-H and O groups along the backbone forming hydrogen bonds between levels in the spiral. The hydrogen bonds each have a relatively small amount of energy, but with many of them working together in a region of alpha-helix they help to stabilize the overall protein structure a great deal. The side chains tend to stick out to the sides of an alpha helix; these play a role in tertiary structure. In a beta-pleated sheet, two polypeptide strands running in opposite directions line up beside each other, forming hydrogen bonds between the strands. This structure is flatter than an alpha helix.

The next level of protein structure is the tertiary structure, the overall three-dimensional shape of a polypeptide chain in space once it is fully folded. The local elements that form the secondary structure are embedded in the tertiary structure. A number of different bonds and interactions work together to form the tertiary structure, including ionic bonds between charged amino acid side chains, hydrogen bonds, covalent bonds (disulfide bridges), and van der Waals interactions between hydrophobic groups. Disulfide bridges in the tertiary structure are a covalent bond between two cysteine side chains that are brought close together by all of the other forces in folding. As a polypeptide folds, amino acids that are far apart in the primary structure along the polypeptide chain can be brought close together in space and interact as part of the tertiary structure. Some components of tertiary structure are more easily reversed than others.

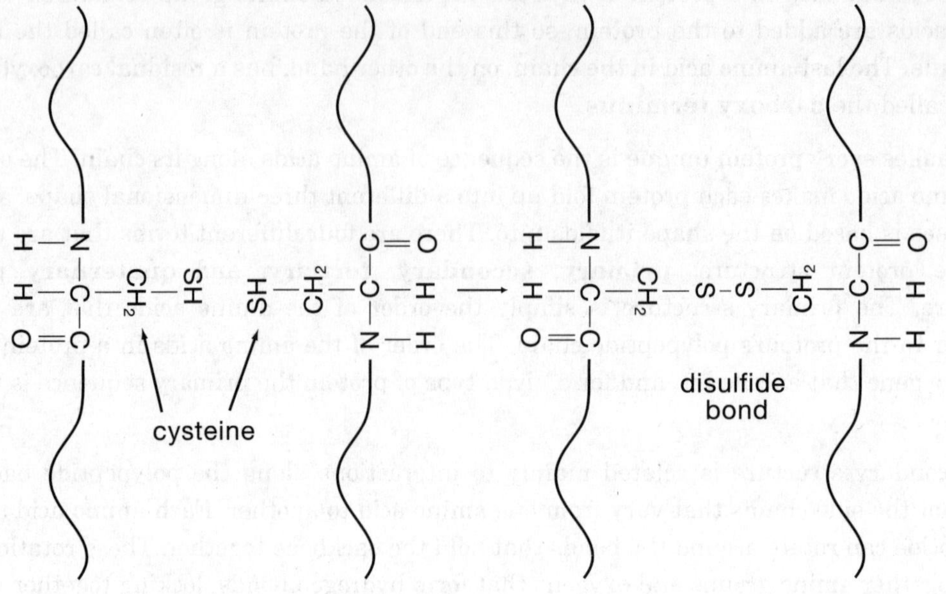

For example, the disulfide bond tends to lock the rest of the tertiary structure in place to keep the structure stable and keep the protein working properly even in an oxidizing environment.

The quaternary structure of a protein involves multiple polypeptide chains that can interact and work together. Each polypeptide is a subunit of the overall protein's structure. The quaternary structure is formed by the interactions of these subunits. Hemoglobin is often described as an important example of the role of quaternary structure in protein function. Hemoglobin is the protein in red blood cells that carries oxygen from the lungs to the tissues. Each hemoglobin protein contains four polypeptide subunits that all work together for hemoglobin to function normally. One of the key characteristics of hemoglobin, cooperative binding of oxygen, cannot occur unless the four subunits are positioned correctly in the quaternary structure to work together properly.

Some proteins have all of the information they need to fold properly contained within the sequence of the amino acids along the protein chain. If the unfolded protein in this case is put into solvent, the interactions of the polypeptide with water and with the rest of the polypeptide chain are enough for the protein to find the correct folding solution and the normal function of the protein. This is not always the case though. Other proteins need help to fold properly, which is provided by protein **chaperones**. Chaperones do not appear to provide any information for specific folding, but they create the right environment a protein needs to fold properly.

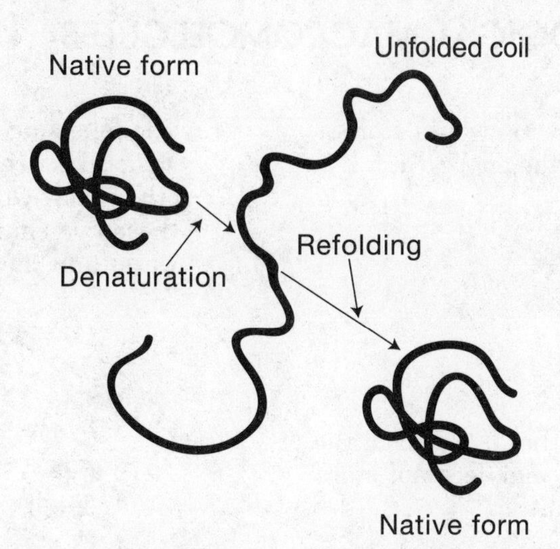

When proteins unfold and lose their properly folded structure, it is called **denaturation**. One factor that causes denaturation is heat. When a protein is heated beyond its normal working conditions, the polypeptide chain moves, breaking the non-covalent bonds that hold the quaternary, tertiary, and secondary structure in place. As a protein unfolds, it may expose patches that were on the interior of the protein, causing it to bind abnormally with other neighboring proteins and become insoluble, forming large clumps and aggregates that cannot easily go back to find their normal folding shape. Other factors that can make proteins denature include extreme pH (too acidic or too basic), salt, solvent (organic solvent, not water), and chemicals that disrupt hydrogen bonds, a key part of protein structure. The primary sequence of a protein is not disrupted when it is denatured. If a protein does not require chaperones and can refold on its own, then, when conditions are returned to normal, it will refold back into its normal shape and resume its normal activity.

EXERCISES: BIOLOGICAL MACROMOLECULES

1. Which of the following would store a larger fraction of the energy in a mammal?

(A) Triglyceride
(B) ATP
(C) Glycogen
(D) Glucose
(E) Sucrose

2. A polysaccharide that is found in the cell wall of plants and does not store energy for the plant is:

(A) Glycogen
(B) Starch
(C) Cellulose
(D) Chitin
(E) Phosphatidylcholine

3. Which of the following best describes the function of DNA in living things?

(A) Conversion of genetic information into proteins
(B) Provide structure of nucleus
(C) Storage of genetic information
(D) Catalysis of essential biological activities
(E) Cofactor for enzymatic reactions

4. The following two carbohydrate structures are enantiomers of each other. How would the solubility of these two enantiomers compare in water and the hydrocarbon hexane?

Enantiomers

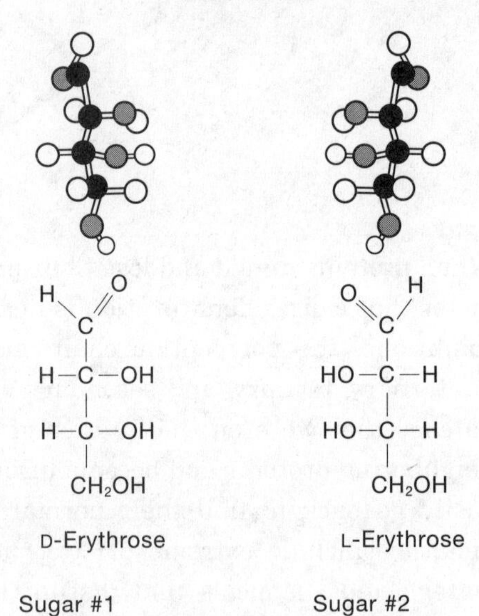

Sugar #1 Sugar #2

(A) Sugar #1 would be more soluble in water, but #2 would be more soluble in hexane.
(B) Sugar #2 would be more soluble in water, but #1 would be more soluble in hexane.
(C) They would both dissolve equally well in water, and neither would dissolve well in hexane.
(D) They would both dissolve well in hexane and poorly in water.
(E) There is not enough information to determine at this time.

5. Which of the following is true about both RNA and DNA?
   (A) Both RNA and DNA contain the same nitrogenous bases.
   (B) The structure of both DNA and RNA is primarily double stranded.
   (C) Both RNA and DNA contain deoxyribose.
   (D) Both RNA and DNA contain phosphodiester bonds.
   (E) Both RNA and DNA store energy in high-energy phosphate bonds.

6. Which of the following reduces the fluidity of lipid bilayer membranes?
   (A) Increasing the percentage of saturated fatty acids in phospholipids
   (B) Increasing the percentage of unsaturated fatty acids in phospholipids
   (C) Increasing the percentage of unsaturated fatty acids in triglycerides
   (D) Reducing the length of fatty acids in phospholipids
   (E) Eating fewer dairy products

7. Which of the following biological macromolecules cannot be hydrolyzed into its component subunits in the human gastrointestinal tract?
   (A) Proteins
   (B) Starch
   (C) Glycogen
   (D) Cellulose
   (E) RNA

8. Ribonuclease (RNAse) is a protein that catalyzes the hydrolysis of RNA molecules into their individual subunits. A sample of ribonuclease (Sample A) is purified in the lab from pig pancreas, and tested for its function, the ability to catalyze the hydrolysis of RNA molecules. To test RNA function, RNAse protein is mixed with RNA in a test tube and incubated at 37°C for 1 hour, followed by measurement of the concentration of free ribonucleotides. Another sample of ribonuclease (Sample B) is boiled first before testing, and then immediately retested for its ability to catalyze the same reaction. A third sample (Sample C) is boiled and then allowed to cool at room temperature for a day before being tested for its function. The following is observed:

Sample A, no boiling: 100 units of RNAse activity

Sample B, boiling and test immediately: 2 units of RNAse activity

Sample C, boiling, cool overnight: 82 units of RNAse activity

Which of the following statements best explains the apparent recovery of RNAse protein function in sample C?

(A) RNA is very stable to heat and is not affected by boiling.
(B) The RNAse protein structure is not affected by heat.
(C) The RNAse does not require a particular tertiary structure to perform its function.
(D) The tertiary and secondary structures of RNAse are denatured by heat and can refold into the original structure at lower temperatures.
(E) The primary structure of RNAse was disrupted by heat and reconstituted with cooling.

ANSWER KEY AND EXPLANATIONS

1. A	3. C	5. D	7. D	8. D
2. C	4. C	6. A		

1. **The correct answer is (A).** Glucose, glycogen, and ATP are more immediate energy sources, but a larger proportion of the total energy stores are found in triglycerides. Sucrose might be eaten by a mammal, but it does not store energy for a mammal.

2. **The correct answer is (C).** Glycogen and chitin are not found in plants, and starch is used for energy, not forming cell walls.

3. **The correct answer is (C).** DNA is the genome, storing the information of genes.

4. **The correct answer is (C).** Water is not stereoselective (it cannot tell the difference between the two enantiomers) and will interact with their polar groups equally well to solvate both types of molecules. Sugars in general have a large number of polar hydroxyl groups and dissolve well in water but would dissolve poorly in a hydrophobic solvent like hexane.

5. **The correct answer is (D).** In both RNA and DNA the nucleotide subunits in the polynucleotides are joined together by phosphodiester bonds. They do not contain the same nitrogenous bases, because RNA contains uracil while DNA contains thymine. DNA is almost always double stranded, while RNA is mostly single stranded, with some regions double stranded as well. RNA contains ribose, while DNA contains deoxyribose. Neither RNA nor DNA stores energy in high-energy phosphate bonds—this describes ATP.

6. **The correct answer is (A).** Saturated fatty acids do not have any double bonds and tend to line up more tightly with each other than unsaturated fatty acids, increasing the number of van der Waals interactions between membrane lipids and reducing membrane fluidity. Triglycerides are not found in membranes and reducing the length of fatty acids will increase, not decrease, fluidity. Changing the diet may affect the body in many ways, but is not likely to change membrane fluidity overall.

7. **The correct answer is (D).** Humans cannot break cellulose down into glucose in their intestinal tract. Starch and glycogen are also polysaccharides built of glucose, but the glycosidic linkages between sugars are different in these polymers, making them easily broken down through hydrolysis into their glucose subunits. Proteins are hydrolyzed by proteases into small peptides and amino acids and RNA is hydrolyzed into individual nucleotides.

8. **The correct answer is (D).** RNAse is a protein that works as an enzyme. The function of enzymes is to catalyze chemical reactions, in this case the hydrolysis of RNA. Like all proteins, RNAse requires a specific folded structure to perform its function. Heat disrupts the non-covalent bonds that are essential for secondary and tertiary protein structure (or quaternary structure), but not the primary structure, which is covalent, the peptide bond. Choice (A) is not correct because it is the RNAse protein and not RNA that is exposed to heat. Choices (B) and (C) are not correct because the information says that the protein structure and function are disrupted by heat.

SUMMING IT UP

- The key biological molecules are: carbohydrates, proteins, nucleic acids, and lipids, all of which include components that are broken apart by hydrolysis.
- Carbohydrates include monosaccharides, like glucose and fructose, that exist in both chain and ring forms, and polysaccharides, like cellulose (plant structure), starch (plant energy), and glycogen (animal energy), with individual glucose subunits polymerized in different ways.
- Nucleic acids (RNA and DNA) store genetic information, encode proteins, and help to synthesize these proteins.
- RNA is generally single-stranded, contains ribose, and contains uracil.
- DNA is generally double-stranded (antiparallel strands), contains deoxyribose, and contains thymine.
- Lipids include triglycerides for energy storage and phospholipids for membranes. Both contain fatty acids of varying length and saturation.
- Proteins are polymerized from twenty different amino acids each of which varies in their side chains. Protein folding has four levels, and proper folding of proteins determines their shape and function.

chapter 5

The Cell

OVERVIEW

- Lipid bilayer membranes
- The eukaryotic cell and organelles
- Cytoskeleton
- Mitochondria
- Plant cells and organelles
- Prokaryotic cells
- The cell cycle and mitosis
- Signal transduction
- Summing it up

All living things are made of cells, from bacteria (single-cell organisms) to humans (trillions of cells). All cells share some fundamental features, while other features are unique to some of the major types of life, such as plants and animals.

LIPID BILAYER MEMBRANES

A key feature of all cells is the membrane that surrounds them, defining and separating the cellular interior and exterior. The membrane barrier is not just a wall—it selectively allows specific substances to move in or out of the cell, while retaining other cellular contents and keeping out unwanted material. The membranes around eukaryotic organelles, described below, play a similar role, isolating and concentrating specific cellular functions inside organelles.

These membranes are composed mainly of phospholipids, those detergent-like molecules with a hydrophobic side and a polar or charged side. The hydrophobic side of phospholipids usually includes two fatty acid side chains with long stretches of hydrocarbons that prefer to interact with each other rather than with water. The polar side of each phospholipid molecule faces out toward water, while the hydrophobic fatty acid side chains face inward toward each other inside the membranes.

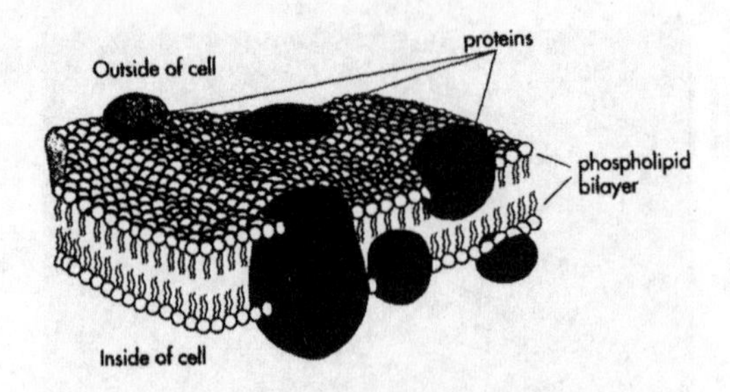

Cross-section of cell membrane
showing fluid-mosaic nature

When mixed with water, phospholipids spontaneously assemble themselves into membranes with two layers of phospholipids that are shaped like sheets or rounded vesicles. The membranes of cells are often called **lipid bilayer membranes** because of the two layers of phospholipids contained in each membrane, with the polar portion of phospholipids facing out toward the water on both sides of the membrane. Within the membrane, the fatty-acid side chains of phospholipids interact with each other mainly through relatively weak van der Waals interactions.

Since the phospholipids are not locked into any specific arrangement and the energy between fatty acid chains is weak, phospholipids are constantly in motion, rapidly diffusing sideways in the membrane, giving membranes a fluid-like nature (the **fluid-mosaic model**). Increasing the temperature of a cell makes lipids (and all other molecules that are not locked in place) move more quickly, increasing the fluidity of membranes. In addition, changing the types of fatty acids or other lipids in membranes affects how fluid they are at different temperatures. If the fatty acids in membrane phospholipids are saturated (no double bonds), they tend to line up straighter in the membrane and pack more tightly with each other. The better the fatty acids pack together, the greater the van der Waals interactions holding the fatty acids together and the less fluid the membrane will be. Similarly, unsaturated fatty acids that have one or more double bonds tend be kinked in the middle, disrupting the packing of the fatty acids. Organisms that live at different temperatures can adjust the fluidity of their membranes by altering their lipid composition. Fish that live in cold water, like salmon, tend to contain more unsaturated fatty acids than are desirable in human diet.

Q Are membrane proteins required for the assembly of phospholipids into lipid bilayer membranes?

A No. Phospholipids alone can spontaneously self-assemble into membranes without any proteins present.

Q How would fatty acids with longer side chains in membrane phospholipids affect membrane fluidity?

A Longer side chains will have more van der Waals interactions with each other, decreasing membrane fluidity.

Cholesterol is another important membrane lipid in animals, including humans, but it is a **steroid** with an entirely different structure from phospholipids. Like other steroids, cholesterol has four rings in its structure and a flat hydrophobic shape.

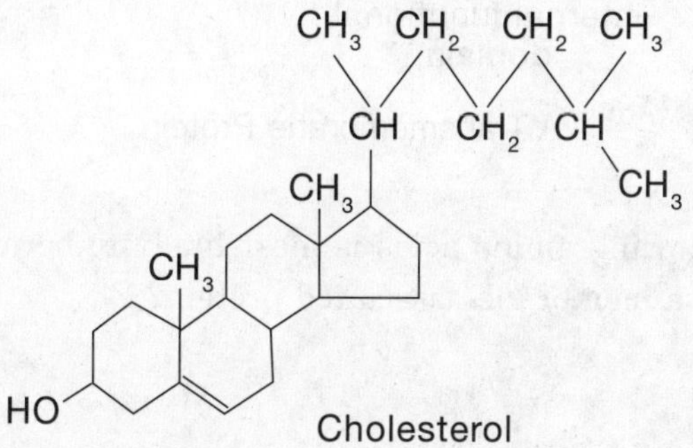

Cholesterol

Because of its different structure, cholesterol packs differently in the membrane between phospholipids, changing how the fluidity of membranes changes with temperature. As temperatures increase, cholesterol buffers the normal increase in membrane fluidity, blocking the movement of phospholipids. At low temperatures, cholesterol breaks up the interactions between fatty acids, while wedging itself between them to keep membranes more fluid.

The energy barrier to sideways movement of membrane lipids is low, but the energy barrier for phospholipids to flip from one surface of the membrane to the other side is high. This flip-flop of phospholipids thus rarely happens. One result of this low rate of lipid movement between sides of a membrane is that, in some cases, the two sides of a lipid bilayer membrane can have different lipid compositions.

In addition to lipids, proteins are another big component of membranes. Many of the functions of membranes are carried out by proteins that are embedded in the membrane. **Transmembrane proteins** cross all the way through the membrane from one surface to the other, contacting and communicating with both sides of the membrane. The portion of the transmembrane protein that pokes into the interior of a cell often has short regions of about fifteen to twenty amino acids called **transmembrane domains.** Transmembrane domains are composed mostly of hydrophobic amino acids and interact with the surrounding hydrophobic fatty acid side chains in the interior of the membrane.

The transmembrane domains thus anchor a protein in the membrane, but do not prevent the membrane from diffusing from side to side in the same manner as membrane lipids. Proteins can have multiple transmembrane domains crossing through a membrane, helping to shape how the protein folds and does its job. For example, GPCRs (g-protein coupled receptors) that transmit signals from the cellular exterior to the cytoplasm are an important class of transmembrane proteins. They have seven transmembrane domains and are often referred to as 7TM receptors.

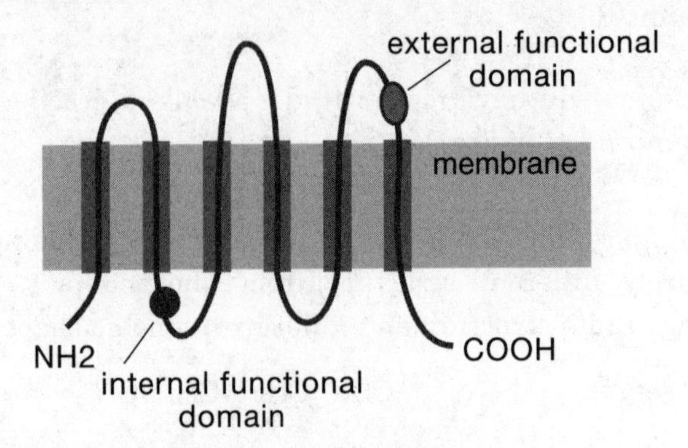

A Transmembrane Protein

Q Which of the following amino acids is most likely to be present in a transmembrane domain of a membrane-associated protein?

(A) Lysine
(B) Serine
(C) Isoleucine
(D) Threonine
(E) Glutamate

Isoleucine, choice (C), is correct. Isoleucine is a hydrophobic amino acid, with an alkyl-group side chain that readily associates with the fatty acid side chains in the hydrophobic interior of the membrane. The other choices are all hydrophilic and less likely to occur in the interior of the membrane in the transmembrane domain.

Since proteins perform essential membrane functions, the type of proteins found in a membrane will vary greatly depending on which membrane in the cell is examined (plasma membrane vs. the inner mitochondrial membrane, for example) and the type of cell examined (muscle, nerve, fat, or liver cell).

Proteins on the cellular exterior are often transmembrane domains that are modified by carbohydrate groups. The modification of these proteins in this way is called glycosylation. The glycosylated groups of proteins are often involved in the binding of cells to each other in tissues and the recognition of cells by each other.

Membrane Transport

One of the key functions of membranes is to act as a selectively permeable barrier, keeping material in and out of the cell or organelle. However, material must move across membranes as needed. There are several methods by which this movement occurs.

PASSIVE DIFFUSION

Some compounds can diffuse on their own directly through membranes. **Passive diffusion** across a membrane always occurs down a concentration gradient, without added energy and without the involvement of transmembrane proteins. The non-polar interior of the lipid bilayer membrane tends to prevent polar molecules from diffusing on their own through membranes, but dissolved oxygen is relatively small and non-polar and readily diffuses across the membrane. In the lungs, oxygen must diffuse from the atmosphere down a concentration gradient through multiple membranes and into the deoxygenated blood that has returned from the tissues. In tissues, oxygen once again diffuses in blood down a gradient from high oxygen to low oxygen. Carbon dioxide diffuses across membranes in the same way, but in the opposite direction—from oxygen, out of tissues where it is produced, into blood at a lower concentration, and finally back into the atmosphere from the lungs.

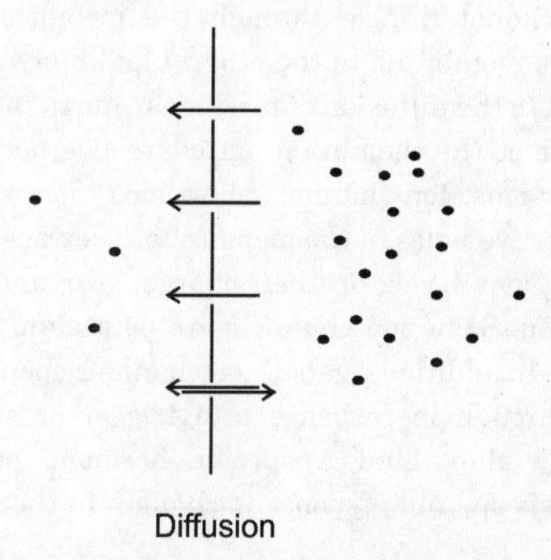

Diffusion

Water can also diffuse to some extent through membranes. When a membrane separates compartments with high and low concentration of solutes (salts, sugars, etc.), water will diffuse across the membrane from the side with the lower solute concentration to the side with the higher solute concentration. This process is called **osmosis** and tends to reduce the difference in concentrations across a membrane. The relative concentration of solutes in cells compared to their surrounding environment are either **isotonic** (solutes are at the same concentration inside and outside of cells), **hypertonic** (solutes are at a higher concentration outside of the cell), or **hypotonic** (solutes are at a lower concentration outside of the cell).

Cells placed in hypotonic solutions will tend to swell as water flows into the cell and can burst if nothing prevents excessive swelling. Cells with a rigid cell wall like plant cells cannot swell to the point of bursting. A plant cell will swell until its cell membrane presses against its cell wall, helping to support the structure of the plant. If a plant does not get enough water, plant cells will not press as hard against their cell walls, causing the plant to sag and wilt.

Q If a red blood cell is moved from an isotonic solution to a hypertonic solution, how will the size and shape of the red blood cell be affected?

A Water will diffuse by osmosis from the region with lower solute to higher solute, which, in a hypertonic solution, would mean water moving out of the cell. As water moves out and the rest of the cellular contents remain the same, the cell will lose volume and shrink.

FACILITATED DIFFUSION

Facilitated diffusion is the diffusion of material across a membrane with the assistance of proteins in the membrane. Some membrane proteins form pore-like channels that allow polar molecules that otherwise cannot diffuse through the membrane to diffuse through the channel from one side of the membrane to the other. Lipid bilayer membranes on their own are essentially impermeable to the diffusion of most ions, due to the energetic barrier imposed by the hydrophobic interior of the membrane on charged groups like sodium, potassium, chloride, calcium, and other ions. Ion channels allow ions to cross through membranes, but they are not just big nonselective holes in the membrane. They are usually highly selective for specific ions, distinguishing ions based on their charge, size, and the water molecules that surround the ions in solution. Some ion channels are potassium channels, while others are sodium channels, and so on. In addition, "gated" ion channels spend some of their time closed, only opening to let ions through in response to a trigger or stimuli. **Ligand-gated ion channels** are closed until they bind a specific hormone or neurotransmitter, while **voltage-gated ion channels** open in response to changes in the voltage across a membrane (more on this later).

Another type of facilitated diffusion involves the movement of water across membranes. Although water can passively diffuse across membranes at a slow rate, transmembrane proteins called **aquaporins** form pores in some membranes allowing water to move more rapidly from one side of a membrane to the other. Aquaporins play a key role in urine formation in the kidney.

Other membrane proteins called **carrier proteins** facilitate transport. Carrier proteins do not simply form channels or pores through a membrane; they bind the material they transport on one side of the membrane and then change their shape to move the material to the other side of the membrane. Carriers transport a variety of substances across membranes but still move material down concentration gradients without any extra energy.

Q Does facilitated diffusion occur more rapidly if more ATP is provided for ion channels?

A No. Facilitated diffusion only occurs down a concentration gradient, from high to low concentration, and does not involve any other source of energy.

Q Will facilitated diffusion or passive diffusion have a maximum rate of diffusion that depends on the presence of proteins in the membrane?

A No proteins are involved in passive diffusion. But the more solute that is present, the greater the concentration gradient, and the faster that diffusion will occur. For facilitated diffusion, there would be a maximum rate of diffusion through the protein involved. The more of the protein that assists in diffusion, the greater the rate would be.

Q If the concentration of solutes in urinary filtrate is very dilute, and the aquaporin channels connect the urinary filtrate to highly concentrated solution in the extracellular fluid of the kidney cells, what direction will water move?

A Water will move by facilitated diffusion through the aquaporins out of the filtrate and into the extracellular fluid of the kidney cells. (See Chapter 13.)

Active Transport

Passive diffusion and facilitated diffusion allow material to move down a concentration gradient across membranes. These processes are energetically favored and will occur on their own without any additional energy. Compounds are also transported across membranes against concentration gradients, from low to high concentration. The movement of material against a concentration gradient is not energetically favored and will not happen on its own. Molecules must be carried up the slope. Because some required materials must move into (or out of) the cell against a concentration gradient, cells perform **active transport,** employing membrane proteins to pump material across the membrane by tapping into a source of energy. This extra source of energy is usually supplied by the hydrolysis of the energy-carrying molecule, ATP. The high-energy phosphodiester bonds in ATP are a common source of energy used by many biochemical processes in cells.

One form of active transport carried out by essentially all cells is the sodium-potassium pump, which uses the energy of ATP to pump sodium out of cells and pump potassium into cells against electrochemical gradients. This form of active transport creates a membrane potential across the plasma membrane and helps cells maintain their shape and osmotic balance.

Another form of energy used in active transport of a given substance is the cooption of the concentration gradient of another substance. A co-transport protein allows one substance to flow down a gradient, but uses this energy to pump another substance against a gradient. For instance, in the intestine, there is more sodium in the interior of the intestine than inside the cells that line the intestine, while glucose is present at a higher concentration inside cells than in the intestine. To absorb glucose from the intestine, a carrier protein binds sodium and glucose and then uses the energy of the sodium gradient to move glucose into cells against a concentration gradient.

Type	Passive Diffusion	Facilitated Diffusion	Active Transport
Transmembrane protein involved?	No	Yes	Yes
Diffusion down a gradient?	Yes	Yes	No
Additional energy required?	No	No	Yes
Example	Oxygen diffusion through membrane	Sodium movement through ion channel	Na+/K+ ATPase

A few additional mechanisms move material in and out of cells without transporting it through a membrane. In **exocytosis**, a membrane vesicle containing cargo (like a protein hormone) fuses with the plasma membrane, releasing its cargo into the extracellular environment. When nerve cells release neurotransmitters, exocytosis is involved. In exocytosis, the contents of the vesicle do not cross through the membrane but are released from the cell when the vesicle fuses with the plasma membrane.

In a reverse process called **endocytosis**, material from outside the cell can be taken inside after being surrounded in a membrane pocket. Receptors on the cell surface bind the material, which triggers a pocket of the membrane to curl inward and form a membrane vesicle with the extracellular material that is now on the interior. The membrane vesicle can at this point target its contents for destruction, or it can be recycled back up to the plasma membrane, allowing its receptors to prepare for another round of endocytosis.

Pinocytosis is a related process carried out by some cells in which material in the extracellular medium is surrounded nonspecifically by patches of the plasma membrane to form vesicles that are internalized. **Phagocytosis** occurs when large particles like pathogenic bacteria are surrounded by a membrane vesicle and internalized to be destroyed. Immune-system cells called macrophages perform phagocytosis of pathogens to neutralize and kill them.

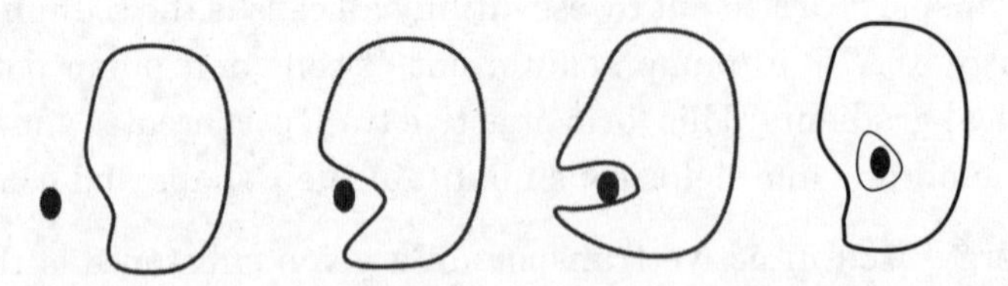

Phagocytosis

THE EUKARYOTIC CELL AND ORGANELLES

One of the major domains of life is the eukaryotes, which includes all life except prokaryotes, bacteria and archaebacteria. All cells (both eukaryotes and prokaryotes) are enclosed by a lipid bilayer membrane that contains the cellular contents. The aqueous intercellular medium is called the **cytosol.** All cells contain ribosomes, relatively small compact structures involved in protein synthesis, and all cells contain a DNA genome.

Eukaryotic cells, however, contain far more than these basic features, including a number of different membrane-bound subcellular **organelles.** Just as the cellular membrane distinguishes the inside and outside of the cell, the membranes surrounding organelles separate organelle functions and contents from the rest of the cell.

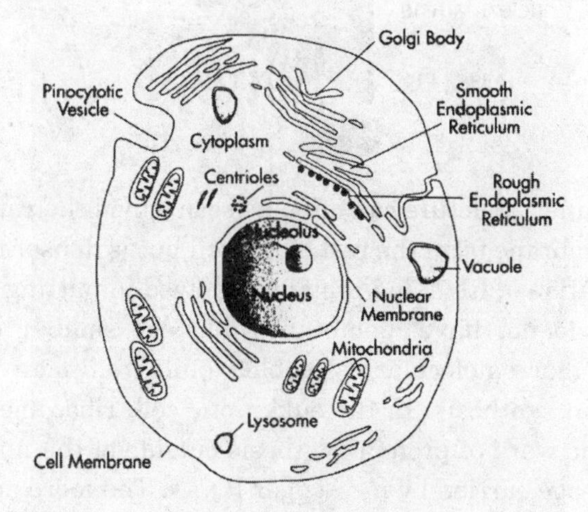

Eukaryotic cell

In eukaryotes, the DNA genome is enclosed in a **nucleus**, which is surrounded by the **nuclear envelope**. The nuclear envelope is not just one membrane, but two layers of membranes, punctuated periodically by nuclear pores that allow material in and out of the nucleus. The DNA genome contained in the nucleus is split among a number of chromosomes.

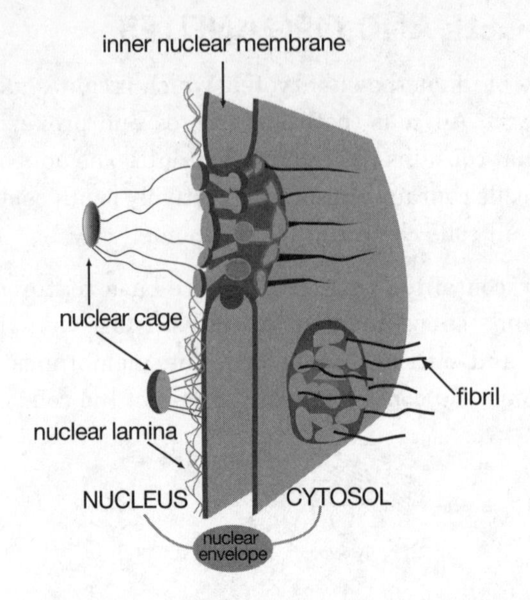

Within the nucleus a smaller structure can often be seen called the **nucleolus**. The nucleolus is not separated by a membrane from the rest of the cell but is denser than surrounding areas due to the high concentration of RNA and proteins involved in putting together **ribosomes** in this region. Ribosomes do not have membranes and are smaller than membrane-bound organelles, but they are macromolecular assemblies built from many different proteins and RNA that conduct protein synthesis. In the eukaryotic cell, ribosomes are assembled in the nucleolus but do all of the work of protein synthesis outside of the nucleus in the cytoplasm, reading the genetic message carried by messenger RNAs. The more active a cell is in protein synthesis, the more ribosomes it will have. Ribosomes can either perform protein synthesis loose in the cytoplasm or bound to the surface of another organelle called the endoplasmic reticulum.

When proteins are synthesized, their fate may be to end up inside one of the specific organelles of the cell, such as the nucleus or mitochondria, to be delivered to the cell membrane, or to be secreted from the cell and released into the extracellular environment. Proteins that are to reside in the cytosol or that are delivered into the nucleus are synthesized by ribosomes in the cytosol. Proteins that are delivered to some of the membrane-bound compartments or that are secreted are first synthesized by ribosomes bound to an organelle called the **endoplasmic reticulum (ER).** Proteins that reside in the cytoplasm are synthesized in the cytoplasm by ribosomes not associated with the ER. Proteins that are targeted to the nucleus are also synthesized in the cytoplasm but contain a nuclear localization sequence about eight amino acids in length. Proteins recognizing this sequence bind to nuclear proteins and assist in their transfer through nuclear pores.

Q If the nuclear localization sequence of a protein is removed, where will the protein probably end up?

A It will be synthesized in the cytoplasm and stay there, rather than getting imported into the nucleus.

The **endoplasmic reticulum (ER)** is not a rounded organelle but a series of membranes that form sheets, tubules, and other shapes that often surround the nucleus, stacked like plates. The sheets and tubes formed by the ER membrane close a narrow space called the ER lumen and, since the ER membrane is joined with the nuclear envelope, the ER lumen is continuous with the space inside the nuclear envelope.

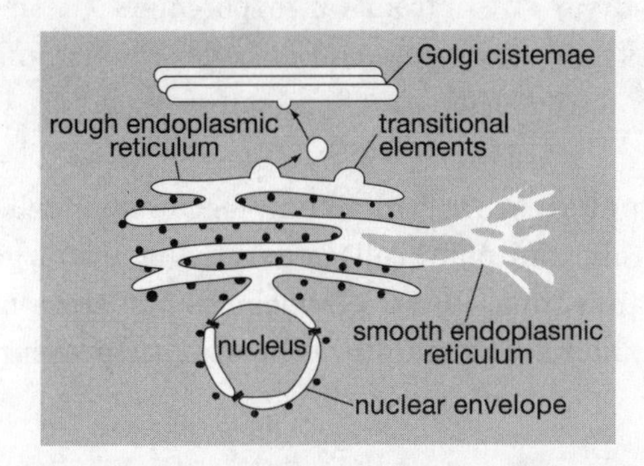

There are two different types of ER, called **smooth ER** and **rough ER** due to their appearance under the electron microscope. Rough ER is named due to its studded appearance in the microscope, with numerous ribosomes attached to its surface synthesizing proteins; thus ribosomes are absent in smooth ER. The smooth ER is not involved in protein synthesis but does perform other functions, like the synthesis of steroid hormones and the detoxification of organic compounds.

Once synthesized on the ER, proteins start to be modified and packaged for delivery to their ultimate destination. Proteins that are destined for the cell surface are often tagged with carbohydrate groups, with this process starting inside the ER lumen. Once proteins are ready for the next step in the secretory pathway, they are packaged into small rounded membrane spheres called vesicles that carry them to the Golgi body.

The **Golgi** body, or the Golgi apparatus, receives transport vesicles that have budded off from the ER. The vesicles fuse with the Golgi membrane and deliver their contents. The Golgi are arranged like a stack of plates, usually between the ER and the cell membrane, with one side of the stack facing the ER and the other side facing outward toward the rest of the cell. The Golgi further modify proteins with carbohydrate groups, as well as tag proteins for delivery to other organelles.

The **lysosome** is an organelle that digests various macromolecules, breaking them up into smaller pieces. Cells can recycle cellular contents in lysosomes, breaking proteins or other material down, and then reusing the subunits to synthesize new macromolecules. The interior of the lysosome is acidic, and the enzymes that digest material inside the lysosome require acidic conditions to work. If the contents of lysosomes were released into the cytoplasm and were still active at the nearly neutral pH found there, these enzymes would begin to destroy the rest of the cellular contents there. Having the activity of lysosomes limited to acidic pH helps to carefully restrict the activity of lysosomal proteins inside this subcellular compartment.

Q Why is it important to restrict the activity of lysosomal proteins to the inside of the lysosome?

A These proteins are designed to hydrolyze a wide range of biological macromolecules, breaking them down from large polymers to small subunits. If these proteins were active in the cytoplasm of cells, they would destroy not just the contents of lysosomes but the rest of the cell as well.

In plants, large central **vacuoles** are prominent features of cells. Vacuoles are not very prominent in most animal cells, but plant cells generally have a large, relatively clear vacuole that performs a variety of functions. Plant cell vacuoles can store material, sequester toxins from the rest of the plant, and help to maintain the osmotic pressure of the plant cell, which physically supports plants.

CYTOSKELETON

The cytoskeleton is a fibrous meshwork of filaments and fibers that span the cytoplasm, giving cells their shape, helping cells to move, moving organelles, and assisting in cell division. The three main types of cytoskeleton are the microtubules, intermediate filaments, and microfilaments.

Microtubules

Microtubules are relatively larger cytoskeletal elements. They are hollow tubes built of the protein tubulin. Microtubules can be very dynamic, with tubulin units added to or removed from the ends of a microtubule to make it grow longer or shorter. Working with the other cytoskeletal elements, microtubules give cells their shape and internal structure. Organelles and membrane vesicles moving between organelles often move along microtubules as if they were railroad tracks, with the movement of the organelles provided by molecular motor proteins.

Microtubules are also associated with centrosomes, which act as microtubule organizing centers near the nucleus. The centrosomes contain two short centrioles, structured like microtubules but as more permanent cellular features. Plant cells have centrosomes but lack centrioles. During mitosis or meiosis, the spindle fibers that separate chromosomes are formed from microtubules, with the two poles of the spindle fiber organized at the centrosomes in animal cells. (See section on mitosis later in this chapter.)

Another specific function of microtubules is to provide the main structural feature and movement of eukaryotic cilia and flagella that project from the surface of some cells. Flagella and cilia move cells like spermatozoa and many protozoans through their surrounding fluid environment, and cilia that line some cells—like those in the human respiratory tract—help to move fluids over the surface of the cells. While cells usually have a small number of flagella (one or a few), cells like paramecia can be covered by large numbers of cilia. Cilia are shorter than flagella. But the interior structure of flagella and cilia, and how they move, is pretty much the same. They both contain microtubules in a hollow tube surrounding central microtubules. Proteins connect the inner and outer microtubules like spokes on the wheel of a bicycle. Motor proteins connect the neighboring microtubules in the outer ring, moving the microtubules past each other to create the motion of cilia and flagella. The whole assembly is connected together, so the motion of the motor proteins on neighboring microtubules causes the cilia to bend back and forth. This motion takes energy, provided by ATP.

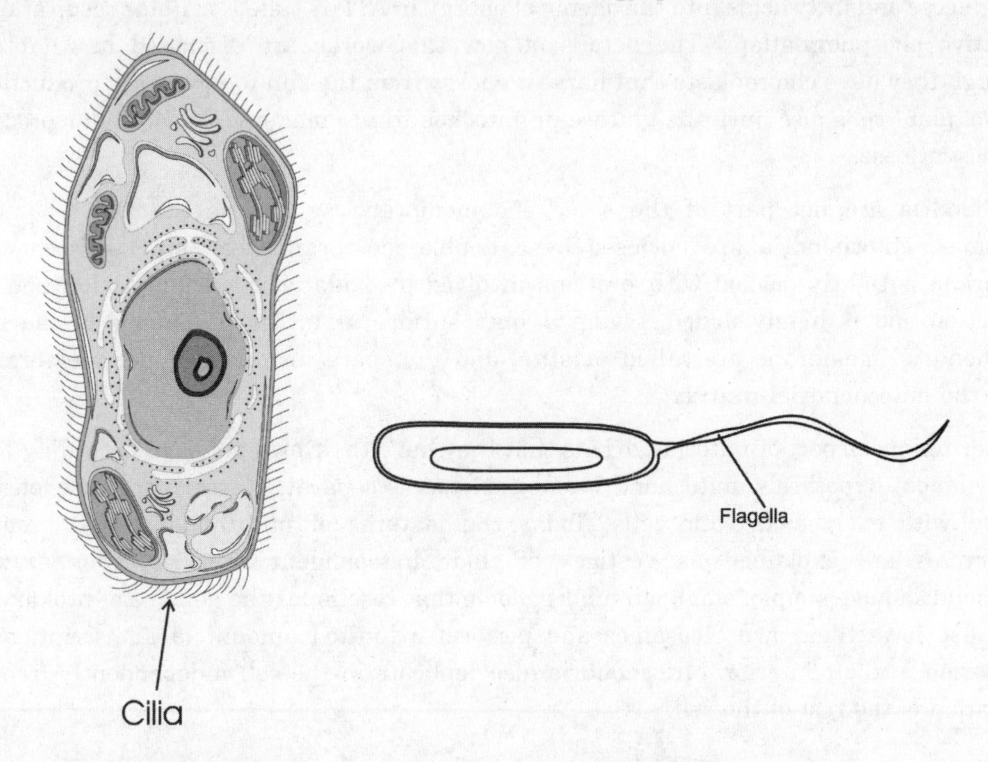

Intermediate Filaments

In the middle of the size range of cytoskeletal elements, intermediate filaments are less dynamic than microtubules, lasting for longer periods of time to maintain the shape of cells and their internal structure. In cells with structures that stay the same most of the time, intermediate filaments probably do a large part of maintaining their structure. One part of the intermediate filament network is the nuclear lamina inside the nucleus, a meshwork that helps support the nuclear envelope and is probably also involved in the organization of chromosomes and the movement of material like mRNA inside the nucleus.

Microfilaments

Microfilaments are relatively small fibers mainly composed of actin. Microfilaments are often involved in cell movement or in the movement of cytoplasm in a cell. The movement of amoeba (or of immune cells that display similar movement) involves changes in the cell's shape and the streaming of cytoplasm driven by microfilaments. Actin alone does not provide movement, but myosin proteins act as motors that bind to actin filaments in order to drive their movement. This is similar to the arrangement of actin and myosin in skeletal muscle cells, as we shall see.

MITOCHONDRIA

Cells need energy for almost everything they do. Biosynthesis, active transport, and cell movement all require energy. **Mitochondria** are the powerhouses of the cell. The energy they produce is chemical energy contained in the form of ATP. The conversion of chemical energy from glucose and fatty acids into the chemical energy in ATP is called "cellular respiration" or "oxidative phosphorylation." The details of how this works are discussed in Chapter 7. Although they have chloroplasts that harvest energy from the sun to power the production of glucose, plant cells also burn the glucose in mitochondria to make ATP for cellular processes like biosynthesis.

Mitochondria are not part of the same endomembrane system as the ER, Golgi, and lysosomes. Mitochondria are enclosed by a double membrane. The inner mitochondrial membrane is tightly packed with proteins involved in oxidative phosphorylation and ATP production and is highly folded, giving it high surface area. The infoldings of the inner mitochondrial membrane are called **cristae,** and the space inside the inner membrane is called the mitochondrial **matrix**.

Another unique aspect of mitochondria is that they have their own genome. According to the endosymbiont hypothesis, mitochondria once existed as separate organisms that eventually merged with early eukaryotic cells. Today the features of mitochondria that resemble prokaryotes are explained as vestiges of their independent existence. For example, mitochondria have simple, small, circular genome that resembles the genome of prokaryotes. They also have their own ribosomes and perform a limited amount of transcription and translation in their interior. Mitochondria also replicate in the cell independently from the replication of the rest of the cell.

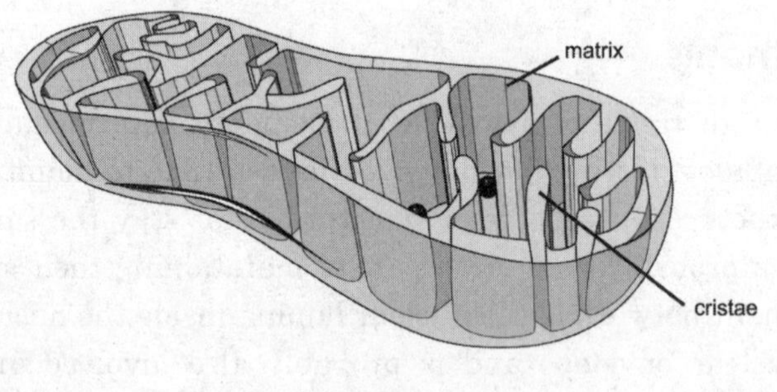

Mitochondrion

PLANT CELLS AND ORGANELLES

Plant cells and animal cells share a great deal in common and also have several obvious differences. Both plant and animal cells are eukaryotic cells sharing a nucleus, ER, Golgi, mitochondria, and cytoskeleton. One of the most obvious differences is the rigid cell wall of cellulose that surrounds and contains plant cells. Giving plant cells their shape, the cell wall also prevents cells from bursting from osmotic swelling. The cell wall provides the strength to counteract physical forces like gravity. The cell wall is made primarily of cellulose, wound into fibers, and cross-linked with other molecules like lignins, making it resistant to digestion or physical stress.

Another obvious difference is the presence of chloroplasts, the organelles that perform photosynthesis, the distinguishing trait of plants. Chloroplasts are also what make plants look green. Although the function of chloroplasts is very different from mitochondria, chloroplasts are similar to mitochondria in several ways. Like mitochondria, chloroplasts are enclosed by a double membrane system, and they both have their own small circular genome, suggestive of a once-independent existence. Like mitochondria, chloroplasts do their own transcription of a limited number of genes in their own genome and have their own protein translation system, including their own ribosomes. Inside chloroplasts are a series of interconnected flattened stacks called the **thylakoids**. The liquid around the thylakoids inside the chloroplasts is called the **stroma,** which contains many of the enzymes involved in photosynthesis. (The means by which chloroplasts perform photosynthesis are described in Chapter 8.)

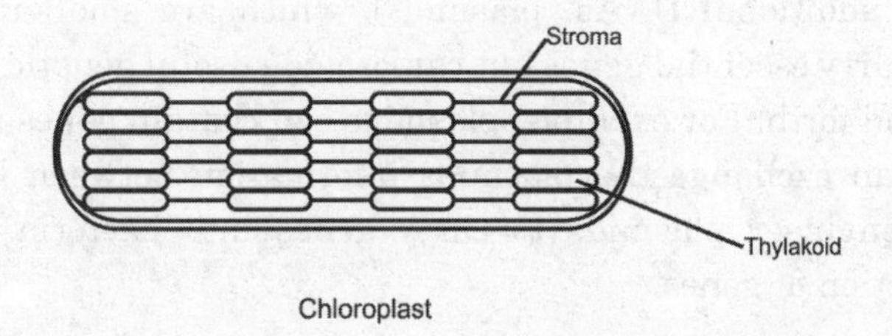

Chloroplast

Plant cells also have unique connections with the cytoplasms of neighboring cells. These **plasmodesmata** allow direct communication through the cytoplasm from one cell to another. Since plant cells are connected in this way, water can flow through the plant from cell to cell to assist in transport of water or other material inside the plant.

Plant cells also have a large, central vacuole for storing material such as food, waste, and ions. In some plant cells, the vacuole can be the largest feature inside the cell. Plants lack some of the features found in animal cells, such as centrioles and intermediate filaments.

Q Which of the following are found only in plant cells?

(A) Centrosomes

(B) Peptidoglycan cell wall

(C) Microtubules

(D) Thylakoids

(E) Lysosomes

A The correct answer is (D). As part of the chloroplast, thylakoids are unique to plants. Peptidoglycan cell walls are found in bacteria, while centrosomes, microtubules, and lysosomes are found in plants, fungi, and animals.

PROKARYOTIC CELLS

At first glance (and probably at second glance), the prokaryotic cell looks far simpler than a typical eukaryotic cell. The prokaryotic cell is surrounded by a lipid bilayer membrane like the eukaryotic cell, but lacks the membrane-bound organelles that populate eukaryotic cells. In prokaryotes, the plasma membrane is surrounded by a cell wall, but not the same type of cell wall found in plants or fungi. The prokaryotic cell wall varies widely in its details but generally includes compounds called peptidoglycan with peptide and sugars groups cross-linked in a massive network of covalent bonds.

Like eukaryotes, prokaryotes have a DNA genome, typically a circular chromosome, unlike the linear DNA chromosomes in eukaryotes. The area around the prokaryotic genome is sometimes called the nucleoid, but it is nothing like the eukaryotic nucleus. Prokaryotes sometimes carry additional DNAs (plasmids), which are smaller circular pieces of DNA. Plasmids do not carry essential genes but can provide useful genetic information that bacteria can trade back and forth. For example, plasmids can contain genes that provide resistance to antibiotics and can exchange the plasmids information between bacterial cells. Molecular biologists often engineer plasmids to carry genes into bacteria, producing proteins and studying the function of genes.

Like eukaryotes, prokaryotic cells carry ribosomes, but the prokaryotic ribosome is smaller and simpler. Also, since there is no nucleus that separates transcription and translation in prokaryotes, prokaryotic ribosomes can start to translate RNA as soon as it is read from the corresponding gene in DNA.

In some prokaryotic cells, the plasma membrane folds inward to form structures that are like eukaryotic organelles, without being entirely distinct organelles. Infoldings of the plasma membrane, called mesosomes, are involved in cellular respiration, increasing the membrane area available to produce energy. Thylakoids are another type of infolding found in some photosynthetic prokaryotes. The thylakoids can fold inside cells extensively, filling much of the cell, but they are still connected to the plasma membrane and do not form a separate organelle.

The method of cell division is also very different in prokaryotes and eukaryotes. Just as prokaryotic cells are much simpler than eukaryotic cells, cell division in prokaryotes is also much simpler, occurring through a process called **binary fission**. The prokaryotic genome is a circular DNA molecule that contains an origin of replication at which DNA synthesis starts when the prokaryotic cell is preparing to divide. When DNA synthesis is complete and the prokaryotic cell contains two complete copies of its genome, the two copies move apart from each other. Once separated, the plasma membrane and cell wall close off in the middle of the cell, creating two identical daughter cells.

THE CELL CYCLE AND MITOSIS

One of the fundamental properties of life is that all life is made of cells. Another fundamental property of life is that all cells come from other cells. Since the origin of the first living cell on earth, all cells in all organisms have been derived from that first cell. The division of a single cell to create two new daughter cells is called cell division. In eukaryotes, this occurs through mitotic cell division as part of the cell cycle. Mitotic cell division is responsible for growth, for healing and regeneration of wounded tissues, and for constant maintenance of tissues in the body. Cells in some tissues are constantly dividing and being replaced, such as cells in the skin, cells that produce hair, and cells that line the intestine.

In the eukaryotic cell cycle, cells first grow larger through biosynthesis. This portion of the cell cycle is called the G_1 phase ("G" stands for "gap," the spaces in the cycle between DNA synthesis and cell division). When the cell reaches a critical size, the S phase ("S" is for DNA "synthesis") of the cell cycle begins in which the DNA genome is replicated until the cell contains twice the normal amount of DNA. Another gap, G_2, follows, in which the cell is prepared for mitotic cell division. Mitosis, the division of the genome, and cytokinesis, the final cleavage of the cell into two daughter cells, complete the cycle.

Mitosis is the part of the cell cycle in which the chromosomes, nuclei, and the rest of the cellular contents are divided between the two daughter cells. The period between mitotic cell divisions is called **interphase**, including G_1, S, and G_2 parts of the cell cycle (see figure below). Mitosis is divided up into specific phases called **prophase**, **prometaphase**, **metaphase**, **anaphase,** and **telophase**. Splitting up the genome between the two daughter cells is a central part of mitotic cell division. Throughout most of the cell cycle, the chromosomes are packaged in chromatin but are still very long and extended and not visible under a microscope. One of the first signs that a cell is approaching mitosis is that the chromosomes become condensed and shortened. The condensation of chromosomes makes it possible for them to be moved in the cell without becoming tangled. Since mitosis occurs after DNA replication, each chromosome is found in the cell in two identical copies, called **sister chromatids,** that are joined in the middle region, called the **centromere**.

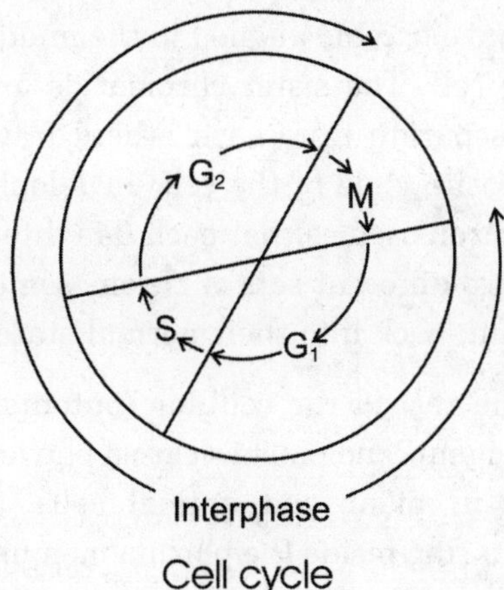

Cell cycle

One of the most visible components of the cell in mitosis is the **mitotic spindle**, an arrangement of fibers made of microtubules that span from one side of the parent cell to the other between the two poles of the cell that will become the two daughter cells. The two poles of the spindle occur at the centrosome, which in animal cells contains the centrioles. During mitosis, microtubules form around the centrosome and radiate out in all directions. This arrangement of microtubules around the centrosome is called the **aster**. The microtubules that span from one aster to another are the spindle fibers.

In prophase, the chromosomes start to condense and the nucleoli region in the nucleus disappears. The spindle starts to form, with two centrosomes separating from each other, asters forming microtubules around them, and spindle fibers extending from one centrosome to the other.

In prometaphase, the nuclear envelope breaks down into small vesicles that will reform the nuclear envelope for the two new cells after mitosis is complete. The centrosomes are fully separated at two opposite ends of the cell, with spindle fibers extending between them from one end of the cell to the other. The spindle fibers attach themselves to the condensed pairs of sister chromatids, hooking on to the chromatids at the central **kinetochore**, a part of the centromere on each chromatid. These spindle fibers attached to the kinetochores start to move the chromatids around, preparing for metaphase.

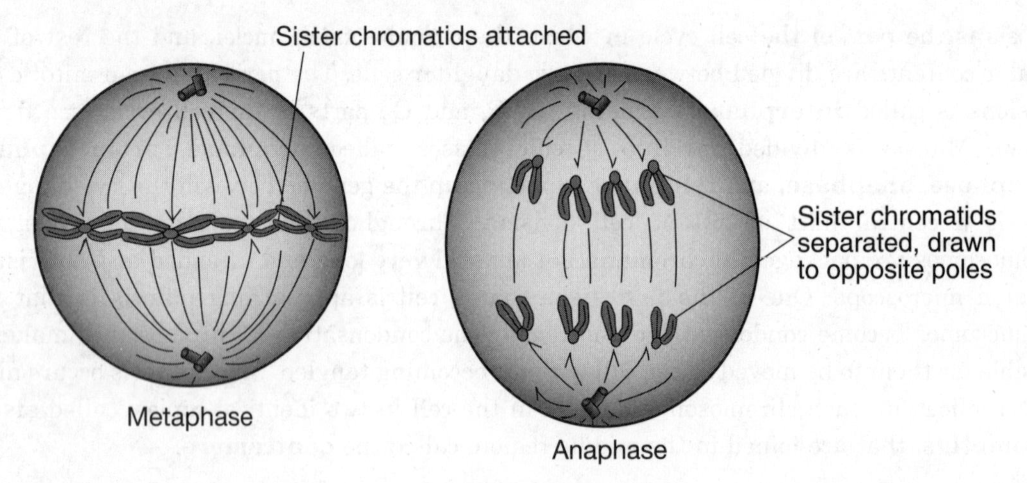

In metaphase, the chromatids have become aligned in the middle of the cell, pulled by spindle fibers toward both ends of the cell. The sister chromatids are still attached. As anaphase begins, the sister chromatids separate from each other, drawn by attached spindle fibers toward the centrosomes at opposite ends of the cell. In telophase, the chromatids are fully separated into the two sets of chromosomes that each daughter cell will receive. The nuclear envelope re-forms around the two different sets of chromosomes, and the chromosomes start to lengthen and spread out again, back into their normal state during interphase.

Finally, through cytokinesis, the rest of the cellular contents are divided between the two daughter cells, and the plasma membrane pinches closed between the two new daughter cells. Cytokinesis occurs differently in plant and animal cells. In animal cells, cytoskeletal elements, microfilaments, form a ring inside the plasma membrane between the two new cells formed after telophase. This contractile ring forms a cleavage furrow that grows deeper until the plasma membrane pinches closed between the two cells, separating them completely. In plant cells, the presence of the cell wall around each cell necessitates a different method of cytokinesis. When plant cells appear near the end of mitosis, vesicles containing components for a new cell wall start to line up between the two new cells. As the vesicles accumulate, they join together creating a cell plate. As the cell plate grows, it becomes a new cell wall isolating the two daughter plant cells from each other.

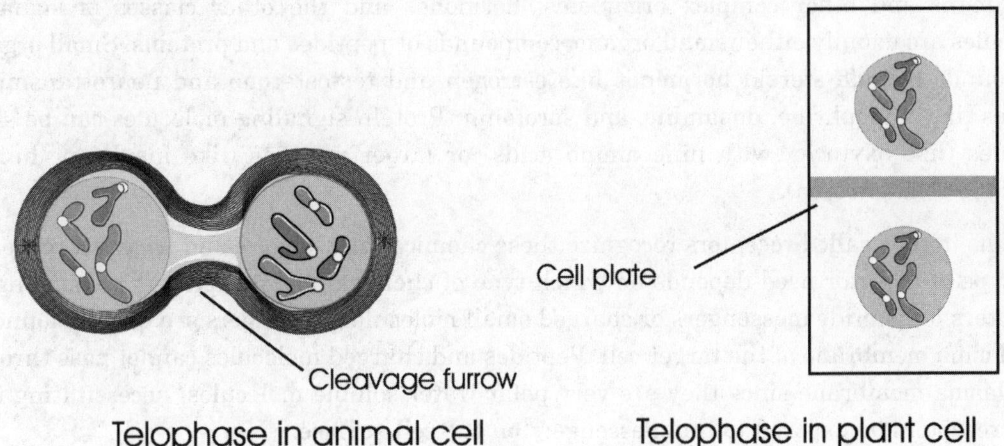

Telophase in animal cell Telophase in plant cell

Control of the Cell Cycle

Mistakes in the cell cycle can have lethal consequences for cells (and their hosts), so it is essential that the cell cycle be tightly regulated. There are checkpoints in the cell cycle that make sure that essential steps are completed before cells move from one stage to the next. Proteins in the cell keep track of the status of events and block progression through the cell cycle until everything is ready. The proteins that control the checkpoints are often enzymes called **protein kinases**. Protein kinases add phosphate groups to the side chains of other proteins, changing the activity of the modified proteins. Cell-cycle regulators identified in molecular and genetic studies of the cell cycle include **cyclins** and **cyclin-dependent kinases**. Cyclin-dependent kinases (cdk) are only active when they are bound to their protein partners, cyclins. Cdk levels don't usually change much as a cell goes through the cell cycle, but cyclin levels rise and fall dramatically. When cyclin levels rise to a certain threshold, their cdk partner is activated, triggering progression through that checkpoint from one stage of the cell cycle to the next.

SIGNAL TRANSDUCTION

Cells do not exist in isolation. The human body contains trillions of cells that must work together and to do so they must communicate with each other. Even single-celled organisms must receive information from their environment and react to it if they are to survive. Cells have evolved signaling systems that receive information and convert it into useful actions. The signaling systems used to communicate between cells and tissues in the body are often chemical messengers. These chemical messengers are released by a cell to either act locally on cells right next door or to act more distantly on other parts of the body. Local messengers are called **paracrine** signals, while distant messengers are hormones, part of the **endocrine** system (see Chapter 13). Specialized messengers released by nerves to act directly on target tissues are called **neurotransmitters.** Paracrine signals and neurotransmitters travel only a very short distance from cell to cell through the extracellular fluid, while hormones are often carried throughout the body in the bloodstream.

In humans and other complex organisms, hormones and the other classes of signaling molecules are usually either small organic compounds or peptides and proteins. Small organic compounds include steroid hormones like estrogen and testosterone and neurotransmitter amines like epinephrine, dopamine, and serotonin. Protein signaling molecules can be short peptides (like oxytocin, with nine amino acids) or larger proteins (like insulin or human chorionic gonadotropin).

Proteins in cells called receptors recognize these chemical messengers and trigger a response. The type of receptor used depends on which type of chemical messenger it is responding to. Receptors for peptide messengers or charged small molecule messengers are usually found on the plasma membrane of the target cell. Peptides and charged molecules cannot pass through the plasma membrane since they are very polar water soluble molecules, necessitating that their receptors be exposed to the messengers on the cell surface.

EXERCISES: THE CELL

1. Which of the following organelles is responsible for hydrolyzing damaged biological macromolecules into their subunit constituents?

   (A) Golgi apparatus
   (B) Rough ER
   (C) Nucleolus
   (D) Lysosome
   (E) Exocytosis

2. Carbon dioxide crosses cellular membranes by which of the following methods?

   (A) Passive diffusion
   (B) Facilitated diffusion
   (C) Active transport
   (D) Endocytosis
   (E) None of the above

3. The movement of cilia does NOT involve which of the following?

   (A) ATP hydrolysis
   (B) Microtubules
   (C) Motor proteins like dynein
   (D) Vesicular transport
   (E) Spoke proteins connecting central microtubules and outer ring of microtubules

4. Which of the following organelles contains its own extranuclear DNA genome?

   (A) Golgi
   (B) Lysosomes
   (C) Endosomes
   (D) Centrosomes
   (E) None of the above

5. Which of the following statements is true regarding proteins found in the mitochondrial matrix?

   (A) All proteins found in the mitochondrial matrix are transcribed and translated within mitochondria.
   (B) Many proteins found inside mitochondria are encoded by nuclear genes and translated in the cytoplasm.
   (C) All proteins found inside mitochondria are transcribed from the mitochondrial genome, but most are translated in the cytoplasm.
   (D) All proteins found inside mitochondria are transcribed from nuclear genes and translated inside mitochondria.
   (E) The mitochondrial genome does not encode any essential protein products.

6. Prokaryotic and eukaryotic cells share which of the following characteristics?

   (A) Linear DNA genomes
   (B) Cytoskeletal elements containing tubulin
   (C) Lipid bilayer membrane providing a selective permeability barrier
   (D) Peptidoglycan cell walls
   (E) Segregation of genome inside a membrane-bound organelle

7. A scientist measuring the DNA content of somatic eukaryotic cells finds that some cells have X quantity of DNA per cell and other cells have 2X DNA per cell. At what stage of the cell cycle are cells with X DNA per cell?

   (A) G1
   (B) S
   (C) G2
   (D) Metaphase
   (E) Prophase

8. In what stage of mitosis do sister chromatids separate from each other and move toward opposite poles of the spindle apparatus?

   (A) Prophase
   (B) Prometaphase
   (C) Interphase
   (D) Metaphase
   (E) Anaphase

ANSWER KEY AND EXPLANATIONS

1. D	3. D	5. B	7. A	8. E
2. A	4. E	6. C		

1. **The correct answer is (D).** The lysosome contains hydrolytic enzymes that break down a variety of biological macromolecules, something the other organelles do not do.

2. **The correct answer is (A).** Carbon dioxide diffuses across cell membranes fairly readily. It moves across membranes down a concentration gradient and without a requirement for membrane proteins, making this an example of passive diffusion.

3. **The correct answer is (D).** Vesicular transport does not have anything to do with cilia; all of the other choices are essential for the structure and function of cilia.

4. **The correct answer is (E).** The only organelles that have an extranuclear DNA genome are mitochondria and chloroplasts.

5. **The correct answer is (B).** The mitochondrial genome encodes many essential proteins, but not all of the proteins found inside mitochondria. Many mitochondrial proteins are encoded by nuclear genes, so they are transcribed in the nucleus and translated in the cytoplasm, before the protein is then imported into mitochondria.

6. **The correct answer is (C).** Prokaryotic and eukaryotic cells all have a lipid bilayer membrane surrounding the cell. Only eukaryotes have linear DNA genomes, a cytoskeleton, and a nucleus, and only prokaryotes have a peptidoglycan cell wall.

7. **The correct answer is (A).** In G1, the cells are growing but not yet replicating their genome. Once the cell enters S phase, it will have more than 1X DNA per cell, until it has 2X DNA throughout G2 and until it fully divides into two daughter cells at the end of mitosis.

8. **The correct answer is (E).** The sister chromatids are the identical copies of each chromosome synthesized during S phase. The two copies remain stuck together until anaphase, ensuring that each daughter cell gets a copy of every chromosome.

SUMMING IT UP

- Cells are the basic unit of living things.
- The principle types of cells are the prokaryotic and eukaryotic.
- Prokaryotic cells lack the membrane-bound organelles found in eukaryotic cells. They generally have a plasma membrane, a single DNA molecule not coupled with protein molecules, small ribosomes, cytoplasm, and a cell wall.
- Membranes enclose the material in the cell and regulate passage of material into the cell from the surrounding environment.
- The fluid mosaic model describes a phospholipids bilayer as the foundation of a plasma membrane. The polar, hydrophilic end lies on the outside of the bilayer, and the non-polar, hydrophobic end lies on the inside. Embedded in the bilayer are cholesterol and a variety of protein molecules.
- Cell organelles are specialized structures within the cell that house specialized functions.
- Cell walls are found only in plants, fungi, some protists, and some bacteria; they consist mainly of cellulose in plants and help control osmotic uptake of water.
- Chloroplasts refer to the site for photosynthesis in plants' cells and they contain photosynthetic pigments, including chlorophyll.
- The cilia and flagella have the same construction and aid in the movement of individual cells or movement of water past cells that are part of tissues.
- Flagella are longer than cilia but of the same material. They aid the cell in movement with a wave-like motion.
- Golgi bodies are also known as Golgi complex or apparatus. They are the site where proteins are modified and other proteins are packaged for release from the cell.
- The mitrochondrion is the powerhouse of the cell and is the site of ATP synthesis only in aerobic respiration.

Energy in Biological Systems

chapter 6

OVERVIEW

- **Thermodynamics**
- **Free energy and equilibrium**
- **ATP and coupled metabolic pathways**
- **Enzymes**
- **Summing it up**

Living things are constantly using energy for practically everything they do, including building biological macromolecules, moving material across membranes, moving and responding to their environment, moving material inside the cell, and going through cell division. Energy takes many different forms, including light and heat and the energy in chemical bonds. Plants get their energy from the sun, transforming the sun's light energy into chemical energy. Animals and decomposers like fungi ingest food containing chemical energy and transform it into other forms of chemical energy that they can store or use immediately to fuel their activities. The conversion of biological compounds from one form to another creates complex networks of chemical reactions and energy flows called **metabolic pathways**. Pathways that take complex molecules and break them down into smaller ones are called **catabolic**. Other pathways that build complex molecules out of simpler ones are called **anabolic,** pathways. The biosynthesis of proteins from smaller, simpler amino acids is an example of an anabolic pathway.

Organisms use the potential energy contained in chemical bonds to do work. Different types of chemical bonds contain different amounts of potential energy that can be released and captured when they are digested by organisms. Ingesting starch and digesting it into individual glucose units that are further metabolized into carbon dioxide is a catabolic process, extracting energy from starch to power other cellular processes.

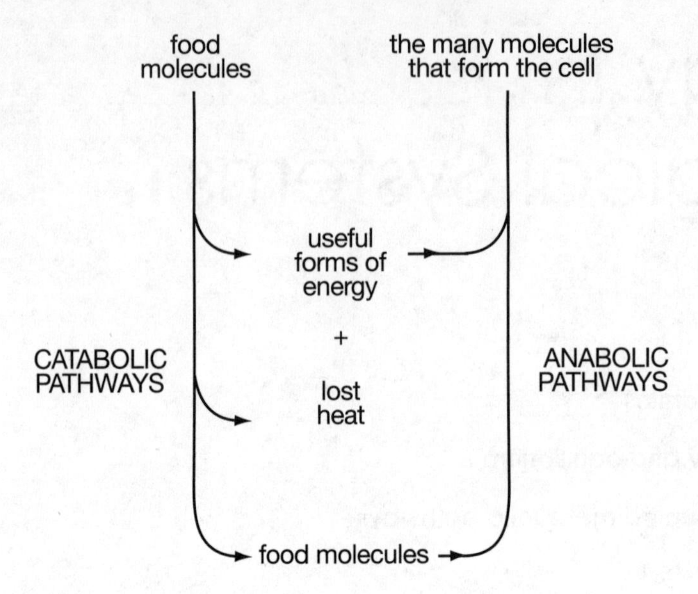

Building blocks for biosynthesis

THERMODYNAMICS

Living things must obey the laws of physics as they convert energy from one form to another. Two laws of thermodynamics are particularly relevant for biological systems. The **first law of thermodynamics** states that the energy of a closed system is constant. Energy can be changed from one form to another, but energy is never created from nothing and it is never lost; energy is always conserved.

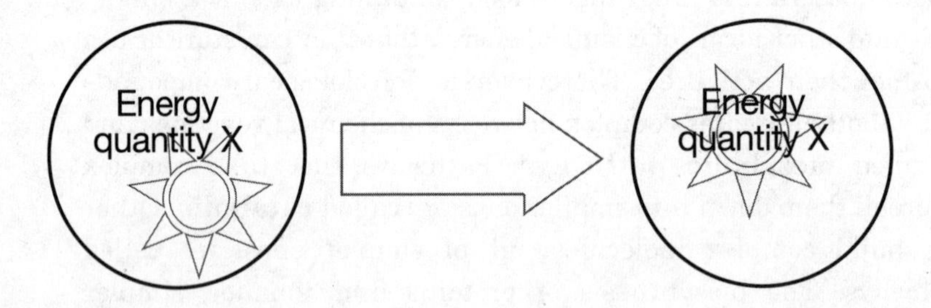

The first law of thermodynamics: Conservation of energy

The **second law of thermodynamics** concerns **entropy**, a measure of the disorder of a system, and states that the entropy of a system always increases over time. Because of the first law, energy is never lost, but because of the second law, the energy used by living things eventually becomes so disordered that it is no longer usable. The reactions involved in the metabolic conversion of energy to useful forms generate heat that is lost to the environment as a by-product, increasing the entropy of the system as a whole (the system that includes the animal and its surroundings). Heat is a disordered form of energy, denoting the level of random kinetic movement of molecules. Over time, as heat dissipates, entropy increases. Yes, energy is conserved, but useful energy decreases and entropy increases over time.

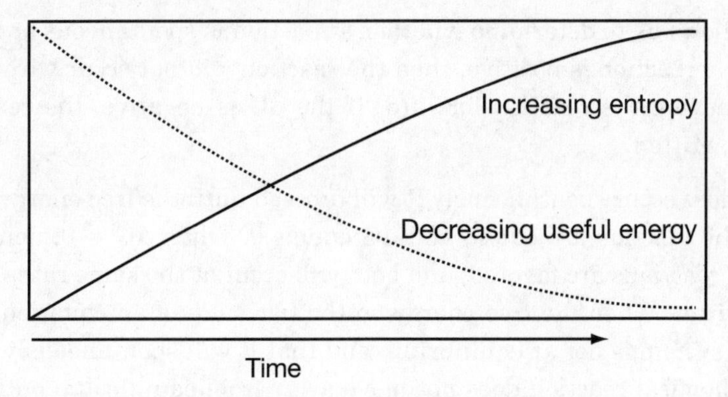

One of the properties of living things is that they are highly ordered, compared to their surrounding environment. The structures of macromolecules like proteins and DNA are much more complex and ordered than any inorganic compounds. This seems to contradict the second law, but when the broader system of which organisms are but a part is taken into account, it is possible to see that this cannot be the case. Living things are able to achieve order and complexity by using energy from their surrounding environment, which is part of the same system as the organism. As an organism uses energy, converting it from one form to another, it decreases its own entropy, but increases the entropy of its surroundings.

Q How do plants convert carbon dioxide into glucose and cellulose to build roots and leaves without breaking the second law of thermodynamics?

A Plants are part of a system that includes not just their local environment but the Earth as a whole and the sun. Plants take useful energy from the sun and convert into chemical energy, releasing heat back into their environment. No energy is lost (the first law is followed), and the overall entropy of the larger system, including the Earth and sun, increases.

FREE ENERGY AND EQUILIBRIUM

The metabolic chemical reactions in cells will either move forward on their own or will require some additional energy to move forward. Reactions that move forward on their own are called **spontaneous**. If a reaction is spontaneous, this does not mean that it happens quickly, only that it will occur on its own without adding any extra energy. Whether a reaction is spontaneous or not depends on the energy of the reactants and the energy of the products. The difference in the energy between the initiation and conclusion of a reaction can be measured by its **free energy**, often called the **Gibbs free energy**. The difference in free energy between the reactants and products of a reaction is given by the equation:

$$\Delta G = \Delta H - T\,\Delta S$$

in which ΔG is the change in free energy, ΔH is the change in enthalpy, a measure of the overall energy, T is the absolute temperature, in kelvin, and ΔS is the change in entropy.

Calculating ΔG allows us to determine whether a reaction is spontaneous or not. If the result is that the ΔG for a reaction is positive, then the reaction will not occur without adding more energy to favor moving the reaction forward. If the ΔG is negative, the reaction will occur spontaneously as written.

When a reaction does occur spontaneously it will proceed until the free energy of the reactants and products of the reaction are equal, and ΔG equals 0. When $\Delta G = 0$, neither the forward nor the backward reactions are favored, and both will occur at the same rate. The reaction has reached **equilibrium**. When the free energy on the two sides of an equation is not zero, this is a sign that the system is not at equilibrium and that it will spontaneously move until it is. A system like a chemical reaction does not move away from equilibrium on its own.

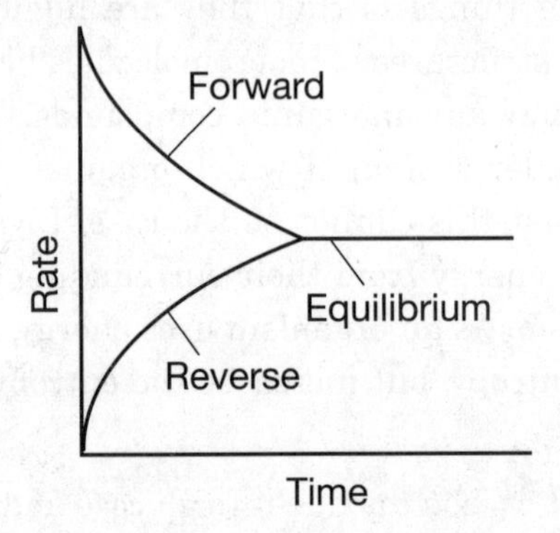

A useful metaphor for thinking of free energy is water flowing downhill. Water will only flow downhill on its own until it reaches the lowest point, a valley. The low point at which water comes to rest is equilibrium. Neither water nor chemical reactions will move uphill without additional energy, like a pump, acting on them.

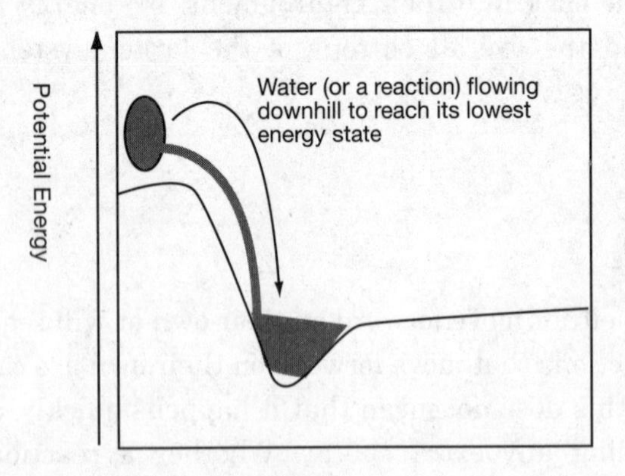

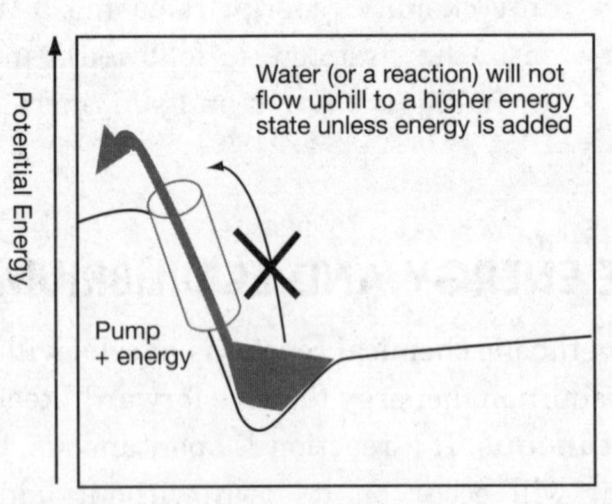

Q In the following reaction, the concentration of A, B, and C are measured in a solution, and the free energy for the reaction is found to be −5 kcal/mol. Does this mean that the reverse reaction C producing A to B does or does not occur?

$$A + B \rightleftarrows C$$

A Both the forward and reverse reaction will occur regardless of the free energy change. What changes is the relative rate of the forward and backward reactions. At equilibrium, both forward and reverse reactions occur at the same rate. With a negative free energy change, the forward reaction will occur faster than the backward reaction, although both are still occurring. If the free energy change is positive, the backward reaction will happen faster than the forward reaction, at least until equilibrium is achieved.

An **exergonic** reaction has a negative free energy, and will proceed spontaneously and release energy, while an **endergonic** reaction has positive free energy and does not occur spontaneously. If an endergonic reaction can absorb energy from its surroundings, it still might move forward.

Q An endergonic process requires energy from its surroundings to move forward, and an endothermic reaction absorbs energy in the form of heat from its surroundings. Some chemicals in solution undergo an endothermic crystallization process. How might an endothermic crystallization affect the temperature of the solvent it occurs in?

A Endothermic processes absorb heat from their surroundings. Absorbing heat from the surroundings means that as crystallization occurs, the solvent will get colder.

As water or a chemical reaction flows downhill from high to low energy, the movement of these systems can be captured to make other things happen. Water flowing downhill can move a turbine in a dam to produce electricity. Similarly, chemical reactions in the body move downhill to lower energy states and release energy that is captured to drive cellular processes. One key fact about living systems is that they are never at equilibrium. At equilibrium, a system has reached the lowest possible energy and cannot do any work, chemical or otherwise. Once water has flowed all the way downhill to the ocean, there is no further energy to be extracted. If a cell or an organism reaches equilibrium, then all of its components have reached the lowest energy and no reactions are proceeding. In other words, it is dead at that point.

ATP AND COUPLED METABOLIC PATHWAYS

One of the key forms of energy currency used by living things is adenosine triphosphate, ATP. ATP contains a 5-carbon sugar, ribose, the purine base adenine, and three phosphate groups. The three phosphates are negatively charged and tend to repel each other but are held together by covalent bonds in a relatively high energy state.

When one of ATP's phosphates is removed, it becomes ADP (adenosine diphosphate) and has less energy. The free phosphate has a much lower energy state, too. This makes the hydrolysis of ATP highly favorable energetically, with a very negative free energy. The following reaction describes the hydrolysis of one phosphate (P_i, inorganic phosphate) from an ATP molecule:

$$ATP + H_2O \rightarrow ADP + P_i$$

This reaction has a $\Delta G = -7.3$ kcal/mol

(This value of ΔG is measured under standardized conditions of specific temperature and concentrations of reagents.)

The energy of ATP hydrolysis is captured to power a wide variety of other cellular processes that are endergonic on their own: chemical reactions, mechanical movement, and the movement of chemicals against electrochemical gradients.

Q An ATPase enzyme is identified that catalyzes the hydrolysis of ATP into ADP and phosphate. If the purified ATPase enzyme is mixed with ATP in solution, how might this affect the temperature of the solution?

A ATP contains energy and its hydrolysis is favored energetically. In the cell, the hydrolysis of ATP is coupled to metabolic processes to make them move forward. Even this process is not 100% efficient, losing some of the energy of ATP hydrolysis to metabolic heat that is lost to the environment. If the ATP is hydrolyzed by the enzyme in isolation, without coupling to other processes, then essentially all of the energy of hydrolysis will be converted into heat.

Like the free energy change for the hydrolysis of ATP, the free energy change for endergonic reactions can be measured as well. If a reaction with a free energy change of +3 kcal/mol is coupled to the hydrolysis of one ATP, with a free energy change of −7.3 kcal/mol, then the free energy change of the overall reaction is −4.3 kcal/mol. In this way, the cell couples energetically unfavorable reactions to ATP hydrolysis to make them move forward:

$$A + B \rightleftarrows C + 3 \text{ kcal/mol}$$

$$ATP \rightleftarrows ADP + P_i \; -7.3 \text{ kcal/mol}$$

$$\overline{A + B + ATP \rightleftarrows C + ADP + P_i \; -4.3 \text{ kcal/mol}}$$

ATP often becomes part of the overall reaction itself. One way this is accomplished is for a phosphate group to be transferred from ATP to another molecule in the reaction and then later removed again by hydrolysis. If the endergonic reaction and the hydrolysis of ATP can be coupled in this way, then the free energy for the coupled reactions can be added together to determine whether the coupled reactions are exergonic, with an overall negative free energy, and then can proceed spontaneously.

The cell must maintain a constant supply of ATP for myriad reactions. As ATP is continuously used by cells for a broad range of reactions, it must also be constantly produced. If the ATP concentration in cells falls much, the result can be deadly for the cell. To keep the ATP concentration constant, ATP is regenerated, adding a phosphate back onto ADP. This reaction is the exact opposite of ATP hydrolysis as described earlier:

$$ADP + P_i \rightarrow ATP + H_2O$$

This reaction has the equal but opposite free energy as ATP hydrolysis, with $\Delta G = +7.3$ kcal/mol rather than -7.3 kcal/mol (see figure below). Producing ATP in this way is not spontaneous; it requires energy from somewhere. One of the most common biological processes performed by almost all living things is the metabolism of glucose, capturing the chemical energy in glucose to produce ATP. Glucose is metabolized, producing carbon dioxide, water, and energy captured in ATP. The ATP is hydrolyzed, releasing this energy again to move other cellular processes forward.

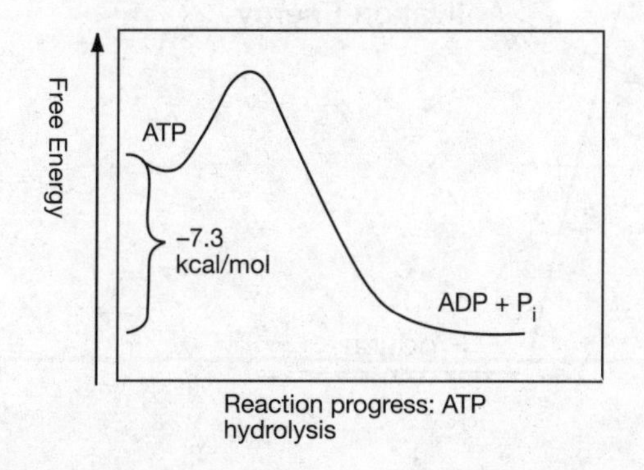

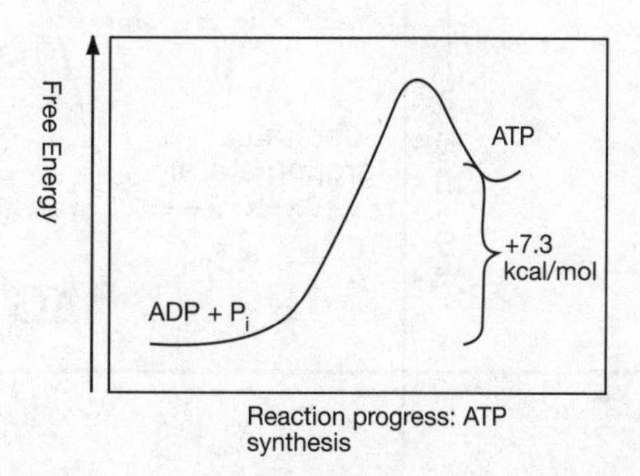

Q If the oxidation of glucose is used by cells to drive the production of ATP, what can you predict about the free energy of glucose oxidation?

A The production of ATP has a large positive free energy change. To move the production of ATP forward, it must be coupled to a reaction with an even larger negative free energy change. The oxidation of glucose must have a free energy change that is negative and significant, and this is just the case.

ENZYMES

Thermodynamic factors determine whatever reactions will occur spontaneously, but they say nothing about how fast they will happen. The rate of reactions is determined by factors other than the energy of the starting material and the resulting products of the reaction.

Some reactions that are quite energetically favorable, with a large negative free energy, occur at a very slow pace if the reaction consists only of reactants mixed in water. The oxidation of a piece of wood is very energetically favorable according to thermodynamics, with a large negative ΔG. A piece of wood is quite stable on its own and will survive intact indefinitely. The reason for the difference between the thermodynamic prediction of spontaneity and the rate at which reactions occur on their own is the result of a barrier to the forward movement of the reaction, a barrier called the **activation energy.**

Thinking of a reaction once again as if it is traversing an energy terrain, a thermodynamically favored reaction starts at a higher point than it ends up, with products having lower free energy. In the middle of the reaction, the chemicals involved in the reaction must pass over a barrier, a hill with higher energy than either reactants or products. This barrier is the activation energy. Even if the reaction is favored, it is generally the case that some energy must be put into the reaction to help the reactants over the activation energy barrier and reach their lower energy state, releasing once again the energy that was put into the reaction. The higher the activation energy barrier, the harder it is for chemicals to get over this barrier, and the more energy that must be put into the reaction to make it move forward.

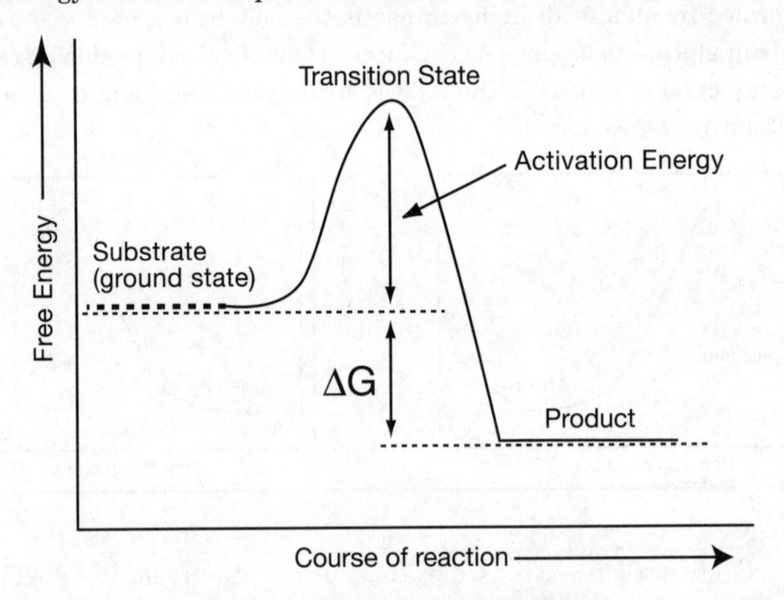

Single-step Spontaneous Reaction

The energy barrier of activation energy results from the high energy intermediate states that compounds move through as they change from reactants to products. In a chemical reaction, some covalent bonds are broken and other new bonds form, all of which takes energy to happen. In the process of breaking old bonds and forming new ones, chemicals briefly assume unstable structures called **transition state intermediates**. The more energy it takes to form the transition state intermediate, the slower a reaction will move forward.

One way to make reactions move more quickly is by heating up a reaction. Heat makes chemicals in solution move more quickly. The more quickly molecules are moving, the more frequently and energetically they will bump into each other, helping them to achieve the transition state more quickly. One form of heat can occur if a substance is burned. A piece of wood is quite stable on its own, but if subjected to the intense heat of a fire, the oxidation of the wood is accelerated enormously, with heating helping the cellulose over the activation energy to reaction with oxygen forming CO_2 and water. The same goes for the oxidation of other molecules like sugar or the hydrocarbons in gasoline. Glucose and gasoline are pretty stable left on their own, but if placed in a flame the heat of the fire increases the rate of the reaction enormously, oxidizing them to release carbon dioxide and water, the lowest energy state of these reactions.

Heat is not the only way to help reactions over the activation energy and move reactions more rapidly forward. Another way to increase the reaction rate is by lowering the activation energy. If the energy required to achieve the transition state intermediate can be reduced, then the activation energy will be reduced, and the reaction can move forward more rapidly even without adding heat. Substances that can perform this feat are called **catalysts**. Catalysts increase reaction rates but are not themselves consumed as part of the reaction they catalyze.

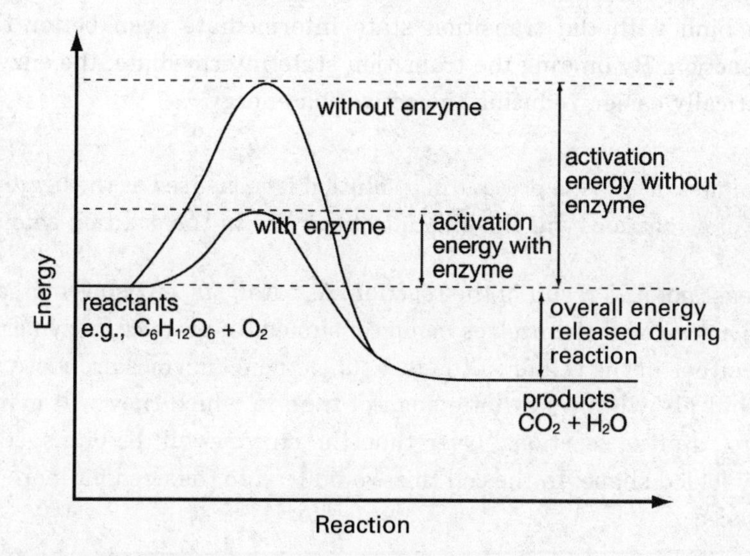

Chemists often use inorganic catalysts like platinum in the lab to speed up reactions, but biological systems use organic catalysts called **enzymes**. Enzymes are usually proteins and are essential for almost all of the complex network of chemical reactions found in metabolism. Enzymes are generally very specific to the reactions they catalyze and the structures of the reactants they will interact with. This specificity is based on the proteinic nature of enzymes.

The reactants in reactions catalyzed by enzymes are called **substrates**, and for an enzyme to catalyze a reaction it first binds to the substrate to form an **enzyme-substrate complex**. Being proteins, enzymes can fold into specific and complex three-dimensional structures. The enzyme-substrate complex does not form just anywhere on the protein but in a specific part of the folded three-dimensional structure called the **active site**, often a cleft in the surface of the enzyme that still has access to the external solvent. The specificity of enzymes is based on the specific recognition of substrate compounds by the enzyme protein in the active site. The active site is essential for an enzyme's activity but usually occupies only a small fraction of the overall structure of an enzyme, with relatively few amino acids in the enzyme actually coming in contact with the substrate.

Enzymes can be *very* specific in binding substrate and catalyzing reactions, differentiating even between closely related structures like fructose and glucose, structural isomers, and even molecules that are stereoisomers (the same in every way except in the orientation of groups around certain carbon atoms).

The specificity of enzymes is based on the shape of the active site and how the substrate fits into the active site. In the three-dimensional structure of enzymes, amino acids fold around the active site to give it a very specific shape that has evolved to interact tightly with the specific substrates for the reaction it catalyzes and not others. The amino acids in the active site interact with the substrate through a variety of bonds, including hydrogen bonds, ionic bonds, and van der Waals reactions, depending on the chemical nature of the substrate—all of which adds up to make the formation of the enzyme-substrate complex more favorable.

Enzymes can increase reaction rates by recognizing substrates, by bringing substrates together in a specific orientation, and by stabilizing the transition state intermediate. In fact, enzymes usually bind with the transition state intermediate even better than with the substrates of a reaction. By binding the transition state intermediate, the enzyme makes its formation energetically easier, reducing the activation energy.

Q If the amount of enzyme present in a solution is measured at the beginning and at the end of a reaction, will the amount of enzyme in the solution change?

A No, at least not as a result of the reaction. As catalysts, enzymes increase the rate of reactions without themselves being consumed. In practice, enzymes do not last forever, either in the cell or in the test tube. Some enzymes are more stable than others, but all will have a finite span of time in which they will maintain their ability to catalyze reactions. Over time the enzyme will be damaged or lose its properly folded shape. In the cell this would lead to the need for more enzymes to be produced.

One way to make an enzyme-catalyzed reaction go faster is by adding more substrate. When substrate concentrations are very low, then only a small percentage of the enzyme molecules present will have their active site occupied by substrate at any given moment. Since enzymes have to bind substrate at their active site as the first step in catalyzing the conversion of substrate to product, the more enzymes that have substrate bound, the faster the reaction will move forward to convert substrate to product. As the concentration of substrate is increased, more enzymes will have substrate bound at their active site, and product will be formed more rapidly. If the substrate concentration is increased higher, then more enzymes will have substrate at their active site, until at very high concentrations, 100% of enzymes will have substrate bound. When this happens, increasing the substrate concentration further does not increase the reaction rate any further, and the enzyme is **saturated**. This maximal enzymatic reaction rate is called the $\mathbf{V_{max}}$.

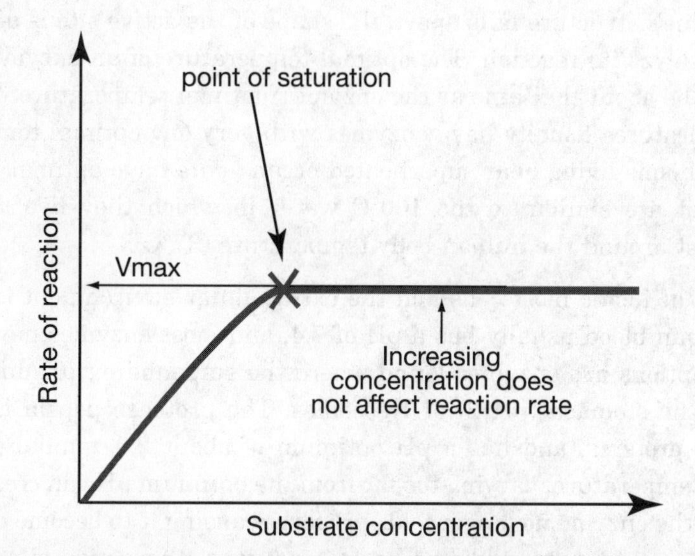

Q A researcher is studying an enzyme-catalyzed reaction in the lab. He finds that a reaction rate for the enzyme increases as the substrate concentration increases until the substrate concentration is increased to 2 micromolar. What will happen to the enzyme if the substrate concentration is increased to 4 micromolar?

(A) The enzyme will denature due to the excessive substrate concentration.

(B) The substrate will bind covalently to the active site, blocking catalysis.

(C) The reverse reaction will start to dominate, reducing the overall reaction rate.

(D) Active sites will be saturated with substrate, and the reaction rate will not increase any further.

(E) The heat of the solution will increase.

A The correct response is (D). The enzyme is saturated with substrate at 2 micromolar, and any further increases in substrate concentration will not change the rate because once the active sites are all full, the enzyme can't work any faster.

The conditions around an enzyme can dramatically affect how well an enzyme works, increasing or decreasing its activity. This can include many different factors, but the most common changes an enzyme might face would be changes in the pH and temperature of the solvent around it. Increasing temperature affects not just the enzyme, but the substrate and solvent molecules around the enzyme. As the temperature increases, the kinetic energy of all of these molecules increases, increasing the rate at which substrate molecules encounter the active site, and increasing the overall reaction rate. The enzyme rate increases with increasing temperature until it reaches a maximum and then decreases with temperatures higher than that which achieved the maximum rate.

The decreased enzyme activity at higher temperatures is caused by stress to the structure of the enzyme at high temperatures, with increased kinetic energy breaking hydrogen bonds, van der Waals, and other interactions that hold the enzyme's three-dimensional structure in place. As the enzyme's structure falls apart, the shape of the active site is disrupted, and the enzyme cannot catalyze its reaction. The optimal temperature of an enzyme for activity has usually evolved to be about the same as the enzyme's optimal temperature. Organisms living at very cold temperatures usually have enzymes with very low optimal temperatures, while extremophile organisms living near superheated ocean vents have optimal temperatures for their enzymes that are similar to the 100°C water in which they live. Human enzymes generally work best around the human body temperature (37°C).

The pH environment inside most cells and the extracellular environment is usually close to neutral (7.0). Human blood usually has a pH of 7.4, and most enzymes have a pH optimum close to this. Exceptions are enzymes found where the surrounding pH differs, such as the acidic interior of the stomach (pH 2) or lysosomes. The protease pepsin is secreted in the stomach to digest proteins, and has a pH optimum of about 2.0, similar to the pH of the stomach. As with temperature, varying too far from the optimum pH can create more extreme conditions so that the enzyme no longer folds properly, causing it to become denatured. As the enzyme's folding is undone by acidic or basic conditions, the active site no longer forms correctly and the enzyme can even form irreversible aggregates with other proteins around it. This is why lemon juice or orange juice makes milk curdle, by making the milk acidic and causing milk proteins to denature, unfold, and stick to each other.

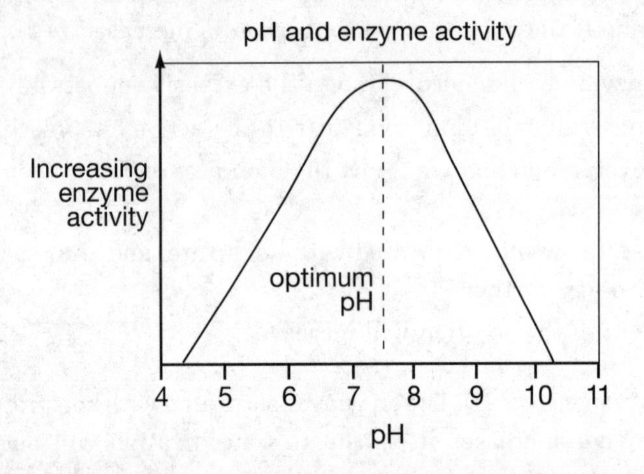

The primary structure of enzymes, consisting of a linear polypeptide of amino acids strung together, is encoded by their genes. The final folded enzyme often includes additional components called **cofactors** that are not encoded in the amino acid sequence of a protein but are essential for enzyme activity. Cofactors can be ions, like zinc or iron, nucleotide-based factors like NADH, or other relatively small organic molecules. Many vitamins are synthetic precursors of enzyme cofactors.

Enzyme Inhibitors

The activity of enzymes can often be inhibited by molecules that do not denature the overall structure of the enzyme but bind to specific sites on the enzyme to block its action. One class of enzyme inhibitors, called **competitive inhibitors,** bind at the enzyme's active site, displacing substrate from the active site. If the inhibitor, not the substrate, is sitting in the active site, then the enzyme cannot act on the substrate to convert it into product. The more active sites that have inhibitors, the fewer active sites that can bind substrate. Competitive inhibitors do not bind covalently to the active site and can be displaced from the active site, although achieving the same level of enzyme activity in the presence of inhibitors requires a higher substrate concentration. Since a competitive inhibitor must fit into the active site, its structure is usually similar to the structure of the substrate for the reaction it is inhibiting.

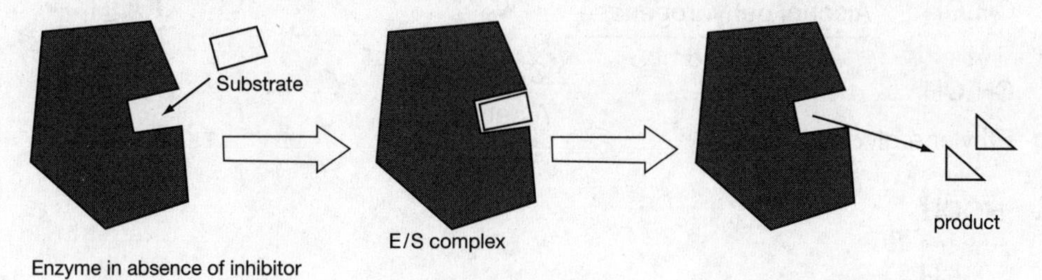

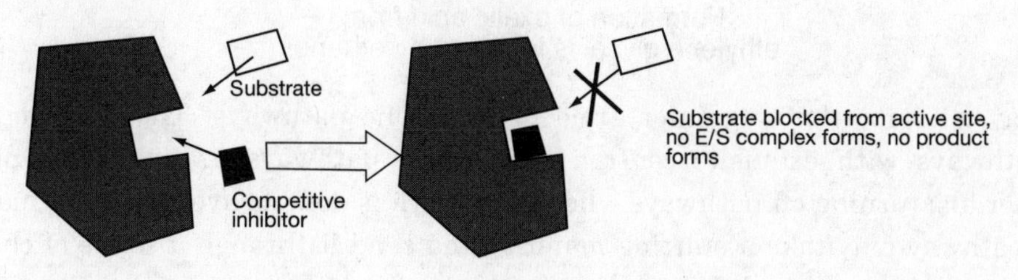

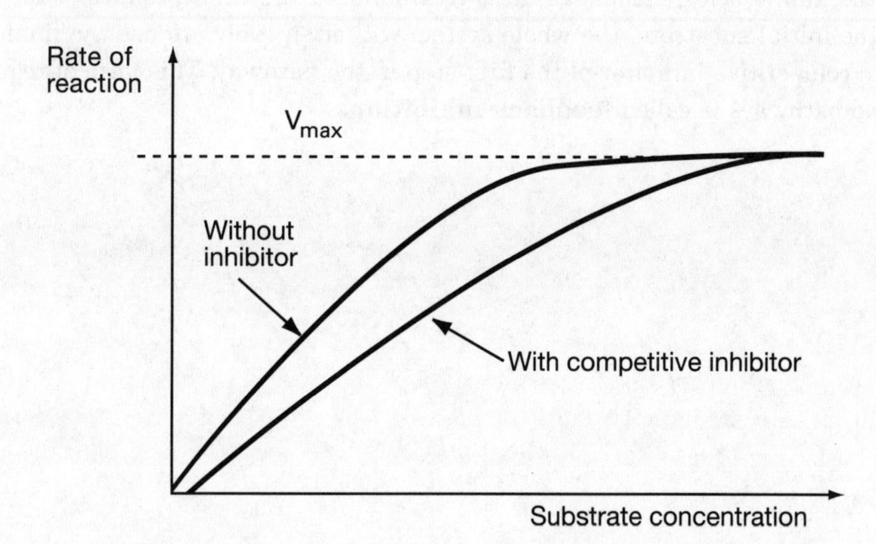

An example of competitive inhibitor is found in the interaction between ethylene glycol and ethanol. Ethylene glycol is found in anti-freeze and is quite toxic; it is converted to oxalic acid in the kidneys in part by the enzyme alcohol dehydrogenase (ADH). Another substrate of ADH is ethanol. If ethanol is bound to the active site of ADH instead of ethylene glycol, it acts as a competitive inhibitor of the conversion of ethylene glycol to oxalic acid, preventing toxic injury. For this reason a classic therapy for ethylene glycol poisoning has been treatment with alcohol. Today another drug called fomepizole is often used instead of alcohol, due to the side effects of treatment with high concentrations of alcohol.

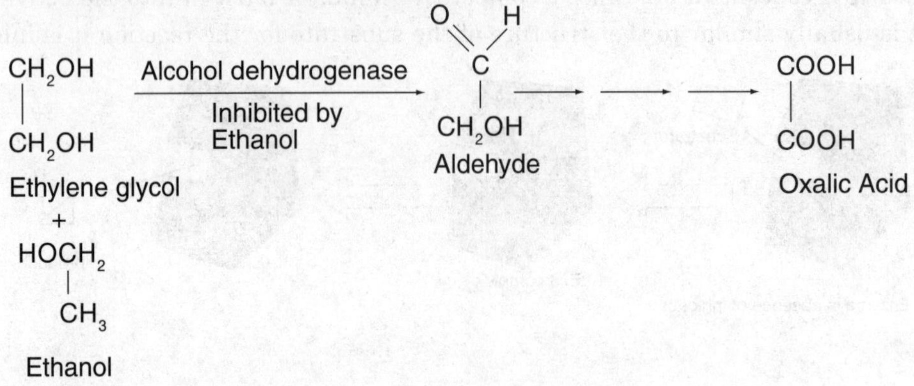

CH_2OH — CH_2OH Ethylene glycol + $HOCH_2$ — CH_3 Ethanol → Alcohol dehydrogenase (Inhibited by Ethanol) → O=C(H) — CH_2OH Aldehyde → → → COOH — COOH Oxalic Acid

Formation of oxalic acid from ethylene glycol is inhibited by ethanol

Inhibitors play an important regulatory role in metabolic pathways. The body regulates metabolic pathways with exquisite control: turning on pathways as needed to produce essential materials; turning off pathways when their work is done to avoid wasting material. A metabolic pathway may take a starting material and send it through a series of chemical reactions, each catalyzed by a different enzyme. If the final product of the pathway is an essential amino acid that is required for proteins to be made, the cell must avoid running out of this essential material. But it also does not want to waste a lot of metabolic energy producing the amino acid if plenty is already around. If the final product of the pathway resembles the initial substrate, the whole system will often evolve to use the final pathway product as a competitive inhibitor of the first step in the pathway. This form of regulation of enzymes and pathways is called **feedback inhibition**.

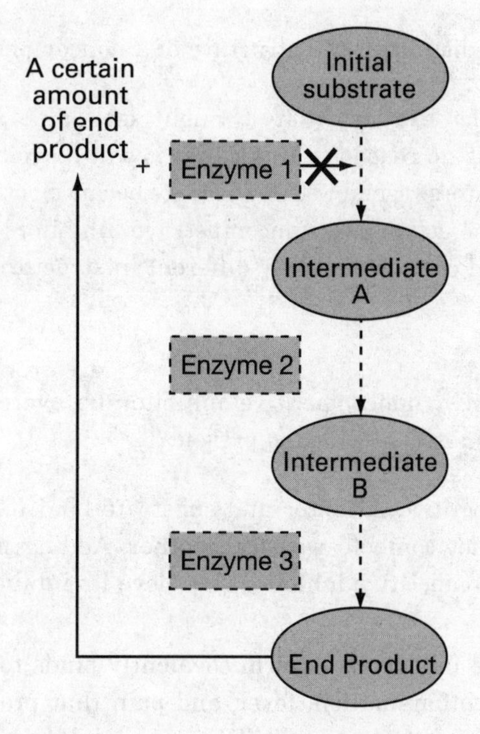

As stated above, competitive inhibitors bind at the active site but do not form an irreversible covalent attachment at the active site and can be displaced from the active site by high concentrations of substrate. **Irreversible inhibitors** also bind at the active site but form a permanent covalent attachment. Once an irreversible inhibitor binds, substrate will be blocked from the active site no matter how much substrate is added.

A third type of enzyme inhibitor binds to enzymes somewhere other than the active site. Since they do not compete with substrate to bind to the enzyme, they are called **noncompetitive inhibitors**. Noncompetitive inhibitors slow the reaction rate by distorting the active site to one degree or another.

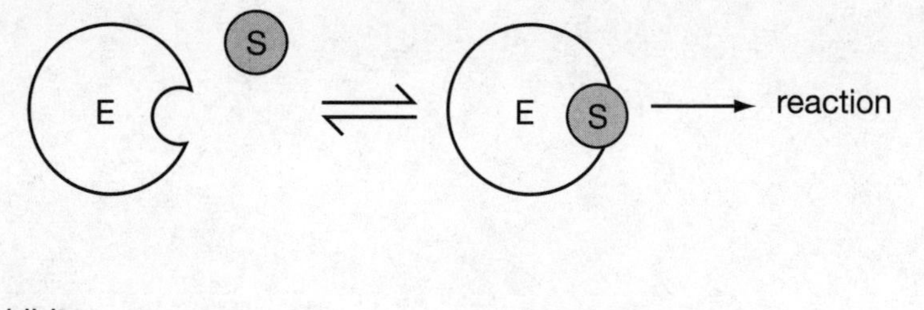

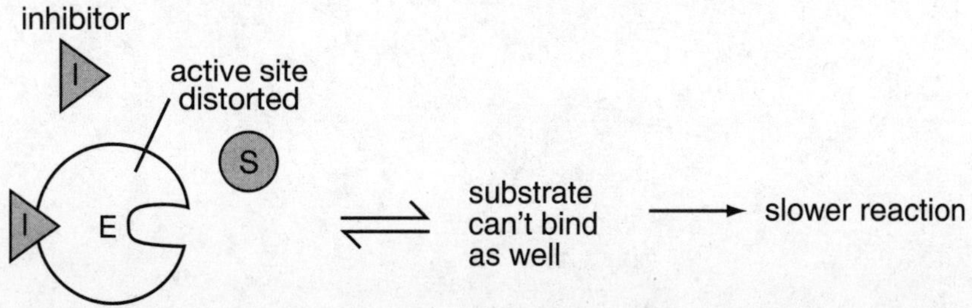

Noncompetitive enzyme inhibitor

Q What can you predict about the structure of a noncompetitive inhibitor?

A You cannot predict exactly what it might look like, but unlike competitive inhibitors, there is no reason to think that it will resemble the substrate. While competitive inhibitors resemble the substrate because both bind to the same place on the enzyme, the active site, noncompetitive inhibitors bind to a different part of the enzyme and could be entirely different in structure.

Q Can the action of a noncompetitive inhibitor be overcome by increasing the concentration of substrate which is present?

A No. The noncompetitive inhibitor and substrate bind to different places on the enzyme and do not compete with each other. Adding more substrate will not displace the noncompetitive inhibitor or relieve its inhibition.

Many drugs act as enzyme inhibitors. Aspirin covalently binds to an enzyme that produces prostaglandins, inhibiting inflammation, fever, and pain that prostaglandin production can create. Antibiotics in the same class as penicillin act as inhibitors of one of the enzymes that produced the peptidoglycan in the bacterial cell wall. Some insecticides called organophosphates or nerve gas agents act as irreversible inhibitors of the enzyme acetylcholinesterase that normally degrades the neurotransmitter acetylcholine at the neuromuscular junction. If the enzyme is irreversibly inhibited, the nerves cannot turn off muscle contraction, causing paralysis.

EXERCISES: ENERGY IN BIOLOGICAL SYSTEMS

1. The ΔG under standardized conditions for the chemical reaction below is +7 kcal/mol. Which of the following statements about this reaction must be true?

$A + B \rightarrow X + Y$

(A) The reaction proceeds quickly forward toward product formation.
(B) The reaction is exergonic.
(C) The reaction is coupled to the hydrolysis of ATP.
(D) The reaction has decreasing entropy and increasing enthalpy.
(E) The reaction will not proceed forward spontaneously as written.

2. A crystal of glucose is placed into a vacuum insulated flask of water and the flask is then sealed. Which of the following statements is true regarding the thermodynamics of the glucose crystal as it dissolves?

(A) The entropy of the contents of the flask increases as the glucose crystal dissolves.
(B) The entropy of the glucose increases but the entropy of the water decreases to the same extent, to conserve entropy.
(C) Entropy increases, but the process for dissolution of the crystal has a positive ΔG.
(D) The mass of the system decreases slightly as the energy decreases.
(E) The mass of the closed system remains the same but the energy decreases.

3. The ΔG for the following reaction is +3.2 kcal/mol.

$A + B \rightleftarrows C$

When a purified enzyme is added to the substrates in a test tube, no product is observed. If ATP is added to the test tube along with the enzyme, product rapidly forms. What is the most likely mechanism by which ATP affects the reaction?

(A) The ATP binds to the enzyme and allosterically alters its conformation.
(B) The hydrolysis of ATP becomes part of the overall coupled reaction, so that the overall reaction is spontaneous although Reaction #1 on its own is not.
(C) ATP makes the enzyme move through the solution faster, increasing its kinetic energy to move the reaction forward.
(D) ATP acts as a catalyst, binding to the substrate to stabilize the formation of the reaction intermediate.
(E) ATP increases the reaction rate but is not itself consumed in the reaction.

4. In the figure below, the inhibitor acts by what mechanism?

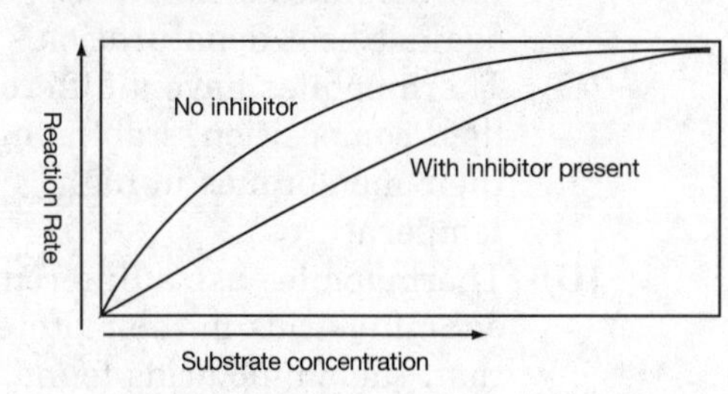

(A) Competitive inhibitor
(B) Irreversible inhibitor
(C) Noncompetitive inhibitor
(D) Feedback inhibitor
(E) It is not possible to determine with the information provided.

5. A researcher studying an enzyme involved in the synthesis of the amino acid valine tests the reaction rate at varying concentrations of substrate and enzyme. How would he expect doubling the enzyme concentration in a reaction working at V_{max} to affect the reaction?

(A) The initial reaction rate and equilibrium will be doubled.
(B) The initial reaction rate will double, and equilibrium will stay the same.
(C) The initial reaction rate and equilibrium will not change.
(D) The initial reaction rate will not change, but the equilibrium will shift toward product.
(E) Initial reaction rate will double, and equilibrium will change toward substrate.

6. Some thermophilic organisms live in temperatures near to or higher than boiling, like those found in volcanic vents. These organisms are often archaebacteria. Which of the following might play a role in the ability of thermophilic enzymes to work at extremely high temperatures?

(A) Polypeptides are joined differently in the primary protein structure of thermophiles, preventing the hydrolysis of peptide bonds.
(B) The tertiary protein structure of thermophiles is stabilized against heat denaturation.
(C) Thermophiles have a different lipid composition, stabilizing their membranes in high temperatures.
(D) Thermophiles use a different set of amino acids in their proteins than the amino acids found in other organisms.
(E) Archaebacteria have alternative metabolic pathways for ATP production.

7. An antibody is identified with a high affinity for the transition state intermediate in the reaction below. When incubated with a solution of A, the antibody is found to catalyze the isomerization reaction converting A to B. Which of the following points are probably changed by the antibody?

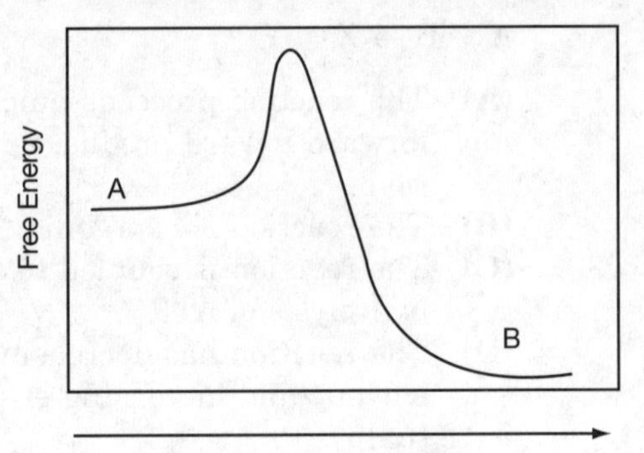

(A) The free energy of A is increased.
(B) The free energy of B is decreased.
(C) The change in free energy from A to B is increased.
(D) The activation energy is reduced.
(E) The entropy for the reaction is increased by the antibody.

exercises

8. A human enzyme tested for activity at 25°C is found to catalyze the formation of 4 mmol/min/mg of enzyme. At 37°C it catalyzes its reaction at 12 mmol/min/mg of enzyme. When the enzyme is returned to 25°C, its reaction rate returns to 4 mmol/min/mg of enzyme. What does this suggest about the activity of this enzyme?

(A) The enzyme is denatured by heat above 37°C.
(B) The enzyme is denatured by exposure to temperatures colder than 37°C.
(C) The enzyme structure is not affected by heat.
(D) The enzyme structure is stable when heated to 37°C.
(E) The enzyme does not require a specific folded structure to catalyze this reaction.

ANSWER KEY AND EXPLANATIONS

1. E	3. B	5. B	7. D	8. D
2. A	4. A	6. B		

1. **The correct answer is (E).** A reaction with a positive ΔG is not spontaneous and will not proceed forward as written. This also means that A is wrong. The definition of exergonic is that a reaction has a negative ΔG. Many reactions that are thermodynamically unfavorable are coupled to ATP hydrolysis, but we don't know if that is true in this particular case. Entropy and enthalpy are two factors that determine the free energy change for a reaction, but we don't know their status in this particular case.

2. **The correct answer is (A).** A glucose crystal is a highly ordered state of matter. As the individual molecules of glucose move into solution, they are moving about randomly and the overall disorder of the system increases. Overall, thermodynamics says that the entropy of any system always increases, if the whole system is taken into account. Entropy is not conserved like mass and energy are conserved. The crystal dissolves spontaneously, so it must have a negative ΔG. Mass and energy will both be conserved. Only the form of energy will change, becoming more and more disordered in the system and less usable for work.

3. **The correct answer is (B).** ATP often plays this role in the cell. ATP can allosterically activate some enzymes, but since we know this reaction has a positive free energy change without ATP, changing this is probably the mechanism involved.

4. **The correct answer is (A).** The inhibition is overcome at higher substrate concentrations to achieve the same V_{max} observed in the absence of inhibitor. This is a hallmark of competitive inhibition.

5. **The correct answer is (B).** Adding more enzyme provides more active sites. If you double the number of active sites working, the initial reaction rate will double. Catalysts increase the reaction rate but do not change equilibria, however.

6. **The correct answer is (B).** All organisms use the same amino and the same peptide bonds in their proteins. Lipid composition and ATP production are irrelevant.

7. **The correct answer is (D).** The antibody is acting as an enzyme, reducing the activation energy by stabilizing the transition state intermediate. As an enzyme, it does not change the free energy of A or B or the difference between A and B. The free energy of A and B is inherent in these molecules.

8. **The correct answer is (D).** The enzyme retains full activity after heating to 37°C, and enzyme function always depends on having the properly folded structure.

SUMMING IT UP

- Enzymes are biological catalysts that accelerate the rate of biochemical reactions.
- Enzymes are not consumed in the reactions they catalyze.
- Enzymes catalyze reactions via action at their active sites.
- The formation of the enzyme's active site, and its catalytic activity, depend on the proper folding of the enzyme protein's structure.
- The addition of more substrate to an enzyme catalyzed reaction will increase the reaction rate up until the substrate concentration is so high that E/S complex cannot be formed any faster. At this point the reaction is saturated.
- Enzymes can increase reaction rates by lowering activation energy, but they do not change reactions' equilibrium or free energy change.
- Enzyme inhibitors are molecules that bind to enzymes and decrease their activity, sometimes to zero. Competitive inhibitors bind at an enzyme's active site and displace substrate. These can be displaced from the active site by high concentrations of substrate. Irreversible inhibitors form a permanent covalent attachment at the active site and block substrate permanently. Noncompetitive inhibitors bind to enzymes at locations other than the active site and do not compete with substrate.

Cellular ATP Production

chapter 7

OVERVIEW

- Metabolism
- Glycolysis and fermentation
- Krebs cycle (citric acid cycle)
- Electron transport or electron transfer phosphorylation
- Other sources of energy (amino acids, fatty acids)
- Coordination of energy pathways
- Summing it up

METABOLISM

A very simplified view of how most life on Earth works is that almost all energy in living things starts with the absorption of solar energy by plants, followed by the conversion of this energy through photosynthesis into the chemical energy of glucose. The energy in glucose is then released through cellular respiration, either by the plants or by other organisms that feed on the plants (fungi, animals, etc.), to produce ATP, with carbon dioxide and water as the end products. The ATP in turn drives the metabolic processes of life, while the water and carbon dioxide can cycle between photosynthesis and cellular respiration.

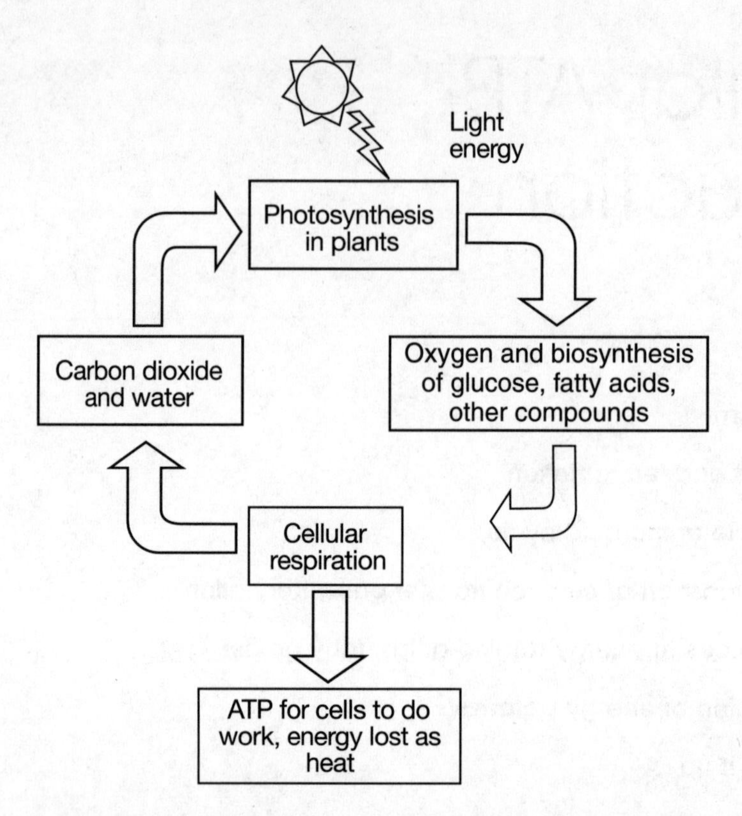

The high-energy chemical bonds in glucose can release their energy in more than one way. If glucose is crystallized and dried out, it will burn if the crystals are placed in a flame, like the caramelized sugar on a crème brulé. When glucose or other sugars burn this way, they react with oxygen in the air, generating heat energy as a visible side product (flames), as well as carbon dioxide and water. Organisms and cells using glucose for energy also oxidize it, but cells use a carefully controlled series of oxidation/reduction chemical reactions to capture as much of the released energy as they can as useful chemical energy rather than the uncontrolled release of energy as heat in a flame.

Oxidation and reduction reactions ("redox" reactions, for short) involve the transfer of electrons between molecules involved in a reaction. A molecule that loses electrons is said to be "oxidized," while a molecule that gains electrons is said to be reduced. The molecule that gives away electrons, also called the electron donor, is a **reducing agent**, while the molecule that gains electrons, an electron acceptor, is an **oxidizing agent**. The oxidizing and reducing agents in a reaction always occur together and cannot occur alone. Redox reactions are like the tango: you really need a pair to do the dance.

In an imaginary reaction, for example, you can track the electrons to see what is being oxidized and what is being reduced:

$$R{-}H\ (e-) + Y \rightleftarrows R + Y{-}H\ (e-)$$

The electron (e−) drawn with the R−H group is the electron associated with the hydrogen atom. Wherever hydrogen goes, this electron goes too.

Q Which of the reactants in the reaction on page 126 is the reducing agent?

A R–H is the reducing agent, the compound that donates an electron to the other reactant, Y.

Q Which of the reactants is oxidized?

A R–H is oxidized. When it reduces Y, it is itself oxidized. You can't have reduction of one molecule without oxidation of another.

Q Which of the reactants is reduced?

A Y is reduced.

Molecules with a lot of hydrogen-carbon bonds are highly reduced, while carbon dioxide is the most oxidized form of carbon. Hydrocarbons, the side chains of fatty acids, are highly reduced, and can release a lot of energy in redox reactions that produce carbon dioxide. These C–H bonds are akin to water at the top of a hill, ready to release their energy by reacting with oxygen or other oxidizing agents. As the high-energy electrons from hydrocarbons or other organic molecules consumed for energy enter metabolism, they are shuttled through a series of redox reactions, that capture their energy. At the end of their path, they reach the most oxidized state of carbon, CO_2.

Molecular oxygen from the atmosphere is usually the ultimate oxidizing agent in cellular respiration, accepting the electrons from these redox reactions until they reduce oxygen to water.

Q Which of the following releases more energy from glucose?

(A) Burning it in a fire, producing water and CO_2

(B) Consumption by a cell, converting it into the same amount of water and CO_2

A Neither. The oxidation of glucose to water and carbon dioxide is the same reaction, with the same ΔG between reactants and products, no matter what path is taken to get from glucose to water and CO_2. The same amount of energy will be released by both. What will be different is the form of energy released. In (A), the energy will be released as heat, and probably lost to the environment without being captured to do work. In (B), some heat will still be lost, but the cell will also capture a great deal of the energy in chemical bonds like those in ATP that can drive work the cell must perform.

One of the key molecules that helps capture energy during the metabolism of glucose or other molecules is called NAD^+, which stands for nicotinamide adenine dinucleotide. The + sign indicates that NAD^+ is a charged molecule. The nicotinamide part of NAD^+ is derived from the vitamin niacin, and if the amount of niacin in a person's diet is not sufficient they can develop the disease pellagra, with reduced metabolic energy production. The rest of the molecule is built from pieces of two nucleotides, making it a dinucleotide.

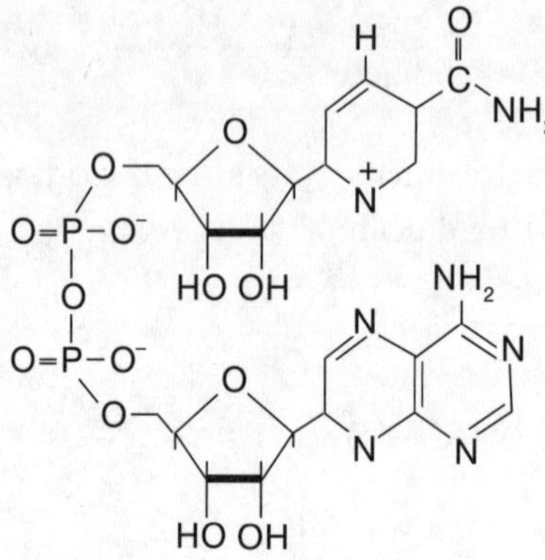

NAD^+

Since living things route electrons through a series of reactions, capturing their energy, they use intermediate molecules to carry these electrons from one stage in the process to the next. NAD^+ accepts electrons, oxidizing the molecule it is capturing energy from. When it oxidizes its partner molecule, it is itself reduced to NADH.

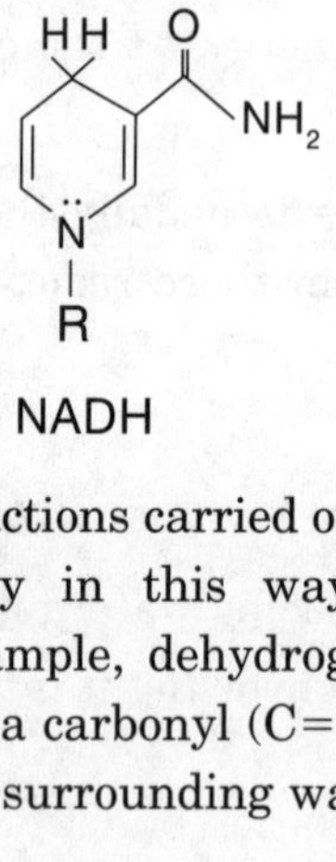

NADH

There are many different metabolic reactions carried out by a general class of enzymes (called dehydrogenases) that capture energy in this way. In the series of steps capturing energy/electrons from glucose, for example, dehydrogenases catalyze reactions in which a hydroxyl group (C–OH) is oxidized to a carbonyl (C=O), producing NADH from NAD^+, and releasing another proton (H^+) into the surrounding water environment.

(Ethanol) CH_3-CH_2-OH

↓ NAD^+ → $NADH + H^+$

(Acetaldehyde) $CH_3-\overset{O}{\overset{\|}{C}}H$

↓ NAD^+ → $NADH + H^-$

(Acetate) $CH_3-\overset{O}{\overset{\|}{C}}-OH$

The three key pathways that oxidize glucose and capture released chemical energy in ATP are: glycolysis, the Krebs cycle (often called "the citric acid cycle" or "the TCA cycle"), and electron transport. Each of these consists of several distinct steps. Although most of the energy captured ultimately ends up as glucose, most of the energy is used to make ATP indirectly.

In glycolysis, a series of reactions convert glucose, a six-carbon sugar, into two molecules of pyruvate, a three-carbon molecule. In eukaryotes, glycolysis occurs in the cytoplasm, and captures only a small part of the overall energy contained in glucose. A small amount of ATP is produced directly by glycolysis, as is some NADH. Most of the energy in glucose is captured in the Krebs cycle, which oxidizes pyruvate all the way down to carbon dioxide, the bottom of the metabolic energy hill. The Krebs cycle takes place inside mitochondria, in the mitochondrial matrix. As pyruvate is oxidized in the Krebs cycle, NADH is produced. This NADH supplies energy in the form of high-energy electrons that enter the electron transport chain, a series of electron carriers that perform redox reactions passing electrons from one carrier to another down the energy hill. The components of the electron transport chain are mostly embedded or closely associated with the inner mitochondrial membrane. As the electrons from NADH move from step to step in the electron transport chain, they move from higher to progressively lower energy levels, and some of their energy is used to pump protons across the inner mitochondrial membrane. This proton gradient is finally used as an energy source to drive the synthesis of the bulk of the ATP that cellular respiration can provide. The production of ATP in this manner, using the redox reactions of electron transport to build a proton gradient, which drives phosphorylation of ATP, is called oxidative phosphorylation or cellular respiration.

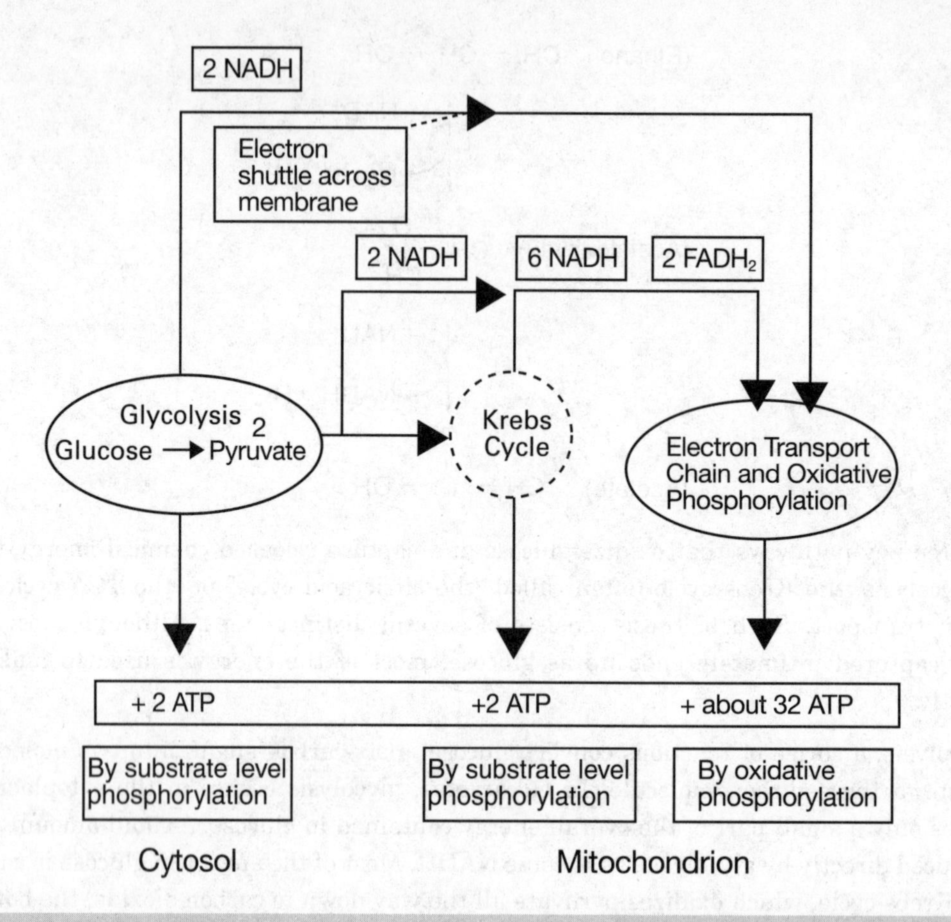

Q If glucose is synthesized with a radioactive carbon atom as the third carbon in the sugar molecule, and this glucose is then fed to cells performing cellular respiration for energy, where will the radioactive carbon atom later be found?

A The carbon atoms in glucose end up as the carbon in carbon dioxide. The radiation would end as carbon dioxide, either in solution as carbonate or carbonic acid or released into the atmosphere as carbon dioxide gas.

Overall, the complete oxidation of one glucose molecule to carbon dioxide through these three processes produces a theoretical yield of thirty-eight molecules of ATP in prokaryotes and thirty-six ATP molecules in eukaryotes. The difference is due to the use of energy to move molecules across the mitochondrial membrane.

GLYCOLYSIS AND FERMENTATION

Glycolysis is used to generate energy by almost all living organisms, from prokaryotes to eukaryotes, and seems to be the most ancient metabolic pathway, arising early in the history of life. In the human body, for some cells (like red blood cells), glycolysis is the only source of ATP. Glycolysis is the starting point for the release of energy from glucose, a metabolic pathway of ten enzyme-catalyzed reactions cutting the six-carbon sugar glucose in two and oxidizing the smaller molecules to create two pyruvates for every glucose that enters glycolysis. These ten reaction steps are broken down into two phases. First, ATP is used to create higher energy intermediates from glucose, then energy is harvested from these intermediates, generating more ATP than was spent in the first phase. Additional energy is captured in NADH. Overall, the first phase uses two ATP for every glucose that enters glycolysis, but the second phase harvests four ATPs and two NADHs, for a net gain of two ATP and two NADH for every glucose.

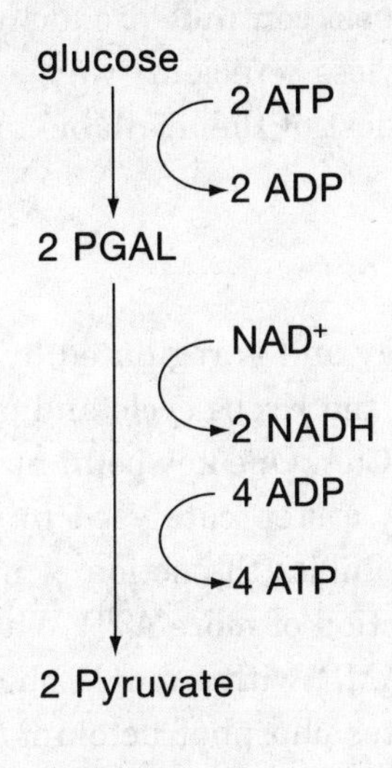

Glycolysis

Even cells that do not perform oxidative metabolism can still generate energy from glycolysis. In the absence of oxygen, cells still need energy, but cannot use the Krebs cycle or electron transport to produce ATP, leaving glycolysis as their only source of ATP production. In these conditions, cells run out of NAD^+ quickly and glycolysis can no longer occur unless the NAD^+ can be regenerated somehow. Cells have found a way around this by using NADH as a reducing agent for pyruvate, generating either ethanol or lactic acid from pyruvate. This process is known as fermentation or anaerobic respiration. Wine and beer contain alcohol as a result of fermentation of sugar by yeast. During strenuous exercise, in which the supply of oxygen to muscle becomes limited, muscle cells will produce lactate to continue generating ATP without oxygen. The sore muscles experienced after such exercise are the result in part of this lactate.

glucose —glycolysis→ pyruvic acid → lactic acid (animal) / ethanol + CO_2 (plant)

Anaerobic respiration (fermentation)

Q If energy can be produced through glycolysis and fermentation alone, what is the benefit to organisms of going through the additional series of steps required for oxidative phosphorylation?

A While a molecule of glucose can produce a total of thirty-six to thirty-eight ATP through oxidative phosphorylation, it produces only two ATP through fermentation. Fermentation can at least continue producing some energy in the absence of oxygen, but it is a far less efficient way to use glucose than oxidative phosphorylation, wasting most of the available metabolic energy.

Regulation of glycolysis

Glycolysis is a key metabolic pathway and is regulated by cells to produce energy when it is needed. When glycolysis slows, then the Krebs cycle and oxidative phosphorylation will slow as well, lacking pyruvate and acetyl-CoA. One key point at which glycolysis is regulated is the third step in this ten-step pathway, a step catalyzed by the enzyme phosphofructokinase. When ATP is abundant in cells, it inhibits the action of phosphofructokinase, slowing down glycolysis as a whole and the production of more ATP. When cells are low in energy, more of their ATP is hydrolyzed, producing AMP, with a single phosphate group rather than three in a row. Increased AMP in cells activates phosphofructokinase, increasing the rate of glycolysis and the amount of ATP produced through glycolysis and oxidative phosphorylation. One more key regulator of the activity of this enzyme is citrate, a molecule that is part of the Krebs cycle. When oxygen is present, the Krebs cycle is the next step in metabolic energy production after glycolysis. Increased citrate levels inhibit glycolysis, keeping these two metabolic pathways working together at just the right rate required by the cell for its energy needs.

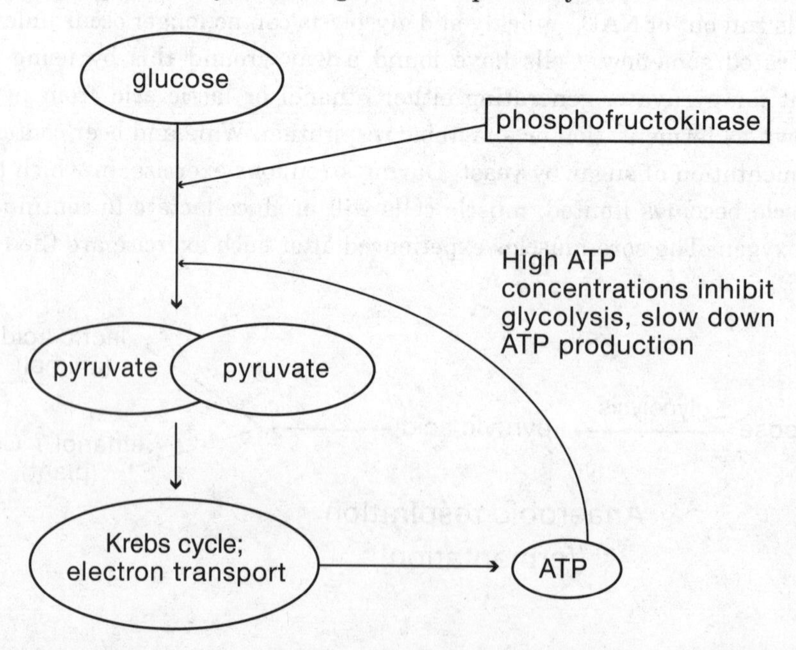

Q When the AMP concentration in cells increases and ATP decreases and oxygen is present, what will happen to the rate of pyruvate formation and the rate of CO_2 production?

A The concentration of ADP and ATP are signs of the energy charge of cells. As ATP decreases and ADP and AMP increase, glycolysis is activated. When glycolysis is activated, more pyruvate is made, which enters in the Krebs cycle. When more pyruvate enters the Krebs cycle, more CO_2 is released by the Krebs cycle.

Glucose is a key metabolic molecule and the most ready energy "currency" for most cells to use, but it is not the only carbohydrate involved in energy production. Other sugars like sucrose, fructose, lactose, and maltose are metabolized and enter into glycolysis to provide energy. The polysaccharide glycogen, which stores glucose in liver and muscle cells, releases glucose when needed, allowing the glucose to enter glycolysis to provide energy. When energy is abundant and glucose levels in the blood are high, cells in the liver and skeletal muscle can import glucose from the blood and use it to add to their glycogen stores. One way that these cells respond to energy levels is from signals received from hormones like insulin. When glucose levels are high after a meal, insulin released by the pancreas acts on liver and muscle cells to increase the import of glucose and activate the enzymes that synthesize glycogen.

Q At which step in glycolysis is molecular oxygen involved?

A None! Glycolysis does not involve oxygen in any way. That is why fermentation can occur in the absence of glycolysis.

Q The protein hormone glucagon is released by the pancreas between meals when glucose levels in the blood are low. How might glucagon be predicted to affect the activity of glycogen synthase?

A When energy levels are low between meals, as measured by plasma glucose levels, then glucagon and other factors will block glycogen synthase and stimulate the release of glucose from glycogen instead, maintaining plasma glucose levels in this way.

In some tissues like liver, when glucose levels in plasma are low, glucose can also be synthesized from other metabolic intermediates, including pyruvate, lactate, and compounds related to amino acids. This metabolic pathway is called **gluconeogenesis**. Gluconeogenesis does use some of the same enzymes as glycolysis but with the reactions occurring in the reverse direction. Where the energetic difference between substrates and products makes a reaction effectively irreversible, gluconeogenesis uses a different enzyme, one unique to this pathway. These two pathways are regulated in opposite manners so that they do not operate in the same cell at the same time. Gluconeogenesis is only activated when blood glucose levels are low and glycolysis is not occurring.

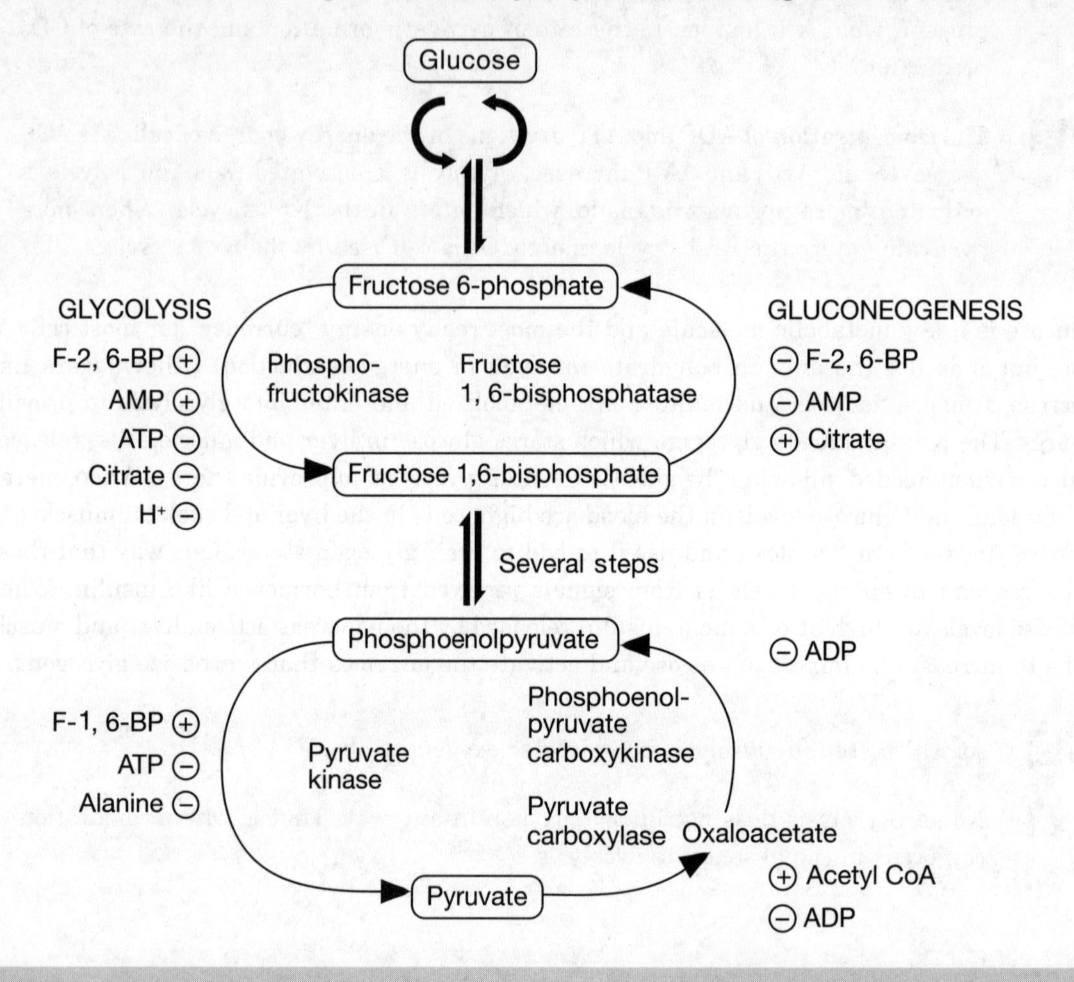

Q What would happen if both gluconeogenesis and glycolysis occurred in the same cell at the same time?

A The cell would be wasting a lot of energy. It takes a lot of energy for the cell to perform gluconeogenesis, much more energy than is produced in glycolysis. If these two pathways are running at the same time, this energy would be spent for nothing.

Q Gluconeogenesis would typically occur during fasting or starvation. Insulin is generally released by the pancreas after meals in response to high glucose levels, stimulating the uptake and use of glucose by tissues. How might insulin be predicted to affect the rate of gluconeogenesis in the liver?

A Insulin inhibits gluconeogenesis. When glucose is abundant in plasma, as occurs after a meal, it would waste energy for the liver to release more glucose into the bloodstream. One of the defects found in diabetes is that insulin does not inhibit hepatic gluconeogenesis, causing the liver to continue releasing glucose and contributing to the abnormally high plasma glucose levels associated with diabetes.

KREBS CYCLE (CITRIC ACID CYCLE)

When oxygen is present, energy production does not stop with glycolysis and fermentation. The pyruvate produced by glycolysis still contains a great deal of chemical energy, energy that is captured through the Krebs cycle and electron transport to produce ATP. The Krebs cycle is a cyclic series of metabolic reactions designed to oxidize pyruvate all the way to carbon dioxide, extracting energy to reduce the electron carriers NAD^+ and FAD.

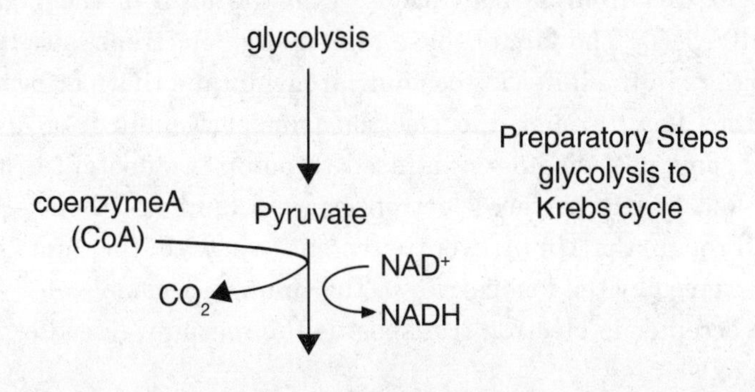

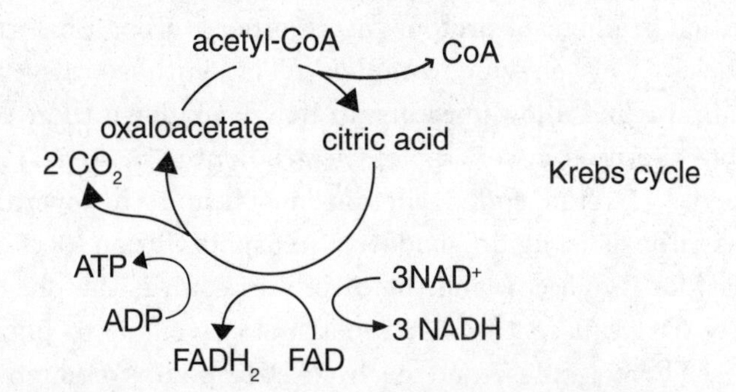

The first step in this process is the transport of pyruvate from the cytosol into mitochondria, after which the three-carbon pyruvate is converted to acetyl-CoA, a two-carbon unit bound to the coenzyme CoA. The enzyme that catalyzes this reaction is the pyruvate dehydrogenase complex, a large multi-subunit enzyme that reduces pyruvate, releasing CO_2 and acetyl-CoA and reduced NADH. In the first of eight reactions that form the Krebs cycle, the acetyl-group in acetyl-CoA is attached to the four-carbon molecule oxaloacetate, creating the six-carbon molecule citrate, or citric acid. The remaining steps of the cycle further oxidize citrate, creating carbon dioxide, ATP, $FADH_2$, and NADH. This loop of reactions is completed by the regeneration of oxaloacetate into acetyl-CoA, ready to begin another loop of the cycle.

Q Although every round of the Krebs cycle adds more acetyl groups to oxaloacetate to form citrate, the net concentration of citrate inside cells does not increase. Why?

A For every acetyl group that comes into the cycle, two carbons leave the cycle as carbon dioxide, keeping the concentration of the intermediates in the pathway roughly constant.

ELECTRON TRANSPORT OR ELECTRON TRANSFER PHOSPHORYLATION

Although glycolysis and the Krebs cycle each generates a small quantity of ATP directly, most of the energy they extract from carbohydrates is in the form of the high-energy electron carriers NADH and $FADH_2$. The fate of these high-energy electrons is to be transferred to a series of energy carriers in the inner mitochondrial membrane that perform a series of redox reactions. As electrons flow through the electron transport chain from a higher energy to lower energy state, some of their energy is used to pump protons out of the mitochondrial matrix. As NADH and $FADH_2$ reduce electron carriers (and are themselves oxidized), the electrons moving through the chain eventually end up with the final electron acceptor, oxygen. Oxygen is reduced by the last carrier in the chain to become water. The use of oxygen as the last electron acceptor in electron transport is the reason we need oxygen in the air we breathe.

Because the inner mitochondrial membrane is impermeable to protons and they cannot diffuse across the membrane passively, pumping protons out of the mitochondrial matrix creates a concentration gradient of protons, or, in other words, a pH gradient. This gradient powers ATP synthesis. The enzyme complex ATP synthase also resides in the inner mitochondrial membrane and allows protons to flow back down their concentration gradient into mitochondria but captures the energy of the gradient to make ATP. The use of the proton gradient to make ATP is often called chemiosmosis, and the overall linkage of electron transport and ATP synthesis is, again, oxidative phosphorylation (because oxidation provided by oxygen is required for it to occur, and the ultimate result is the phosphorylation of ADP to make ATP). Each NADH causes the electron transport chain to pump enough protons to create about three ATP molecules, and each $FADH_2$ pumps enough protons to drive the synthesis of two ATP molecules.

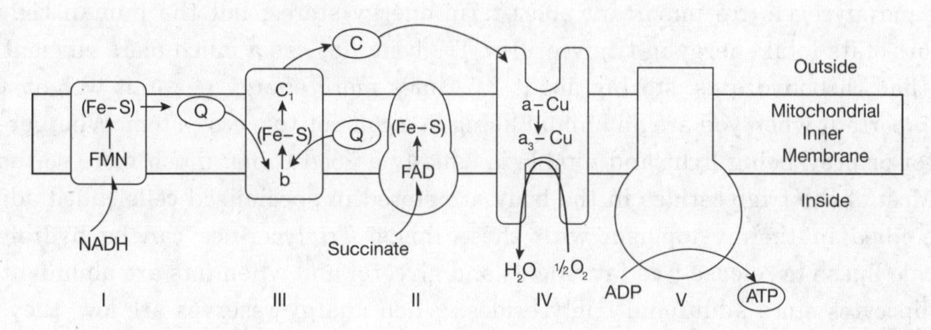

In the study of oxidative phosphorylation, scientists use a variety of methods to figure out the connection between electron transport and ATP synthesis. One way to do this is to isolate mitochondria and change the pH of their surrounding environment. Given oxygen, pyruvate, and other reagents, mitochondria isolated from the cell will continue making ATP. If the mitochondria are deprived of pyruvate or oxygen, they will stop making ATP. If acid is added to their external environment, they will start making ATP again, even if no pyruvate or oxygen is provided. Why is this? Adding external acid bypasses electron transport and the Krebs cycle, providing the pH gradient directly to drive ATP synthesis.

In another scenario, what will happen to ATP synthesis by isolated mitochondria if pyruvate, oxygen, and all other necessary substrates for oxidative phosphorylation are provided, and small holes are introduced into the inner mitochondrial membrane? The answer is the holes will allow protons to passively diffuse across the inner mitochondrial membrane, bypassing ATP synthase.

The principle of chemiosmosis occurs in other settings as well. Chloroplasts generate a pH gradient across their membrane as part of photosynthesis, using it to produce ATP, and bacteria produce a pH gradient across their cell membrane to make ATP or drive other cellular functions.

Overall, cellular respiration is a very efficient process. For every glucose molecule that enters glycolysis, as many as thirty-eight ATP molecules can be produced. These thirty-eight ATP molecules contain about 40% as much energy as a molecule of glucose, making the process 40% efficient at capturing the energy in glucose in a useful form. The internal combustion engine in your car, for comparison, captures about 20–25% of the energy in gas in a useful form.

Fermentation on the other hand, only produces two ATP per every glucose consumed. Fermentation is essential for some organisms to grow in the absence of oxygen, but it is far less efficient at ATP production than oxidative phosphorylation.

OTHER SOURCES OF ENERGY (AMINO ACIDS, FATTY ACIDS)

Glucose and glycogen are important short-term energy stores, but the human body stores much more of its total energy in triglycerides. Triglycerides are a much more efficient energy reserve than carbohydrates, storing about 2.5 times more energy per unit weight. Staying light is important when you are an animal living in a competitive ecosystem. Whether you are a predator or prey, being light and nimble is usually a good thing, much more so than for a potato. Most of the triglycerides in the body are stored in specialized cells called adipocytes that are filled in their cytoplasm with these lipids. Triglycerides can be hydrolyzed by triglyceride lipase to release free fatty acids and glycerol and when fats are abundant after a meal, adipocytes store additional triglycerides. When energy reserves are low, they release fatty acids and glycerol. Fatty acids can be released into the blood to be used as an energy store elsewhere in the body.

To use fatty acids for energy, cells import them into mitochondria and perform a pathway called beta oxidation, in which fatty acids are progressively shortened two carbons at a time to create acetyl-CoA and to reduce the energy carriers $FADH_2$ and NADH. This shortening process continues in cycles until a fatty acid is completely broken down into smaller units ready to burn for energy. The acetyl-CoA units enter into the Krebs cycle inside mitochondria to generate energy, and the $FADH_2$ and NADH transfer their electrons to the electron transport chain. The fatty acids found in triglycerides are on average sixteen to eighteen carbons long, so several rounds of beta oxidation are required for them to be fully prepared for use in oxidative phosphorylation. Ultimately, metabolic oxidation of a single 18-carbon fatty acid can result in the production of as many as 146 ATP molecules, making the energy yield from fatty acids much higher than the yield from glucose.

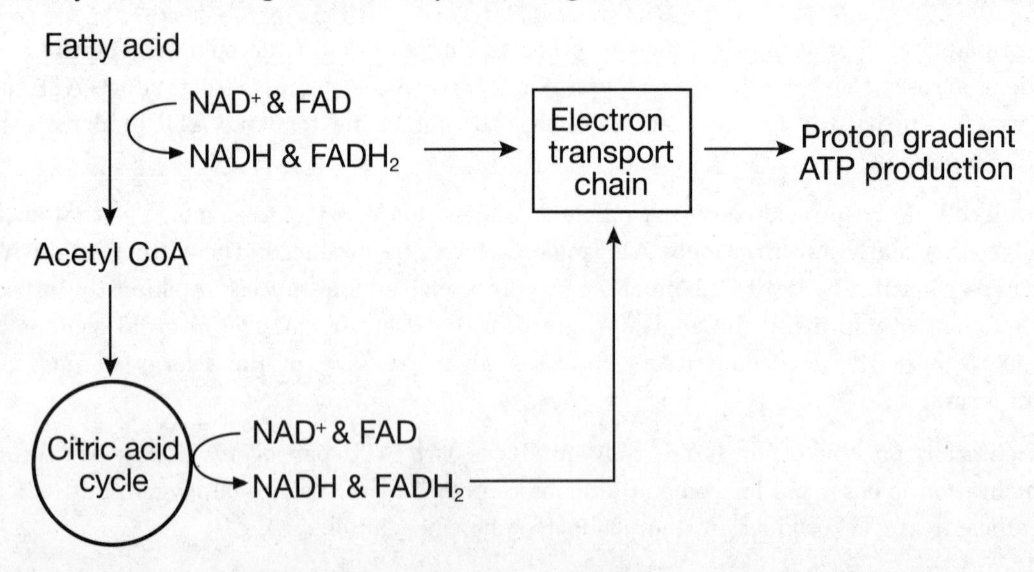

The pathways that synthesize fatty acids look a lot like the beta oxidation pathway, but fatty acid synthesis occurs in the cytoplasm whereas beta oxidation occurs inside mitochondria. To build fatty acids, the pathway starts with two acetyl-CoA groups and joins them to create a four-carbon molecule. Several rounds of joining additional acetyl groups results in a fatty acid. The acetyl groups used to build fats can come from carbohydrates, since acetyl-CoA is one of the products of carbohydrate metabolism from glycolysis and the Krebs cycle. Thus, even a low-fat meal can result in fat storage in adipocytes through the synthesis of fatty acids from acetyl-CoA. The reverse however does not occur; glucose cannot be produced from the acetyl-CoA derived from beta oxidation of fats. The reaction producing acetyl-CoA from pyruvate is irreversible.

Q What is one reason that beta oxidation and fatty-acid synthesis might occur in different cellular compartments?

A If they both happened in the same place at the same time, energy would be wasted in a futile loop.

Several amino acids found in proteins are closely related to intermediates in the Krebs cycle and, in addition to building proteins, they are broken down to create energy or metabolic intermediates used for the synthesis of other molecules. For this to occur, the nitrogen in the amino group is first removed by one of two reactions. In oxidative deamination, the amino group is replaced by oxygen, releasing ammonia. Ammonia is quite toxic so after leaving cells it is transported to the liver where it is converted into the less toxic molecule urea and is then excreted by the kidneys in urine.

In transamination, the amino group is moved from an amino acid to another molecule to create a new amino acid. These reactions can produce intermediates in the Krebs cycle such as alpha-ketoglutarate, which can in turn lead to a variety of metabolic products.

COORDINATION OF ENERGY PATHWAYS

All of the energy pathways described are coordinated with each other and tuned to the energy requirements of the cell. Glycolysis is regulated to move forward no faster at pyruvate production than the Krebs cycle can use pyruvate. Similarly, the Krebs cycle will only move forward at the rate required to generate NADH and $FADH_2$ as fast as they are consumed by electron transport, and electron transport will only move forward at the rate required to maintain ATP at a constant level inside cells. Cells and organisms operate to maintain homeostasis, of which having a constant ATP concentration is one important part. If the ATP concentration inside cells falls, cells will be unable to maintain their essential activities and may even die.

Q If oxygen, glucose, and other essential substrates are provided in excess to cells in culture, will their rate of ATP production increase as well?

A Cells operate to maintain homeostasis, with constant ATP concentration inside the cell. Cells do not increase the rate of oxidative phosphorylation to produce as much ATP as possible, only as much as they need. If excess oxygen and glucose are present, the factor regulating the rate of ATP production is how much ATP is used. If the rate of consumption increases, the cells will increase the rate of production to match.

EXERCISES: CELLULAR ATP PRODUCTION

1. If radioactive oxygen is delivered in the atmosphere of a mouse, where will this oxygen be found a few hours later?

   (A) Carbon dioxide
   (B) Glucose
   (C) Water
   (D) Citrate
   (E) Pyruvate

2. The metabolic poison dinitrophenol was once used as a diet drug. This compound makes the inner mitochondrial membrane permeable to protons, and too much of it can prove toxic. If a low, sublethal dose of dinitrophenol is given to an animal, which of the following will occur?

   (A) Glucose consumption will increase.
   (B) Oxygen consumption will decrease.
   (C) ATP production will increase.
   (D) The pH gradient across the inner mitochondrial membrane will increase.
   (E) The rate of glycolysis will decrease.

3. Yeast cells are facultative anaerobes, able to continue growing in conditions where there is little or no oxygen. If yeasts are grown in two flasks of culture medium containing glucose as their food source and oxygen is deprived from one of the flasks, which of the following will occur in the flask lacking oxygen?

   (A) The yeast will stop growing.
   (B) The rate of glucose consumption will increase.
   (C) Alcohol production will decline.
   (D) ATP production will halt.
   (E) Glycolysis will stop.

4. Which of the following is secreted into the bloodstream by cells to be used as a source of energy by other cells in the body?

   (A) Glucose
   (B) Glucose 6-phosphate
   (C) Insulin
   (D) ATP
   (E) Glycogen

5. Which of the following can be metabolized to produce ATP through oxidative phosphorylation?

   (A) Glycogen
   (B) Triglycerides
   (C) Amino acids from proteins
   (D) Free fatty acids
   (E) All of the above

6. According to the first law of thermodynamics, energy is always conserved. In the oxidative phosphorylation of glucose, about 40% of the energy in glucose is captured in ATP. Where does the rest of the energy go?

   (A) It is converted to decreased entropy.
   (B) It is the free energy of the reaction products.
   (C) It is transferred to the surroundings as heat.
   (D) It is converted into the chemical energy in carbon dioxide.
   (E) It is stored in the mitochondrial proton gradient.

7. Beta oxidation of fatty acids is similar to fatty acid synthesis in which of the following ways?

   (A) They both occur in the mitochondrial matrix.
   (B) They both involve two carbon acetyl-CoA units.
   (C) Both occur without additional energy from an outside source.
   (D) They both require molecular oxygen as an electron acceptor.
   (E) Both processes use NADH as a cofactor.

8. Phosphoglucoisomerase catalyzes the isomerization of glucose 6-phosphate to fructose 6-phosphate. Although this enzyme greatly accelerates this reaction, it fails to catalyze the isomerization of galactose 6-phosphate. Which of the following is the most likely explanation for this observation expressed below?

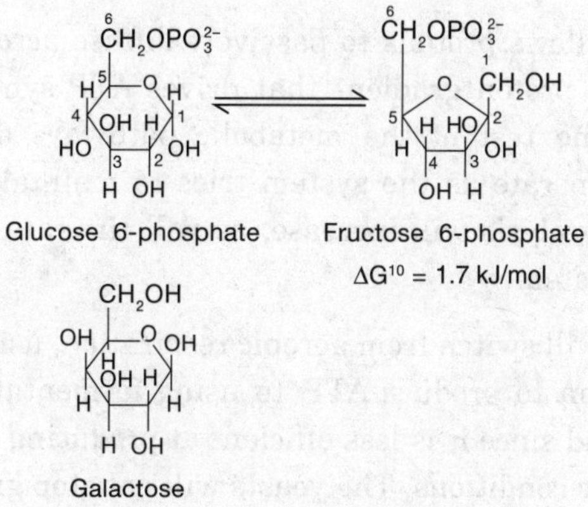

Glucose 6-phosphate ⇌ Fructose 6-phosphate

$\Delta G'^{\circ} = 1.7$ kJ/mol

Galactose

(A) Galactose 6-phosphate is a competitive inhibitor of this reaction.
(B) The active site of phosphoglucoisomerase fails to bind diastereomers of glucose 6-phosphate.
(C) Isomerization reactions can only occur in the context of the full metabolic pathway.
(D) The free energy galactose 6-phosphate is much higher than the free energy of glucose 6-phosphate.
(E) Folding of this enzyme is very sensitive to changes in the surrounding solvent environment.

9. The nitrogen found in urea in humans comes from which of the following?

(A) Acetyl-CoA
(B) Atmospheric nitrogen
(C) Phospholipids
(D) Polysaccharides
(E) Amino acids

10. One of the reactions in the Krebs cycle is:

Isocitrate + $NAD^+ \rightleftarrows$ alpha-ketoglutarate + CO_2 + NADH + H^+

In this reaction, which of the following occurs?

(A) Isocitrate is oxidized and NAD^+ is reduced.
(B) Isocitrate is reduced and NAD^+ is oxidized.
(C) Carbon dioxide is oxidized and isocitrate is reduced.
(D) Both isocitrate and NAD^+ are reduced.
(E) No electrons are transferred in this reaction, so it is not a redox reaction.

ANSWER KEY AND EXPLANATIONS

1. C	3. B	5. E	7. B	9. E
2. A	4. A	6. C	8. B	10. A

1. **The correct answer is (C).** As the ultimate electron acceptor, molecular oxygen is reduced by the electron transport chain to water.

2. **The correct answer is (A).** The drug allows protons to passively diffuse across the mitochondrial membrane, disrupting the proton gradient that drives ATP synthesis. ATP synthesis will slow or halt, but the rest of the metabolic pathways driving oxidative phosphorylation will increase in rate as the system tries to maintain ATP production. Glucose consumption and glycolysis will increase, as will the rate of the Krebs cycle and the electron transport system.

3. **The correct answer is (B).** The yeasts will switch from aerobic respiration, using the Krebs cycle and oxidative phosphorylation to produce ATP, to using fermentation to make ATP. Fermentation uses glucose, and since it is less efficient at producing ATP it will use even more glucose than in aerobic conditions. The yeasts will not stop growing since they are facultative aerobes and can use fermentation to grow (A is wrong). Alcohol production will not decline. Alcohol is the result of fermentation, so alcohol production will increase (C is wrong). ATP production cannot halt. If it does, the cells will die (D is wrong). Glycolysis will not stop, because this is part of fermentation, and it will actually increase in rate due to the lower efficiency of fermentation in ATP production.

4. **The correct answer is (A).** Some cells like liver cells can release glucose into the blood to be used by other cells for energy as needed. Glucose 6-phosphate is found inside cells as an intermediate in glycolysis, but it is not secreted for use by other cells in the body. Insulin is a hormone, not an energy source. ATP is found inside cells for energy, not secreted into the blood. Glycogen is an important energy store, but it is a large polysaccharide found only inside cells. It is not secreted by one cell to be used by another.

5. **The correct answer is (E).** All of these can enter glycolysis and the Krebs cycle to produce ATP through oxidative phosphorylation.

6. **The correct answer is (C).** The remaining energy is heat. This heat of metabolism allows mammals to maintain a constant body temperature but does not contribute to ATP production.

7. **The correct answer is (B).** Beta oxidation breaks down fatty acids into acetyl-CoA groups that can be oxidized in the Krebs cycle, while fatty acid synthesis adds acetyl-CoA groups to make fatty acids grow two carbons at a time. Although the two pathways look superficially similar, just run in reverse, they have several important differences. Beta oxidation occurs in mitochondria and produces NADH while fatty acid synthesis occurs in the cytoplasm and uses NADPH (A and E are wrong). Also, beta oxidation releases energy while fatty acid synthesis uses energy (C is wrong). Neither pathway involves molecular oxygen directly (D is wrong).

8. **The correct answer is (B).** Galactose and glucose are diastereomers of each other, differing in the configuration around one chiral carbon. Enzyme-active sites are complex three-dimensional arrangements of precisely positioned amino acids evolved to allow very specific substrates to fit in the active site, even to the point of excluding molecules that differ from each other in a small way like this. Pathways handling sugars like glycolysis must be very precise in which sugars enter the pathway to prevent the production of waste material.

9. **The correct answer is (E).** The denitrification of amino acids produces ammonia, which is then converted to urea by the urea cycle.

10. **The correct answer is (A).** Isocitrate is oxidized, releasing a molecule of CO_2, with some of the energy used to reduce NAD^+ to NADH. Choice C is not correct; CO_2 is already oxidized and isocitrate is oxidized by this reaction, not reduced. In a redox reaction like this, only one member of each pair is oxidized while the other is reduced (D is wrong). Electrons are transferred in this reaction, although they are not drawn out by themselves (E is wrong).

SUMMING IT UP

- The energy currency of the cell is ATP.
- ATP provides the energy for many biological chemical reactions. It is composed of three distinct parts:
    - A five-carbon sugar called ribose
    - An adenine base
    - Three phosphate groups aligned in a linear fashion, extending from the central, ribose sugar
- Autotrophs carry out photosynthesis and respiration—they still need energy when the sun goes down—and heterotrophs carry out respiration only.
- Anabolism is the constructive part of metabolism during which larger molecules are formed.
- Catabolism is the destructive phase of metabolism during which energy is released and complex molecules are broken down.
- Respiration can be both aerobic and anaerobic.
- Cellular respiration is a catabolic reaction that releases the stored energy in glucose so it can be stored in ATP.
- There are three stages in cellular respiration:
    - Gycolysis
    - Krebs (citric acid) cycle
    - Electron transport system

Photosynthesis

chapter 8

OVERVIEW

- **Light into glucose**
- **Photosystems**
- **The Calvin cycle for glucose production**
- **Other photosynthetic routes**
- **Summing it up**

The vast majority of life on Earth ultimately derives energy from the sun. Plants get their energy directly from the sun, capturing the radiant energy of the sun through photosynthesis and converting the sun's energy into the chemical energy of glucose and ATP. Animals use the sun's energy indirectly by eating plants to extract their chemical energy, moving this chemical energy through food webs. As organisms that produce their own food energy from the sun, plants are **photoautotrophs**. In addition to plants, some protists, algae, and some bacteria also perform photosynthesis. All of these organisms are the primary producers of almost all of Earth's biosphere. Organisms that cannot make their own food are **heterotrophs**, the consumers.

The site of plant photosynthesis is the leaf. Two of the requirements for photosynthesis, in addition to sun, are carbon dioxide and water. Pores in the surface of leaves called **stomata** open to allow gases to be exchanged between the interior of the leaf and the atmosphere. Plants require carbon dioxide for photosynthesis to occur, so stomata open to allow carbon dioxide to enter leaves. Oxygen, a by-product of photosynthesis, is also a waste product, so the opening of stomata also allows oxygen to diffuse out of the leaf.

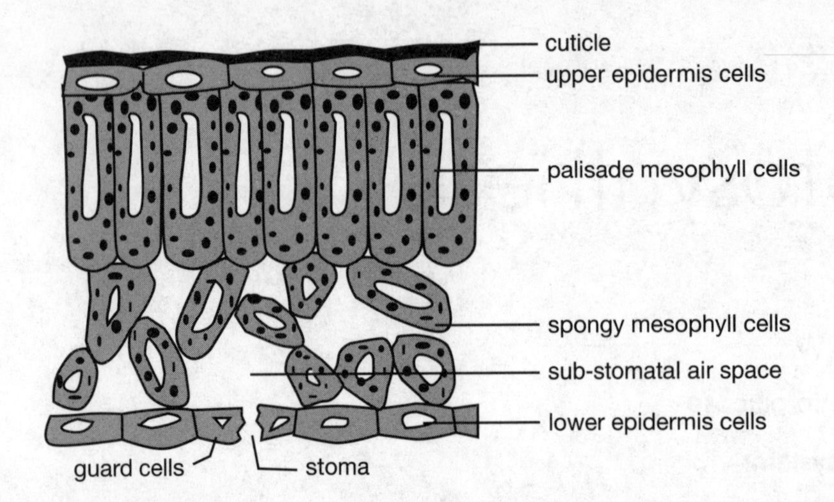

Q Other than photosynthesis, what biological process produces significant quantities of the oxygen found in the Earth's atmosphere?

A Really, photosynthesis is pretty much it. The oxygen in Earth's atmosphere today is the result of billions of years of photosynthesis. If plants and other photosynthetic organisms like algae went away, the oxygen would pretty much go away with them.

Within plant cells, photosynthesis happens in chloroplasts. Chloroplasts resemble small photosynthetic prokaryotes, and like mitochondria chloroplasts are the remnant of cells that probably lived independently early in evolution but started to live symbiotically inside early eukaryotic cells.

PLANT CELL STRUCTURE

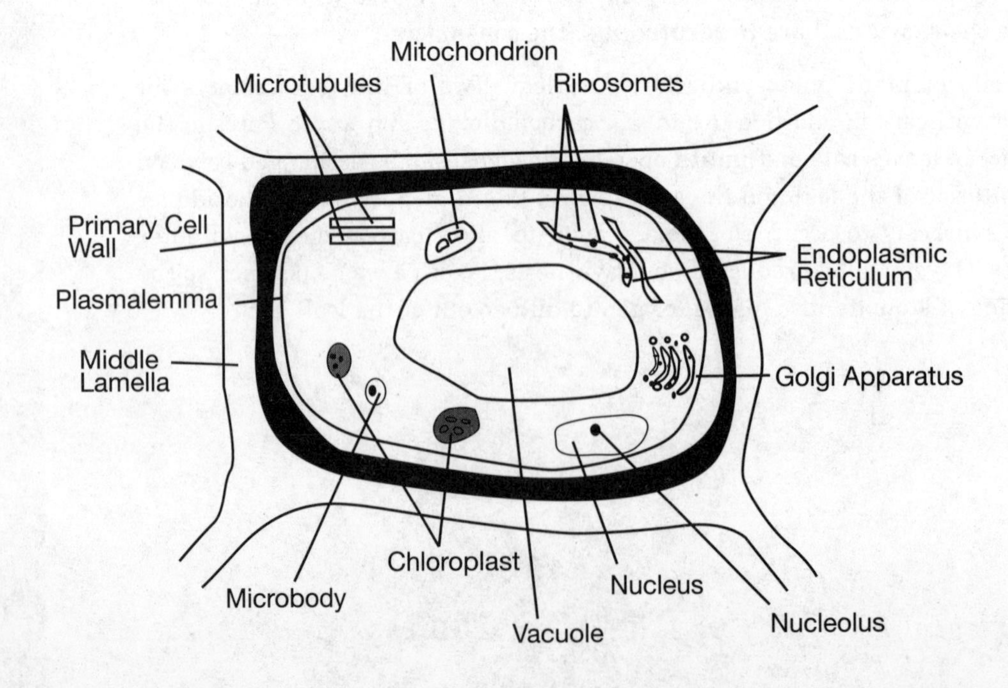

The series of stacked disc-like structures inside the chloroplast are called **thylakoids**, and the fluid inside chloroplasts around the thylakoids is the **stroma**. The light absorption pigments of chlorophyll are in the membranes of the thylakoids.

Q Under a light microscope, which part of the plant cell will appear green?

(A) Stroma

(B) Thylakoid

(C) Cytoplasm

(D) Cell wall

(E) Outer chloroplast membrane

A The correct answer is (B). The pigments of chlorophyll are in the thylakoid and give the plant cell its green color.

LIGHT INTO GLUCOSE

Overall, the reactions of photosynthesis can be described by this equation:

$$\text{Sunlight} + \underset{\text{Water}}{H_2O} + \underset{\text{Carbon dioxide}}{CO_2} + \longrightarrow \underset{\text{"Carbohydrates"}}{\text{"}CH_2O\text{"}} + \underset{\text{Oxygen}}{O_2}$$

Using light energy, carbon dioxide from the air, and water, plants make carbohydrates and oxygen. Of course, this does not happen in one reaction step but in a number of reactions in different pathways. The overall result, though, is the opposite of cellular respiration. In cellular respiration, glucose is fully oxidized to carbon dioxide and water, releasing energy. In photosynthesis, carbon dioxide and water are reduced to form glucose, with energy input.

Photosynthesis is usually viewed as two sets of reactions called the **light-dependent reactions** and the **light-independent reactions.** A few subsequent products result from the light-dependent reactions. One of these is that light energy from the sun energizes electrons, and these energized electrons reduce the high-energy electron carrier $NADP^+$ to make NADPH. The light-dependent reactions also split water to release oxygen and create a proton gradient that drives ATP synthesis. In chloroplasts, the proton gradient is created across the thylakoid membrane, pumping protons from the stroma around the thylakoids into the interior of the thylakoids. The synthesis of ATP in chloroplasts in this manner is called **photophosphorylation**.

Q If water containing a heavy isotope of oxygen (O_{18}) is provided to plant cells, where will the oxygen be found after the cells are exposed to light?

(A) Oxygen
(B) Carbon dioxide
(C) Glucose
(D) ATP
(E) Cellulose

A The correct answer is (A). The light-dependent reactions split water, releasing molecular oxygen.

The light-independent reactions are also called the **Calvin cycle**. This part of photosynthesis usually occurs in the light, coupled to the light-dependent reactions. In the Calvin cycle, the energy from the light-dependent reactions is used to build glucose from carbon dioxide. To do this, carbon dioxide is absorbed from the atmosphere, reduced, and joined with smaller molecules that are present in the chloroplasts. The Calvin cycle occurs in the stroma.

PHOTOSYSTEMS

Let's look at photosynthesis in more detail. First, light comes in from the sun. The sun's energy covers a broad range of the electromagnetic spectrum, of which plants selectively absorb and use a relatively narrow band of wavelengths in the visible spectrum. Light is absorbed in plants by compounds called pigments, mainly chlorophyll. Different pigments in plants absorb somewhat different wavelengths, including chlorophyll a, chlorophyll b, and carotenoids. These pigments usually absorb in the blue and red part of the spectrum, not in the middle green region. Thus, plants are green because they do not use green light for photosynthesis, reflecting it back instead into the eye of the observer.

When light is absorbed by the pigments, it is not lost, but changed into a different form. Tracing the energy from the photon through photosynthesis allows us to trace how its energy is converted into other forms, like the chemical energy that eventually produces glucose.

Q If one group of bean plants is placed in a container with a blue light and another group is placed in a container with a green light, which beans will grow more quickly?

A The beans that absorb more light will perform more photosynthesis and will have more energy produced, helping them grow more quickly. Blue light is better absorbed by photosynthetic pigments than green light, so the plants in the blue lights will grow faster.

Chlorophyll pigments in chloroplasts have one end with a porphyrin ring that absorbs light and another end with a hydrophobic tail that makes chlorophyll interact with membranes.

When a photon with the right wavelength hits chlorophyll, an electron is kicked up to a higher energy orbital. Chlorophyll molecules are not isolated on their own inside mitochondria. They are found in large assemblies of proteins and pigments in the chloroplast membrane called **photosystems**. Some of the pigments in the photosystem gather light like antennas, passing energized electrons from pigment to pigment until they reach a **reaction center**. In the reaction center, specially positioned chlorophylls in protein complexes take photons from the other accessory pigments and transfer these electrons to a primary electron acceptor.

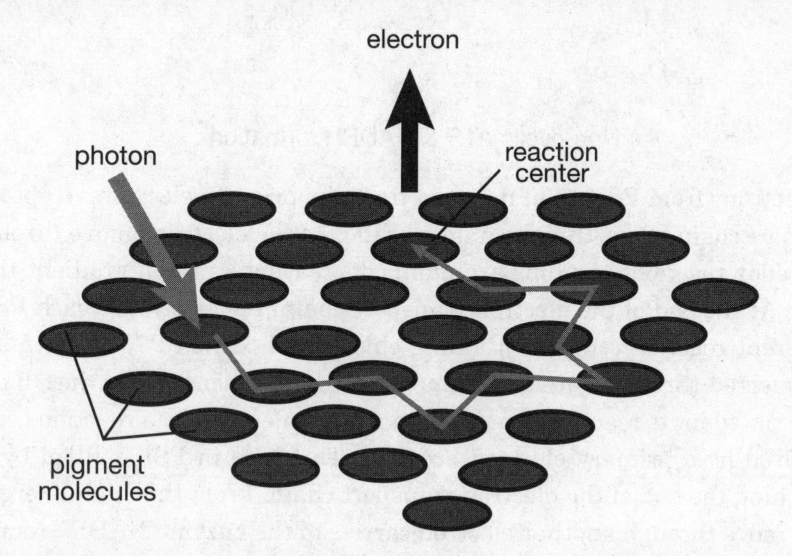

There are two different types of photosystems in the thylakoid membrane of chloroplasts, photosystem I (PSI) and photosystem II (PSII), each of which performs somewhat different parts of photosynthesis called cyclic and non-cyclic photophosphorylation.

Cyclic and Non-cyclic Photophosphorylation

NON-CYCLIC ELECTRON FLOW

Non-cyclic photophosphorylation involves both photosystem I and photosystem II and is responsible for most of the photosynthetic activity of plants. In this pathway, photons are first absorbed by one of the antenna pigments in PSII and then moved through a series of antenna pigments until they reach the reaction center. At the reaction center in PSII, two chlorophyll a molecules are positioned to absorb light best at 680 nm; they are called the P680 chlorophylls. When the photons reach the P680 reaction center in PSII, they excite a PSII electron that is captured by the primary electron acceptor in PSII. Water is split at PSII, releasing electrons, along with hydrogen and oxygen atoms. The electrons fill the hole left by the excited electron in the reaction center, and two oxygen atoms react with each other to form molecular oxygen, O_2.

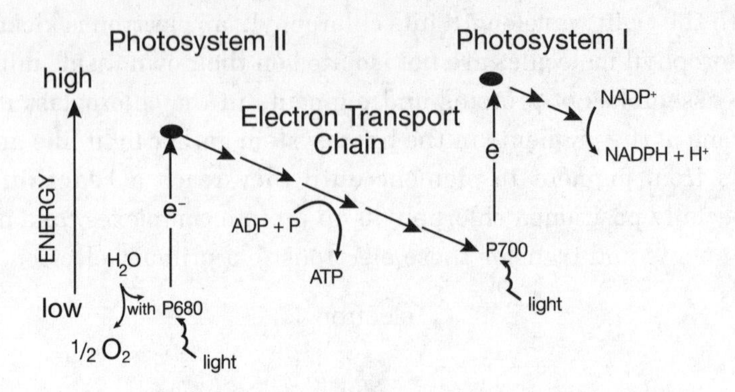

Non-cyclic ATP & NADPH formation

The excited electrons from PSII then move on from the primary electron acceptor through an electron transport chain in the thylakoid membrane. As the electrons move through the chain in a series of redox reactions, protons are pumped, creating a proton gradient that can drive ATP synthesis. At the end of the electron transport chain, the electrons reach PSI, which has a slightly different reaction center containing chlorophylls called P700. When a photon hits PSI, it is transferred through antenna pigments to the reaction center, just like in PSII. In PSI, a hole is created in its reaction center when a photon energizes a reaction center electron, which is captured by a primary electron acceptor. This hole in PSI is filled by the electron from PSII reaching the end of the electron transport chain. From this high-energy carrier, the PSII electrons move through another electron carrier to the enzyme $NADP^+$ reductase, which makes the reduced high-energy electron carrier NADPH, another one of the key products of photosynthesis.

Q What happens to the photon which initially is absorbed by antenna pigments in photosystem II?

A The photon moves through the pigments until it gets to the PSII reaction center. There its energy is transferred to an electron, which moves to a higher energy state.

Q Where does the PSII photoexcited electron eventually end up?

A First, it goes to the PSII primary acceptor, then through the electron transport chain to PSI. It fills the hole left in PSI when a photon excites a PSI electron to jump to a high-energy state.

Cyclic Electron Flow

Although photosynthesis usually proceeds through the non-cyclic system, sometimes it only involves PSI in what is called cyclic photophosphorylation. In the cyclic system, electrons excited in PSI by light are still transferred to the primary electron acceptor, but instead of passing to the $NADP^+$ reductase, they flow back to the electron transport chain between PSII and PSI. From there they are transferred through the chain, using their energy to pump protons and produce ATP. At the end of the chain, they fill the hole left in the PSI reaction center when an electron is excited by a photon, beginning the cycle again.

In the non-cyclic system, the movement of electrons through PSII, the electron transport chain, PSII and the $NADP^+$ reductase produce about the same amount of ATP and NADPH. In the cyclic system, electrons bypass the $NADP^+$ reductase, producing ATP but no NADPH. This all fits together with the needs of the Calvin cycle for glucose production. The Calvin cycle needs more ATP than NADPH, and if it relied entirely on non-cyclic photophosphorylation it would run short of ATP.

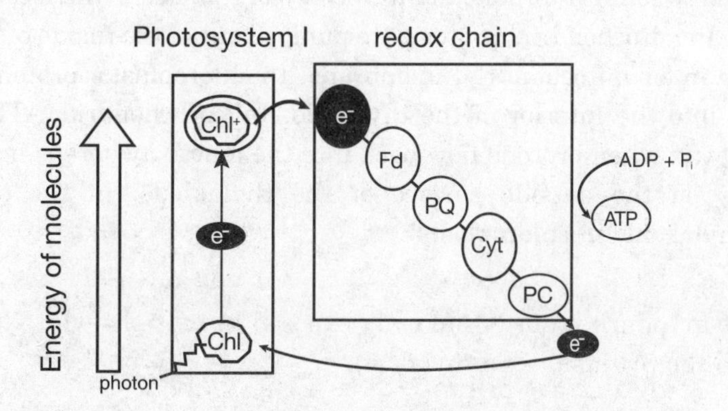

Cyclic ATP formation

Both the cyclic and non-cyclic systems use an electron transport chain to pump protons across a membrane (the thylakoid membrane), generating a pH gradient that drives ATP synthesis. Sound familiar?

Q What other organelle generates a proton gradient for ATP synthesis?

A The mitochondrion.

While there are a lot of similarities between the way chloroplasts and mitochondria produce ATP, there are also some differences. Both take high-energy electrons and pass them through a series of oxidation/reduction reactions with electron carriers in a membrane-associated electron transport chain. In both organelles, this electron transport chain pumps protons, converting the energy of the electrons and using it to create a different form of energy, the pH gradient across a membrane. The use of the pH gradient to make ATP from ADP and phosphate by chemiosmosis occurs by the same mechanism in both cases.

Q What would happen to ATP production in chloroplasts if the chloroplast membrane is permeabilized to allow protons through, but the thylakoid membrane remains undisturbed?

A ATP production would continue. The pH gradient that drives ATP production in chloroplasts is developed across the thylakoid membrane, not the chloroplast membrane.

One big difference between ATP production in the two organelles is the location and direction of the pH gradient. In mitochondria, protons are pumped out of the inside of the organelle (the matrix), across the inner mitochondrial membrane. In chloroplasts, protons are pumped by electron transport into the interior of the thylakoid. In mitochondria, ATP synthesis takes place inside the matrix when protons flow back into the matrix, whereas in chloroplasts ATP synthesis happens on the outside surface of the thylakoids, in the stroma, the fluid surrounding the thylakoids in chloroplasts.

Q If a plant in a dark room is suddenly exposed to a bright white light, what will happen to the pH inside the thylakoid?

A Since plants pump protons into the interior of the thylakoids as part of photosynthesis, turning on the light will pump protons into the thylakoids, lowering the pH inside the thylakoids.

THE CALVIN CYCLE FOR GLUCOSE PRODUCTION

Photosynthesis is all about capturing the sun's radiant energy and using this energy to drive biosynthesis. The cyclic and non-cyclic photophosphorylation systems, the so-called "light-dependent reactions," are only half of the story. The light-dependent reactions capture the sun's energy and convert it to chemical energy in the form of ATP and NADPH, but the Calvin cycle takes the energy from ATP and NADPH and uses it to build sugar from carbon dioxide. Carbon dioxide is a much lower energy, more oxidized form of carbon than glucose, so energy must be put into the system for this reaction to happen. The energy of ATP and NADPH meets this need.

While the light-dependent reactions take place in the thylakoids, the Calvin cycle takes place in the stroma, outside the thylakoids.

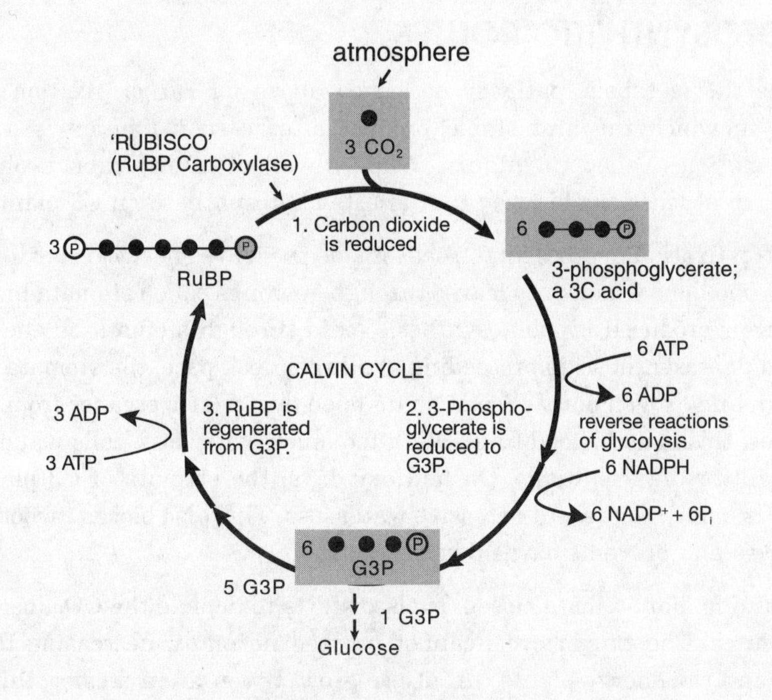

THE CALVIN CYCLE

The first thing to happen in the Calvin cycle is that carbon dioxide is absorbed from the atmosphere, and through carbon fixation one carbon is attached to a five-carbon sugar to create an unstable six-carbon sugar that breaks in half to make two 3-carbon sugars, two molecules of 3-phospholglycerate (3-PG). **Carbon fixation** in plants by this mechanism is catalyzed by an enzyme called **rubisco.** Next, the 3-PG molecules are phosphorylated using ATP and reduced using NADPH, creating a three-carbon sugar called glyceraldehyde-3-phosphate (G3P). Finally, more reactions regenerate the five-carbon sugar that is the acceptor for CO_2 at the start of the cycle.

Q What will happen to the production of carbohydrates in a leaf if the sun is shining but no carbon dioxide is available?

A The Calvin cycle is where carbon dioxide from the atmosphere is used to build carbohydrates, using energy from ATP and NADPH produced in the light-dependent reactions. If no carbon dioxide is available, the Calvin cycle cannot proceed and no sugar is produced, no matter how much light, ATP, and NADPH are available.

To make one molecule of 3-PG takes about nine molecules of ATP and six molecules of NADPH. The Calvin cycle does not produce glucose directly, though. The 3-PG produced by the Calvin cycle is the starting point for reactions leading to synthesis of glucose and other carbohydrates.

OTHER PHOTOSYNTHETIC ROUTES

Most plants use the metabolic pathway described above for carbon fixation, catalyzed by rubisco. Plants in which the first stable product after carbon fixation is a three-carbon molecule are sometimes called **C3 plants.** While a wonder of molecular evolution, rubisco seems imperfect in at least one big way that affects photosynthesis in C3 plants.

In order for carbon fixation to proceed, the cells inside the leaves need carbon dioxide from the atmosphere. Carbon dioxide gets into leaves through openings called stomata in the surface of leaves, and oxygen produced by photosynthesis exits through stomata on the leaves. On a sunny day when photosynthesis is proceeding at a breakneck pace, the stomata open to allow carbon dioxide in and oxygen out. When stomata open, water also escapes from leaves, and if too much escapes, this is not a good thing. With too much water loss, cells will dehydrate and the plant will wilt or die eventually. On hot, dry days, the stomata of C3 plants will close again, even if it is sunny, to prevent excessive water loss. This also blocks carbon dioxide from getting into leaves and prevents oxygen from exiting.

With the sun shining, but stomata closed, rubisco starts to deplete the CO_2 inside leaves and oxygen accumulates. The Calvin cycle cannot proceed normally, decreasing the yield from photosynthesis and the energy plants can use to grow. If you are a farmer, this means your crops will not produce as well. In these conditions, rubisco will bind oxygen instead of carbon dioxide, producing a two-carbon molecule that is split into carbon dioxide rather than entering the normal Calvin cycle. Producing CO_2 rather than building sugar is called **photorespiration** and leads to a further decline in photosynthetic productivity of C3 plants in hot, dry conditions.

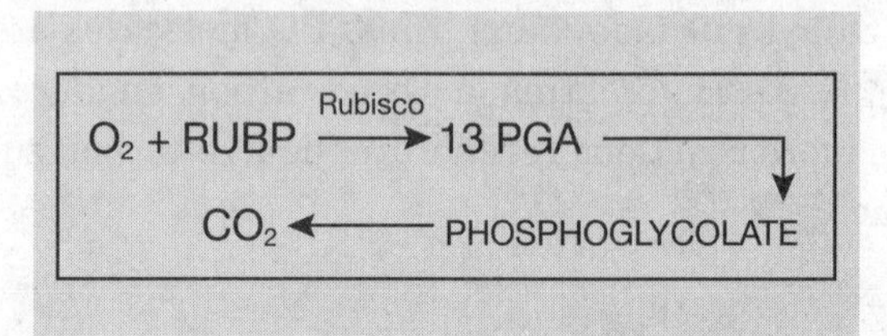

Photorespiration in C3 plants

Q When photorespiration produces a two-carbon intermediate, how do plants capture the metabolic energy of this compound?

A They don't. This molecule is wasted and is split into carbon dioxide just to get rid of it. This wasted molecule is a drain on the energy of the plant, reducing the overall energy supplied by photosynthesis for other metabolic activities.

One mechanism plants have evolved to grow in hot, dry conditions is found in **C4 plants**. C4 plants use a different mechanism for carbon fixation and have a different structure of cells inside their leaves, allowing them to avoid photorespiration and to thrive in hot climates with lots of sun. The corn plant is one example of a C4 plant that is grown on millions of acres of farmland in the U.S.

In C4 plants like corn, carbon fixation is carried out by the enzyme **PEP carboxylase,** rather than rubisco, adding CO_2 to a three-carbon molecule (PEP) to create a four-carbon molecule, oxaloacetate. This four-carbon molecule produced by carbon fixation in C4 plants gives them their name. PEP carboxylase binds carbon dioxide more tightly than rubisco so it can fix carbon even when CO_2 concentrations are low. And it does not bind oxygen, avoiding the unproductive diversion of energy produced by photorespiration in C3 plants.

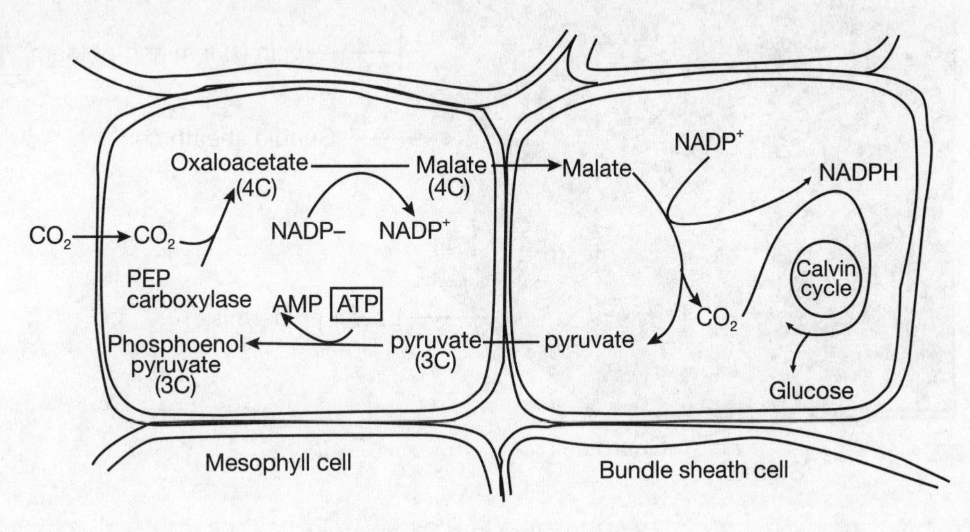

C4 plant carbon fixation

In addition to using a different enzyme for carbon fixation, C4 plants split the activities of carbon fixation and the Calvin cycle into different cells organized around vascular tissue running through leaves. **Mesophyll** cells closest to the surface of the leaf contain PEP carboxylase and do carbon fixation, then transfer four-carbon molecules to neighboring **bundle sheath** cells directly through cytoplasmic connections called plasmodesmata. Bundle sheath cells are where the Calvin cycle takes place. In bundle sheath cells, the four-carbon carrier is broken down again into a three-carbon unit and carbon dioxide. The carbon dioxide is then used by rubisco and incorporated into the Calvin cycle. The three-carbon molecule, pyruvate, is recycled back into mesophyll cells to make more PEP for carbon fixation.

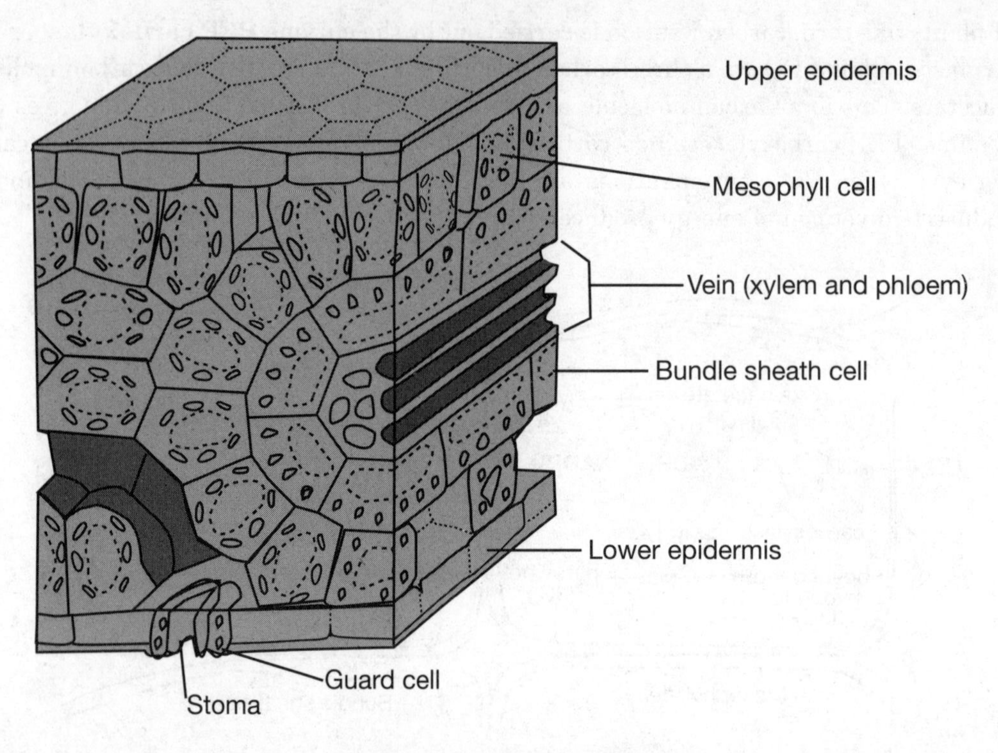

Leaf structure in a C4 plant

Q Do C4 plants lack the gene for rubisco?

A No. They still have rubisco and it is required for the Calvin cycle, but it does not fix carbon dioxide directly. It fixes carbon dioxide that is released from four-carbon molecules in bundle sheath cells after being fixed in mesophyll cells by PEP carboxylase.

Q How are carbon fixation and the Calvin cycle separated in C3 plants?

A They aren't, and that's the problem for C3 plants on a hot, sunny day.

Another mechanism for carbon fixation is found in **CAM plants**, such as cacti. CAM plants are adapted to living in very hot and dry conditions in which water conservation is essential for survival. C4 plants will partially close stomata during the day to prevent excessive water loss, while in hot, sunny conditions the CAM plants will close their stomata altogether. In a C3, or even in a C4, plant this would lead to a severe loss of photosynthetic efficiency, while in a CAM plant photosynthesis can still proceed. Closing their stomata during the day, CAM plants will open them at night. CAM plants fix carbon at night into various organic acid compounds that are stored in vacuoles. During the day, while the stomata are closed, CO_2 is released from these organic acids for use by the Calvin cycle in the presence of the sun.

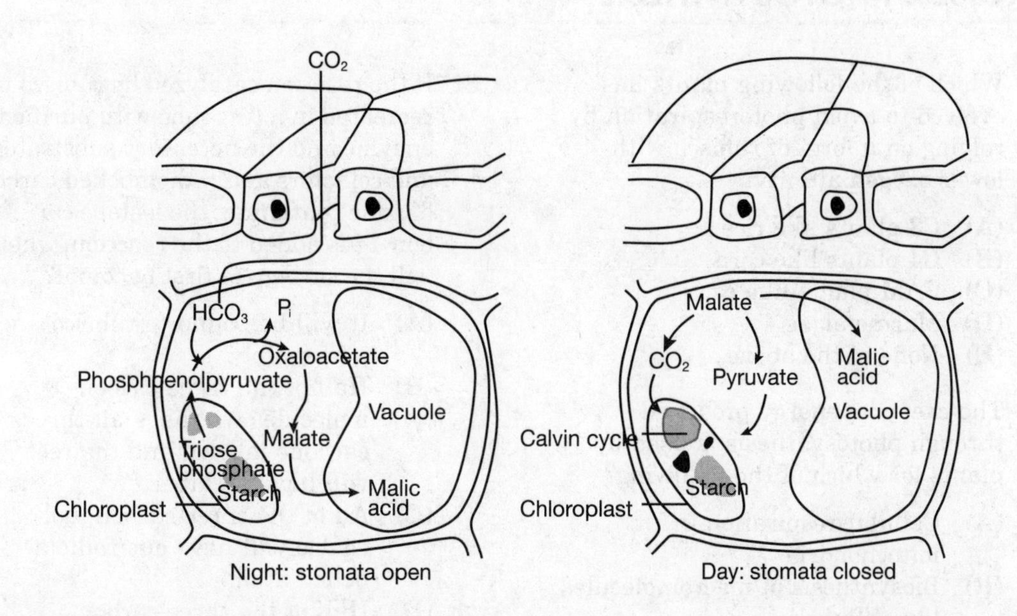

Q Are there any plants that have adapted to water loss by performing the Calvin cycle at night?

A No. The Calvin cycle always happens in concert with the light-dependent reactions. Only carbon fixation has changed.

EXERCISES: PHOTOSYNTHESIS

1. Which of the following plants has evolved to avoid photorespiration by relying on a form of rubisco with lower oxygen affinity?

   (A) C3 plants like rice
   (B) C4 plants like corn
   (C) CAM plants like cacti
   (D) Marine algae
   (E) None of the above

2. The chemical energy produced through photosynthesis is used by plants for which of the following?

   (A) Cellular respiration in mitochondria
   (B) Biosynthesis of macromolecules like cellulose
   (C) Maintenance of electrochemical gradients across cell membranes
   (D) Transport of cytoplasmic contents within cells
   (E) All of the above

3. If the reaction catalyzed by rubisco is recreated in a test tube with purified enzyme and the necessary substrates and cofactors and radiolabelled carbon dioxide containing the isotope carbon-14 is added to the reaction, where will the carbon-14 first be found?

   (A) It will be bound to rubisco irreversibly.
   (B) Half of the three-carbon molecules will have all three carbons labeled and the rest will have no label.
   (C) All of the three-carbon molecules will have one radiolabelled atom.
   (D) Half of the three-carbon molecules will have one labeled atom and the rest will not have any radiolabel.
   (E) The radiolabel will be found in glucose and no other intermediates.

4. In an experiment, a researcher exposes sections of leaves from C3 plants to different wavelengths of light. One leaf segment is exposed to light with a wavelength of 550 nM and another leaf segment is exposed to an equal intensity of light with a wavelength of 450 nM. Both leaf segments are exposed to the same temperature and gas concentrations. Which of the following will be observed?

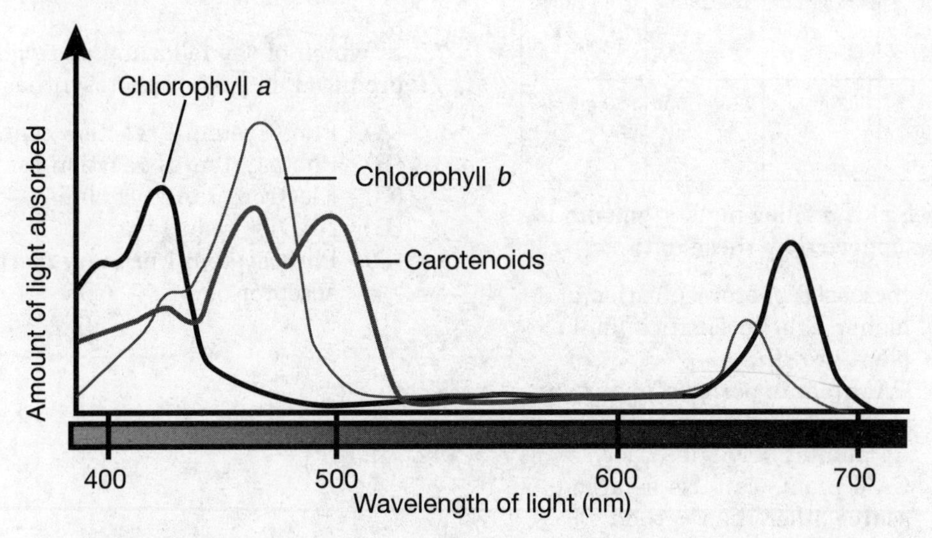

(A) At 450 nM, more organic acids will be produced.
(B) More carbohydrate will be produced at 450 nM.
(C) More carbohydrate will be produced at 550 nM.
(D) Fewer organic acids will be produced at 450 nM.
(E) Both leaf segments will have the same productivity.

5. CAM plants are weighed, exposed to different intensities of light in test chambers for several months, and then weighed again at different time intervals. There are ten plants in each group.

	Time 0	1 month	3 months
Light intensity 1×	300 grams +/− 20	350 grams +/− 15	460 gram s +/− 40
Light intensity 2×	450 grams +/− 33	528 grams +/− 40	675 grams +/− 57
Light intensity 3×	600 grams +/− 60	720 grams +/− 50	920 grams +/− 75

Which of the following statements is best supported by these data?

(A) Increased photorespiration at higher light intensities limits plant growth.
(B) CAM plants perform photosynthesis more efficiently than C4 plants.
(C) CAM plants can use a carbon source other than carbon dioxide.
(D) The growth of CAM plants is proportional to both time and light intensity.
(E) The growth of CAM plants is not proportional to light intensity.

6. In which of the following conditions will the most organic acids accumulate in a plant cell?

(A) On a hot night in a C3 plant
(B) On a hot, sunny day in a CAM plant
(C) On a cool summer night in a CAM plant
(D) On a hot, cloudy day in a C4 plant
(E) On a hot, sunny day in a C3 plant

7. In which of the following is oxygen produced as part of photosynthesis?

(A) Photosystem I reaction center
(B) Photosystem II reaction center
(C) Electron transport chain
(D) $NADP^+$ reductase
(E) Photosystem I primary electron acceptor

ANSWER KEY AND EXPLANATIONS

1. E	3. D	5. D	6. C	7. B
2. E	4. B			

1. **The correct answer is (E).** Some groups of plants have evolved alternative ways to fix carbon dioxide to avoid this problem with rubisco, but the solution is not generally found in rubisco itself.
2. **The correct answer is (E).** Everything that a plant cell does that requires energy gets that energy from photosynthesis ultimately. All of these activities are performed by cells of plants and all of them require chemical energy.
3. **The correct answer is (D).** The first step in the Calvin cycle is carbon fixation from carbon dioxide, catalyzed by rubisco. Carbon dioxide is joined to a five-carbon sugar, creating a six-carbon intermediate that quickly breaks down into two equivalent three-carbon molecules. In an isolated system, this would mean that for every carbon dioxide that is fixed by rubisco, two three-carbon molecules will be made, but only one of these will be labeled.
4. **The correct answer is (B).** First, C3 plants do not use organic acids as part of carbon fixation so choices A and D are out. The absorbance spectrum indicates that the leaf pigments absorb light best at 450nM, much more efficiently than at 550 nM. Most of the light at 550 nM will be reflected by the leaf and will not contribute to photosynthesis. At 450 nM more of the light energy is captured by photosynthetic pigments and converted into chemical energy.
5. **The correct answer is (D).** Answer choice (A) is true for C3 plants, although it does not apply to CAM plants. Choice (B) is true, but it does not relate to the data shown, particularly since no C4 plants were tested, so this is not the correct answer.
6. **The correct answer is (C).** CAM plants open their stomata at night and do carbon fixation at night to avoid water loss on hot, sunny days. Carbon is stored as organic acids during the night so that carbon dioxide can be released during the day. The highest concentrations of organic acids would be present at the end of the night.
7. **The correct answer is (B).** Water is split to feed electrons into the hole left in the PSII reaction center when an electron is excited by a photon and leaves the reaction center.

SUMMING IT UP

- Photosynthesis uses the energy of sunlight to produce carbohydrates from water and carbon dioxide.
- The light-dependent reactions of photosynthesis absorb sunlight and produce ATP and NADPH.
- The Calvin cycle uses ATP and NADPH to produce carbohydrates.
- There are two types of photophosphorylation for ATP production: cyclic and non-cyclic. Non-cyclic moves electrons through both photosystem I and II and produces both ATP and NADPH. Cyclic photophosphorylation produces only ATP.
- ATP production in the light-dependent reactions requires the production of a proton gradient, with protons pumped into the interior of thylakoids. This ATP production resembles ATP production in mitochondria in many ways.
- In C3 plants, rubisco fixes carbon dioxide, attaching one carbon to a five-carbon sugar to create a transient six-carbon molecule. This molecule quickly breaks down into two 3-carbon molecules (thus the term "C3") that go on to make other sugars like glucose.
- Photorespiration occurs in C3 plants when stomata close, depleting carbon dioxide and causing oxygen to accumulate. Photorespiration wastes metabolic energy.
- C4 plants avoid photorespiration by having carbon fixation in an outer layer of cells called mesophyll, while the Calvin cycle occurs in different cells called bundle sheath cells. C4 plants use PEP carboxylase to fix carbon in mesophyll cells and create a four-carbon molecule.
- CAM plants avoid photorespiration in a different way: They fix carbon at night into various organic acid compounds that are stored in vacuoles. During the day, while the stomata are closed, CO_2 is released from these organic acids for use by the Calvin cycle in the presence of the sun.

Genetic Inheritance

chapter 9

OVERVIEW

- **Chromosomes, meiosis, and sexual reproduction**
- **Inheritance patterns in humans**
- **Summing it up**

Let's go back to one of the basic principles of living things: ***all living things on Earth today come from other living things.*** When living things reproduce, they give their progeny characteristics, or traits, that determine many aspects of how their progeny will develop and grow. Offspring inherit from their parents many of the traits that distinguish individuals within a species. Your eye color, hair color, general height, and general appearance are derived all or in part from traits that your parents gave to you. The science of how traits are passed from parents to progeny is called **genetics**.

Genes are the fundamental unit of inheritance. Genes are contained in the genome of every living thing, and the sum total of an organism's genes defines all of its inherited features. Genes are located at discrete locations on chromosomes.

Organisms that reproduce asexually produce progeny that are exact genetic copies (or clones) of themselves. Prokaryotes can exchange genetic information in the form of extrachromosomal genetic elements like plasmids, but prokaryotes reproduce asexually, copying the genome of the parent cell exactly and giving each of two daughter cells an identical copy of the genome when the parent cell splits in two through binary fission. The only way in which the daughter cells will have a genome that is different from the parent cell is through random mutation, a relatively rare event.

Other organisms can reproduce asexually or sexually, depending on conditions. For example, at times, some plants can reproduce vegetatively, starting new individuals from shoots of a parent. The vast majority of plants and animals reproduce sexually, however, and sexual reproduction is the focus of most of genetics. Sexual reproduction is the dominant form of reproduction because it provides for greater variation in progeny than mutation alone, which gives populations and species a greater capacity to adapt to a variable environment.

CHROMOSOMES, MEIOSIS, AND SEXUAL REPRODUCTION

In sexual reproduction, progeny are created when gene cells, called **gametes,** from two different parents join to create the first cell of a new individual. The union of two gametes in this way is called **fertilization**. In humans and other animals, the two gametes are an ovum and a spermatozoon (sperm). The single-celled progeny, a **zygote**, then grows and develops, until it becomes sexually mature and creates its own gametes. Every generation alternates in this way, between the gametes that pass genes from one generation to the next, and the rest of the cells of the body, or the **somatic cells**. Somatic cells include all cells of the body that are not gametes, that are not part of the germ-line of the animal.

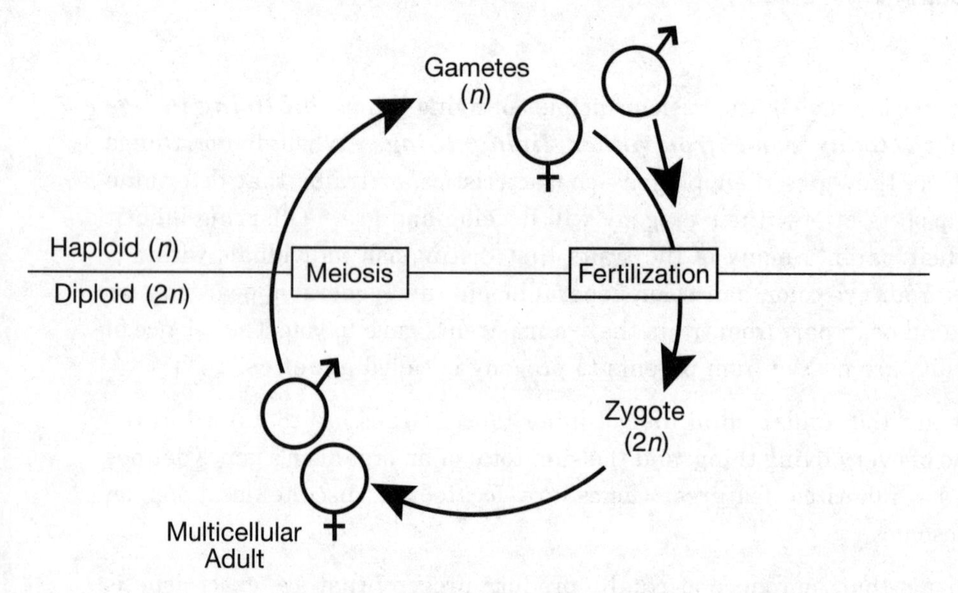

The information about meiosis and sexual reproduction above applies to animals. The mechanisms involved in transmitting genetic information between generations are somewhat different in other types of organisms like fungi and plants.

The human genome is split into linear pieces of DNA called chromosomes. There are twenty-three different chromosomes in the human genome and the number of different chromosomes is characteristic for each species. Human somatic cells are **diploid,** containing two copies of each chromosome, or forty-six chromosomes total. The two copies of each chromosome are inherited, one copy from each parent. Gametes (ova and sperm) only have one copy of the genome: a single copy of each chromosome. Having a single copy of the genome, gametes are called **haploid**. To create gametes, cells go through a form of cell division called **meiosis** (also called "reductive cell division"), which divides the number of chromosomes that the resulting gametes carry. When two gametes join in the act of fertilization as part of sexual reproduction, the resulting cell, the first cell of the new offspring individual, receives two copies of each chromosome, making it diploid again. The number of copies of the genome a cell has is often called "n". Haploid cells are $1n$, diploid cells are $2n$, triploid cells are $3n$, and tetraploid cells are $4n$.

Q During the mitotic cell cycle, how many copies of each chromosome does a human somatic cell have during the G2 phase of the cell cycle?

A It has two copies of each chromosome during interphase, since it is diploid, and it has four copies of each chromosome during G2 phase up until the daughter cells separate in mitosis and cytokinesis.

Of the twenty-three pairs of human chromosomes found in diploid somatic cells, twenty-two of the pairs are the same in men and women. These chromosomes are called **autosomes** in genetics, and traits found to be associated with genes on these chromosomes are called **autosomal**. The other pair of chromosomes is the **sex chromosomes**. Human females have two X chromosomes (XX), whereas human males have an X and a Y chromosome (XY). The Y chromosome is quite a bit smaller than the X and contains unique genes involved in sex determination. In a technique called **karyotyping**, technicians can take the condensed chromosomes from a cell and photograph them under a microscope to examine them for chromosomal abnormalities. Karyotyping can reveal when an individual carries too many or too few chromosomes, such as the trisomy (three copies) of chromosome 21 found in individuals with Down syndrome.

Q Can karyotyping reveal the changes in nucleotide sequence that usually distinguish different versions of genes?

A No. Karyotyping can only reveal big changes affecting whole chromosomes or big chunks of chromosomes. Most gene changes that distinguish different versions of genes involve very small changes to one or a few nucleotides that are not visible under a microscope.

Q If an organism is triploid, how many copies of the genome will its gametes carry?

A This is sort of a trick question. Meiosis reduces the number of chromosomes in half from somatic cells to create gametes. A triploid organism carries three copies of the genome. Going through meiosis does not work well, since you can't have 1.5 copies of the genome. If its cells try going through meiosis, some cells would have one copy of one chromosome and other cells would have two. It's all pretty messy and basically does not work. As a result, triploid organisms are usually sterile. Some common fruits like bananas and seedless watermelon are triploid—that's why they are seedless: they lack gametes (seeds).

Meiosis

Meiosis is the form of cell division that creates gametes, reducing by half the number of chromosomes that the gamete carries. This is in contrast to mitosis, which creates cells carrying the same number of chromosomes as the parent cell. Although the overall result of meiosis is quite different from **mitosis**, many of the features of mitosis and meiosis are similar. Meiosis, however, involves one round of DNA replication followed by two rounds of cell division, whereas mitosis involves DNA replication followed by a single round of cell division. The two rounds of cell division in meiosis are called **meiosis I** and **meiosis II.**

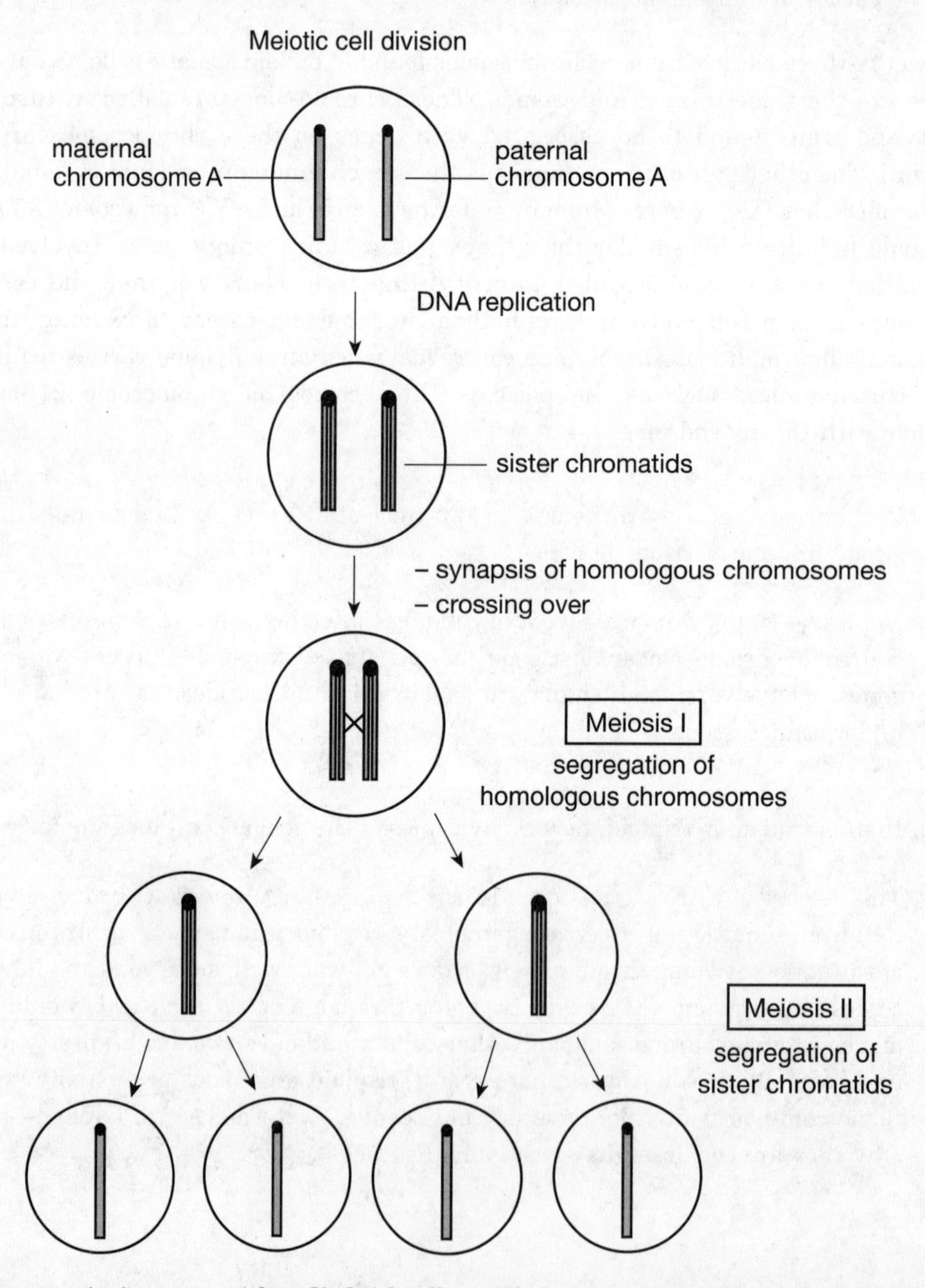

In humans, meiosis starts with a **diploid** cell, a cell that is a precursor to gametes in men or women. The pair of matching chromosomes in the diploid cell is called **homologous**

chromosomes. Since the two homologous chromosomes originated from two different parents, they represent the same genetic information, but they are not identical. They contain the same genes along their length, but small variations in the "copies" of genes (called "alleles") in each individual result in the unique inherited differences every human displays.

Q When a cell goes through DNA replication, every chromosome is replicated. Does this mean that the original and copied versions of a chromosome are homologous chromosomes?

A No. The homologous chromosomes are different copies of the same chromosome but they are not identical to each other. For example, we all carry two versions of human chromosome number 4 in all of the diploid somatic cells of our body and all of these carry the same genes, but they are not identical to each other. The two versions of the chromosome carry numerous but small differences in the sequence of DNA along their length, including differences in genes.

Before meiosis, the diploid cell goes through DNA replication. Every chromosome is copied exactly, producing a cell that has 4n copies of the genome, 92 chromosomes in humans. With four copies of the genome, every chromosome is present in the four copies as well. Each of two homologous chromosomes at this point has an identical copy. These identical copies are called **sister chromatids**. The sister chromatids are physically connected in the middle region of the chromosome, helping to keep the chromosomes organized.

Having completed DNA replication, the cell can enter the first meiotic cell division, meiosis I. The different phases of meiosis I are similar to the stages of mitosis, and they have the same names.

Meiotic prophase I: Chromosomes condense and homologous chromosomes line up together. When homologous chromosomes line up, they form complexes of matching DNA that can be exchanged between them. This process is called **meiotic recombination**. The junction at which DNA is broken from an existing strand and new bonds are made is called a **chiasmata**. In addition, in prophase, the spindle forms and the nuclear envelope breaks down.

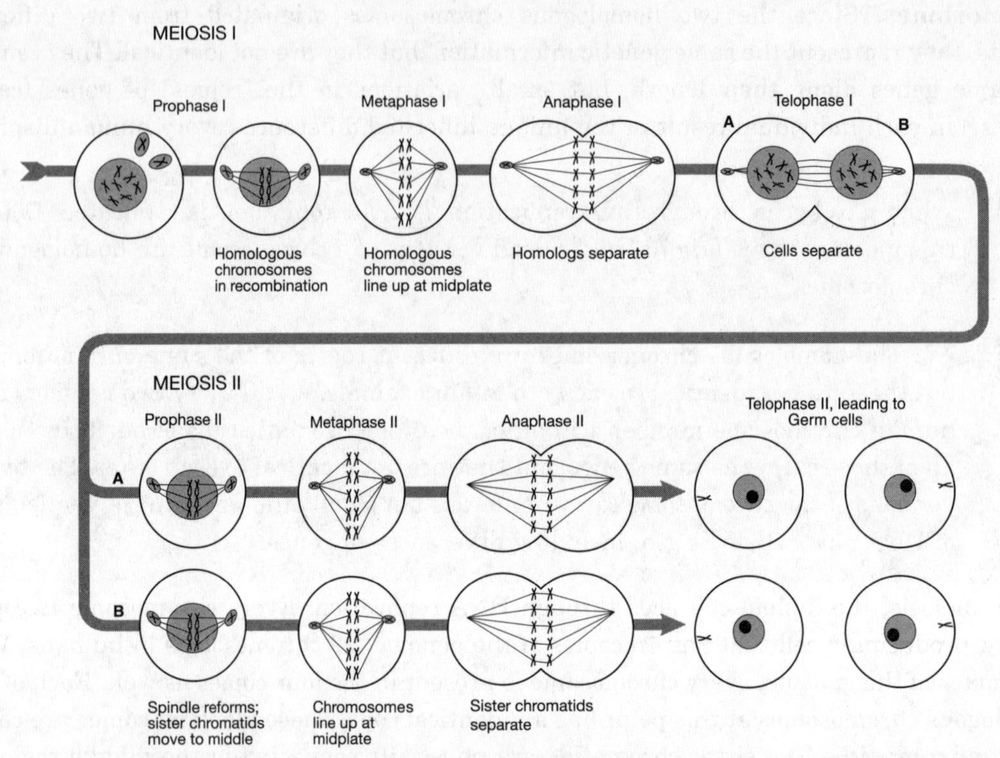

Phases of Meiosis I and II

Meiotic metaphase I: The pairs of homologous chromosomes line up in the middle of cell, equidistant from the two poles of the cell. The homologous chromosomes are each attached at a kinetochore in the middle to spindle fibers.

Meiotic anaphase I: Homologous chromosomes are separated, drawn toward opposite ends of the dividing cell by the spindle. The sister chromatids are still attached.

Meiotic telophase I: The two daughter cells divide, each containing the pair of sister chromatids for every chromosome.

Meiosis II: Similar to meiosis I (and mitosis), except that there is no recombination in meiosis II and the end result is a haploid cell with 1*n* copies of the genome.

Meiotic prophase II: Spindle reforms, and sister chromatids move toward the middle of the cell.

Meiotic metaphase II: Chromosomes are lined up in the middle of the cell, with each sister chromatid attached to kinetochores in the spindle.

Meiotic anaphase II: Sister chromatids are separated from each other, drawn toward opposite ends of the cell.

Meiotic telophase II: Chromosomes decondense, the nuclear envelope forms, and the two daughter cells divide. Each daughter cell at this point contains only a single copy of the genome, one copy of each chromosome.

Q In meiosis II, are the two sister chromatids exactly identical to each other in their DNA sequence?

A No. Sister chromatids are identical to each other immediately after DNA replication, but in meiosis I, recombination occurs between homologous chromosomes. As a result, the two sister chromatids exchange material with matching homologous chromosomes, but they would each do so differently and with a different homologous chromosome.

The key advantage of sexual reproduction is the great diversity it creates in a population through the constant mixing of alleles and traits in new combinations. As we will see later, having greater diversity in a population's gene pool allows that population to be more likely to adapt to environmental challenges. Meiosis increases the genetic variation of individuals in a population in a few ways.

One way meiosis creates variation is that homologous chromosomes randomly separate from each other in meiosis I. For every chromosome, there are two varieties in a diploid organism like you and me, one coming from your father and the other originating with your mother. Every gamete will get one copy of every chromosome, but it could be either the copy from the father or the copy from the mother. Since there are twenty-three chromosomes in humans, there are 2^{23} different potential ways these chromosomes can be combined in gametes—more than 8 million gametes can be generated in a person from this one mechanism alone! In genetics, the random movement of homologous chromosomes into gametes during meiosis in this manner is called **independent assortment** (more on this later).

Another way that meiosis increases diversity is through meiotic recombination between homologous chromosomes. Every chromosome contains many genes along its length, usually ranging from hundreds to thousands of genes on a chromosome. While every copy of chromosome number 10 in humans has the same genes along its length, the exact version of the genes in any individual may vary slightly. These differences make the two homologous chromosomes different from each other. If two genes lie on the same chromosome, and if there was no recombination, then these genes would move through meiosis as if they were joined. If your mother had a gene encoding the hairy arms trait, and she had on the same chromosome a gene conferring blue hair, then any gamete of hers containing this version of this chromosome would always carry these two traits together. Meiotic recombination means that even genes on the same chromosome need not stay together in the same combination on the same chromosome from one generation to the next.

(Although not part of meiosis, another way sexual reproduction creates diversity is through random combination of gametes during fertilization.)

Q How does mitotic recombination differ from meiotic recombination?

A Mitotic recombination does not exist. Recombination only occurs in meiosis, in prophase I.

Mendelian Genetics

Although the fact that kids inherit certain traits from their parents may seem obvious, how this occurs is less obvious. Prior to our current knowledge of genes at the molecular level, **Gregor Mendel** and others inferred the behavior of genes from their visible impact on organisms from one generation to the next. Mendel was a pioneer with his pea plant breeding experiments, and any introduction to genetics usually starts in his garden in the 1800s. The basic patterns of genetic inheritance that Mendel found do not explain everything, but they are still very relevant for living things (not just peas).

In studying his peas, Mendel first identified specific, clear features of the plants, the inheritance of which he could study. These inherited features are called **traits**. Traits are not caused by the environment, but by the inborn genes that an organism carries. The visible manifestation of these traits is called the **phenotype**. The combination of genes that produces a phenotype is called the **genotype**. Figuring out how genotypes and phenotypes relate is a crucial part of genetics.

To start his experiments, Mendel would pick a particular trait and breed together plants that had this trait from one generation to the next. If some peas had smooth seeds and others had wrinkly seeds, he would breed the plants with smooth seeds together for several generations until they produced only offspring with smooth seeds. These plants are called **true-breeding**, and using them in subsequent breeding experiments made experiments simpler to set up and interpret.

Mendel created true-breeding peas for both smooth seeds and for wrinkled seeds, then bred these peas together and examined their offspring. The offspring of such a cross between true-breeding lines are called the **F_1 generation**. When Mendel did a cross between true-breeding smooth seed peas and true-breeding wrinkled seed peas, he found that all of the offspring in the F_1 generation all looked the same, producing round smooth seeds. If traits are caused by genes, where did the genes causing the wrinkled trait go in the F_1 generation?

To look at this question further, Mendel bred these F_1 generation plants against each other to produce another generation, the **F_2 generation**. In the F_2 generation, Mendel saw a curious thing. While none of the F_1 plants had wrinkled seeds, the wrinkled seed trait reappeared in about one-fourth of the plants in the F_2 generation. The other three-fourths of the plants produced smooth seeds. This leads to another question—where did the wrinkled trait come from in the F_2 generation if it was not seen in the F_1 generation?

Some fundamental concepts about the behavior of genes help to explain this behavior. One of these concepts is that traits are caused by discrete genes that organisms carry, and that a given organism carries two copies of a gene. These two copies are inherited one from each parent, and they can either be the same or different from each other. When there are two or more different versions of a gene, these are called **alleles** of each other. The true-breeding peas that have smooth seeds and those that have wrinkled seeds have different alleles of a gene involved in seed formation. We now know that alleles of a gene differ in their DNA sequence, and these changes in a gene relate to different functions in the appearance of an organism.

A diploid organism can carry two copies of a gene that are the same or two that are different. If the two copies of a gene are the same, the organism is called **homozygous** in its genotype for this gene. If the two copies are different, it is called **heterozygous**.

Q Does being heterozygous describe the genotype or the phenotype of an organism?

A The genotype. Being heterozygous or homozygous will impact the phenotype, but really these words describe the genes an organism carries.

Another aspect of gene behavior derived from Mendel's data is that different types of alleles seem to behave differently in their impact on an organism's appearance, the phenotype. In the cross of plants with smooth and wrinkled seeds, the F_1 plants must have had one copy of the allele that causes smooth seeds and one copy of the allele that causes wrinkled seeds, but all of the plants made smooth seeds. Where is the allele that causes wrinkled seeds in the F_1 generation? It must be there in the plants, but not expressed in the presence of the other allele causing smooth seeds. Alleles that can lie silent and unobserved in this way in the F_1 generation are called **recessive alleles,** while alleles causing traits that are always observed in the F_1 regardless of the other copy of the gene are called **dominant alleles.**

Q According to the description above for the behavior of the smooth and wrinkled seed traits in peas, which allele is recessive, and what is the phenotype of heterozygous plants carrying both the recessive and dominant alleles?

A The F_1 generation plants produce only smooth seeds, although they must all carry the allele for wrinkled seeds as well. This would make the allele for wrinkled seeds recessive, and the smooth seed allele dominant. The phenotype of heterozygous plants in this example is the phenotype of the F_1 plants, which all carry both alleles, one from each parent. These plants all express the dominant allele, the one for smooth seeds.

Mendel devised two laws to describe the behavior of alleles in his breeding experiments: the **law of segregation** and the **law of independent assortment**. The law of segregation says that the two copies of a gene that an organism carries can separate from each other when they are transmitted from one generation to the next. We now know that this is the result of the mechanics of meiosis. The two alleles are carried on homologous chromosomes that are randomly segregated into different gametes as part of meiosis.

The law of independent assortment applies to the behavior of more than one trait. Mendel examined not just one trait in peas but seven different traits, and sometimes he would examine the behavior of more than one trait in a single cross. If two traits are controlled by two different genes that lie on different chromosomes, then these two different traits will be passed from generation to generation independently of each other.

One very useful way of visualizing and predicting the results of cross between plants is a tool called the **Punnett square**. Let's look at a cross involving a single trait, the smooth/wrinkled pea seed trait again. If two F_1 generation plants are crossed with each other, how can we predict how the F_2 generation will look? We know that the F_1 generation plants are all heterozygous for the gene controlling seed shape, containing a copy of both the recessive and dominant alleles. Lets call the dominant allele *S* (for smooth) and the recessive allele *s*. Upper case letters are used to represent dominant alleles; lower case letters represent recessive alleles. To set up a Punnett square, the two alleles for the two individuals in the cross are separated to represent the two possible gamete types that may be created by each parent plant in meiosis for this trait. Every gamete can contain one allele or the other, but not both and not zero. The possible gametes are lined up on the two sides of the square, and the internal squares represent possible combinations of gametes that a cross between these two individuals may create.

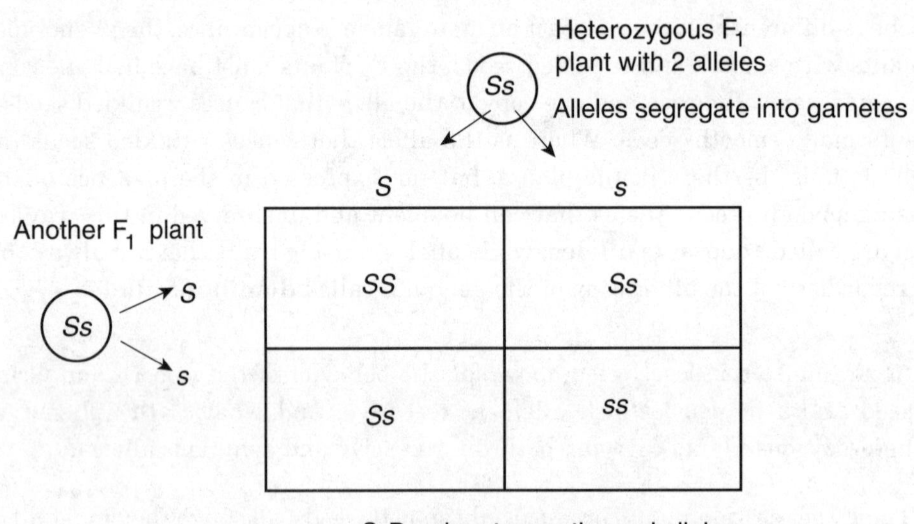

Punnett Square for cross of two F_1 heterozygous plants for smooth/wrinkled trait

First, let's look at the genotypes that the square predicts—the various combinations of alleles in offspring of the cross. One square contains a homozygous dominant result, two squares are heterozygous, and one square is homozygous recessive. From the genotypes we can then predict phenotypes. Since the allele for smooth seeds is dominant, the homozygous dominant and heterozygous combinations of alleles will all produce smooth seeds. Only the homozygous recessive plants will produce wrinkled seeds.

Does the Punnett square predict anything about the relative abundance of genotypes and phenotypes in the offspring of the cross? The answer is yes. The number of squares occupied by a given result leads to the probability of that result in the cross. There are 3 out of 4 squares in the cross that would produce the smooth seed phenotype, and this mirrors the probability of seeing this phenotype in the cross: 75% of offspring will have smooth seeds.

In fact, you don't really need the Punnett square to solve this sort of problem, because the answers can also be determined using some simple calculations. There is a 50% probability that a given gamete will have one allele or the other. The probability of two events happening is obtained by multiplying the probability of each event together. For a specific combination of gametes coming together, first find the probability of each gamete type, then multiply these together. The probability of an individual offspring being homozygous recessive is: ½ × ½ = ¼, or 25. The Punnett square makes it easier to figure things out though by visualizing what is happening, particularly when a cross is more complicated or you are still working on getting comfortable with problems like this.

The Punnett square can also be used to see what happens in a more complicated cross looking at two different traits at the same time. Another trait that Mendel looked at was the color of the peapod. Some plants were green and others were yellow. These plants were also bred to be true-breeding and, further, plants were crossed to be true-breeding for both seed shape and pod color. Green pod color is determined by a dominant allele (let's call it *P*), while yellow pods are caused by a recessive allele (let's call this one *p*). Let's say that we bred the peas to create one true-breeding line with all green pods and smooth seeds, and another line with all yellow pods and wrinkled seeds. These two lines are crossed to create an F_1 generation.

Q What will the phenotype of the F_1 generation be?

A All of the plants in the F_1 will be heterozygous for both traits, so they will all express only the dominant alleles, making them plants with green pods and smooth seeds.

Let's use a Punnett square to predict the result of a cross between two of these F_1 plants. The first step is to create all of the possible allele combinations on the two sides of the square as shown. In this example we assume that the alleles are on different chromosomes and will follow the law of independent assortment. If this were not the case, it would quickly become obvious in the results of the cross. It is also possible to create a Punnett square to predict what would happen if the two traits are encoded by genes closely linked on the same chromosome. After populating the two sides of the square with all possible allele combinations in gametes, the next step is to match up the various resulting combinations that would occur through fertilization.

F_1 plants: *PpSs*

All possible gametes:

	S	*s*
P	*PS*	*Ps*
p	*pS*	*ps*

	PS	*Ps*	*pS*	*ps*
PS	*PPSS* Green pods, smooth seeds	*PPSs* Green pods, smooth seeds	*PpSS* Green pods, smooth seeds	*PpSs* Green pods, smooth seeds
Ps	*PPSs* Green pods, smooth seeds	*PPss* Green pods, wrinkled seeds	*PpSs* Green pods, smooth seeds	*Ppss* Green pods, wrinkled seeds
pS	*PpSS* Green pods, smooth seeds	*PpSs* Green pods, smooth seeds	*ppSS* Yellow pods, smooth seeds	*ppSs* Yellow pods, smooth seeds
ps	*PpSs* Green pods, smooth seeds	*Ppss* Green pods, wrinkled seeds	*ppSs* Yellow pods, smooth seeds	*ppss* Yellow pods, wrinkled

First, figure out all of the possible gametes, then all of the possible genotypes created by matching up these gametes together as happens during fertilization and from this predict the resulting phenotypes. In this case, looking at either trait alone, there is a 3:1 ratio of the dominant phenotype compared to the recessive phenotype, since the dominant phenotype is observed with both the homozygous dominant genotype and the heterozygous plants. The square predicts that 75% (9⁄12) of the peas will have smooth seeds and that 75% (9⁄12) of the peas will have green pods. Because the genes are not linked, displaying independent assortment from each other, it is not always the same plants that have smooth seeds and green pods, however.

The phenotypic ratio observed is:

9 Smooth seeds, Green pods :
3 Smooth seeds, yellow pods :
3 wrinkled seeds, Green pods :
1 wrinkled seeds, yellow pods

The four different phenotypes represent various combinations of the dominant and recessive phenotypes for the two traits, including

9 Dominant, Dominant :
3 Dominant, recessive :
3 recessive, Dominant :
1 recessive, recessive

This 9 : 3 : 3 : 1 phenotypic ratio is a classic F_2 result, given the assumptions we've laid out.

As with experiments looking at just one trait, another way to solve a problem dealing with a cross involving two traits is to calculate the probability of an outcome. First, what is the probability that a pea plant will have smooth seeds in this cross?

The smooth seed result is produced with two genotypes, dominant homozygous (SS) and heterozygous (Ss). The probability of the dominant homozygous genotypes is $\frac{1}{2} \times \frac{1}{2}$, or $\frac{1}{4}$. The heterozygous genotype can happen two different ways (sS and Ss), so the probability of this is added: $(\frac{1}{2} \times \frac{1}{2}) + (\frac{1}{2} \times \frac{1}{2})$, or $\frac{1}{2}$. The overall probability of having a smooth seed is found by adding the fraction of dominant homozygous and heterozygotes together: $\frac{1}{2} + \frac{1}{4}$, for an overall probability of $\frac{3}{4}$ that a plant will have smooth seeds.

The same rules can be applied for the double cross. First, find the probability for each trait on its own, then multiply the probability of both traits happening together. If there is a $\frac{3}{4}$ probability of a plant having smooth seeds and a $\frac{3}{4}$ probability that it will have green pods, then there is a $\frac{3}{4} \times \frac{3}{4} = \frac{9}{16}$ probability that a plant will have both smooth seeds and green pods.

Q In genetic crosses looking at real plants, the numbers of each phenotype observed are often close to the predicted result, but not exactly the same. Does this deviation from the exact predicted ratio indicate something is wrong with the experiment or the rules of genetics?

A Not at all. This is to be expected in a real experiment. The real numbers will almost never exactly match the predicted numbers. Statistics would say that if we had an infinitely sized population then the numbers should match the predicted ratios, but you don't find infinite populations in the real world.

INHERITANCE PATTERNS IN HUMANS

If the rules of genetic inheritance that Mendel found applied only to peas then Mendel would probably not be all that famous. In fact, it was only when 3 independent researchers found these Mendelian ratios in other organisms that Mendel's work was "rediscovered." As it is though, his observations about genetics apply to other organisms that reproduce sexually, including humans. The rules of genetic inheritance based on the work of Mendel have been used to study many common human traits like eye color, as well as a variety of genetic diseases.

Since human geneticists do not generally have the opportunity to pick who mates with whom, the human geneticist must use what information is available, usually by studying patterns of inheritance within families. A **pedigree** is often constructed to do this, drawing on paper the individuals and relationships within a family and charting who is affected by a trait and who is not. In a pedigree, each level represents a generation in the family. Males are represented by squares and females by circles. Horizontal lines connecting a male and female indicate marriage, and vertical lines indicate children resulting from this marriage. Individuals affected by the trait being studied have their box or circle filled in; others have open circles or boxes.

Let's say, for example, that there is a trait causing short thumbs that is found in some members of a family. To see how this trait is transmitted, a geneticist collects information about as many family members as he can to reconstruct the history of the disease in this family. He finds the following:

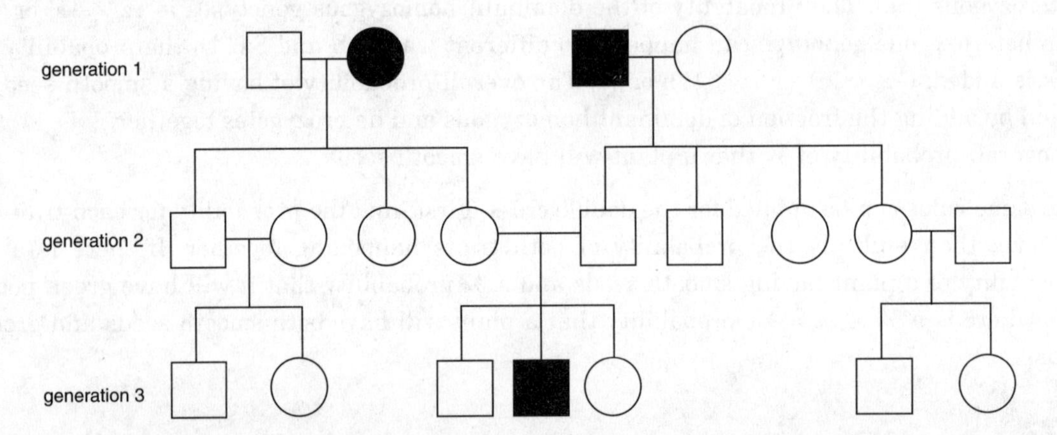

In this pedigree, the disease is found in two individuals in the first generation, then skips the second generation, and appears again in the third generation. A dominant trait can never skip a generation like this, because a dominant trait will always be expressed if it is present, and it must be present in the second generation if it appears in the third. Since it is not dominant, the trait must be recessive. Knowing this, you can start to fill in some of the missing information in the pedigree, like the probable genotype of the people in the second generation who were parents of the affected individual in the third generation. Let's call the dominant allele for this trait *T* and the recessive allele *t*. If the trait is recessive, and neither parent expresses the trait but they have kids that do, then the parents must both be heterozygous for the recessive allele (*Tt*).

Recessive traits are often associated with alleles that produce a nonfunctional protein. The reason the allele is recessive is that the dominant allele in the heterozygote still produces the normal functional protein, so that there is still at least 50% of the normal protein around. Having 50% of the normal protein present is often enough to still produce the normal function. An example of this is albinism. Albinism is seen in people who are homozygous for a defective version of a gene involved in producing pigment. Heterozygous individuals appear normal, however. Heterozygous individuals who carry a recessive allele for such traits without expressing them are called **carriers**.

The impact of recessive traits on individuals expressing them ranges from innocuous to lethal. Some genetic diseases like cystic fibrosis are only observed in individuals who are homozygous for a recessive disease allele. The gene associated with cystic fibrosis is autosomal recessive and encodes a defective version of an ion channel expressed in cells lining the lungs and digestive tissues. Individuals carrying one copy of the allele causing cystic fibrosis appear normal overall but are carriers of the disease. The single normal copy of the gene that they have is enough to provide normal function. Individuals with two copies of the defective recessive allele don't have any properly working ion channels and secrete large quantities of chloride, plugging their airways with excessive mucus. Even with modern medical treatments, individuals who are homozygous for the cystic fibrosis allele have a reduced life expectancy.

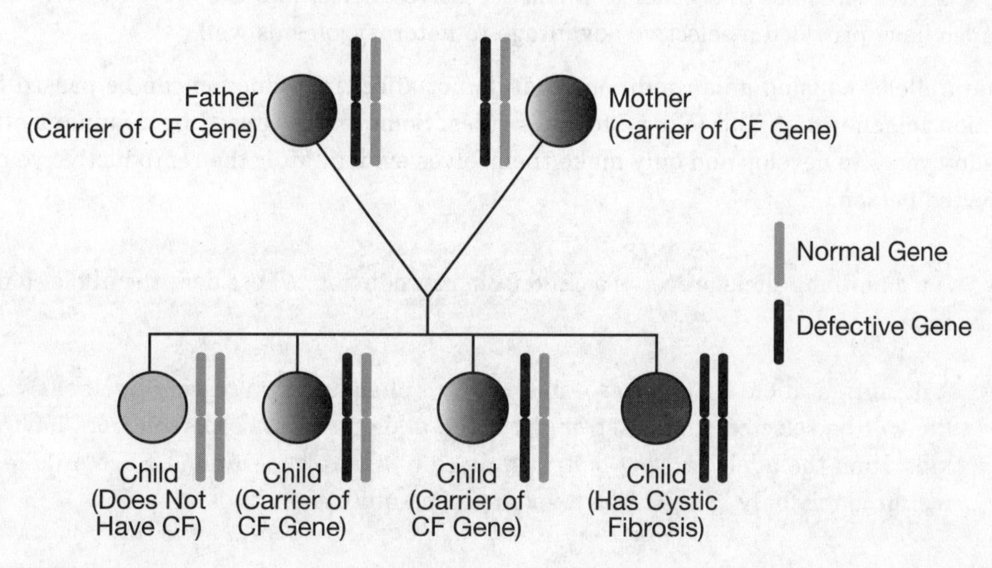

Individuals who express recessive alleles (like cystic fibrosis) that cause severe or lethal genetic conditions will often have no children or few children, meaning that they will not pass their alleles as frequently to future generations. Nonetheless, many genetic conditions such as cystic fibrosis persist in the population. Why is this the case? Heterozygous carriers are much more abundant in the population than the homozygous recessive individuals and heterozygotes have no decrease in the number of children they produce, overall. As a result, any selective pressure against the disease allele is only observed in homozygotes and the overall effect on the larger population is often very small.

Q If the number of cystic fibrosis heterozygotes in the population is 5% of the population and marriage is random, what are the odds that two heterozygotes will marry?

A The odds are $\frac{1}{20} \times \frac{1}{20}$, or $\frac{1}{400}$ marriages. In these marriages, the odds of a child being homozygous for the allele are $\frac{1}{2} \times \frac{1}{2}$, or 1 out of 4.

Sickle-cell anemia is a genetic disease that is most common in parts of Africa and in the U.S. among African Americans. It is caused by a specific allele of one of the subunits of hemoglobin, the protein that carries oxygen in red blood cells. In individuals who carry two copies of the sickle cell allele, the hemoglobin inside their red blood cells is prone to forming large aggregates that deform and damage red blood cells, causing anemia. The condition in homozygous individuals can be severe with a variety of health problems, particularly in children if they are not identified and treated. Despite the severe phenotype associated with homozygotes, the allele causing sickle cell anemia is abundant in this population. Why? Shouldn't it be selected against over longer periods of time? Heterozygous individuals usually express no symptoms or mild symptoms, but in addition, heterozygous individuals are somewhat protected against malaria, a severe health problem in those parts of Africa where the sickle-cell allele is common. In the region of Africa where the allele arose, the resistance of heterozygotes to malaria probably gave them a selective advantage over sickle-cell homozygotes or normal homozygotes, preserving the sickle-cell allele in this population at high levels. It is possible that other common recessive alleles, like the cystic fibrosis allele, might also have provided a selective advantage to heterozygotes as well.

Dominant alleles causing mild conditions that do not affect reproduction can be passed from generation to generation and follow Mendel's rules. Some traits caused by dominant alleles take many years to develop and only make themselves evident after the reproductive years of the affected person.

Q If a dominant allele prevents a child from reproducing, where does the allele come from?

A A dominant allele that causes a more severe phenotype expressing itself early in life will be selected against strongly. If the allele prevents a person from having kids, then the allele will not be transmitted to future generations. Such an allele will probably only arise through spontaneous mutation.

Beyond Mendelian Genetics

Mendel's rules are based on the molecular and cellular processes that form gametes and they apply broadly to a wide variety of traits across the kingdoms of life. However, they do not universally explain the genetic inheritance of traits. Mendel was fortunate to pick the traits he studied in peas. These traits are associated with only one gene, and for which alleles behaved as purely recessive or dominant alleles. Things are not always this simple, though. Fully understanding patterns of inheritance requires expanding your understanding of genes beyond Mendel's original rules.

INCOMPLETE DOMINANCE AND CODOMINANCE

In Mendel's peas, recessive alleles were not seen at all in heterozygotes, and dominant alleles produced a phenotype that looked the same in homozygotes or heterozygotes. This situation is called **complete dominance**. Sometimes, however, heterozygotes have a phenotype that is somewhere in the middle of the phenotype of homozygotes of either recessive or dominant alleles. An example of this would be a plant whose flower color is controlled by two alleles in which heterozygotes have an intermediate color. If a true-breeding individual of this plant with yellow flowers is crossed with a true-breeding plant with red flowers, the F_1 progeny have orange flowers. This phenotype result indicates **incomplete dominance.**

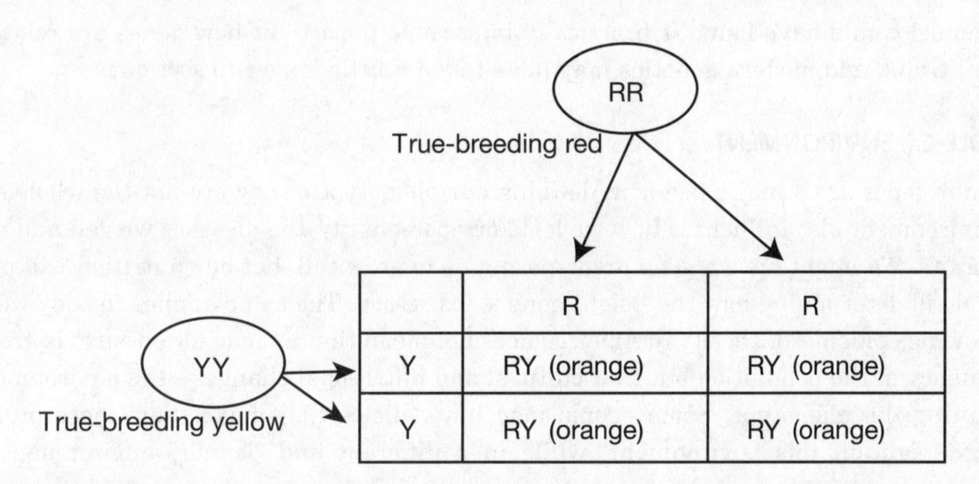

Codominance is a form of gene expression in which multiple dominant alleles are each expressed at the same time without any "blending" of the dominant phenotypes (unlike incomplete dominance). The most common example of this involves the alleles that control blood type. Blood types are associated with proteins expressed on the surface of red blood cells, and there are three different common alleles, *A, B,* and *O*. *O* is a recessive allele, in which no protein is expressed. *A* and *B* are both dominant to the *O* allele, but codominant to each other. If an individual has both *A* and *B* alleles, they express both dominant phenotypes.

Q A researcher finds that the height of sunflowers is controlled by two alleles, one for short plants (h) and one for tall plants (H). Homozygous hh plants are 1 meter tall, and homozygous HH plants are 2 meters tall. In a cross between true-breeding short plants and true-breeding tall plants, the F_1 generation is all the same height, 1.7 meters. Is this an example of complete dominance, incomplete dominance, or codominance?

A This is incomplete dominance. Neither allele is expressed fully in heterozygotes, and the phenotype is somewhere in the middle of what is produced by either allele alone.

POLYGENIC TRAITS

Traits often lack discrete, neatly defined characteristics. Rather, some traits fall somewhere along a continuous range. Although the height of people is at least partially inherited from parents, people do not grow to a distinct number of different heights. Any sufficiently large group of people, if analyzed statistically, will display a bell-shaped curve of heights with every possible height between the extremes. Other traits like skin color also fall in a broad range. Many such traits are controlled by multiple genes, or are **polygenic**. The more genes that are involved, the more complex the relationships among genes that cause the trait, and the more difficult it is to relate the trait to any one specific gene. The susceptibility of people to diseases like obesity and schizophrenia is thought to have a genetic component. But with many genes interacting in a complex fashion, the exact pathway from genotypic interactions to phenotypic trait is hard to determine. If the color of pea flowers were controlled by five genes instead of one, Mendel would have found it practically impossible to sort out how genes are related to physical traits, and modern genetics may have taken a little longer to sort out.

THE ROLE OF ENVIRONMENT

While our genes are a major factor in defining our phenotypes, they are not the whole story. Our environment also influences how we look, our personality, the diseases we get, and many other traits. We might have a gene predisposing us to grow tall, but our nutrition will play a large role in determining how the height gene is expressed. The same applies to body weight. The growing epidemic of obesity in America does not mean that a "gene for obesity" is growing in frequency in the population but that cultural and nutritional changes—the environment—are shifting the phenotype. Some people may have alleles that make them more prone to gaining weight in this environment, while in a different and "leaner" environment this phenotype would not have been observed.

LINKAGE AND RECOMBINATION

Mendel's law of independent assortment describes the independent segregation of alleles into gametes. This law holds if two traits are controlled by genes on *different* chromosomes. But what if this is not the case?

Earlier, a cross was described for peas in which two traits were examined that displayed independent assortment. Now let's imagine a scenario in which these two traits are encoded by genes next to each other on the same chromosome. If there is no meiotic recombination between the genes (as we will discuss shortly), then the results of the F_1 and F_2 cross will play out differently than before. Let's use a Punnett square to visualize this scenario. First, true-breeding lines for peas with green pods, smooth seeds (GGSS) and for yellow pods, wrinkled seeds (ggss) are established. Since we are assuming that the genes are linked, we need to move the alleles through the Punnett square in a different fashion—one that replicates the cytological situation. There are only two—not four—possible types of gametes in any pea:

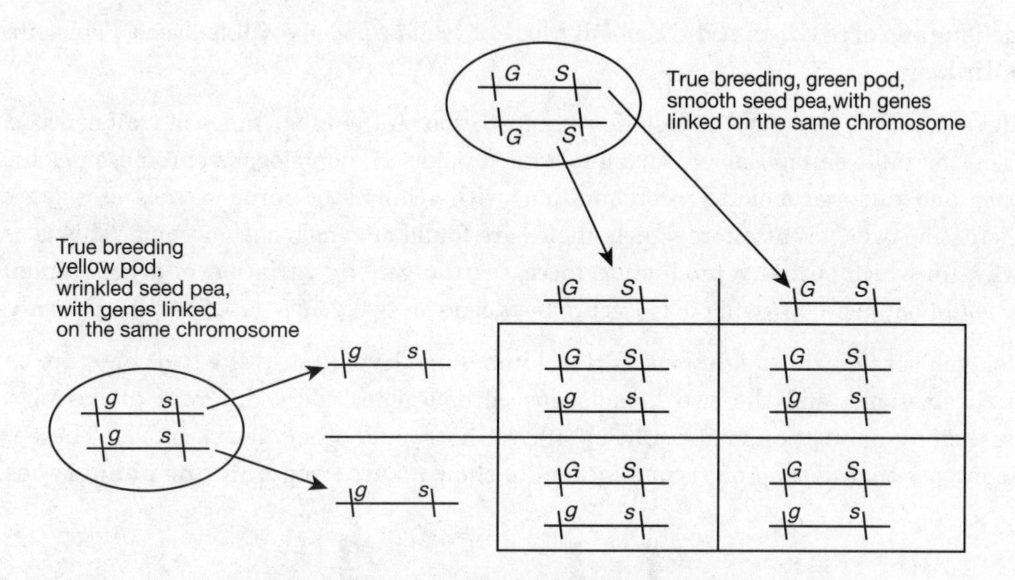

Punnett Square for cross between parental true breeding lines in which both genes are linked on the same chromosome

In the F_1 generation resulting from this cross, all of the peas are heterozygous for both traits, and all of them have green pods and smooth seeds. So far, the phenotypic result is identical to a situation in which the genes are on different chromosomes. In the cross between F_1 plants though, something peculiar happens.

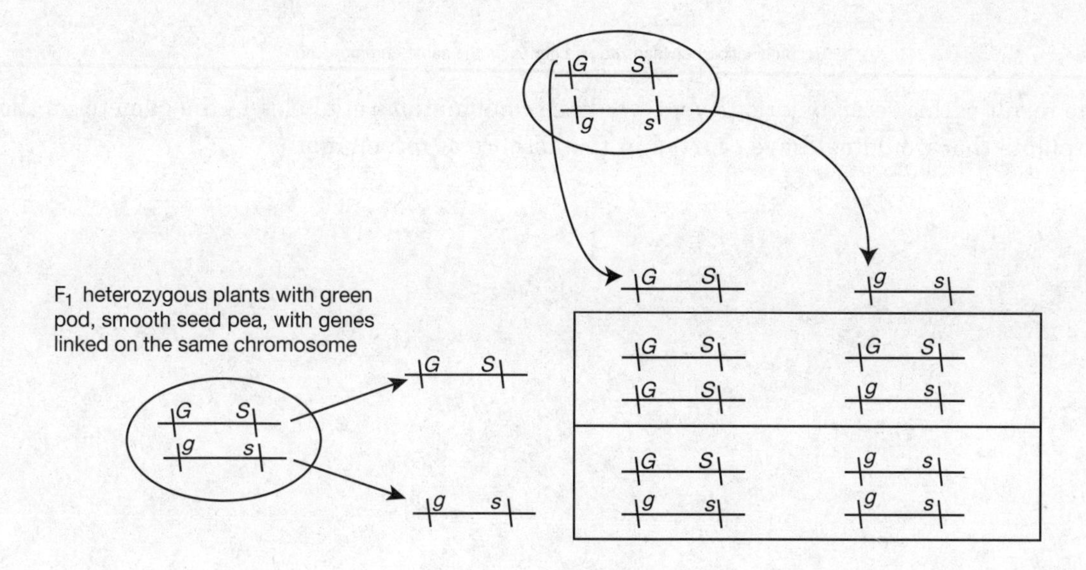

The F_2 generation

Again, since the genes sit close by on the same chromosome and cannot be separated, there are only two possible gametes. Any plant getting the dominant allele for one gene will also get the dominant allele for the other gene, and the same is true for recessive alleles. These were the combinations found in the true-breeding parents. These results resemble those of the F_1 of a monohybrid cross: only four possible genotypes and two possible phenotypes, along with the

Mendelian ratio of 3 (Green pod + Smooth seed) : 1 (yellow pod + wrinkled seed). This effect is called **linkage**.

Now let's introduce the effect of meiotic recombination on the inheritance of traits encoded by genes on the same chromosome. During meiotic prophase I, homologous chromosomes line up together and undergo meiotic recombination, with sections of chromosomes crossing over. This "crossing over" event alters which alleles are found on which chromosome. This is one of the ways in which sexual reproduction increases the genetic variation of a population. So, what would happen if crossing over occurs between the two genes described for peas above?

Let's sketch out how this might look at the chromosome level first. If the two genes are on the same chromosome, and the two homologous chromosomes carry different alleles of these genes, then recombination might split up alleles that would otherwise be linked. These types of phenotypes that depend on recombination to show up are **recombinant phenotypes**.

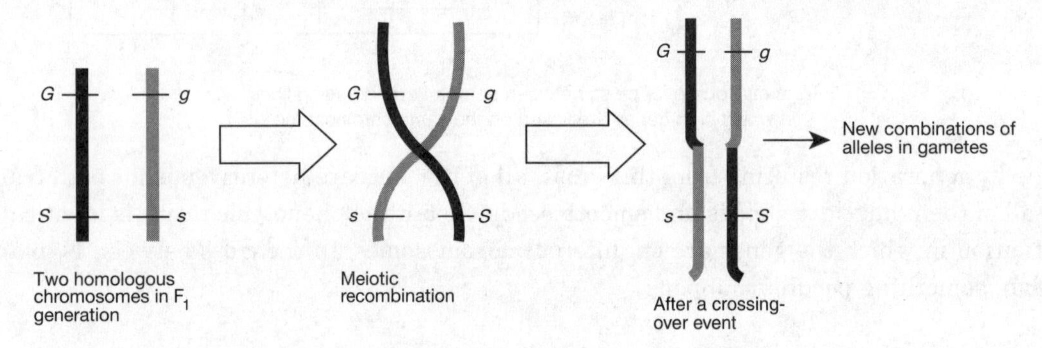

Recombination between two pea genes on the same chromosome

The result of this recombination event would be combinations of alleles in the gametes of the F_1 plants that would not have ocurred in the absence of recombination.

A. Linkage with no crossing over

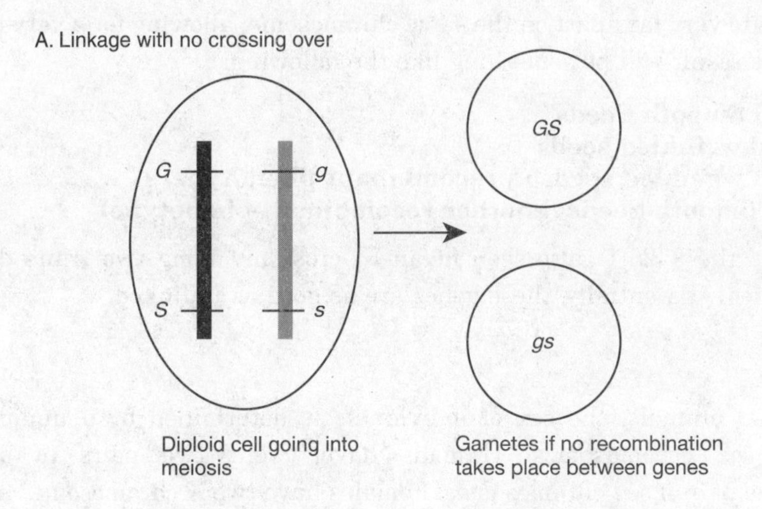

B. Linkage with crossing over

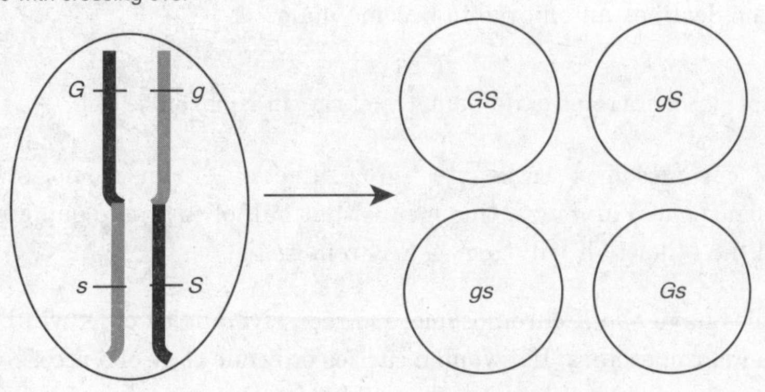

Some gametes have new combinations of alleles due to recombination

The shorter the distance between two genes on a chromosome, the less tightly linked they are, and the more likely crossing over will occur. The distance between genes on a chromosome is proportional, therefore, to the frequency of crossing over events. Genetic or genomic maps can be created by determining these relative frequencies among all genes on a chromosome. If the genes are so close that no recombination is observed between the peas, then the ratio of offspring between two F_1 plants will be:

75 Green Pods and Smooth Seeds
25 yellow pods and wrinkled seeds
or 3:1.

However, if the genes are slightly farther apart, allowing a low level of recombination between the genes, the result will be something like the following:

73 Green Pods and Smooth Seeds
24 yellow pods and wrinkled seeds
2 Green Pods and wrinkled seeds (a recombinant phenotype)
1 yellow pod and Smooth Seed (another recombinant phenotype)
Not quite 3:1.

Finally, if the genes are very far apart on the same chromosome, allowing for a very high level of recombination, the result will be something like the following:

55 Green Pods and Smooth Seeds
18 yellow pods and wrinkled seeds
17 Green Pods and wrinkled seeds (a recombinant phenotype)
6 yellow pods and Smooth Seeds (another recombinant phenotype)

This ratio is close to the 9:3:3:1 ratio seen in an F_1 cross involving two traits displaying independent assortment. Essentially, these genes are as good as unlinked.

SEX LINKAGE

In humans and other animals, the sex of individuals is determined by a unique pair of chromosomes, the sex chromosomes. Humans have twenty-two pairs of autosomal chromosomes and one pair of sex chromosomes. Females have two X chromosomes and males have an X and a Y chromosome. The Y chromosome in males is much smaller than the X chromosome and thus contains relatively few genes. One of the few genes it *does* have is called SRY. This gene destines an embryo to become male.

Q What percentage of male gametes carry an X chromosome?

A Sex chromosomes segregate into gametes during meiosis just like other chromosomes. In males, this means that half of the male gametes will have an X and the other half will carry a Y chromosome.

Human females have two X chromosomes, so recessive alleles on X will behave like those on all of the other chromosomes. If a woman carries only one copy of a recessive allele, she will be a heterozygous carrier and express the wild-type phenotype. A female would need to be homozygous for the recessive allele to express it.

The story for men is different. Men only have one X, and their Y lacks most of the genes on the X. If a man inherits a recessive allele on his X chromosome, he will express it since there is no other copy of the gene to mask the recessive allele. So, when it comes to genes located on the X chromosome, men can't hide anything. Genes located on the X chromosome are called sex-linked genes, because their inheritance pattern depends on the sex of the individual. We can use a Punnett square to examine the inheritance pattern observed for a sex-linked trait. One of the most common examples of a sex-linked trait is for red-green color blindness. Color blindness is caused by a defect in a gene on the X chromosome for one of the pigments responsible for color vision. Let's call the dominant healthy allele *C* and the recessive defective allele *c*.

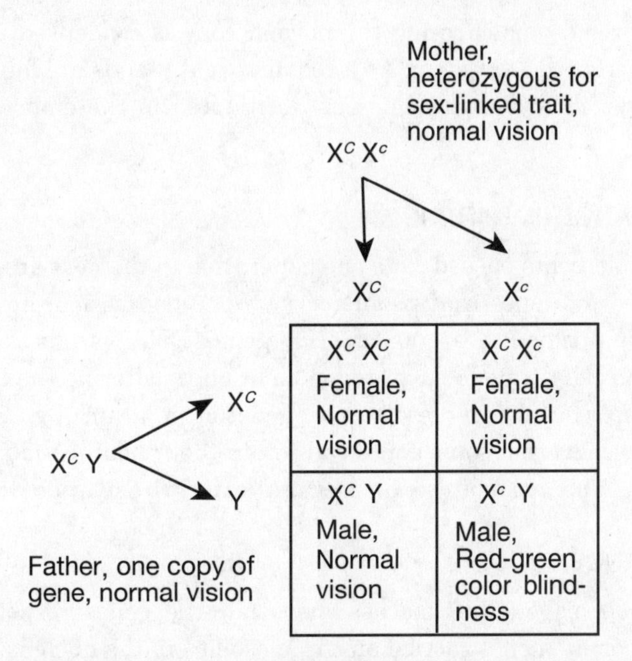

In this cross, the female carries the dominant allele on one X chromosome and the recessive allele on the other X. The male only has one copy of the gene, on his sole X. Neither of the parents is color blind. What about their kids? How many of the males and how many of the females will be color blind? According to the Punnett square, females always get an X from their father, so they always have at least one dominant allele and are never color blind. Things work out differently for male children though. The males get the Y chromosome from their father (this is why they are male) and they receive their X from their mother, with a 50% chance that it will be either the X with the recessive allele or the X with the dominant allele. Males will have a 50% chance of being color blind in this cross.

Q If a color blind man marries a woman who is homozygous for the normal visual pigment allele, what percentage of their male children will be color blind?

A Zero. The boys will get a normal allele on the X from their mother, and they will get their Y chromosome from their father. They will only have one allele, but they will have normal color vision.

X CHROMOSOME INACTIVATION

Although women have two X chromosomes in every diploid cell, they are not both expressed. During embryonic development of female mammals, one of the X chromosomes is condensed and inactivated in each embryonic cell so that it expresses few of its genes. This condensed X chromosome is called the Barr body. The other X chromosome expresses genes along its length normally. As the female continues development and the embryonic cell divides mitotically, all of the daughter cells will also have the same inactivated X. This results in patches of tissue in females that have one or the other X chromosome inactivated, or a mosaic pattern of expression. A gene involved in sweat gland formation, for example, is found on the X chromosome. Women carrying one copy of a defective allele of this gene will have patches of skin where the X chromosome carrying the normal copy is condensed and other patches of skin where the other copy is expressed. As a result, sweat glands are found in patches on the skin depending on which X chromosome was inactivated in the embryonic development of each patch of skin.

EPIGENETIC INHERITANCE (IMPRINTING)

Generally the heritable traits passed from one generation to the next are based on differences in the DNA sequences of genes that people carry. Gene expression, however, is affected by more than just the nucleotide sequence of a gene. DNA can be chemically modified. Methylation of a gene can turn gene expression on or off by altering the chromatin. It is often thought that gamete formation resets the genetic program by fixing any chemical modifications, but it seems that some genes can retain their modification and pass it on to the next generation. This phenomenon is called **epigenetic inheritance** or **imprinting**.

MATERNAL INHERITANCE

So far, our discussion of genetics has assumed that the genes responsible for traits are encoded on nuclear genes—genes carried on chromosomes in the nucleus. There are, however, a few genes encoded in the DNA found in mitochondria. When gametes form in humans or other animals, ova are usually quite large, with extensive cytoplasm, including mitochondria, that is contributed to the zygote after fertilization, whereas sperm cells contribute only their genome to the zygote. As a result, all of the mitochondria in the zygote and the resulting organism come from the ova, and thus the mother.

There are a few cases of diseases that are associated with mutations in genes encoded in the mitochondrial genome. These diseases pass from mothers to all children and are not passed from fathers to their children. The inheritance pattern is different from sex-linked traits.

EXERCISES: GENETIC INHERITANCE

1. In what stage of meiosis are sister chromatids separated from each other?

(A) Metaphase I
(B) Prophase II
(C) Anaphase I
(D) Telophase I
(E) Anaphase II

2. If a man and a woman are both heterozygous carriers of the allele causing cystic fibrosis, what percentage of their male children will have the disease?

(A) 0%
(B) 25%
(C) 33%
(D) 50%
(E) 100%

3. A man and a woman are both heterozygous carriers of an autosomal recessive allele. In homozygotes, this allele is associated with a neurodegenerative disorder that leads to progressive loss of muscle control. This couple has had one male child affected by this disorder. What is the probability that their next child will be affected?

(A) 0%
(B) 12%
(C) 25%
(D) 50%
(E) 75%

4. A young individual is observed who suffers from a severe genetic condition caused by a dominant allele that causes sterility. Examination of their genes reveals that the allele is found in both somatic cells and gametes. What is the most likely source of an allele that is carried on an autosomal chromosome?

(A) X chromosome of maternal grandfather
(B) Meiotic recombination
(C) X chromosome of paternal grandmother
(D) Mutation occurring in oogenesis
(E) Mutation arising in somatic cells of affected individuals

5. A geneticist is studying two traits in fruit flies, a common genetic model organism. One of the traits determines if flies will have normal long wings or short, stubby wings. Another trait determines if flies will have red eyes or white eyes. True-breeding flies with long wings and red eyes are created and true-breeding flies with stubby wings and white eyes are created. In the F_1 generation, all of the flies have long wings and red eyes, but in the F_2 generation crossing F_1 flies, the following is seen:

670 flies with long wings and red eyes

220 flies with stubby wings and white eyes

38 flies with long wings and white eyes

22 flies with stubby wings and red eyes

This information best supports which of the following?

(A) The traits are codominant.
(B) The two traits are determined by genes on the same chromosome.
(C) The traits segregate independently.
(D) Meiosis is abnormal in these flies.
(E) Flies with white eyes are selected against in the F_1 and F_2 generation.

exercises

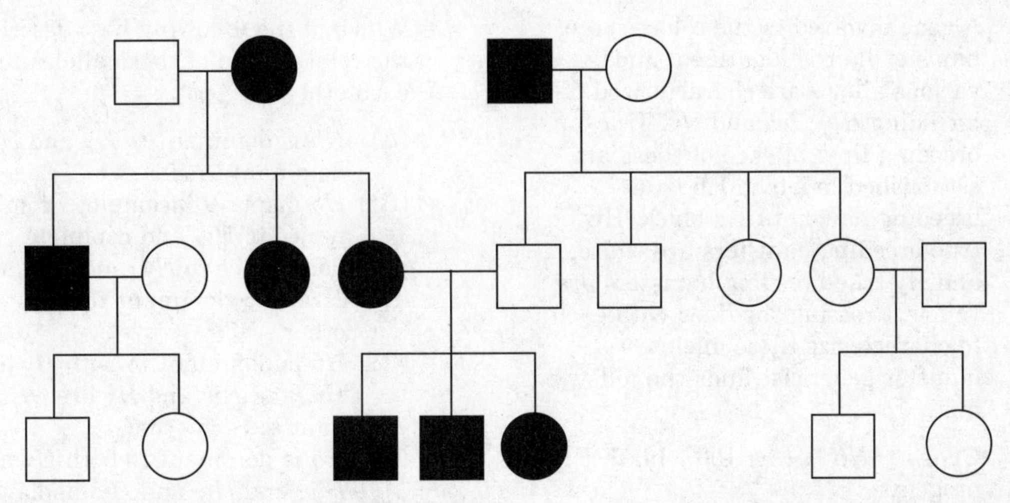

6. In the pedigree above, black circles indicate that a muscle-wasting disorder is observed in that individual. Which of the following is the best explanation for the inheritance pattern that is observed?

   (A) Imprinting
   (B) Incomplete dominance
   (C) Sex-linked trait
   (D) Caused by a gene encoded in mitochondrial genome
   (E) The trait is polygenic.

7. One hypothesis about aging is that the deterioration of tissues with increasing age is caused by damage to genes. A protein involved in energy metabolism is encoded in the nuclear genome, translated in the cytoplasm, and then transported into the mitochondrial matrix where it performs its function. The gene that encodes this protein is found to have the following number of mutations per 1,000 base pairs of DNA from blood cells in various age groups, with the average values reported out of 20 people examined in each age group:

   10–20 years: 0 mutations
   20–30 years: 0
   30–40 years: 0
   40–50 years: 1
   50–60 years: 5
   60–70 years: 10

   Which of the following statements is best supported by this information?

   (A) Changes in mitochondrial function are related to aging.
   (B) More rapid aging will be maternally inherited.
   (C) More mutations are found in the mitochondrial genome of older people.
   (D) Blood cells are more susceptible to mutation than other tissues.
   (E) The mutations observed in this gene will not be related to a heritable trait.

8. The Law of Segregation that Mendel observed in genetics is based on which of the following molecular and cellular events?

   (A) The separation of homologous chromosomes in meiotic anaphase I
   (B) Meiotic recombination
   (C) The separation of sister chromatids in meiotic anaphase I
   (D) The blending of traits affected by more than one gene
   (E) The union of gametes in fertilization

9. A gene involved in the coloration of hamster hair is identified, and various alleles are characterized, including *Hb, Hw,* and *Hy*. True-breeding lines of each of these are established in which Hb true-breeding hamsters are black, Hw true-breeding hamsters are white, and Hy true-breeding hamsters are yellow. Crossing the lines with each to characterize these alleles, a hamster geneticist finds the following:

Cross 1: *Hb* × *Hw*: 100% black F_1 progeny

Cross 2: *Hb* × *Hy*: 100% black F_1 progeny

Cross 3: *Hw* × *Hy*: 100% light yellow F_1 progeny

Cross 4: F_1 progeny from Cross 1 are crossed with true-breeding black: 100% black animals

Cross 5: F_1 progeny from Cross 1 are crossed with true-breeding white: 50% black and 50% white animals produced

Cross 6: F_1 progeny from Cross 3 are crossed with true-breeding black: 100% black animals are produced

Cross 7: F_1 progeny from Cross 3 are crossed with true-breeding yellow: 50% yellow and 50% light yellow hamsters are produced

Which of the following best describes the relationship of these alleles to each other?

(A) *Hb* is dominant to *Hy* and *Hy* is dominant to *Hw*.
(B) *Hb* displays incomplete dominance to *Hw* and complete dominance to *Hy*, and *Hw* and *Hy* are codominant to each other.
(C) *Hb* is dominant to both *Hy* and *Hw* and *Hw* and *Hy* are codominant to each other.
(D) *Hb* is dominant to both *Hy* and *Hw*, and *Hw* and *Hy* display incomplete dominance toward each other.
(E) *Hb* is dominant to *Hy, Hy* is dominant to *Hw,* and *Hw* is dominant to *Hb*.

ANSWER KEY AND EXPLANATIONS

1. E	3. C	5. B	7. E	9. D
2. B	4. D	6. D	8. A	

1. **The correct answer is (E).** Sister chromatids stay paired together, attached, until anaphase II. Homologous chromosomes separate in meiosis I and sister chromatids separate in meiosis II.

2. **The correct answer is (B).** The inheritance of this gene has nothing to do with the sex of the children, since the gene is carried on an autosome and not a sex chromosome. For both the man and the woman, ½ of their gametes will carry the allele causing the disease. Using either a Punnett square, or probability, you can find out that the odds of any child getting the disease depend on the various combinations of possible gametes, in this case $\frac{1}{2} \times \frac{1}{2} = \frac{1}{4}$ probability, or 25% of the children, regardless of their sex.

3. **The correct answer is (C).** The odds are the same for every child, regardless of history. The probability is determined by the genotypes of the parents, not by history. There is a 25% chance that any child will get two copies of the recessive allele, one from each parent.

4. **The correct answer is (D).** A dominant allele causing a severe or lethal condition in young people will not often pass from one generation to the next. Meiotic recombination can create new combinations of alleles on a chromosome, but it does not create new alleles unless a mistake occurs. Mutations in somatic cells can be ruled out since the allele is found in their gametes as well.

5. **The correct response is (B).** The two traits must be determined by genes that lie on the same chromosome. If the two traits sat on different chromosomes and displayed independent segregation, then the F_2 generation would look a lot different. The four phenotypes would be in the ratio of 9:3:3:1 in the F_2 generation. If the two genes were so closely linked on the same chromosome that no recombination occurred between them, then there would only be two phenotypes in the F_2 generation, long wings with red eyes and stubby wings with white eyes, in the ratio of 3:1. As it is, the predominance of the dominant:dominant phenotype and the recessive:recessive phenotype, with a small number of flies with recombinant phenotypes, suggests that these genes are linked on the same chromosome, although with at least enough distance between them to display some recombination.

6. **The correct answer is (D).** The trait is passed from mothers to all of their children in every case and never from fathers to their children. This is what is observed for traits that are encoded by genes in the mitochondrial genome. The ova contribute all of the mitochondria at fertilization, so mitochondrial traits are always inherited from the mother.

7. **The correct answer is (E).** Some of the answer choices may seem reasonable but are either not related to the data or interpret them incorrectly. Change in mitochondrial

function may indeed be related to aging, but there is no cause and effect relationship established here. The mutations are in the nuclear genome, not the mitochondrial genome, so even if there is a cause and effect relationship (that has not been established), it would not result in maternal inheritance. Although blood cells are examined in this experiment, there is no comparison to other cells. The mutations that are observed, those that occur almost entirely in people past their reproductive years and occur in somatic cells, will not be found in gametes and will not be passed to future generations.

8. **The correct response is (A).** Diploid organisms contain two copies of every gene, one on each of two homologous chromosomes. The Law of Segregation reflects the separation of these two copies of each gene during meiosis, one copy into each haploid gamete.

9. **The correct answer is (D).** The evidence indicates that *Hb* is dominant to both of the other alleles. The relationship of the other two alleles toward each other, however, takes more thinking. The light yellow phenotype sounds like an intermediate heterozygote phenotype between the true-breeding white or true-breeding yellow phenotypes, which is consistent with incomplete dominance. **Cross 7** further supports this: the yellow hamsters in this cross are *HyHy* homozygotes, while the light yellow animals are *HwHy* heterozygotes.

SUMMING IT UP

- The science of how traits are passed from parents to progeny is called genetics.
- Genes are the fundamental unit of inheritance and are contained in the DNA genome of all living things.
- The vast majority of plants and animals reproduce sexually.
- Meiosis is reductive cell division in which a round of DNA replication in a diploid cell leads to two rounds of cell division creating haploid gametes.
- In sexual reproduction, progeny are created when cells (called gametes) from two different parents join to create the first cell of a new individual. Fertilization is the union of two gametes in this way.
- The human genome is split into linear pieces of DNA called chromosomes. There are twenty-three different chromosomes in the human genome and the number of different chromosomes is characteristic for each species.
- Recombination between homologous chromosomes in meiotic prophase I creates new combinations of alleles to increase genetic diversity during sexual reproduction.
- Alleles are slightly different versions of the same gene.
- When an individual carries two different alleles for a gene, sometimes one is expressed and the other is silent. In this circumstance, the expressed allele is called dominant and the silent allele is called recessive.
- The genotype of an individual is the combinations of alleles that it carries in its genome. The phenotype is the way the genotype is expressed in appearance and behavior.
- Maternal inheritance is a pattern in which a mother passes a trait to all children. It is caused by transmission of traits in the mitochondrial genome.
- Sex-linked traits are caused by genes on the X chromosome. Men express recessive sex-linked traits more often than women since men have only a single X chromosome.
- Human somatic cells are diploid, that is, containing two copies of each chromosome, or forty-six chromosomes total.
- In his law of independent assortment, Mendel described different traits being segregated into gametes, independently of each other.
- Understanding many genetic traits requires expanding your understanding of genes beyond Mendel's original rules.
- Linkage is the effect of having genes on the same chromosome. Genes far apart on the same chromosome display less linkage due to recombination between the genes during meiosis.

The Genetic Code

chapter 10

OVERVIEW

- The structure of DNA and genes
- DNA replication
- Gene expression
- Transcription
- RNA processing
- Translation
- The effect of mutation
- Genome sequencing
- Current biotechnology
- Viral and bacterial genetics
- Summing it up

Knowledge of the role of DNA as the genetic material, the "stuff" of life, is so common today that it is hard to believe that scientists were still debating the nature of the genetic substance only a few decades ago. Many thought that DNA was too simple chemically to be responsible for the enormous task of providing genetic instructions. They believed that genetic material had to stem from more sophisticated molecules such as proteins. Fortunately, several experiments that are now considered classics helped steer this debate in the right direction.

Many of these experiments were performed on simple genetic systems such as viruses. Although the role of DNA as genetic material was unknown in the 1930s and 1940s, scientists did know at the time that viruses could transmit genetic instructions. Alfred Hershey and Martha Chase, for example, radioactively labeled bacteriophage viruses either with S^{35} proteins or P^{32} DNA, and then looked to see which label they would find inside bacteria that had been infected by these viruses. Whatever was acting as genetic material and reprogramming bacteria with viral genes needed to enter the bacteria. They discovered that the viral nucleic acids, not the viral proteins, entered the bacteria—and therefore that the proteins could not have been the gene carriers. The genes had to reside in the DNA.

THE STRUCTURE OF DNA AND GENES

When scientists began to realize that DNA is genetic material, they sought to determine exactly how it played this role. The answer lies in DNA's construction. The fundamental building blocks of DNA are nucleotides. Each nucleotide contains three parts:

1. a sugar (deoxyribose)
2. phosphates
3. nitrogenous bases adenine (A), guanine (G), thymine (T), and cytosine (C)

DNA is a polymer built out of nucleotide subunits strung together. The sugar and phosphate groups form the backbone of a strand of nucleotides, attached 3' to 5' on each sugar, with the nitrogenous bases projecting from the sides.

Although DNA is made of polymerized strands of nucleotides, the strands do not resemble a string of beads as you might imagine. Rather, each strand in DNA is directional, and its direction is indicated by the orientation of the sugars in its backbone. The deoxyribose sugars in the backbone of each strand are connected to the rest of the chain by their 3' and 5' hydroxyl groups. At one end of the DNA strand, the 5' group in the last sugar is exposed; at the other end of the strand the last sugar in the 3' group is exposed. The direction of the strand is often described as 5' to 3' based on the orientation of the ends of the strand (and all the sugars in between). A key fact (and a very testable one) about DNA polymerases is that they only polymerize the new addition of nucleotides in one direction, from 5' to 3'. This means that they can only add nucleotides to a free 3' group.

Q Once a string of nucleotides is synthesized into a polynucleotide, is it a complete DNA molecule?

A No. DNA is a double helix and almost always consists of two polynucleotide strands wrapped around one another.

A single polynucleotide strand of DNA does not usually exist on its own. Each strand of DNA pairs with another strand, and the two wrap around each other in the famous double helix shape. This structure was elucidated by James Watson and Francis Crick with the help of Rosalind Franklin's X-ray diffraction photos.

In the double helix, each of the nitrogenous bases in one strand is always paired precisely with another base in the other strand: adenine is paired with thymine and guanine with cytosine. These pairs hold onto each other via hydrogen bonds: Two hydrogen bonds hold AT together and three hydrogen bonds hold GC together. These matching pairs of bases in DNA are often called **base pairs.** For example, you would count the number of bases in the length of a gene as "2000 base pairs" rather than as single bases. The two strands in the DNA double helix run in opposite directions (i.e., are "antiparallel") and are not covalently attached. The hydrogen bonds holding the strands together can be "melted" and separated, although it takes energy (ATP or heat) to do so.

Q If a strand of DNA contains the nucleotide bases ATCGTA, reading from 5' to 3', what is the sequence of its partner strand, from 5' to 3'?

A DNA strands in the double helix run antiparallel to one another. The partner strand would be:
5'-ATCGTA-3'
3'-TAGCAT-5' (partner, read 3' to 5')
To read it 5' to 3', turn it around: 5'-TACGAT-3'

The double helix shape of DNA helps to ensure the proper pairing of bases and maintains the integrity of the genetic information DNA holds. Bases that do not pair correctly (for instance, A and G) do not fit into the double helix and will not form the correct hydrogen bonds.

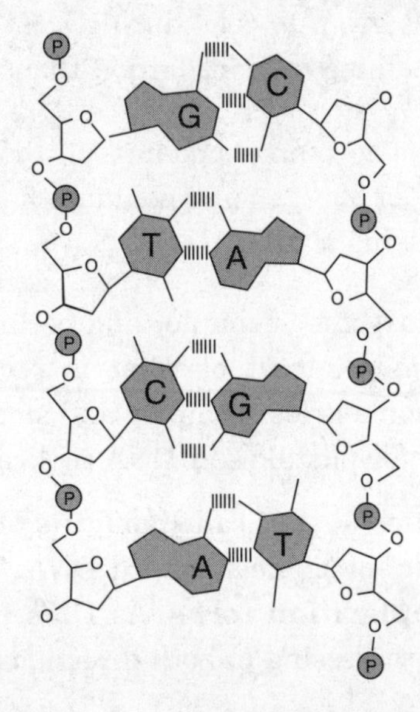

When Watson and Crick published their description of the configuration of DNA in 1954, the structure immediately suggested how DNA contains genetic information and how the information is copied as part of DNA replication during mitosis or meiosis the behavior of which was already known through cytological studies. Since each base in one strand of DNA partners precisely with a matching base in the other strand, the DNA double helix contains two copies of the information in each gene. If the two strands are separated, each DNA strand is used as a "template" to assemble a new copy of the partner strand, producing two complete and exact copies of the original.

DNA REPLICATION

To start replication, a DNA double helix unzips to create a small, single-stranded region. This single-stranded DNA is used as a template for the assembly of new DNA by pairing bases with those in the older single strand and then connecting the new base links. After DNA replication, if the new strands are labeled, you can see that each double helix contains one-half old DNA and one-half new DNA. This form of DNA replication is called **semiconservative**, because one-half of the old DNA has been "conserved."

Q After replication is complete, in double-stranded DNA, the amount of guanine matches the amount of which nitrogenous base: cytosine, adenine, or thymine?

A It matches the amount of cytosine, because G always base-pairs with C.

Q If radioactive thymine is given to cells during DNA replication, will one copy of the genome end up containing more radiation than the other copy?

A They will both contain the same amount of radiation. In semiconservative replication, both copies of the genome contain the same amount of new DNA synthesized during DNA replication.

Because DNA replication in prokaryotes is simpler and better understood than that of more complex organisms such as humans, most of what is presented here will be based on prokaryotes. The prokaryotic genome is less complex and smaller than that of more complex organisms; thus, the "price" paid for mistakes in DNA replication is perhaps not as dire.

All organisms begin DNA replication at sites called **origins of replication**. At the origin, the DNA opens up to create a "bubble" of single-stranded DNA. The *y*-shaped junctions on both sides of the bubble are called **replication forks**. As DNA replication progresses, it moves outward, with replication forks progressing in both directions.

Q When bacteria replicate their genome, is the entire bacterial genome single-stranded before replication begins?

A No. Because replication starts at a replication fork and moves progressively through the genome, only one section of a genome is single-stranded at any given moment during DNA replication.

Enzymes that synthesize DNA using single-stranded DNA as a template are called **DNA polymerases**. A DNA polymerase adds new nucleotides one at a time to a DNA polynucleotide chain, making sure that each new base matches its partner base in the template strand. The polymerization reaction uses the hydrolysis of high-energy phosphate groups from the nucleotides as energy to keep the reaction going. Each nucleotide monomer added to DNA includes three high-energy phosphate groups, such as those in the energy carrier ATP. When the monomer is added, two phosphates are released together as pyrophosphate, and the pyrophosphate is further cleaved to become inorganic phosphate. The hydrolysis of these phosphodiester bonds is the energy that drives DNA polymerization.

Q Without ATP hydrolysis, is DNA hydrolysis thermodynamically spontaneous?

A Yes. DNA hydrolysis has a negative ΔG, and occurs spontaneously without additional energy (an enzyme would immensely speed up this reaction). DNA synthesis, the reaction going in the other direction, is not spontaneous with extra energy from ATP hydrolysis.

Q Can DNA polymerase fill in the following single stranded region:
5'-AACTGTCGATTAGT-3'
3'-TTGACA-5'

A No. This would require that the polymerase add nucleotides from 3' to 5', which it cannot do.

Not only does each strand in DNA have a direction, but the two strands partnered in DNA are also oriented in opposite directions. While one strand in each double helix is oriented from 5' to 3', the other strand has an **antiparallel orientation** relative to its partner. This has important ramifications for how DNA replication occurs. When a replication fork opens up as part of DNA replication and separates the two strands of DNA, one of the strands is oriented 3' to 5' and the other strand is oriented from 5' to 3'.

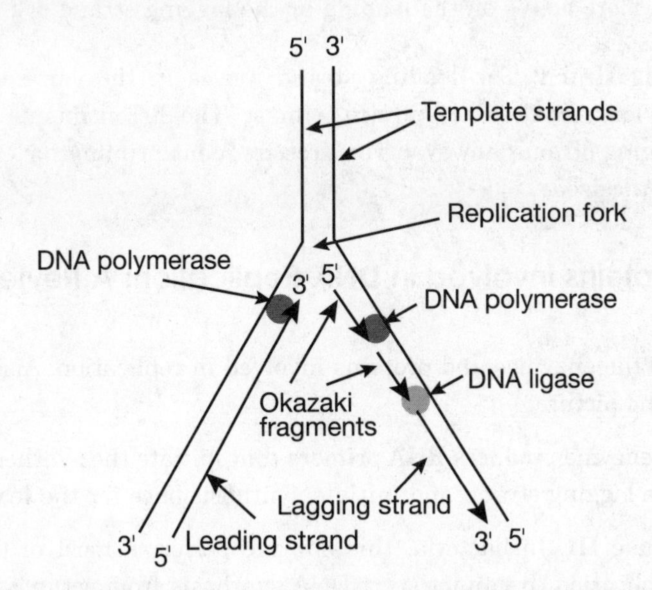

On the **leading strand** at the replication fork, synthesis of new DNA moves from 5' to 3', in the same direction as the replication itself. Synthesis on the other strand, the **lagging strand**, presents a complication, however. Since this strand is oriented in the direction opposite the leading strand, 5' to 3' DNA synthesis must move in the direction opposite that of the replication fork. Once the replication fork opens up and exposes a new stretch of template on the lagging strand side, DNA polymerase fills in this side as well, moving away from the fork. In this way, synthesis of the lagging strand creates discontinuous short stretches of DNA called **Okazaki fragments**. After DNA polymerase has completed its job, another enzyme, **ligase,** fuses the Okazaki fragments to make the lagging strand a single long piece of DNA, just like the leading strand.

Q Do Okazaki fragments occur on the leading strand or lagging strand of DNA replication?

A They occur on the lagging strand, where synthesis is discontinuous in short stretches of DNA.

Although DNA polymerases are responsible for extending DNA along a single-stranded piece of template, they cannot start DNA synthesis without an existing region of double-stranded nucleic acid. An enzyme called **primase** synthesizes RNA on top of the DNA template to create a short region of RNA/DNA hybrid. On the lagging strand, primase must start every new fragment that is created as the replication fork progresses. This "primes" the process for DNA polymerase to continue.

The RNA primers are not permanent, however. After primase and **DNA polymerase III** processes are complete, another enzyme—**DNA polymerase I** in prokaryotes—replaces the RNA primers with DNA. Ligase completes the job by "sealing" the last remaining link between nucleotides in the lagging strand. The fully replicated stretch of double-stranded DNA is then complete.

Q Is primase more active on the leading or the lagging strand of DNA replication?

A The lagging strand. The leading strand moves in the same direction as the replication fork and does not require primase. The discontinuous DNA replication on the lagging strand, however, requires frequent priming for each new stretch that is synthesized.

Enzymes and Proteins Involved in DNA Replication: A Review and Extension

Let's pause to recap the enzymes and proteins involved in replication. Also, we'll introduce a few more to fill in the picture.

- **Primase.** This enzyme produces RNA primers that initiate the synthesis of each Okazaki fragment on the lagging strand and initiate synthesis once for the leading strand.
- **DNA polymerase III**. In bacteria, this enzyme performs most of the DNA synthesis during DNA replication. It cannot start DNA synthesis from scratch on single-stranded DNA, and it only produces new DNA from 5' to 3'.
- **DNA polymerase I**. In bacteria, this enzyme removes RNA primers and replaces them with DNA. DNA polymerase I also works from 5' to 3'; it requires ligase to seal the last link between nucleotides.

- **Ligase.** This enzyme seals phosphodiester bonds between neighboring nucleotides in double-stranded DNA.
- **Topoisomerase.** DNA consists of two strands wrapped around one another in a double helix. When DNA opens to create single-stranded regions that are templates in DNA replication, the opened region must be unwound. This unwinding in one region of DNA creates tighter coiling in neighboring regions. The enzyme topoisomerase relieves this torsional stress around the replication fork so that replication can proceed.
- **Helicase.** Replication requires energy to unzip each new stretch of DNA. The protein helicase unwinds double-stranded DNA to make it single-stranded.
- **Single-stranded binding proteins.** After helicase unwinds a piece of DNA into a single strand, the single-stranded sections must be prevented from rebinding and becoming double-stranded before replication is complete. Single-stranded binding proteins bind to the single-stranded regions to stabilize them until DNA polymerase can do its work.

Q Does ligase work before or after DNA polymerase I in each stretch of DNA being produced on the lagging strand during DNA replication?

A Ligase catalyzes the last step in DNA replication. First, primase makes the RNA primer, then DNA polymerase III synthesizes a stretch of DNA. Then DNA polymerase I removes the RNA primer and replaces it with DNA, after which ligase fuses the DNA fragments together into one long DNA strand.

Proofreading and Repair

DNA has perhaps the most important function of all molecules. It must correctly instruct the cells of the body, direct the production of all RNA and proteins, and accurately pass down genetic instructions from generation to generation. The consequences of errors can be deadly, so accurate DNA replication is essential for a healthy organism. For this reason, organisms have evolved a complex set of molecular machinery to "proofread" DNA when it is created and to repair errors in DNA when they elude the proofreading process or occur after DNA replication is complete.

Ensuring accurate DNA replication is a two-step process:

1. At the time of replication, DNA polymerase adds each new nucleotide at the 3' end of the lengthening polynucleotide chain and then checks the newly added nucleotide for errors. It ensures that As are matched with Ts and Cs are matched with Gs. When all is correct, the polymerase moves to the next nucleotide.

   If an error is detected, however, the polymerase can repair the mistake. In addition to having 5' to 3' polymerase activity, DNA polymerase III has 3' to 5' exonuclease activity—this means that it can remove the last nucleotide in the event of an error and can try again to add the correct base. This is the process called **proofreading.**

2. Even with the proofreading process, errors may slip through on occasion—once every 100,000 base pairs or so. The cell has DNA repair systems that scan the double helix for irregularities in shape in locations where bases have not paired correctly. To repair these

errors, the nuclease enzyme first cuts the DNA at the site of the problem and then removes the bases that are damaged or have been paired incorrectly. DNA polymerase fills in the gap, and ligase reseals it.

Q Does DNA polymerase III back up ten nucleotides to fix a mistake made during DNA replication?

A No. It checks the last nucleotide during proofreading, rather than scanning the whole DNA strand for errors.

Aside from errors that occur during DNA replication, DNA damage by other factors may also lead to errors in the DNA sequence. Chemical mutagens can react with DNA to alter its structure, preventing genes from being copied correctly. Radiation can damage DNA by breaking bonds with energized particles that strike the DNA molecule. The same repair mechanism we just reviewed can repair this type of damage as well. If the repair mechanism fails, however, additional errors can occur and in some organisms this could lead to cancer. If DNA damage is severe, eukaryotic cells will halt the cell cycle, preventing damaged DNA from being replicated and perhaps causing the cell to enter **apoptosis**—programmed cell death.

Telomerase and Telomeres

One last detail you'll need to remember about DNA replication concerns the ends of eukaryotic chromosomes. Each eukaryotic chromosome is an enormously long, linear DNA molecule with two free ends. If DNA polymerases function in only one direction (from 5' to 3'), how are the ends of chromosomes replicated? DNA polymerase alone can't do the job, and if the chromosome ends aren't replicated, the chromosomes will grow shorter with each round of replication.

The ends of chromosomes are replicated by a unique enzyme called **telomerase**. Special DNA sequences called **telomeres** at the ends of eukaryotic chromosomes are also involved. Telomeres contain repeated DNA sequences, with the same short sequence of bases repeated many times at the end of each chromosome. Telomerase acts as a unique DNA polymerase to synthesize additional repeats using its own RNA, which it carries as a template. This allows telomerase to accomplish what other DNA polymerases cannot.

Telomerase is active in germ-line cells, where it extends telomeres during gamete formation, ensuring that telomeres retain their length from generation to generation. Most somatic cells lack telomerase, however, so with increasing age and cell divisions, their telomeres shorten. Degraded ends of chromosomes that lack telomerase place a limit on the number of divisions cells can undergo. Division of many cancer cells, for example, is limited by the loss of DNA at the ends of their chromosomes, although some cancer cells do express telomerase and thus overcome this barrier. Future cancer treatments could involve the use of telomerase inhibitors.

Q If the expression of telomerase could be blocked in cancer cells, how might this affect the progression of the disease?

A The telomeres of the cancer chromosomes would become progressively shorter until the chromosomes would become too short for the cancer cells to continue proliferating.

Q Do prokaryotes need a telomerase-like enzyme?

A No. Prokaryotes have circular genomes, not linear chromosomes. A circular DNA has no ends that will degrade with repeated rounds of replication, so no telomerase is necessary.

GENE EXPRESSION

Genes guide the activities of living things and pass traits from one generation to the next. They are inherited instructions written in the sequence of bases along the length of each DNA molecule. Each gene carries instructions for producing a protein (or in some cases, an RNA), leaving it to proteins to execute the message written in the DNA. In **transcription,** the message carried by a gene in DNA is copied into an RNA version. The RNA copy of the message is then read by a different set of molecular machinery and is used to produce the protein encoded by the gene. This process is called **translation**.

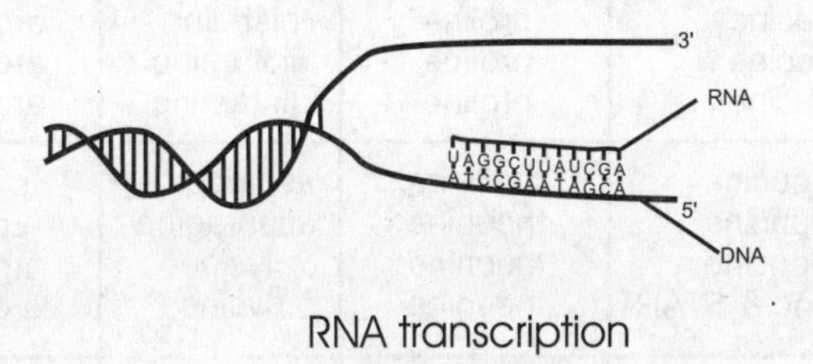

RNA transcription

One of the major functions of a cell is to read genes and use them to produce proteins. But this must happen at the right time and in the right place. Not all genes in human DNA produce their corresponding protein at all times; that would be chaos. What distinguishes a heart cell from a liver cell, for example, is that unique sets of genes are active in each tissue type, producing different sets of proteins in heart, liver, and all other tissue types. The selective reading of genes—turning them "on" and "off"—is called the **regulation** of gene expression. This occurs at many levels, beginning with transcription of genes and including RNA processing control, RNA stability, RNA translation into protein, and protein stability.

How is a ribosomal RNA translated?

Ribosomal RNA is not translated into proteins. The RNA itself is built into ribosomes.

The Genetic Code

In what is called the **genetic code**, each gene in DNA encodes a protein using an arrangement of nucleotide bases along its length. The genetic code is written using groups of three nucleotides that encode each amino acid added to a protein. For example, the three nucleotides AGG (adenine-guanine-guanine) in a gene indicate that the amino acid arginine is to be added at this position in a protein. Each three-nucleotide group of bases in a gene is called a **codon**. The codons for each amino acid can be determined from a table that includes all possible combinations of the three nucleotides. ("U" is often used instead of "T" in such tables to represent the uracil that is incorporated in RNA when the gene is transcribed and translated.)

	Second Letter				
First letter	U	C	A	G	Third letter
U	phenylalanine phenylalanine leucine leucine	serine serine serine serine	tyrosine tyrosine STOP STOP	cysteine cysteine STOP tryptophan	U C A G
C	leucine leucine leucine leucine	proline proline proline proline	histidine histidine glutamine glutamine	arginine arginine arginine arginine	U C A G
A	isoleucine isoleucine isoleucine methionine & START	theonine theonine theonine theonine	asparagine asparagine lysine lysine	serine serine arginine arginine	U C A G
G	valine valine valine valine	alanine alanine alanine alanine	aspartate aspartate glutamate glutamate	glycine glycine glycine glycine	U C A G

mRNA codons and accompanying amino acids

Twenty different amino acids go into building proteins, so there must be at least one codon for each. In addition, the genetic code includes codons that indicate the position where production of a protein starts and where it stops.

Q If the genetic code is written in groups of three nucleotides, how many codons can there be?

A There are four different bases in DNA, so with three nucleotide codons, the number of possible codons is $4 \times 4 \times 4$, or 64.

Q Are there any possible combinations of nucleotides that do not have meaning in the genetic code?

A All 64 possible combinations carry meaning.

With 64 codons and 20 amino acids, as well as the codons used to start and stop translation, some redundancy exists in the genetic code, and some amino acids can be encoded by more than one codon. As many as six different codons can code for the same amino acid—arginine. This redundancy is also referred to as **degeneracy**. When multiple codons are used for the same amino acid, they often differ only at the third position. For example, the codons CGT (or CGU in RNA), CGC, CGG, and CGA all signal the addition of arginine, and they differ only in the last position. As a result, mutations in the final nucleotide of codons often do not change the sequence of proteins. This reduces the negative impact of mutations. In other cases, amino acids that are structurally related—such as the hydrophobic amino acids leucine, isoleucine, and valine—also have similar codons. Mutations may result in a similar amino acid being substituted in a gene that might not greatly impact the protein's structure. Such mutations are called **conservative mutations**.

Since the genetic code is written in groups of three, the location where the gene begins determines the way the message is read in a given string of nucleotides. This results in different **reading frames**. For example:

5'-AAAGGGCCCAAAGGGCCC-3'

If the reading frame in this gene starts with the first nucleotide, then the first three codons are AAA, GGG, and CCC, and the corresponding amino acids that will be added to a protein will be lysine, glycine, and proline. If the reading frame starts with the third nucleotide, however, then the gene will be read entirely differently, and the first three codons would be AGG, GCC, and CAA, adding arginine, alanine, and glutamine to a protein. Having the correct reading frame is essential for an accurate reading of a gene's message. Reading the wrong reading frame can produce no protein, or a completely different protein, from the same piece of DNA.

Q A single nucleotide is inserted into a gene. How does this affect the reading frame of the gene? What if three nucleotides are inserted?

A Inserting a single nucleotide muddles the reading frame. Inserting three nucleotides restores the original reading frame, with an extra amino acid inserted in the protein (unless the extra codon is a stop codon).

With a few minor exceptions, all living things use an identical genetic code. The highly conserved nature of the genetic code throughout all domains of living creatures suggests that it evolved early in the history of life on Earth and was probably used in its present form in the progenote—the genome of the organism that gave rise to all of life.

TRANSCRIPTION

The first step to reading a gene is producing an RNA copy of the gene's message. Enzymes called **RNA polymerases** perform this activity, using DNA as a template to produce an RNA copy of the gene. RNA polymerases are similar to DNA polymerase in that they use DNA as a template and synthesize nucleic acids in a strictly 5' to 3' direction. Unlike DNA polymerase, however, RNA polymerases do not need a primer to begin, and they have no proofreading capacity. RNA polymerases also use nucleotides containing ribose instead of deoxyribose (found in DNA) and use uracil instead of thymine. In general, prokaryotes have one type of RNA polymerase, while eukaryotes have three: RNA polymerase I produces ribosomal RNA, RNA polymerase III produces tRNAs involved in translation, and RNA polymerase II transcribes genes that produce proteins.

Q How are mistakes in RNA transcripts corrected?

A They are not corrected. While mistakes in DNA replication may lead to severe consequences, errors in RNA are not as dire. This is because RNA molecules are constantly created and degraded, so RNA polymerase requires less "fidelity" and need not be as accurate as DNA replication.

The quantity of DNA in a cell can be enormous. In human cells, 3 billion nucleotides of DNA make up the human genome. RNA polymerase detects gene locations among this giant population of nucleotides by seeking special DNA sequences called **promoters**. These sequences are much more complex in eukaryotes than in prokaryotes. In prokaryotes, promoters are clearly defined, and they operate near the transcriptional start site. Many, but not all, eukaryotic promoters contain a TATA sequence just "upstream" of the point where RNA polymerase begins transcription.

Promoters often contain DNA sequences that direct the binding of sequence-specific transcription factors; these attract RNA polymerase to the gene. These proteins binding on promoters "flag" the genes for the polymerase to transcribe. Before a eukaryotic RNA polymerase will begin transcription, several factors must bind on a promoter near the transcriptional start site. Once this is done, RNA polymerase opens the DNA double helix, accesses the DNA strand it reads as a template, and begins creating an RNA transcript of the gene.

Q If radioactive thymine is given to cells, will RNA be radiolabeled?

A No. RNA contains uracil, not thymine.

Q Does RNA polymerase synthesizing RNA move in both directions away from a promoter, like the two replication forks that move away from an origin of DNA replication?

A No. The promoter instructs the polymerase where to start transcription and in which direction to go on the gene being transcribed. It travels in only one direction from the promoter and start site.

Eukaryotic genes have promoters near the transcriptional start site and DNA elements farther from the start site. The more distant elements, called **enhancers**, also regulate transcription. Enhancers probably interact with promoters by bending the intervening DNA and bringing them closer, even if they are distant from one another along the sequence of bases in DNA. The protein binding to DNA in promoters and enhancers regulates gene expression at specific times in development, in specific cells and tissues, and in response to specific stimuli such as hormones.

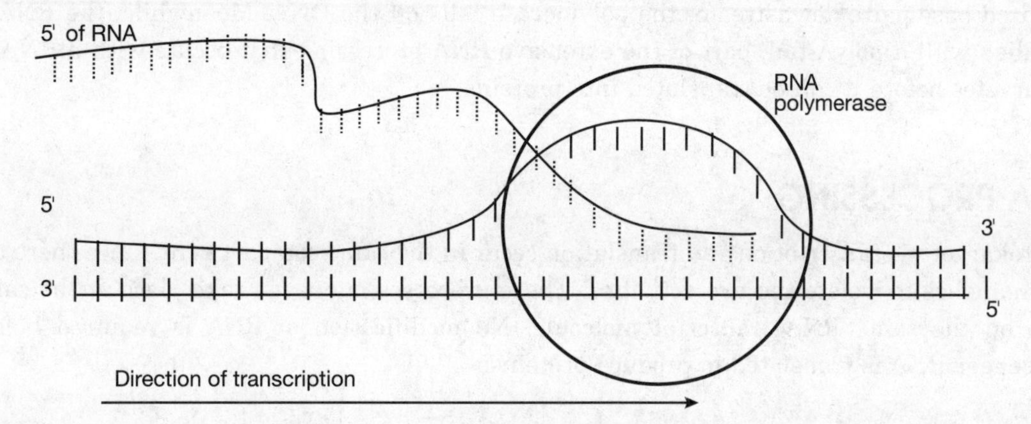

Although genes occur in double-stranded DNA, only one of the DNA strands in a gene is actually used as a template by RNA polymerase. When the sequence of bases in RNA is written, it is commonly written from the 5' end of the RNA toward the 3' end of the molecule, in the same order in which it is made by RNA polymerase. Other than the fact that it contains uracil instead of thymine, the RNA made by RNA polymerase is complementary to the DNA template strand and will be read with the same sequence of bases as the nontemplate strand. The nontemplate strand in DNA is also called the **coding strand** because it lines up with the RNA produced from the gene. Another name for it is the **sense strand**, in contrast to the template strand which may be called the **antisense strand**.

Example:

RNA: 5'-ACUUGGCCUUAA-3' RNA transcript

DNA: 5'-ACTTGGCCTTAA-3' nontemplate strand (also called sense or coding strand)

3'-TGAACCGGAATT-5' template strand (also called antisense strand)

Q Is the nucleotide sequence of the sense strand the same as the mRNA sequence?

A They line up together and are the same, except that RNA contains uracil instead of thymine.

Once RNA polymerase has started reading a gene and producing its RNA copy of the gene's message, it continues to move along the DNA and read the rest of the gene, creating an increasingly longer **transcript** of the gene (the RNA produced by RNA polymerase during transcription). As the polymerase moves along DNA, it transiently opens a new stretch of double-stranded DNA, which closes again after the polymerase passes. The RNA copy trails behind the polymerase. Transcribing genes with more than one RNA polymerase at a time produces many transcripts of a gene and increases protein production.

In prokaryotes, RNA polymerase continues lengthening the RNA transcript until it reaches a DNA sequence called the **terminator**. It then falls off of the DNA template, ready to start another cycle of transcription. In eukaryotes, the process is more complex. When eukaryotic RNA polymerase II transcribes through a polyadenylation sequence, AAUAAAA, it keeps transcribing endlessly, but the mRNA is cleaved behind it by another enzyme. Several hundred base-pairs downstream, the polymerase falls off the DNA. Meanwhile, the RNA is modified with a polyA-tail, part of the extensive RNA processing that occurs with mRNA in eukaryotes before it can be translated into proteins.

RNA PROCESSING

In prokaryotes, transcription and translation occur in the same compartment, since there are no organelles to isolate the two activities. These processes may even take place at the same time on the same RNA transcript molecule. No modification of RNA is required before messenger RNA is translated to produce proteins.

Q Is mRNA double-stranded or single-stranded?

A mRNA is primarily single-stranded, without a specific secondary structure.

In the case of eukaryotes, RNA is extensively modified and processed in the nucleus before it is ready for translation in the cytoplasm. One of the first modifications takes place at the front end of the pre-mRNA soon after transcription begins. A methylated guanine is added to the RNA as a 5'-cap. This modification protects RNA from degradation and helps the translation machinery recognize mRNA. The 5'-cap also helps export RNA from the nucleus to the cytoplasm.

Q Can ribosomes start translating a eukaryotic RNA while it is still being transcribed?

A No. Prokaryotes can do this because translation and transcription are not compartmentalized, but in eukaryotes, transcription and translation are entirely separate processes.

Another modification of pre-mRNA is the addition of a poly-A tail at the other end of the transcript. After the nascent RNA is released by RNA polymerase II, a different enzyme adds a long string of A nucleotides onto the 3' end of the RNA. This string follows the poly-adenylation signal. The poly-A tail also helps stabilize messenger RNA against degradation via **ribonucleases**—enzymes that are constantly chopping RNA into small bits. Like the 5'-cap, the poly-A tail may also help with recognition of mRNA in translation and export from the nucleus.

Q Are the long string of As on the 3' end of eukaryotic mRNA encoded in the genome?

A No. They are added after transcription is complete and the RNA is no longer being read from the DNA template in genes.

Q Where does poly-A tail addition take place in the cell?

A All eukaryotic RNA processing takes place in the nucleus, before mature mRNA is exported to the cytoplasm for translation.

The next big step in eukaryotic RNA processing before translation can occur is called **splicing**. Eukaryotic genes are often split into sections that code for proteins, called **exons**, and intervening sections that do not, called **introns**. The DNA sequence of a gene, and the RNA transcript it first produced by RNA polymerase, can contain a number of alternating exons and introns. All of the introns must be removed from the pre-mRNA before the pre-mRNA can be translated. Complex assemblies of RNA and protein called **spliceosomes** help to splice introns from RNA, excising each intron and rejoining surrounding exons. The RNA component of spliceosomes helps recognize junction sequences in the pre-mRNA on both sides of an intron that marks it for removal.

Transcribing introns may seem like a waste of a cell's precious resources, because it produces a long RNA only to cut out and throw away whole sections of it. Although prokaryotes lack introns today, they may have had them in their evolutionary past. The introns may have evolved away from what appears to be a wasteful use of resources.

Introns and splicing may help provide greater flexibility in gene expression and evolution in higher organisms. Over many generations, exons may accidentally connect in fortuitous new ways in the genome, creating new genes and proteins. The number of genes in the human genome is smaller than many scientists had anticipated, but the number of proteins may still be high, in part because of the production of multiple versions of a protein from the same gene via alternative splicing.

A surprising finding that emerged from studies of RNA splicing is that some organisms have self-splicing introns, capable of catalyzing their own removal from genes. The observation of enzyme activity in RNA molecules, both natural and artificial, has since been expanded.

 How can RNA molecules act as enzymes instead of proteins?

 Enzymes are defined not by what they are made of, but by what they do. If an RNA molecule can assume a certain conformation that forms an active site, allowing it to reduce the activation energy for a reaction, then it qualifies as an enzyme.

Degradation of mRNA provides another means of regulating gene expression. RNA does not last in the cell forever. Even as new mRNA is being made, old mRNA is being destroyed by enzymes called nucleases. The amount of mRNA in the cell at any moment is a balance between how fast mRNA is made and how fast it is degraded. mRNA lacking a 5' cap or poly-A tail is broken down quickly. Specific mRNAs can also be broken down more rapidly than average. A mechanism called small interfering RNA (siRNA) has recently been identified and used by biologists to block the expression of a gene by causing selective degradation of its mRNA.

TRANSLATION

In protein translation, the genetic message in messenger RNA is used to produce the protein encoded by a gene, translating the language of nucleotides into the language of amino acid sequences. In prokaryotes, translation occurs in the same cellular compartment as transcription (in the cytoplasm) and translation of an mRNA can begin while the mRNA is still being transcribed. A string of ribosomes on an mRNA, called a **polyribosome**, can even translate an mRNA while it is still being transcribed. In eukaryotes, transcription and mRNA processing occur in the nucleus, but translation occurs in the cytoplasm and mRNAs must be exported from the nucleus before they are translated. Polyribosomes can form on mRNA in eukaryotes as well, although they do not begin until the mRNA is fully transcribed and exported from the nucleus.

To translate an mRNA, the cell must match each three-nucleotide codon in mRNA with the correct amino acid the particular codon specifies. A special type of RNA molecule called **transfer RNA (tRNA)** matches amino acids with the correct codon. tRNAs are relatively small RNAs, about 80 nucleotides long, and they have complex secondary structure, including base-pairing within the RNA molecule to form regions of RNA-RNA double helices.

Q How might heat affect the structure of tRNA molecules?

A With enough heat, the double-stranded regions holding the tRNA folded structure together would melt, and the structure would unfold.

Each tRNA includes a three-nucleotide **anti-codon** complementary to one or more codons in mRNA. Each tRNA is also activated to have a specific amino acid covalently bound to it; the amino acid matches the codon that the tRNA translates in the genetic code. During translation, the pairing of codons with anti-codons helps position the amino acids linked to tRNAs to a growing polypeptide chain.

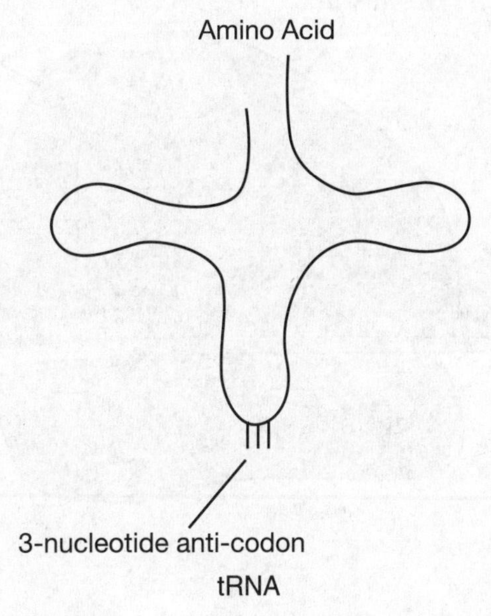

tRNA

Enzymes called **aminoacyl-tRNA synthetases** activate tRNAs, attaching the correct amino acid to each tRNA. There are twenty different aminoacyl-tRNA synthetases that attach the correct amino acid to each tRNA and forty-five different tRNA molecules. Not counting the three stop codons, a total of sixty-one codons encode amino acids, so some tRNAs recognize more than one codon. The ability of some tRNAs to recognize more than one codon exists because some codons differ at the last (third) position for the same amino acid. This ability of tRNAs to bind more than one codon through flexible recognition of the third nucleotide in some codons is called **wobble**.

Q Can an aminoacyl-tRNA synthetase attach more than one amino acid to a given tRNA?

A No. It would result in an inappropriate amino acid being inserted into a protein during translation. Much ATP is burned to prevent this from occurring. There are more tRNAs than there are aminoacyl-tRNA synthetase enzymes to activate them, so some of the synthetases can activate more than one tRNA. However, a given synthetase does not recognize more than one amino acid.

Ribosomes are large molecular assemblies of protein and RNA that bind all of the elements of translation and help them move correctly through protein translation. The RNA component of ribosomes is called **ribosomal RNA or rRNA**. Each ribosome is built from two subunits, one large and one small. The ribosome has a binding site for mRNA and three different positions binding tRNAs, matching anti-codons on tRNA with codons in the mRNA. The ribosomal P site binds a tRNA with the growing polypeptide, the A site binds the next aminoacyl-tRNA to be added, and the E site contains tRNAs that have already transferred their amino acid to the growing protein and are ready to be released from the ribosome.

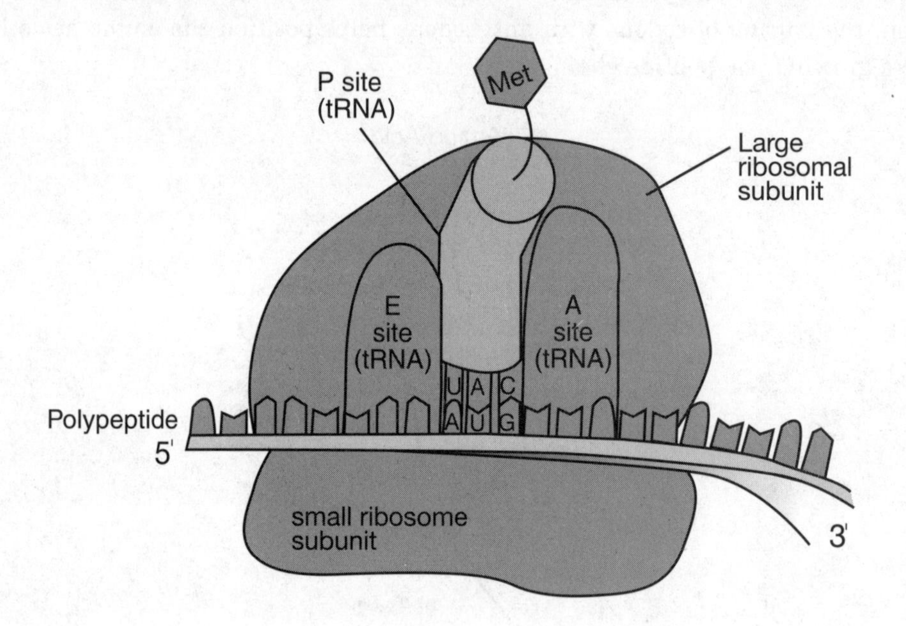

Translational Initiation

To start translation, an mRNA molecule binds to a small ribosomal subunit and to the aminoacyl-tRNA recognizing the translation initiation codon in the mRNA. Ribosomes recognize the codon AUG in mRNA as the initiation codon, which codes for methionine. Therefore, most proteins are translated with methionine as the first amino acid. This methionine is sometimes removed from the protein later on.

The first amino acid in a protein is called the **N-terminus** because it has a free amino group. Protein translation always proceeds from the N-terminus toward the **C-terminus**, with a free carboxy group at the other end of a polypeptide.

 Is the C-terminus of a protein generally charged at physiological pH?

 Physiological pH is about 7.4, which is almost neutral, and the C-terminus is a carboxylic acid group that will be deprotonated at this pH, so this gives it a –1 charge.

Q Is the methionine encoded by the translation initiation codon found at the N-terminus or the C-terminus of proteins?

A The methionine is the first amino acid in the newly synthesized protein, so it is found at the N-terminus.

Once the small ribosomal subunit is positioned with the mRNA and the tRNA at the initiation codon, the large ribosomal subunit binds to form the translation initiation complex. Additional translation initiation factors and GTP hydrolysis help form the translation initiation complex in preparation for making the first peptide bond (GTP, or guanosine-5'-triphosphate, is a purine nucleotide that acts as a substrate for RNA synthesis during transcription). In the translation initiation complex, the first tRNA with methionine is positioned at the P site. The next tRNA then binds to the second mRNA codon, positioning itself at the ribosomal A site. Two GTP molecules are hydrolyzed to ensure the accuracy of every aminoacyl-tRNA positioned at the A site and in line to be added to the growing protein chain. Once tRNAs are positioned at the P and A sites, the ribosome catalyzes the formation of a peptide bond between the amino acids bound to tRNAs at the P and A site. This transfers the growing peptide chain from the tRNA at the P site to the amino acid bound to a tRNA on the A site.

In **translocation**, the mRNA and associated tRNAs move one codon on the ribosome after the formation of each peptide bond, using hydrolysis of another GTP to move forward. The mRNA always translocates one codon at a time—from 5' toward the 3' end of the mRNA—and is translated in the same direction.

Q At which ribosomal binding site does tRNA first bind?

A The A site. From there it moves to the P site, with the growing polypeptide attached, and then to the E site to leave the ribosome after it no longer has an amino acid bound.

Q During translation, how many nucleotides does the mRNA translocate with each cycle of translation?

A The ribosome moves the mRNA three nucleotides with each cycle of translation, because there are three nucleotides in each codon.

Translation of an mRNA continues in this manner, through codon after codon, until a stop codon is reached in the mRNA. There are no tRNAs that bind to stop codons, so when a stop codon appears in the A site, a protein called a **translation release factor** binds to the stop codon and releases the growing polypeptide from the tRNA at the P site. Once the final polypeptide is released, the mRNA is released as well. The two ribosomal subunits dissociate from one another and prepare to start another cycle of translation.

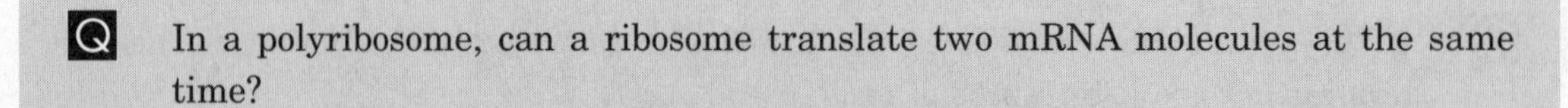

Q In a polyribosome, can a ribosome translate two mRNA molecules at the same time?

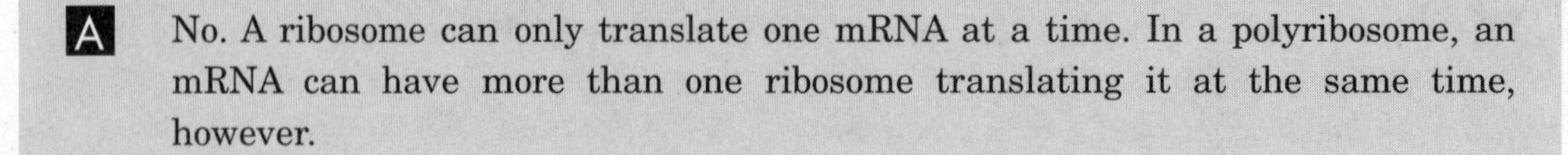

A No. A ribosome can only translate one mRNA at a time. In a polyribosome, an mRNA can have more than one ribosome translating it at the same time, however.

Regulation of translational initiation provides another opportunity to regulate gene expression. Proteins binding to an mRNA can prevent the gene from being translated, and modifying translation initiation factors can accelerate or decelerate the overall rate of translation of all mRNA molecules.

In both prokaryotes and eukaryotes, ribosomal RNA is a critical part of ribosomal structure. It also plays a key role in translation, apparently by carrying out the catalytic activity of ribosomes while the proteins in ribosomes stabilize the organism's structure. Prokaryotic ribosomes are smaller than eukaryotic ribosomes and are similar to organelle ribosomes in mitochondria and chloroplasts. Some antibiotics act specifically on prokaryotic ribosomes, inhibiting bacterial translation to reduce bacterial growth.

To function correctly, the finished protein must fold into its proper structure. Some proteins fold on their own into the proper structure; others require assistance in the form of **chaperonins** or **molecular chaperones**. Chaperonins do not force proteins to assume a specific folded structure, but they do prevent proteins from denaturing as they are folding. This creates an environment that allows the protein to find the correct folded structure.

Newly synthesized proteins may also be modified in several different ways. Some proteins are cleaved by protease enzymes at one or more sites in the primary structure. Many peptide hormones are produced by the cleavage of larger precursor proteins. Carbohydrate groups, phosphates, and coenzymes such as metal ions, which many enzymes require for function, are not encoded in the primary protein sequence in genes. Instead, they are often added to proteins after they are synthesized.

Q A modified form of tyrosine with a phosphate group on the hydroxyl (phosphotyrosine) is often found in proteins. Is there a codon for phosphotyrosine in the genetic code?

A No. Phosphate is added to tyrosine after protein translation by tyrosine protein kinases. Phosphorylation is a common form of post-translational protein modification that regulates protein activity in signal transduction cascades. It is reversible; other enzymes called protein phosphatases remove the phosphate from the protein.

Like mRNA, proteins do not last indefinitely in the cell. Over time, they wear out and must be replaced by newly synthesized proteins. In addition, proteins can be selectively tagged for destruction, so that their function is eventually switched off. Proteins linked with a small peptide called **ubiquitin** are targeted for destruction in proteosomes.

THE EFFECT OF MUTATION

DNA replication is an impressively accurate process but, even so, mistakes can occur. DNA damage by chemicals and radiation happens fairly frequently, and although most of the errors are eventually repaired, some remaining damage may cause **mutations**—heritable changes in the sequence of nucleotides in the genome. Mutations can affect one or more nucleotides, or they can change even larger sections of genes and chromosomes.

Q Is a change in the methylation pattern of a gene a mutation?

A No. This does not change the sequence of nucleotides in the genome.

Q Is a change of an adenine to a thymine in a nontranscribed and nontranslated region of a gene a mutation?

A Yes. A mutation can happen anywhere, and it does not have to change the protein encoded by a gene to qualify as a mutation.

Changes that affect a single base-pair in DNA are called **point mutations**. Most of the time, mutations in the genome that occur outside of genes and their regulatory sequences will have no consequence to cells. Mutations that change the nucleotide sequence of genes are called **silent mutations** if they have no effect on the amino acid sequence of the protein produced by the gene. For example, if the mutation changes one codon in a gene to another codon for the same amino acid, the mutation will be silent because it has no effect on the resulting protein. If a mutation causes a change in the encoded protein of a gene and changes one amino acid to another, however, the mutation is called a **missense mutation**. If a mutation creates a premature stop codon, translation of an mRNA will terminate early, thereby truncating the protein. This is called **nonsense mutation** and will usually create a nonfunctional protein.

Q What type of point mutation creates a gene associated with a recessive allele and a protein with a significantly reduced molecular weight?

A Recessive alleles are often associated with versions of a gene that lose function. A premature stop codon in a gene causes reduced molecular weight, truncating the resulting protein and causing it to lose function, so it is a nonsense mutation.

Another type of mutation involves inserting or deleting nucleotides in a gene. Either of these occurrences throws the translation machinery out of synch with the normal reading frame. This causes a **frame-shift mutation**. If a deletion removes three nucleotides, or an insertion adds three, then the normal reading frame is preserved but with one fewer or one more amino acid in the protein than before the mutation.

GENOME SEQUENCING

Scientists are currently engaged in characterizing not just single genes or proteins but the whole genome of some organisms, including humans. The Human Genome Project has already revealed that we have about 25,000 genes in our genome—far fewer than had been expected, considering the complexity of the human body. But the degree of complexity that our genes can author relies not just on the number of genes, but on mechanisms such as alternative splicing and proteolytic processing. These increase the diversity of gene products and their functions.

Q If humans have, on average, three alternative splicing versions of a gene and there are two ways that the protein translated from each transcript can be proteolytically cleaved, then how many different gene products do humans have?

A Three different splice versions of 25,000 genes would produce 75,000 different transcripts. Two different proteins produced from each transcript would produce 150,000 different gene products from these 25,000 genes. Other modifications to the human genome further increase the diversity of gene products.

Some geneticists have focused primarily on trying to identify the genes all humans carry. Others are sorting out what each specific gene does. Studying the genome has already been useful in helping identify how genes are involved in specific diseases and how individual genome variations make every human being different. On a molecular level, each of us varies at many single base-pairs from every other human. These variations are called **single nucleotide polymorphisms** or **SNPs**. It has been discovered that SNPs in genes alter the associated protein function and cause disease, and that they are also responsible for some of the many traits that make us individuals.

The sequencing of the human genome has thus far revealed much about human beings and how we evolved and spread across the planet. Examining mitochondrial DNA sequences in people around the world has allowed scientists to trace the maternal lineages of proto-humans through centuries of migration, from hominids to the origin of *Homo sapiens* in Africa and beyond.

Comparing genomes of different types of organisms can also reveal where genes themselves may have come from. Some genes are members of multigene families, related in sequence and function. One such family involves a great number of identical repeats of the same gene. The genes encoding ribosomal RNAs are an example. Hundreds of copies of these genes are repeated in clusters alongside one another on five different chromosomes. Each repeat in these clusters encodes three ribosomal RNAs that are transcribed together in one RNA and cleaved to produce the three RNAs in ribosomes. Because they are transcribed together, these RNAs are produced in the equal ratios necessary for building ribosomes.

Other gene families encode similar but nonidentical genes. For example, in the human genome, about 500 genes encode protein kinases and about 800 genes encode G-protein coupled receptor genes. The related genes in these two families did not evolve independently, converging on a common sequence. Instead, various versions of genes probably arose through a process of gene duplication, followed by mutation and evolution of various copies for varying functions. Mistakes in meiotic recombination or DNA replication can create gene duplication. When multiple copies of genes exist, mutations in one copy are better tolerated than they would be if only a single copy existed.

Gene clusters can include **pseudogenes** as well. These have strong sequence similarity to genes, but they cannot create functional protein because of mutations, such as a stop codon in the middle of the sequence. Pseudogenes are probably produced through a gene duplication event, which is followed by mutation and degradation of one nonessential copy of the gene.

Transposons

Genes and regulatory sequences take up only about 25 percent of the human genome. What about the other 2.2 billion base pairs out of the 3 billion in the genome? Much of this genetic material are repetitive DNA elements that apparently do not encode proteins but nevertheless are present in many copies. A majority of repetitive DNA—and a great proportion of the human genome—resembles transposable DNA elements, or **transposons**. Transposons can move within the genome from one site to another. The movement can occur via different mechanisms: (1) by cutting the transposon out of one place in the genome and reinserting it at another location, or (2) by copying the transposon to produce another one that can insert itself somewhere else in the genome. Most of the copies of apparent transposable elements appear to be degraded in sequence over long periods, however, and only a small fraction of transposable elements in the human genome appear to be active. Nevertheless, several cases of apparently recent transpositions and resulting mutations have been documented in humans.

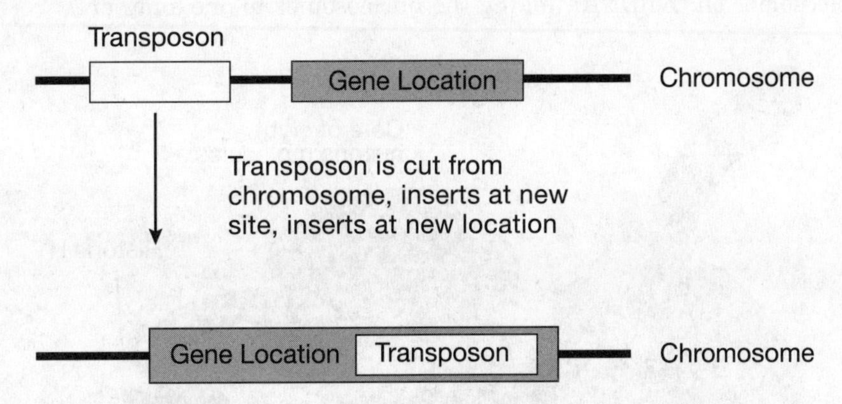

Transposons can be hundreds or thousands of base pairs long, and in some cases they are transcribed and even encode a protein that is translated. One protein product encoded within some transposons is a form of reverse transcriptase. If a transposon is transcribed, and the resulting mRNA is translated into the enzyme reverse transcriptase, this enzyme can in turn make a DNA copy of the transposon RNA. This DNA copy might then insert a new transposon elsewhere in the genome. This form of transposon is called a **retrotransposon**.

The reverse transcriptase produced in this manner might be related to the reverse transcriptase encoded in the genome of retroviruses. Retrotransposons resemble retroviruses in some ways, but they never leave the cell in which they are located. The movement of transposons in the genome probably plays a role in genomic changes occurring during evolution, including changing regulatory sequences, inactivating a gene by inserting itself into the middle of the gene, or creating new copies of genes. Usually these changes in the genome, like any mutation, will produce a selective disadvantage for the individual carrying them. Occasionally, a change that produces no change in fitness or one that even increases fitness will be propagated from one generation to the next.

Chromatin Revisited

Eukaryotic genomes cram a lot of material into the nucleus. The human genome contains about 3 billion base pairs of DNA, split into twenty-three pairs of chromosomes. This much DNA would be about 2 meters long if stretched out, but the typical diploid cell must fit this mass into a nucleus that is only a few microns wide. Not only must DNA fit into this limited space, but it must also remain available to be transcribed. This requires careful packing of the cell's genomic luggage.

To accomplish this, eukaryotic genomes are packaged with proteins in a multilevel organization called **chromatin**. The first level of organization in chromatin is the wrapping of DNA around clusters of proteins called **nucleosomes**. When chromatin is isolated from a cell and viewed under an electron microscope, it resembles a series of beads on a string. Each nucleosome has 200 base pairs of DNA wrapped around it in two loops and contains a cluster of eight positively charged **histone** proteins. There are two each of four histones in each nucleosome: H2A, H2b, H3, and H4. The positive charge of histone proteins neutralizes the negative charge of the phosphate groups in the DNA backbone. Another histone, H1, binds between nucleosomes on the DNA, linking the nucleosomes to one another.

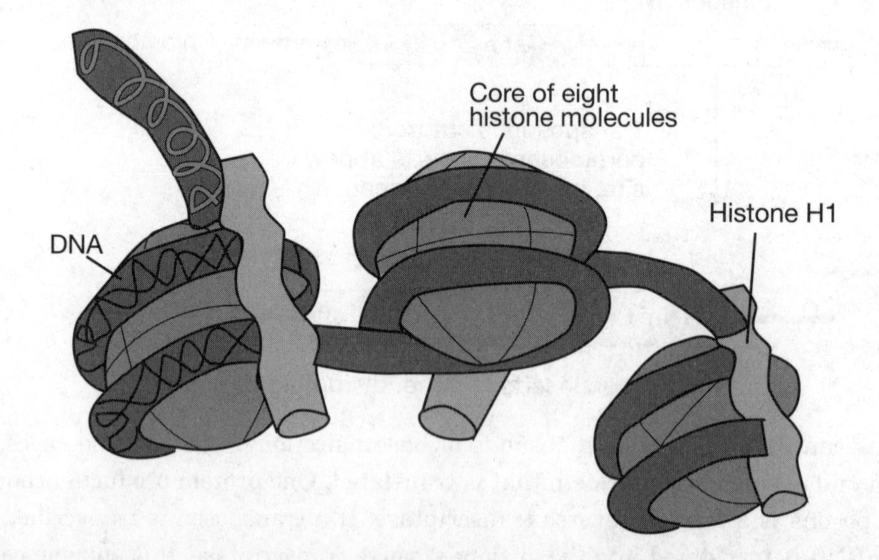

Q Which amino acids are histones likely to have in abundance?

A Lysine and arginine, amino acids with side chains that are basic and have a positive charge at neutral pH.

Nucleosomes are folded into more compact chromatin structures such as the 30 nM fiber, a condensed string of nucleosomes that are bundled up closely with one another. The 30 nM fibers condense further into large, looped domains, anchored at their base during interphase to proteins that provide nuclear structure.

In addition to regulating RNA polymerase and promoter interaction, opening chromatin to access genes is an important part of regulating gene expression. Some regions of the genome are even more tightly packed into a form known as **heterochromatin**, in which most genes are not expressed. Histones are often modified to change how tightly they bind DNA, modifying the positively charged amino acids in their tails with acetyl groups that change their charge from positive (+1 per lysine or arginine) to neutral (no charge). Without as many charges, histones bind DNA more loosely, making the DNA more accessible for proteins involved in gene expression such as RNA polymerase. DNA binding proteins that bind to promoters and enhancers will often recruit histone deacetylases to that region of DNA in the gene, helping to loosen up the DNA structure for other proteins.

Q Would an actively transcribed gene have histones that are more acetylated or less acetylated than a gene that is inactive?

A An actively transcribed gene would generally have more loosely associated histones, giving RNA polymerase and other transcription factors better access to the gene. This means the gene would have more acetylated histones.

Q Does histone acetylation regulate gene expression in bacteria?

A No. Bacteria don't have histones or chromatin.

CURRENT BIOTECHNOLOGY

Modern biology has provided a molecular understanding of genetics and evolution for biologists. It has also given them the tools to experiment with genes on a molecular level. Biologists can now identify a gene from an organism, determine which amino acid sequence it encodes, and then insert the gene into another organism. A gene from humans that encodes a protein hormone, for example, can be removed from the human genome and inserted into bacteria, which produce the hormone in large quantities so it can be used as a medicine or therapy. The manipulation of DNA and genes in this way is often referred to as **recombinant DNA technology** or **genetic engineering**.

Q How can bacteria produce proteins from a human gene if bacterial ribosomes differ so greatly from human ribosomes?

A There are differences in ribosomes, but the similarities between human and bacterial translation are more important than those differences. Most important, the genetic code of both organisms is the same. The gene must use bacterial transcription initiation sequences when inserted into bacteria and it cannot contain introns, but the gene sequence itself can be translated in almost any species.

Recombinant DNA

Bacteria are convenient to work with in the lab because they grow rapidly and are much easier to cultivate than eukaryotic cells. To work with genes in bacteria, biologists will often insert the gene into extra-chromosomal DNA elements called **plasmids**. Plasmids are small, circular DNA fragments that usually consist of only a few thousand base pairs. Bacteria exchange plasmids with one another and use them to "trade" antibiotic resistance genes, for example. To insert a gene in a plasmid, a researcher must first isolate his gene in a DNA fragment, then insert this fragment into the plasmid. **Restriction enzymes** are commonly used to help with this process.

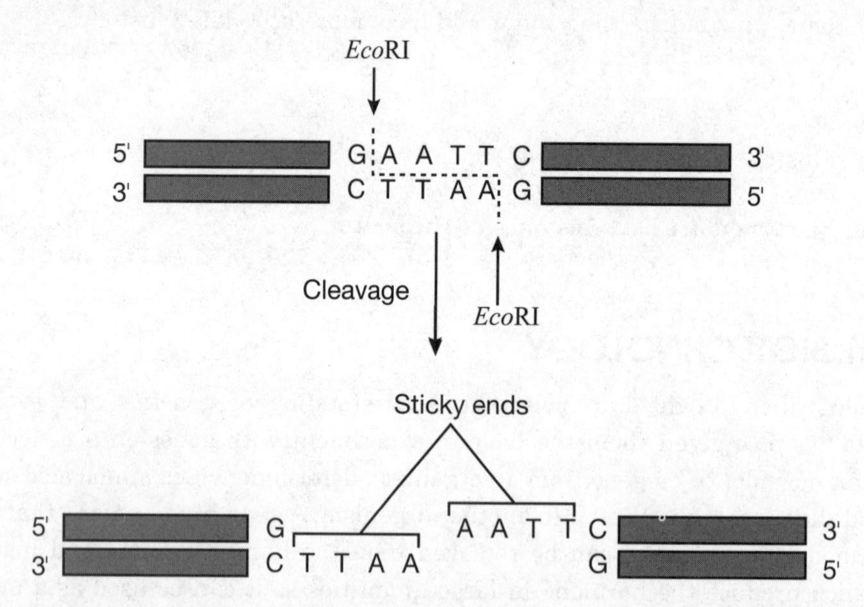

Restriction enzymes have been described as molecular scissors that recognize and cut DNA at specific sequences. A great variety of restriction enzymes have been characterized that cut DNA at several different sequences called **restriction sites**. If a researcher knows which DNA sequence must be cut to move a favored gene, it's easy to check the catalog of a molecular biology company for an enzyme that cuts in that location.

Bacteria probably produce restriction enzymes in order to cut the DNA of viruses that might infect them. Many restriction enzymes leave one end of a DNA protruding at the cut, creating a short, single-stranded region called a **sticky end**. This sticky end can hybridize with another one if the end of the other region matches. If the same restriction enzyme is used to cut two different DNAs, both will have the same sticky ends. They can then anneal (i.e., zip together) to form a new double-stranded region that brings the two different DNAs together. Another enzyme, **DNA ligase**, seals the DNAs together to create one DNA.

Q Some restriction enzymes cut DNA so that both strands end at the same place, creating blunt ends. Would it be easier to ligate two DNAs that have blunt ends or to ligate two DNA fragments with matching sticky ends?

A It is easier to ligate DNA fragments that have sticky ends, because the sticky ends form base pairs that hold the two ends together and prepare them for ligase to seal them together.

Q The restriction enzymes used to create DNA fragments must be removed or their activity must be stopped before ligase is used. Why?

A If the restriction enzyme is still present, it will cut up the DNA molecules that the ligase is sealing together.

If a DNA fragment is ligated into a plasmid, the plasmid can then be reinserted into bacteria, in a process called **transformation.** The bacteria can then be grown to produce more of the plasmid. Plasmids are often used to carry a gene that provides resistance to an antibiotic, such as ampicillin, along with the new DNA being inserted into the plasmid. After transformation, the biologist can grow bacteria in the presence of the antibiotic, allowing only bacteria carrying the plasmid to grow.

Q Bacteria are often used to grow plasmids and produce proteins. Some proteins, however, are inactive when expressed in bacteria and require eukaryotic cells for proper protein expression. What activities might eukaryotic cells supply that bacteria lack?

A Bacteria have all the basic translation apparatus and use the same genetic code as other organisms, but they have a few key differences that affect how certain proteins are expressed and function. Bacteria do not add the same post-translational modifications as eukaryotic cells—for example, glycosylation—and they do not have the same molecular chaperones as eukaryotic cells. Cell surface transmembrane proteins are not inserted in membranes in bacteria in the same way as they are with eukaryotic cells.

What might prevent a gene from the human genome from producing a protein if placed in bacteria? Eukaryotic genes are far more complex than bacterial genes. Eukaryotic messages usually have introns; bacterial genes do not. If a eukaryotic gene containing an intron is inserted into bacteria, the bacteria cannot process the RNA properly and the correct protein is not made. A solution to this problem is to create a different version of the human gene that is lacking introns before inserting it into bacteria. If the human mRNA for the gene is discovered and copied into DNA, this DNA copy of the gene will lack introns, which would have been spliced from the mRNA before it was exported from the nucleus. Another factor differentiating prokaryotes from eukaryotes is the transcriptional regulatory sequences and promoter regions. Since a plasmid is being inserted into bacteria, the gene it carries must use bacterial promoters and transcriptional start sites if it is to be expressed.

Q Which of the following RNA processing steps do bacteria perform: splicing, 5'-cap addition, polyA tail addition, or none of these?

A Bacteria do not perform any of these RNA processing steps, so RNA can be translated even while it is being transcribed.

Q Do bacteria have more than one protein translated from the same RNA?

A Yes. All genes in an operon will be transcribed together in one RNA and then translated into multiple different proteins, starting from internal translation initiation sites. This is not generally observed in eukaryotes. There are exceptions, however, such as in some viruses that infect eukaryotes.

Polymerase Chain Reaction

Another major innovation in biotechnology has been the **polymerase chain reaction (PCR)**. PCR allows biologists to amplify a gene rapidly if it is present in a sample. A scientist starts with a double-stranded sample of DNA, such as DNA in blood that has been taken from a crime scene. The DNA is heated to separate the two strands, and then short pieces of DNA called **primers** are incubated to become double-stranded (anneal) with each of the single-stranded pieces. Allowing the primers to bind requires that the sample be cooled somewhat, because too much heat denatures DNA (breaks up base pairs between strands). Next, a scientist would use a heat-stable DNA polymerase to copy each of the two single-stranded DNAs, starting with the annealed primers and extending them from 5' to 3', as DNA polymerases always do. This process is then repeated for multiple cycles, doubling the amount of DNA present with each cycle.

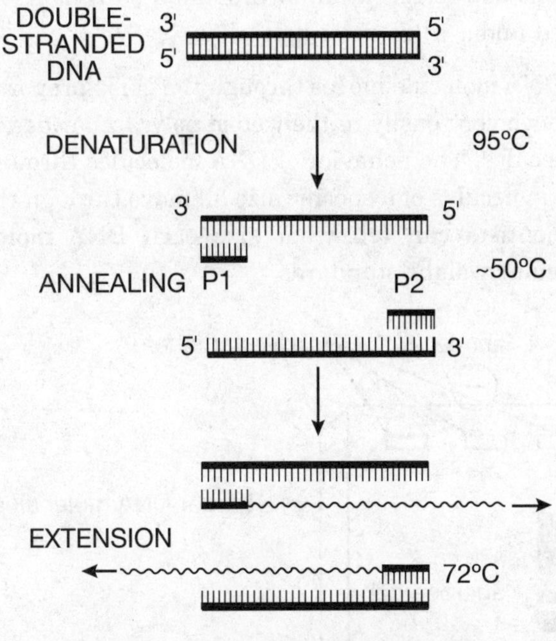

PCR. P1 and P2 are DNA primers

Q Why are primers needed for the DNA polymerase used in PCR?

A Primers initiate the synthesis of DNA.

The heat-stable DNA polymerases used in PCR have been discovered in organisms that thrive in high-temperature environments such as thermal vents and hot springs. Since the DNA polymerase is heat-stable, many rounds of DNA amplification can be performed with a single sample of DNA polymerase. Once PCR reactions are established, the whole sequence of heating, annealing, and elongating can be programmed in a rather elaborate PCR machine that runs without supervision.

Q If no heat-stable DNA polymerases were known, how could scientists perform PCR?

A Biologists would have to add a new batch of DNA polymerase for each round of the process, adding it to the cooled reaction to lengthen DNAs from the primers. When the DNAs are heated, the polymerase would be denatured and would then require more DNA polymerase for the process to continue.

Another commonly used method of working with DNA is **agarose gel electrophoresis**, a process that separates DNA into fragments of varying length and molecular weight. Agarose gels are thick, clear, gelatinous substances made up of a tangled web of polysaccharides derived from seaweed. DNA is placed in an agarose gel along with buffer salts, and the culture is then placed in an electric field. Negatively charged molecules, including DNA, migrate through the gel toward the positively charged electrode (the cathode). DNA in the gel is stained with a fluorescent dye (e.g., ethidium bromide), so that its fragments can be seen clearly and photographed under ultraviolet light.

The rate at which each DNA molecule moves through the gel is proportional to its size. Larger DNA molecules do not move very easily in the web of polysaccharides in the gel, so they don't go as far as smaller molecules. The behavior of DNA molecules through electrophoretic gel is so predictable that DNA molecules of a specific size all move through the gel together, forming a fluorescent band. Scientists can tell what size each DNA molecule is by comparing fragments against molecular weight standards.

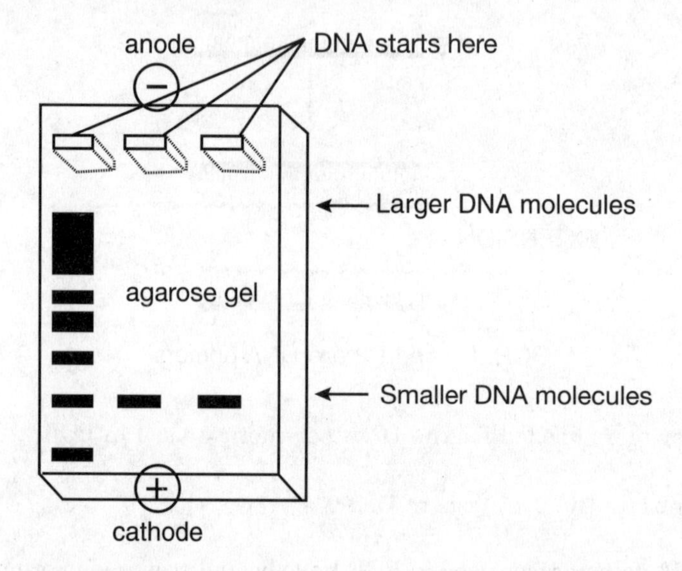

Q Why is DNA negatively charged?

A DNA is negatively charged because the phosphate groups in the backbone of DNA are negatively charged.

Biotechnologists use agarose gels to purify pieces of DNA and examine the structure of plasmids. The scientist can tell when insertion of a specific DNA fragment into a plasmid has been successful by taking a sample of plasmid DNA, digesting it with restriction enzymes, and then subjecting the digested DNA to agarose gel electrophoresis. DNA of a specific size can also be purified from agarose gels by cutting out "bands" from a gel. Once purified, the DNA can be mixed with a plasmid or other fragments to create a new DNA.

For example, a biotechnologist digests a plasmid with the restriction enzyme *Eco*RI and digests a DNA fragment containing a gene with *Eco*RI. Then these DNAs are ligated, the bacteria are transformed with the ligated plasmid, a large quantity of the plasmid is grown, and the biotechnologist checks to see whether the fragment containing the gene was successfully inserted into the plasmid. The technologist knows that the digested plasmid is 3,000 base pairs long and that the DNA fragment containing his gene has *Eco*RI restriction sites about 1,000 base pairs apart. To examine the new plasmid that has been created, the technologist digests the DNA with the *Eco*RI and then subjects the digested DNA to agarose gel electrophoresis. When the gel is run under UV light, the following pattern appears:

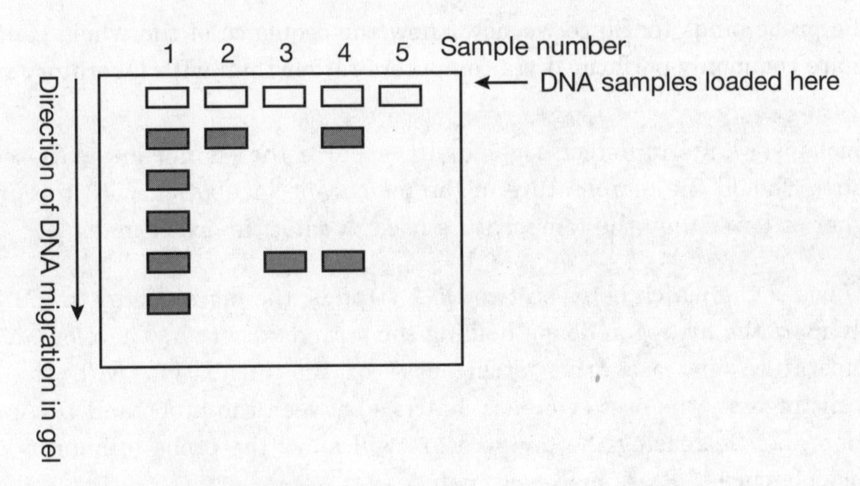

The samples loaded on the gel were:

1. Molecular weight standards
2. The original plasmid digested with *Eco*RI
3. The purified *Eco*RI-digested DNA fragment containing the gene before ligation
4. The plasmid recovered from bacteria after ligation, transformation, and *Eco*RI digestion
5. No DNA loaded

Q Did the researcher succeed in ligating the DNA fragment into the plasmid?

A Yes. The DNA purified from bacteria that had been transformed reveals the more rapidly migrating *Eco*RI fragment that matches the size of the *Eco*RI fragment the biotechnologist had purified and ligated into the plasmid.

Another use of agarose gels is to characterize DNA after PCR amplification. If successful, a PCR reaction produces a DNA fragment of a specific desired size. Running the gel helps the researcher see whether the reaction is successful. If it is, the agarose gel can again be used to purify the DNA.

A method called the **Southern Blot,** named after British biologist M. E. Southern, was once a key method for looking for genes in DNA, but it is less common today. To perform a Southern Blot, a researcher first separates DNA fragments of varying sizes in an agarose gel. These are then transferred out of the gel onto a flat membrane. The DNA sticks tightly to the membrane, but it can still be denatured to make it single-stranded. The denatured DNA on the membrane is incubated with a "probe," another single-stranded piece of DNA labeled with a fluorescent dye or a radioactive isotope designed to detect a specific DNA. If a DNA on the membrane matches the probe and forms double-stranded DNA with it, then that fragment is detectable by the fluorescence or radiation at that position on the membrane.

One of the uses of Southern Blot, in addition to seeking a specific gene, is to detect genes that are not identical but are related closely enough in sequence that the probe can bind to them. The Southern Blot has been used to seek new members of gene families. Altering the conditions in which the probe is incubated with DNA on the membrane can affect what sort of sequences the probe binds to. Since we now know the sequence of the whole genome, such experiments are commonly performed via computers rather than with laboratory work.

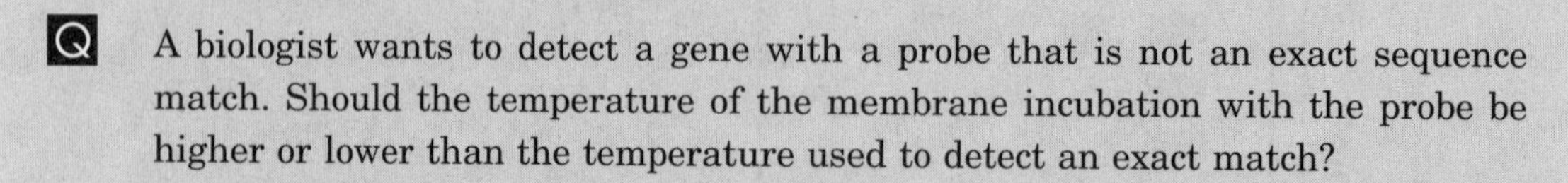

Q A biologist wants to detect a gene with a probe that is not an exact sequence match. Should the temperature of the membrane incubation with the probe be higher or lower than the temperature used to detect an exact match?

A The better the match between two DNA strands, the more energy it will take to pull apart the hydrogen bonds holding the strands together. Therefore, a higher temperature increases the "stringency" of the incubation with the probe, permitting only the best sequence matches between the probe and DNAs on the membrane. Decreasing the temperature will allow the probe to bind to a wider range of genes that are not exact matches.

VIRAL AND BACTERIAL GENETICS

Many of the studies of life at the molecular level have initially focused on bacteria and viruses. It is assumed that learning the basics of these simpler systems and then applying and expanding what we learn helps us to better understand more complex organisms such as humans.

Viruses

In general, viruses are not considered living organisms. Technically, viruses are obligate intracellular parasites that perform no metabolic activity of their own and cannot reproduce on their own. However, they do carry genes and they reproduce in other living organisms. Viruses infect essentially all living organisms, including bacteria, fungi, protists, plants, and animals. They use a variety of genomes, double-stranded and single-stranded, as well as RNA or DNA.

The genome type of a virus tells scientists a great deal about its life cycle. In viruses with a single-stranded RNA genome, the genome can be used either directly for translation or indirectly by copying into mRNA or DNA. Retroviruses, such as HIV, have a single-stranded RNA genome and carry reverse transcriptase. When a retrovirus infects a cell, reverse transcriptase copies the viral RNA genome into double-stranded DNA. The double-stranded DNA is inserted into the cellular genome, where it can be transcribed for the most part by using cellular transcription machinery. RNA transcribed from the inserted virus serves as mRNA for translating viral proteins, or as new copies of the viral genome.

DNA viruses often rely on a mix of viral and cellular proteins for transcription and translation of viral genes. Viral replication sometimes depends on cellular DNA replication factors, so viruses sometimes have mechanisms that turn on cellular DNA replication activity and then use that activity to copy their own genome.

Q Will reverse transcriptase inhibitors prevent HIV from gaining entry into cells?

A No. Entry into cells does not involve reverse transcriptase, and the reverse transcriptase cannot be active unless the virus is inside a cell. Cell entry involves protein binding on the surface of the viral envelope, with receptors on the target cell's membrane. Blocking reverse transcriptase with drugs can block progression of the viral life cycle after cell entry, however, and can slow viral infection.

The viral genome is packaged inside a shell of viral proteins called the **capsid**. The variety of virus shapes depend on the type of proteins in the capsid. An example is the complex "lunar lander" shape of some **bacteriophages**, which are viruses that infect bacteria.

Some viruses that infect animals also have a lipid **envelope** surrounding them when they bud from an infected cell. The viral envelope is derived from the cell's membranes and can include cellular proteins and viral proteins. Envelope proteins produced by the viral genome can play important viral functions, such as recognition and binding to proteins on the surface of cells, helping the virus enter and infect the cell.

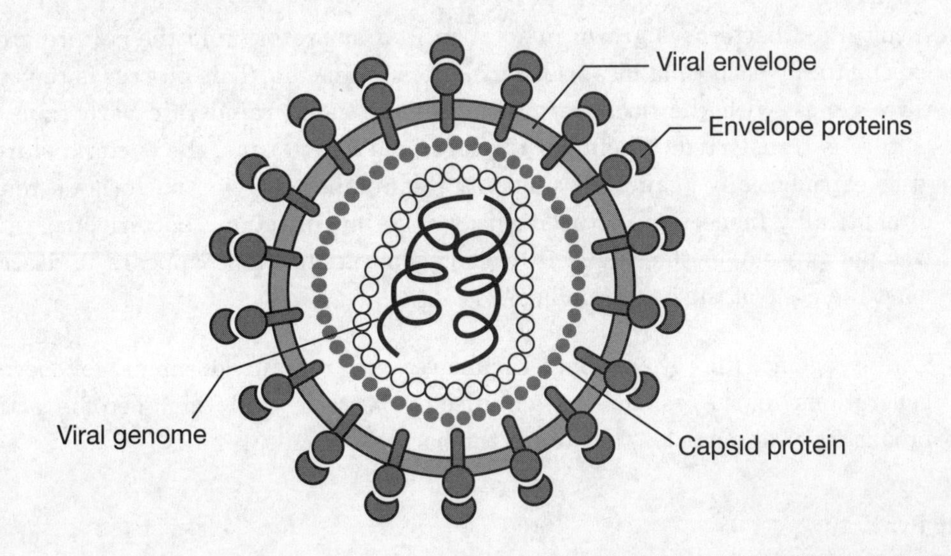

Q Do viruses synthesize the lipids in the viral envelope?

A No. In fact, viruses don't really synthesize anything. They simply use the infected cell's own biosynthetic capabilities and redirect them to produce more copies of themselves.

The viral life cycle typically follows this five-step pattern:

1. The virus infects a cell by inserting its genes into the cellular genome, directing the cell's systems to express viral genes.
2. New copies of the viral genome are produced.
3. Viral capsid proteins are produced.

4. New viruses are assembled in the infected cell.
5. The infected cell releases the new viruses, which are ready to infect other cells.

Many steps of the viral life cycle occur inside infected cells, which creates a challenge for the human immune system. Fortunately the immune system, mediated by T-cells, monitors what happens inside cells and kills virus-infected cells (see the section on the immune system in Chapter 13).

Bacteriophages have two different types of life cycle: a **lytic cycle** and a **lysogenic cycle**. In the lytic cycle, viruses replicate as soon as they infect bacteria. After new virus copies have been assembled, the infected cell is lysed (destroyed by antibodies called lysins), thereby releasing the new copies of the virus. In the lysogenic cycle, viruses do not immediately reproduce inside an infected cell. Instead, the virus inserts its genome into the bacterial genome. Once they are integrated, the viral genome may remain silent for one or more rounds of DNA replication and bacterial reproduction. At some point, the virus may become active again, replicate, and lyse the infected cell. It has been suggested that bacterial infections that are resistant to antibiotics might be treated with therapy involving bacteriophage viruses.

Q A culture of bacteria is grown in solution in a laboratory and the culture grows exponentially each time it is transferred to fresh media. This process is repeated three times, with the bacteria growing at the same rapid rate each time the culture is transferred. When it is transferred a fourth time, the bacteria start to grow exponentially again, but suddenly the number of living bacteria decreases dramatically. Inspection of the media reveals an infectious bacteriophage that was not present in the media in earlier generations. This appears to describe what life cycle of the bacteriophage?

A The lysogenic life cycle. Bacteriophages can remain dormant for several generations in the genome of an infected bacteria and then become active, producing viral particles that infect the bacteria in culture.

Bacteria

Bacteria reproduce asexually, replicating the genome and producing two genetically identical daughter cells (with the exception of mutations that may be introduced in the process). However, there are three ways in which bacteria can exchange DNA and genetic information, even without sexual reproduction:

1. **Transduction.** Bacteriophages are designed to move viral genes from bacteria to bacteria. In transduction, bacteriophages move bacterial DNA and viral DNA and transfer bacterial DNA into other bacteria. How does bacterial genomic DNA get into the virus? During a bacteriophage infection, some of the new viral particles that are assembled may include DNA from the bacterial genome instead of simply the viral genome. When such a virus infects a new cell, it carries the bacterial DNA with it, and the bacterial gene might become part of the genome of the new infected cell. Since the space inside the viral package is limited, the bacterial DNA will generally crowd out part or all of the

viral DNA. In this way, the viruses may gain entry into other bacteria but not produce infectious viruses.

2. **Transformation.** This process involves the uptake of DNA by bacterial cells from the external environment. If purified DNA is incubated with specially prepared bacteria, for example, some of the bacterial cells will take in the foreign DNA. Some bacteria even have specialized mechanisms to transport DNA from the external environment into cells.

3. **Conjugation.** In bacterial conjugation, DNA from one bacterium moves in one direction through a **pilus**, a narrow, temporary tube connecting two cells. The bacterium that is the DNA donor and which forms the pili is often called the male, although this is not technically sexual reproduction. A plasmid called an **F factor**, encoded genes involved in forming pili, is usually required for conjugation. In most cases, conjugation moves only the genes on the F factor plasmid between cells, but if the F factor integrates in the bacterial genome, it can drag bacterial genes along with F factor DNA between cells. The sequential movement of bacterial genes and viral DNA during conjugation has been used to map the relative positions of bacterial genes in the bacterial chromosome.

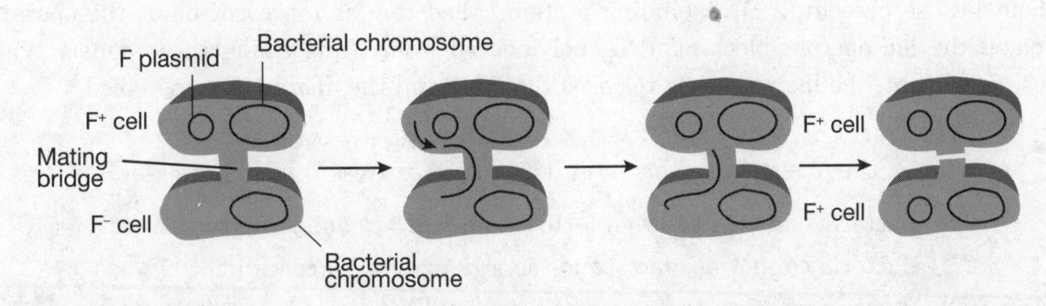

Q How do viral particles inject bacterial genes into infected cells if the viral genome in the viruses is not complete?

A The viral particle carries any type of DNA, whether or not the packaged DNA is a complete copy of the viral genome. Infection with such viruses cannot create new infectious viruses because one or more viral genes are missing. However, it is still capable of inserting into cells whatever DNA it carries.

Q Viruses are involved in which of the following forms of introducing new genes into bacteria: transformation, transduction, or conjugation?

A Transduction, in which bacterial genes are packaged into viruses. Transformation involves naked DNA; conjugation involves bacteria passing plasmids from one to another without virus involvement.

Bacterial Gene Expression

The basics of gene regulation were originally studied in *E. coli*, in which it was found that groups of genes with a related function were often regulated together. Groups of genes that occur together in the bacterial chromosome next to one another and that have related functions are called an **operon**. All genes in a bacterial operon are often controlled from a single bacterial promoter and transcribed together in one long mRNA. This mRNA can be translated into multiple different proteins encoded by the genes transcribed along its length. (Eukaryotes don't normally follow this process.)

The lac operon is a common example. Bacteria need to produce several genes to use lactose as an energy source—including proteins that hydrolyze lactose into monosaccharides and a transporter that moves the sugars into the cell. Without lactose, producing these proteins would be a waste of metabolic energy, and bacteria for the most part do not produce them under such conditions. Their production is rapidly and greatly induced by the presence of lactose. The genes for these proteins are located side by side on the bacterial chromosome and are transcribed together, so transcription of all genes in the operon is induced at once. Without lactose present, a DNA-binding protein called the lac repressor binds the operator region of the lac operon, blocking RNA polymerase from transcribing these genes. When lactose is present, the lac repressor releases the gene and the operon is transcribed.

Q What is the phenotype of bacteria that do not express the lac repressor?

(A) Bacteria can only grow on lactose and will not use other sources of energy.

(B) Bacteria cannot use lactose for energy under any conditions.

(C) Bacteria express genes in lac operon at all times, whether or not lactose is present.

(D) More RNA from the lac operon is transcribed but not translated.

A The correct answer is (C). Without the lac repressor, bacteria will express the genes of the lac operon constitutively, whether or not lactose is present. This creates a wasteful diversion of bacterial resources and places them at a selective disadvantage compared to normal bacteria when no lactose is available.

EXERCISES: THE GENETIC CODE

1. Which of the following is not a mechanism for the regulation of gene expression?

   (A) Altering the frequency of transcriptional initiation by RNA polymerase II
   (B) Decreased expression of telomerase
   (C) Switching splicing from one splice junction site to another
   (D) Increased degradation of mRNA
   (E) Altered acetylation of histones

2. A researcher identifies a gene encoding a protease enzyme expressed in the pancreas. The researcher also finds, nearby on the same chromosome, a stretch of nucleotides that are closely related to the sequence of the protease gene, but with an extra nucleotide inserted soon after the translation initiation codon. Which of the following best describes the nature of the related region of the genome near the protease gene?

   (A) A pseudogene created by a gene duplication event, followed by mutation
   (B) A transposable element inserted into the protease gene
   (C) A pseudogene created by mutation, followed by a gene duplication event
   (D) Convergent evolution
   (E) Non-disjunction that created an extra copy of one chromosome in pancreas cells

3. Which of the following best describes the role of primase in DNA replication?

   (A) It creates frequent short RNA stretches that DNA polymerase extends on the leading strand.
   (B) It does away with the need for a template in telomeres.
   (C) It synthesizes short RNA polynucleotide that DNA polymerase extends into Okazaki fragments on the lagging strand.
   (D) It seals DNA regions between breaks in the double helix.
   (E) It initiates DNA replication by joining the first few nucleotides of DNA.

4. Which of the following takes place in the eukaryotic cytoplasm?

   (A) Transcription
   (B) RNA processing
   (C) Ribosome assembly
   (D) Splicing
   (E) Translation

5. When linear DNA fragments are subjected to agarose gel electrophoresis, the distance that the fragments migrate in the gel is proportional to:

   (A) The molecular weight of the DNA fragments
   (B) The charge density of the DNA
   (C) The folded shape of the fragments
   (D) The percentage of adenine and thymine base pairs in the DNA
   (E) The compaction of the DNA into nucleosomes

6. Which of the following is *not* a step in the life cycle of a retrovirus?

   (A) Packing of genome into viral capsid proteins
   (B) Reverse transcription of viral RNA into DNA
   (C) Insertion of double-stranded DNA copy of viral genome into genome of host cell
   (D) Replication of viral RNA by RNA polymerase into new copies of RNA genome
   (E) Budding of mature virus with membrane envelope from infected cells

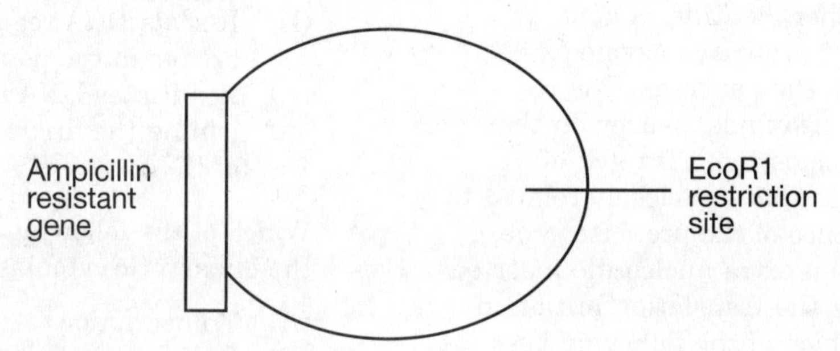

7. A biologist has a plasmid containing a gene providing resistance to the antibiotic ampicillin. The plasmid is digested with the restriction enzyme *Eco*RI, as is a human gene expressed in the immune system. This creates a linear fragment of DNA containing the gene. After purifying the *Eco*RI digested plasmid and the fragment containing the gene to remove the *Eco*RI enzyme, the biologist incubates the two DNAs with DNA ligase, then transforms bacteria with the ligated plasmid. The transformed bacteria are plated on growth media on one agar plate containing ampicillin and on another agar plate lacking ampicillin. The bacteria produce a large number of colonies on the plate lacking ampicillin but no colonies on the plate containing ampicillin. Which of the following best explains this observation?

   (A) The gene inserted in the wrong place in the plasmid.
   (B) The gene inserted into the plasmid was not transcribed properly in bacteria.
   (C) The gene inserted into the plasmid was not translated properly in bacteria.
   (D) The gene product is not active in bacteria.
   (E) The transformation of bacteria with the recombinant plasmid was not successful.

8. The DNA polymerase generally used in the polymerase chain reaction (PCR) has which of the following properties?

   (A) Extends DNA polynucleotides without a template
   (B) Performs 3' to 5' DNA synthesis
   (C) Synthesizes DNA without a primer
   (D) Is heat-stable during the denaturation step of the PCR cycle
   (E) Produces its own RNA primers in each reaction cycle

9. A researcher isolates an animal virus that buds from infected cells and treats the virus with a protease enzyme. The researcher discovers that after protease treatment, the viruses are no longer able to infect other cells. Which of the following is the most probable explanation for this result?

- **(A)** The protease was contaminated with a nuclease.
- **(B)** The virus uses proteins as part of its genome.
- **(C)** Proteases sometimes degrade DNA and RNA.
- **(D)** The protease inactivated one or more viral envelope proteins.
- **(E)** The proteases were not active under these conditions.

10. A mutation in a gene that changes a valine residue to a leucine residue at the same position in the encoded protein is best described as:

- **(A)** Nonsense mutation
- **(B)** Silent mutation
- **(C)** Conservative mutation
- **(D)** Frame shift mutation
- **(E)** Antisense mutation

ANSWER KEY AND EXPLANATIONS

1. B	3. C	5. A	7. E	9. D
2. A	4. E	6. D	8. D	10. C

1. **The correct answer is (B).** Telomerase maintains the ends of chromosomes, but it is not normally expressed in most somatic cells and it does not regulate the expression of specific genes. The other answer choices are all examples of gene regulation at various stages of mRNA production and processing.

2. **The correct answer is (A).** A pseudogene is a region near a functional gene that resembles the gene in its sequence but deviates enough that it no longer produces a functional gene product. The most likely mechanism by which this would occur would be via duplication of the gene in the evolutionary past, followed by mutation of one copy until the gene became a nonfunctional pseudogene. Inserting an extra nucleotide in the gene would change the reading frame and probably create a nonfunctional protein.

3. **The correct answer is (C).** Primase is an RNA polymerase that creates short regions of RNA using DNA as a template. DNA polymerase cannot initiate DNA synthesis without a primer, so it extends from the end of the RNA primer. Primase does not play a major role in the leading strand since DNA replication is continuous; however, it is important in lagging strand synthesis, where DNA replication is discontinuous.

4. **The correct answer is (E).** Transcription and RNA processing occur in the cell nucleus. Ribosome assembly occurs in the nucleolus of the nucleus.

5. **The correct answer is (A).** DNA fragments migrate in agarose gel electrophoresis according to their molecular weight. The bigger the DNA fragment, the smaller the distance it travels through the gel. The charge density of DNA is always the same. Linear pieces of DNA on their own, without proteins, virtually always have the same general snake-like shape when moving through a gel. The number of AT base pairs does not change migration in a gel, and nucleosomes are not relevant here.

6. **The correct answer is (D).** The viral RNA genome is not replicated directly from RNA. A DNA copy of the genome is made and this inserts itself into the infected cell's genome. From the DNA copy, viral genes are transcribed, and a long RNA is produced to serve as viral genome in the next generation of mature viruses. All of the other steps are normal steps in the life cycle of retroviruses.

7. **The correct answer is (E).** The failure of bacteria to grow in the presence of ampicillin has nothing to do with the gene being inserted in the plasmid. If the plasmid enters the bacteria in any form, with or without the inserted gene, it provides ampicillin resistance to the bacteria. The bacteria are not ampicillin-resistant after the transformation because the transformation with plasmid DNA did not work.

8. **The correct answer is (D).** The DNA polymerase used in PCR is remarkably heat-stable, so the enzyme is added just once for many reaction cycles. It cannot, however, perform the activities mentioned in the other answer choices. Just as with

other DNA polymerases, it needs a primer, only synthesizes DNA from 5' to 3', uses a template, and does not make RNA.

9. **The correct answer is (D).** Viral proteins in the envelope help the virus bind to new cells to infect and gain entry into these cells.
10. **The correct answer is (C).** Mutations that change one amino acid to another, chemically similar amino acid are conservative.

SUMMING IT UP

- Nucleotides are the fundamental building blocks of DNA. Each nucleotide contains three parts: a sugar (deoxyribose); phosphates; and nitrogenous bases adenine (A), guanine (G), thymine (T), and cytosine (C). The sugar and phosphate groups form the backbone of a strand of nucleotides, attached 3' to 5' on each sugar, with the nitrogenous bases projecting from the sides.
- In the double helix, each of the nitrogenous bases in one strand is always paired precisely with another base in the other strand, in base pairs that are held together in hydrogen bonds. In replication, the double helix opens to create a small, single-stranded region, which provides the template for new DNA assembly by pairing bases in the older single strand with new base links.
- Accurate DNA replication is essential for a healthy organism. DNA has a "proofreading" mechanism to catch errors during replication and to repair errors that elude the initial proofreading process.
- In the genetic code, each gene encodes a protein using an arrangement of nucleotide bases along its length. The genetic code is written using groups of three nucleotides that encode each amino acid added to a protein. During transcription, the message carried by a gene in DNA is copied into an RNA version; in gene translation, the RNA copy of the message is then read by a different set of molecular machinery and is used to produce the protein encoded by the gene.
- RNA is a critical part of ribosomal structure in prokaryotes and eukaryotes. It also plays a key role in translation by carrying out the catalytic activity of ribosomes while the proteins in ribosomes stabilize the ribosomes' structure.
- Recombinant DNA technology involves identifying a gene from an organism, determining which amino acid sequence it encodes, and then inserting the gene into another organism. A gene from humans that encodes a protein hormone, for example, can be removed from the human genome and inserted into bacteria, using bacteria to produce the hormone in large quantities so it can be used as a medicine or therapy.
- PCR allows rapid amplification of a gene that is present in a sample. After heating to separate DNA strands, primers are incubated to anneal with each single-stranded piece. After the sample is cooled, a heat-stable DNA polymerase is used to copy each single-stranded DNA. The process is repeated for multiple cycles.
- Viruses can infect essentially all living organisms. Those with a single-stranded RNA genome can use the genome either directly for translation or indirectly by copying into mRNA or DNA. Retroviruses, such as HIV, use reverse transcriptase to copy the viral RNA genome into double-stranded DNA and insert it into the cellular genome, where it is transcribed using cellular transcription machinery and serves as mRNA for translating viral proteins or as new copies of the viral genome.
- Bacteria reproduce asexually, replicating the genome and producing two genetically identical daughter cells. They can exchange DNA and genetic information, even without

sexual reproduction, in one of three ways: transduction, in which bacteriophages move bacterial DNA and viral DNA and transfer bacterial DNA into other bacteria; transformation, in which DNA is taken up by external bacterial cells; and conjugation, in which DNA from one bacterium moves through a pilus connecting two cells.

[illegible]

Evolution and Diversity

chapter 11

OVERVIEW

- **The evidence of evolution**
- **Population genetics**
- **Speciation**
- **Classification**
- **The origin of life on Earth**
- **Summing it up**

Mendel and geneticists of subsequent generations outlined the mechanism by which traits are passed from one generation to the next, explaining a great deal about the living world. Another unifying concept in modern biology was put forward initially by Darwin and his contemporary Wallace, who saw that the Earth is not populated by a multitude of unrelated organisms, but that living things are all interrelated, derived from a common ancestor in the remote past, and that each is adapted to a unique ecological niche. One simple idea brought this all together and made sense of life on Earth—evolution.

Darwin's ideas were based on observations of the world around him. One observation was that living things often produce great numbers of offspring that vary in many different traits. Some of these traits are inherited, but not all of the offspring produced in each generation survive and reproduce. Those with traits affecting how well individuals survive and reproduce will pass on those traits to the next generation. The selective inheritance of traits will change a population over a number of generations. This change in populations over time is **natural selection**.

Natural selection is a key part of evolution, and it is tied to the concept of fitness. **Fitness,** a relative, not absolute, measure of an organism's adaptation to the environment at a given moment in time. Evolution does not engineer the best possible solution that we might think of but only selects the highest fitness out of the variations present in a population.

Q Does having greater fitness mean that an organism lives longer?

A No. Fitness has nothing to do with living longer or "being healthier" in the normal way we use the word. It is a measure of an organism's adaptation to its environment at any given time.

Populations must vary for natural selection to occur. If a population is homogeneous, like a clone army, then there cannot be any selection for variants that provide increased survivability and reproductive success. The genetic variation that natural selection acts on comes from mutation and sexual reproduction. Mutation is the only way new alleles can be created in a population, and sexual reproduction creates new combinations of alleles. Natural selection then exerts pressure on the population, increasing the abundance of some alleles over others.

THE EVIDENCE OF EVOLUTION

There are signs of evolution throughout the living world and in traces left from living things of the past. Scientists study evolutionary relationships through the comparison of anatomical structures, embryonic development, the fossil record, and molecular studies of genes.

In order to discern the evolutionary relationships between organisms, we have often relied on an examination of anatomical differences and similarities. Mammals might be grouped as having a common evolutionary descent because they have the same basic body structure, with four limbs, hair, and similarities in skeletal structure and internal organs. Similar body structures stemming from a common evolutionary ancestor are called **homologous structures**.

Internal structures can reveal common origins where they might not be apparent based on surface appearances alone. The fore-limb bone structure of whales, bats, and humans share similarities that might not be expected based on external appearances and function of these structures.

Despite the different functions of a whale fore-limb (none), a bat wing (flight) and a human arm (varied), the underlying structure is essentially the same, indicating a common ancestor. On the other hand, organisms with different ancestral lines and function have evolved similar structures due to similar ecological pressures driving evolution. Such structures are called **analogous** rather than homologous. For example, dolphins may look like fish, but it is not because of a relatively recent ancestral connection. Both lines have independently evolved a streamlined body plan to better enable movement in a common aquatic environment.

The distinction between homology and analogy is one of degree. All life shares a common ancestor—in that sense, all traits are technically homologous. However, it is useful and accurate to consider similar forms in quite distantly related organisms stem from an adaptive response to similar environmental pressures—that is, as analogous structures.

Q If crabs and dragonflies have similar jointed appendages that have been adapted for different functions, is this an example of homologous structures or analogous structures?

A Crabs and dragonflies are both arthropods and thus share many structural features, such as jointed appendages and a chitin exoskeleton. These structures arising from a relatively recent (or, as in this case, well-conserved) divergence are homologous.

Examination of the genes, and proteins derived from genes, also reveal evolutionary relationships. Genes that rapidly evolve can be useful for looking at recent evolutionary events while genes that evolve slowly are useful for longer spans of evolutionary time. Ribosomal RNA genes mutate slowly, as does the cytochrome c gene. Mitochondrial DNA changes more rapidly and has proven useful for comparison of migration of *Homo sapiens* around the world. While comparison of genes has proven useful in many cases, this method assumes that the rate of mutation is always the same from species to species, an assumption that has been shown to be false.

The fossil record provides additional pieces of the puzzle of life on Earth, although it is also imperfect. The fossil record is biased for organisms that have hard body parts, that are abundant, and that die in sedimentary deposits that preserve the record of the animal. Evidence for species that do not fill all of these requirements is rare or perhaps nonexistent, but this does not mean that the species did not exist.

POPULATION GENETICS

To the geneticist, a **population** is not just a bunch of animals that live together. The more precise definition is this: **A population is a group of sexually reproducing and interbreeding organisms that can produce fertile offspring.** A population is the basic unit of evolution: organisms do not evolve. Only populations evolve. Natural selection changes the relative abundance of alleles in a population over generations, increasing the abundance of alleles that confer higher fitness and decreasing the abundance of those that confer lower fitness.

Q If an individual organism in a population has a spontaneous mutation producing pigmentation with a better match for their surroundings and that mutation is passed to their offspring, has the organism evolved?

A No. The population is the basic unit of evolution. Looking at one individual or their offspring is not enough to make conclusions about evolutionary trends. If the mutation provides greater fitness, and the allele associated with the mutation increases in abundance in the population over generations, then evolution will have occurred.

The sum of all the alleles of all genes are present in population is called the **gene pool.** Each allele is present at a certain frequency in the population. The **allele frequency** is calculated as follows:

$$\text{allele frequency} = \frac{\text{\# of specific alleles}}{(\text{total \# of alleles})}$$

Consider a simplified situation in which there are only two alleles for a gene. Call the alleles *p* and *q*. The allele frequency of each must be a fraction of 1, or 100%.

Now, imagine there is a population of 1,000 diploid wombats, whose eye color is controlled by two alleles, *E* and *e*. Since there are 1,000 wombats and they are all diploid, there will be 2,000 copies of the gene (2,000 alleles) in the population. If you know the breakdown of genotypes, you can see how many copies of each allele are present:

100 homozygous *EE* wombats (200 *E* alleles)

700 Heterozygous *Ee* wombats (700 *E* alleles and 700 *e* alleles)

200 homozygous *ee* wombats (400 *e* alleles)

If you add these, you'll see that there are 900 copies of the *E* allele and 1,100 copies of the *e* allele in the population. The allele frequency of the *E* allele is 900/2000 = 0.45 (45% of alleles are *E*)

The allele frequency of the *e* allele is 1100/2000 = 0.55 (55% of alleles are *e*)

To see if the calculations are correct, add the two allele frequencies:

$$0.45 + 0.55 = 1.0, \text{ or } 100\%$$

Evolution stems, in large part, from changes in the allele frequencies in a population over time. In fact, **microevolution** is defined as the change in allele frequencies in a population over time. Two scientists, Hardy and Weinberg, created a set of rules describing an ideal mathematical situation that population geneticists have used as a standard of comparison with actual populations. To visualize their world, imagine that the gene pool is like a jar of marbles, where each marble is a copy of one allele, and differently colored marbles represent different alleles. In each generation an individual will have a random selection of the marbles, but the overall makeup of the marbles in the gene pool jar stays the same. This situation is called the **Hardy-Weinberg equilibrium**.

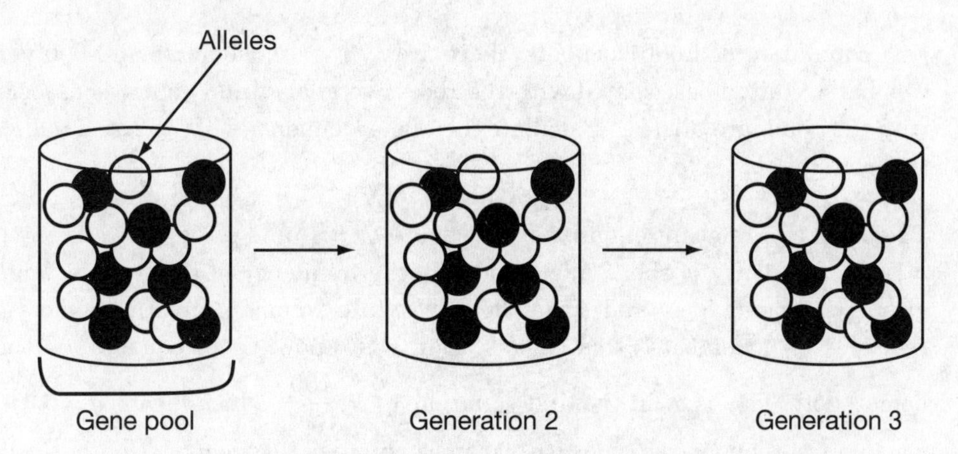

Hardy-Weinberg equilibrium maintaining constant allele frequency in a population from one generation to another

When a population is at Hardy-Weinberg equilibrium, the relative abundance of genotypes can be calculated based on allele frequencies. We know that the two allele frequencies must equal 1. Call the frequency of one allele p and the frequency of the other allele q. Thus, $p + q = 1$. However, we know that alleles combine in three ways: homozygous dominant, homozygous recessive, and heterozygous. If the dominant allele is T and the recessive is t, then:

p^2 = the frequency of the TT homozygote

$2pq$ = the frequency of the Tt heterozygote

q^2 = the frequency of the tt homozygote

How do we know this to be the case? Take a look at this Punnett square:

	T	*t*
T	*TT*	*Tt*
t	*Tt*	*tt*

This shows the familiar 1:2:1 genotypic ratio. Since p = the frequency of T and q = the frequency of t, we can derive an equation that describes the relative frequency of T and t:

	p	q
p	$p \times p = p^2$	$p \times q = pq$
q	$p \times q = pq$	$q \times q = q^2$

What do we now see? We have p^2, a q^2, and 2 pq's or:

$$p^2 + 2pq + q^2 = 1$$

We know that $p + q = 1$—that's why we were able to set this new equation to 1 (and also, incidentally, why the operation is addition). This equation, which follows logically from the 1:2:1 genotypic ratio of a monohybrid cross and from the fact that probabilities must add up to 100% (i.e., must equal 1), is called the **Hardy-Weinberg equilibrium**.

Q In a population of 1,000 wombats, there are 100 white wombats and 900 yellow wombats. White is associated with the recessive allele t and yellow is associated with the dominant allele T. Calculate the allele frequencies of the T and t allele in this population.

A There are two relevant equations: $p^2 + 2pq + q^2 = 1$, and $p + q = 1$. Let's say that p is the frequency of the T allele and q is the frequency of the t allele. You are given the phenotypes, and since the only white wombats are the homozygous recessive (tt) wombats, this means that you know the frequency of the tt homozygote. This is q^2 in the first equation. $q^2 = \frac{100}{1000}$, which means $q = 0.316$. If you know q, then since $p + q = 1$, you can calculate p, the frequency of the T allele:

$$p + 0.316 = 1$$

$$p = .684$$

The allele frequency of T is 0.684 (68%) and the allele frequency of t is 0.316 (32%).

You can also calculate the frequency of the Tt heterozygotes and the dominant TT homozygotes, by the way, since at this point you know p and q. The frequency of the TT yellow homozygotes is $p^2 = 0.47$ and the frequency of the heterozygotes is $2pq = 0.43$.

In another variation, you can be given the allele frequencies and asked to determine the phenotypes in a population. Let's say the frequency of the T allele is 0.7 and the frequency of the t allele is 0.3. Plug these into the $p^2 + 2pq + q^2 = 1$ equation to get the various genotypes, and from this you can get the phenotypes.

$$p^2 = \text{the frequency of the } TT \text{ homozygote} = 0.7^2 = 0.49$$

$$2pq = \text{the frequency of the } Tt \text{ heterozygote} = 2(0.7 \times 0.3) = 0.42$$

$$q^2 = \text{the frequency of the } tt \text{ homozygote} = 0.3^2 = 0.09$$

The TT and Tt wombats are yellow, which means that the frequency of yellow wombats is:

$$0.49 + 0.42 = 0.91 = 91\%$$

White tt wombats are therefore present with a frequency of 9%.

Natural populations are virtually never in Hardy-Weinberg equilibrium. In order to see why this is the case, let's explore the conditions that must hold for a population to be in Hardy-Weinberg:

- There is no mutation. The main effect of mutation is the introduction of new alleles. Mutation is a relatively rare event, so it will not exert an appreciable effect on allele frequencies between generations, but the rare new alleles introduced by mutation may over many generations have an impact when combined with other influences like natural selection.

- There is no migration or gene flow. If there are two different populations that are geographically close enough for animals to move and interbreed between the populations, then they will take their alleles along with them, taking them out of one gene pool and into another. It's like jumping between gene pools.
- The population is large enough to prevent random drift. Part of the inheritance of alleles between generations is determined by statistics. In a large population, this is not very noticeable. If your jar of marble-alleles has 4,000 marbles, 50% blue and 50% yellow, and you are taking a thousand out for the next generation, the odds are that the thousand marbles you select will be pretty evenly divided between blue and yellow, reflecting the gene pool they came from. What if only ten marbles were selected for the next generation? Statistics is better at predicting the long-term effects in large groups than small ones; small groups tend to deviate more from the larger group they are selected from. If you pick ten marbles, you might get three yellow marbles, or none, and if all future generations are derived from these ten founders, the future generations will be skewed based on this.
- Mating is random. It is often seen that mates are quite selective, looking for specific phenotypes like the length of tail feathers or size of antlers in the preferred mate. The more this is true, the more it will affect the way that alleles are transmitted from one generation to the next and upset Hardy-Weinberg equilibrium.
- There is no natural selection. One of the assumptions for Hardy-Weinberg equilibrium to be established is that there is no natural selection. If the frequency of alleles is altered because of varying fitness between different genotypes and phenotypes, then allele frequencies will change.

Q If there is no natural selection observed in a population of fruit flies, but some members of the population move back and forth with another population and interbreed, can Hardy-Weinberg equilibrium be established?

A No. It's not enough for one of the requirements for Hardy-Weinberg to be true for equilibrium to hold. The situation must meet all of the assumptions, and since there was migration in this case, at least one requirement was not met.

So, if the Hardy-Weinberg equilibrium never occurs in nature, what's the point of it? It is used as an ideal "reference point" against which the behavior of real populations can be profitably compared.

Humans hunted California sea otters for their pelts in the 1800s, reducing the population to a small colony of survivors. Today the sea otter has recovered much of its range and numbers, but genetic analysis shows that it has very little genetic variation for the size of its population. What happened? When the population was reduced to a small number of animals, it lost the genetic variation that originally existed in the larger population. Subsequent generations only had the alleles that were present in this smaller population. This effect is called the **bottleneck effect**. Such limited genetic variation can make it difficult for the population to respond to new environmental challenges like climate change, pollution, or disease.

Genetic bottlenecks also occur when new environments are colonized by a small number of organisms. This has been observed in islands many times, and is called the **founder effect**. Once again, the small number of animals provides all of the genetic variation initially available in this new population. However, as in the case of mutation and recombination, what is usually a drawback can, under certain circumstances, be a benefit. A fortuitous marriage of founder gene pool with a new environment can cause rapid evolution of new forms, including new species.

Q Why is having low genetic variation in a population's gene pool usually a problem?

A Having a variety of alleles in a population increases the chances that some individuals will enjoy increased fitness in the face of new environmental challenges. One example is bananas: the world's commercial banana crops are essentially clones, lacking genetic variation. If a new strain of fungus arises that overcomes the genetic defenses of the current banana crops, lack of genetic variation places all bananas at risk. Similarly, while California sea otters appear to be thriving in current conditions, they have may lost alleles that would have helped them adapt to new food sources as climate change warms the ocean and their prey.

There are several types of natural selection, depending on how selection affects a population. Traits are usually distributed in a more or less bell-shaped curve. In **disruptive selection**, organisms with a trait near the middle of the distribution are selected against. The effect on a population over time is that the distribution tends to spread out or perhaps even diverge into two different groups. In **directional selection**, organisms with a trait to one side of the distribution are selected against, pushing the average of the population over time in the opposite direction. In **stabilizing selection**, animals at both extremes are selected against, driving the population to have a narrower range of the trait in the middle.

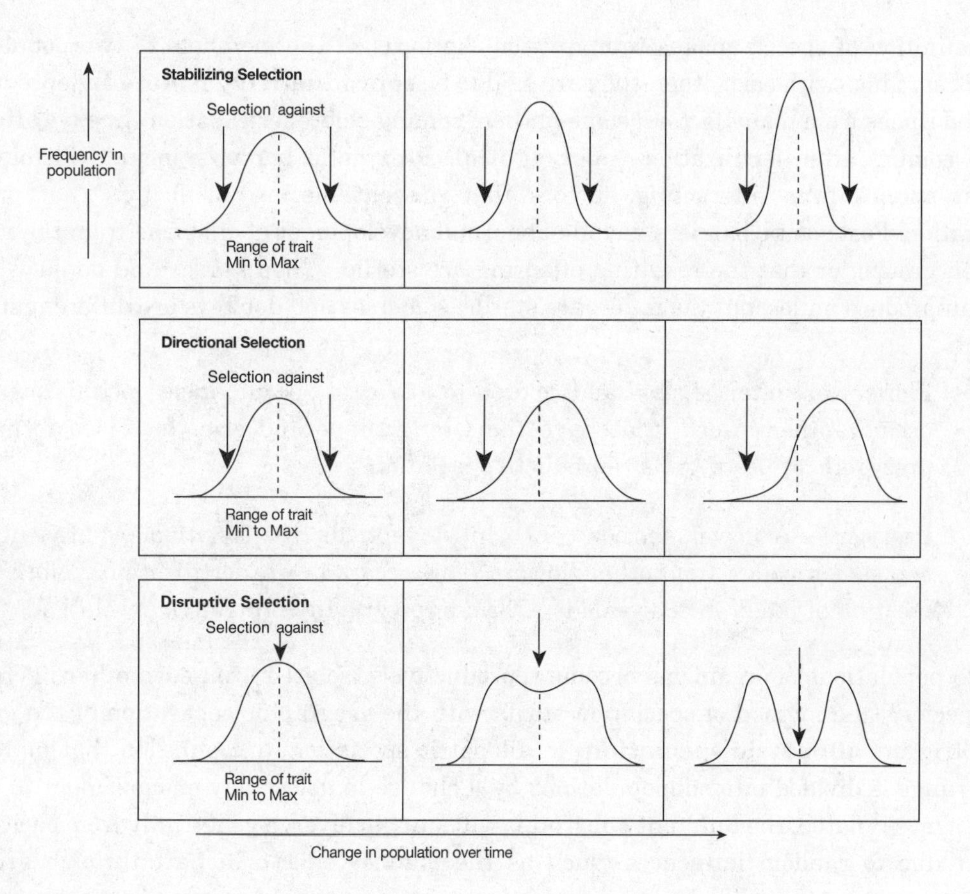

SPECIATION

What is a species? To the biologist, a species is more than just a group of animals that look the same. Animals with similar appearances are sometimes different species, like insects that evolve to mimic each other. To most biologists today, a **species** is a population that can interbreed and produce fertile offspring. If two groups cannot interbreed and produce fertile offspring, they are not the same species. We humans are all of the same species, *Homo sapiens,* not because we all look the same, but because we can interbreed and produce fertile offspring.

Q Two populations of a hummingbird are recently geographically separated, one on an island and another on the mainland, and they do not appear to interbreed. The birds on the island have a much smaller average size. When a biologist takes birds from the island to the mainland in an aviary, the birds mate and produce offspring that are fertile and breed to produce a second generation as well. Are these populations two different species?

A No. Although they are geographically separated, they are not reproductively isolated and therefore are not two different species. With additional time and continued separation, evolution may create reproductive isolation and speciation.

This definition of species makes reproduction the focus. If the members of two populations cannot reproduce with each other, they are said to be **reproductively isolated**. Reproductive isolation comes from many factors, some of them coming before fertilization (**prezygotic**) and others coming after fertilization (**postzygotic**). Prezygotic barriers include factors that prevent species from interacting, factors that prevent mating, or factors that prevent fertilization. Postzygotic barriers include abnormal development of offspring from the mating between groups, or that the resulting offspring are sterile. When horses and donkeys mate, they can produce mules, but the mules are sterile, so horses and donkeys are different species.

Q Flowers are often shaped and colored to attract a specific insect pollinator and exclude other insects that have the wrong shape and size. Is this a form of prezygotic or postzygotic reproductive isolation?

A Prezygotic. A flower species can achieve reproductive isolation by preventing transfer of pollen from other flowers. This occurs before fertilization, before the creation of the zygote, so this is a form of prezygotic reproductive isolation.

How do populations of organisms become reproductively isolated from each other to create a new species? One method of speciation starts with the geographic separation of the groups, which triggers **allopatric speciation**. In allopatric speciation, a population that inhabits a broad range is divided into subpopulations by a change in geography or movement to a new island. Once isolated, the different subgroups will start to diverge genetically from each other, in part due to random influences (such as the founder effect); in part through selective pressure.

Sympatric speciation is another mechanism. In this case, speciation occurs within a population that is still geographically intermixed. One way this can occur is through spontaneous polyploidy within the population. Once a polyploid strain arises, it cannot interbreed successfully with other members of the population, even if they inhabit the same region.

A population of polyploid salamanders lives in a wetland environment. A freeway is built through the middle of the wetlands, splitting the population in two. On one side of the freeway, over a period of fifty years, the wetlands grow dryer and the salamanders are found to change their life cycle to live more under wet leaves than in the water. At the end of the fifty years, both groups are still polyploid, and in vitro fertilization of one group by the other creates viable offspring, but when a researcher mixes individuals of the two groups he finds that their mating behavior has changed so that they no longer interbreed. Is this an example of speciation and, if so, what type of speciation?

 Yes, this is speciation. The two groups have become reproductively isolated from each other and no longer interbreed. It does not matter if they produce offspring by artificial methods. If they do not interbreed because of changes in mating behavior, then the altered mating behavior forms a prezygotic barrier, and the populations are reproductively isolated from each other. The type of speciation is allopatric speciation. The event leading to speciation was the change in geography—the road being built. The two populations were first isolated from each other by the road, after which reproductive isolation could occur. The fact that the salamanders were polyploid was irrelevant in this case.

Models of speciation have traditionally assumed that most species evolve gradually over time, through small, continuous, incremental changes. The fossil record is taken to reflect only certain "snapshots" of this process, which is an extrapolation from "microevolution" (that which can be observed in a human lifetime) to "macroevolution" (the history of life). However, the hypothesis of **punctuated equilibrium** takes the fossil record at face value. That record shows that most species remain nearly unchanged once they appear—and their appearance is itself relatively sudden (in geological terms). The relative frequency of sudden versus gradual change, as well as the mechanisms by which punctuated equilibrium could occur, are still much debated.

Associated with the debate between punctuated equilibrium and "gradualism" is not only the importance of mass extinctions, but also the importance of "mass speciations." **Adaptive radiation** is a particular type of speciation observed when a few organisms colonize a new habitat like an island. If the island has very few organisms early on, there will be a great diversity of ecological niches that are unoccupied, and the colonizers will often diversify into species that occupy these niches. Whereas the existence of adaptive radiation at relatively local levels is uncontroversial, the notion that post-mass-extinction adaptive radiations have determined the history of evolution more than the operation of selection, drift and so on is, like the punctuated equilibrium hypothesis with which it is associated, still a matter of intense debate.

Q After a volcanic eruption, a number of species from the surrounding environment colonize the rocky surface, causing rapid growth of plants and animals in a previously barren region. Is this an example of adaptive radiation?

A No. There is no information given suggesting that speciation is occurring. The species from the surrounding environment colonized the area.

CLASSIFICATION

Assigning scientific names to classify living things in a systematic manner is called **taxonomy**. This process started with the Swedish scientist Carolus Linnaeus and continues today as new species continue to be discovered. The system uses hierarchical categories to classify each organism, starting with domains as the highest possible category and ending with each individual species. Each species is often referred to using its genus and species names together, like *Homo sapiens.* This is called **binomial nomenclature.** In the following example, the bolded terms refer to *Homo sapiens'* full classification through all taxonomic levels.

Domains: Bacteria, Archae, **Eukaryotes**
Kingdom: **Animals**, plants, fungi
Phylum: Arthropods, **chordates**, echinoderms
Subphylum: **Vertebrate**
Class: **Mammals**, reptiles, birds fish
Order: **Primates**, rodents, insectivores
Family: Feline, canine, bovine, *hominidae*
Genus: *Homo*
Species: *sapiens*

In addition to these major groupings, additional subgroupings are often used, such as subphylum (vertebrates) or subclass. Taxonomy is interrelated with **phylogenetics**, the study of evolutionary relationships among organisms. Phylogenetics is used to construct trees depicting the relative relationships among organisms.

The phylogenetic groupings of organisms change as more is learned about evolutionary relationships through new fossil finds and molecular tools like genome sequencing. The current system with the three domains of life (archae, bacteria, and eukaryotes) is fairly recent. The changes in phylogenetic groupings reflect the continuous advancement of our understanding of life on Earth.

THE ORIGIN OF LIFE ON EARTH

Based on a variety of evidence, the Earth formed about 4.5 billion years ago, along with the rest of our solar system. The first fossil evidence of life on Earth, prokaryotic cells preserved in stone, comes from about 3.5 billion years ago. This leaves about a billion years about which we are not quite sure. During part of this billion years the early Earth was still being violently bombarded by bodies of various sizes. There is much about the origin of life on Earth that we will probably never know for sure. Any trace of this process has probably been destroyed in the intervening time and geologic change. We can, however, make hypotheses that can be tested. Some clues are buried in the living things that populate the Earth today, remnants of the first life that gave rise to all living things. Other clues come from laboratory experiments that aim to reproduce early conditions to see whether and how life arises.

Life on Earth started with the gradual assembly of the chemicals of life from the compounds that formed spontaneously on the early Earth. The Earth's atmosphere in the first billion years was quite different from today, and it has been suggested that it was composed of compounds such as nitrogen, carbon dioxide, methane, ammonia, and water vapor, with virtually no molecular oxygen. As a result, the early Earth's atmosphere may have been reducing, while today's atmosphere is oxidizing. In this reducing (or at least nonoxidizing) atmosphere, with the addition of solar radiation and energy from lightning strikes, prebiotic compounds like amino acids may have spontaneously formed. Laboratory experiments have shown this to be possible.

The next step toward the first living cell was the spontaneous assembly of these compounds into polymers, such as joining amino acids into peptides. Protein-like polymers have been shown to spontaneously form from amino acids mixed in solution. Another key step would have been the formation of membranes that created the first cell-like structures. Phospholipids mixed in water spontaneously assemble into membranes in rounded structures with the same size and shape as small simple cells. In the prebiotic world, such membranes could have surrounded prebiotic molecules, concentrating and protecting them, helping prebiotic chemistry to occur more efficiently. Early proteins assembled randomly may even have acted as primitive enzymes, speeding the rate of reactions slightly.

Q Is it possible that living organisms could arise again independently of other living things on the modern Earth?

A Not too likely. The prebiotic world is postulated to have been greatly enriched with organic compounds that provided precursors and were free in the environment. Soon after life arose, this situation changed. Prior to the advent of photosynthesis, the first living things were probably heterotrophs living off of these free organic compounds for energy and building blocks. The world is different today, altered by living organisms and subject to interaction with them. Also, the oxidizing atmosphere of today's Earth would prevent some of the reactions or degrade some molecules that would have been able to spontaneously form in the early Earth.

For systems to be living systems, they need some sort of genetic material. A variety of clues suggest that RNA may have played this role in the first living things. Some of these clues are "molecular fossils" in living things today, including the intimate role of RNA in gene expression. RNA is not the genetic material in most living things, but mRNA carries the information encoded in DNA genes for it to be translated, RNA molecules in ribosomes are responsible for the catalysis of peptide synthesis, and tRNAs bridge the gap between mRNA and the ribosome. RNA plays an essential role in catalysis of splicing, and the self-splicing nature of some introns inspired biologists to view RNA as perhaps a relic of early living things. Single-stranded, and a flexible polymer with lots of potential for hydrogen bonds, RNA may have played a role both as genome and catalyst in early living things. If an RNA molecule could copy itself, this may have been the first genome, and it could also evolve, as RNAs that were more efficient at copying themselves would come to dominate the RNA world. RNAs that could also assemble other RNAs that helped the process would also have a selective advantage. These may have been the first genes.

To piece together the first living cell from which all life on Earth today is derived, biologists look for characteristics that all living things today share. The three domains of life are the bacteria, the eukaryotes, and the archae. These three groups share the following characteristics that were probably also found in the "progenote," the last common ancestor from which all life on Earth since then derived. These characteristics probably included the following:

- lipid bilayer membrane
- amino-acid-based proteins
- DNA genome
- Chemiosmotic ATP production
- Glycolysis, and other basic biosynthetic and metabolic pathways

Q Was the progenote photosynthetic?

A Probably not. Photosynthesis arose early in the history of life, but the precursor to all other living things was probably not photosynthetic, but rather chemosynthetic, as undersea vent microorganisms are today.

One of the challenges for the first cells would have been finding energy to fuel the reactions of living things. At first, living things probably used molecules free-floating in their local environment to supply energy. But as the supply of spontaneously formed compounds ran low another energy source would have been needed. The evolution of photosynthesis was a major event in the history of life, harnessing the energy of the sun for living things. Interestingly, photosynthesis in many prokaryotes does **not** produce molecular oxygen by splitting water. But as prokaryotes developed photosynthesis that splits water, the oxygen released by this process eventually affected the entire biosphere. The chemistry of the oceans was changed, and as oxygen entered the atmosphere, the chemistry of Earth's surface changed as well. The availability of oxygen in the atmosphere drove some organisms that could not tolerate oxygen into ecological niches where oxygen remained scarce while making the evolution of metabolic reactions reliant on oxygen possible.

Q Would organisms carrying out oxidative phosphorylation have an advantage over those carrying out fermentation for energy in an oxidative atmosphere?

A Yes. Oxidative phosphorylation is much more efficient at extracting energy from food molecules than fermentation, and extracting energy more efficiently is a big selective advantage.

The three domains of life did not evolve independently of each other. Early in their joint history, their paths were intertwined. The first eukaryotes probably evolved about 2.5 to 2.7 billion years ago, based on fossil evidence. Eukaryotes on Earth today are a lot different from prokaryotic cells, with a proliferation of distinct subcellular organelles bounded by membranes. Eukaryotes have a cytoskeleton and linear chromosomes and divide by mitosis and meiosis to reproduce sexually. The endoplasmic reticulum, nuclear envelope, Golgi, and lysosomes may have originated from infoldings of the cellular membrane. Some prokaryotes today still have similar infoldings of their membranes.

One molecular clue to the origin of eukaryotes comes from mitochondria. Essentially all eukaryotic cells have mitochondria—or signs that they once had mitochondria. Mitochondria are organelles in eukaryotic cells, but they resemble prokaryotic cells in many ways:

- Small, circular DNA genome
- Ribosomal translation system resembling prokaryotes'
- Transcription-like prokaryotic genes
- Replication that resembles the binary fission of prokaryotic cell division

It is now believed that mitochondria were once independent prokaryotic organisms that existed either as internal parasites of early eukaryotic cells or were internalized by those cells. Once internalized, the mitochondrial ancestor and the eukaryotic precursor developed a symbiotic relationship. The eukaryote got energy from the mitochondria, which were able to use oxygen to drive oxidative phosphorylation, something the eukaryotic precursor could not do on its own. The early mitochondria got a variety of metabolic needs satisfied as well, as metabolic material is plentiful inside of a cell. The origin of eukaryotes in this manner is called **endosymbiosis**. Over time, most of the genes of the mitochondria were transferred to the nuclear genome. In fact, many mitochondrial proteins today are transcribed from nuclear genes and translated in the cytoplasm and then imported into mitochondria.

Q If a parasitic organism has a nucleus but no mitochondria and has genes in its nuclear genome that appear to have originated from mitochondria, what does this suggest about the evolutionary history of this organism?

A In most organisms, many mitochondrial genes appear to have been transferred to the nuclear genome over the billions of years of eukaryotic evolution. Parasites often rely on the host organism for much of their metabolic needs. It seems that as a parasite this organism lost its mitochondria, leaving behind the molecular relic of mitochondrial genes in the nuclear genome.

Chloroplasts appear to have arisen in an anlogous manner, originating as an independent photosynthetic prokaryotic cell that was internalized to become an endosymbiont, giving rise to eukaryotic algae and, later, plants. Chloroplasts, like mitochondria, have their own genome, and carry out their own translation and transcription.

EXERCISES: EVOLUTION AND DIVERSITY

1. In examining the ribosomal RNA genes of bacteria, archae, and eukaryotes, a researcher found that the ribosomal RNA of archae and eukaryotes were more related to each other than to bacteria. Which of the following does this suggest?

   (A) Bacteria evolved translation independently of the other branches.
   (B) Archae and eukaryotic ribosomes share the same sensitivities to antibiotics.
   (C) The last common ancestor of archae and eukaryotes existed more recently than the last common ancestor of bacteria and archae.
   (D) Eukaryotes evolved from archae.
   (E) Life arose on Earth on more than one occasion.

2. Which of the following best describes the role of fitness in natural selection?

   (A) Organisms with higher fitness live longer.
   (B) Organisms with higher fitness are less subject to natural selection.
   (C) The more advanced an organism is in evolutionary terms, the greater its fitness will be.
   (D) Organisms with greater fitness have greater strength, ensuring their survival.
   (E) Organisms with greater fitness pass on more of their alleles to future generations than other organisms.

3. A population of diploid gerbils has two alleles of a gene affecting tail length, a dominant allele L that produces a long tail and a recessive allele l that produces a short tail. The population has 5,000 gerbils, of which 1,000 have short tails and 4,000 have long tails. What percentage of the population is heterozygous for the two alleles?

   (A) 10%
   (B) 20%
   (C) 25%
   (D) 37%
   (E) 50%

4. Which of the following will NOT disrupt Hardy-Weinberg equilibrium in a population of rabbits?

   (A) Migration from another population
   (B) Introduction of a new allele through mutation
   (C) Meiotic recombination that generates new combinations of alleles in a population
   (D) Sexual selection of mates with greater disease resistance
   (E) Natural selection of rabbits that produce more young

5. Which of the following is generally agreed must occur for speciation to occur?

   (A) Geographic isolation
   (B) Reproductive isolation
   (C) A discernible change in anatomical structure
   (D) Bottleneck effect
   (E) Adaptive radiation

6. Which of the following were probably NOT found in the last common ancestor of the three domains of living organisms on Earth today?

   (A) Lipid bilayer membranes
   (B) Glycolysis
   (C) Peptidoglycan cell walls
   (D) Electron transport system
   (E) L amino acids in proteins

7. Which of the following best describes a situation involving allopatric speciation?

   (A) Slow accumulation of mutations, followed by a period of rapid evolution
   (B) Organisms living in the same geographic region achieve reproductive isolation through a genetic change like polyploidy.
   (C) Organisms are geographically isolated but are not reproductively isolated.
   (D) Organisms living in different geographic regions become reproductively isolated from each other.
   (E) The founder effect leads to adaptive radiation on an island habitat.

8. Volcanic islands often have unique species that are found nowhere else. The best explanation for this is:

   (A) Adaptive radiation from a few founders
   (B) Genetic bottlenecks
   (C) Convergent evolution
   (D) Sympatric speciation
   (E) Nonrandom mating

ANSWER KEY AND EXPLANATIONS

1. C	3. E	5. B	7. D	8. A
2. E	4. C	6. C		

1. **The correct answer is (C).** This is an example of a molecular clock. Molecular clocks are not exact and are subject to problems, but they suggest evolutionary relationships like this, suggesting which species shared a common ancestor most recently.

2. **The correct answer is (E).** When discussing natural selection, fitness relates to the ability of organisms to pass on their alleles and traits to future generations. The greater the fitness, the more alleles an organism passes on to future generations.

3. **The correct answer is (E).** Given the phenotypes in this problem, it is necessary to calculate the allele frequencies and then to calculate the genotype frequency. Call p the frequency of the L allele and q the frequency of the l allele in the population. The only genotype producing short-tailed gerbils is the ll homozygote, so the abundance of short-tail gerbils will reveal the frequency of this allele.

   $q^2 = 1000/5000$
   $q = 0.45$
   If $q = 0.45$, then $p = 0.55$, since $p + q = 1$

   The heterozygote frequency = $2pq$, or $2 \times 0.45 \times 0.55$, or 0.50, which is 50%.

4. **The correct answer is (C).** Meiotic recombination will not change the abundance of alleles from one generation to the next, while all of the other factors will, disrupting Hardy-Weinberg equilibrium.

5. **The correct answer is (B).** Speciation involves the reproductive isolation of two populations from each other, one way or another. It does not always involve geographic isolation, such as when sympatric speciation occurs. It does not necessarily involve a change in appearance.

6. **The correct answer is (C).** Only bacteria have this feature, not archae or eukaryotes, so this feature probably arose after this group split and was probably not present in the progenote. All of the other features are found in all of the domains, including sophisticated molecular features like electron transport systems for energy production.

7. **The correct answer is (D).** Allopatric speciation begins with geographic isolation of two populations, which then allows changes in the genotype and phenotype of the populations to develop, which cause reproductive isolation.

8. **The correct answer is (A).** Islands formed from volcanoes start with no terrestrial life and are colonized by terrestrial species that reach the island over the ocean. The few organisms that arrive at the island find an environment with little competition and many potential ecological niches. Through adaptive radiation, animals and other organisms can rapidly evolve to fill a variety of ecological niches, and this is what is often seen on islands.

SUMMING IT UP

- Natural selection is the differential survival and reproduction of organisms with heritable variation over time. Natural selection alters the genotypes and phenotypes of a population over time.
- The variation that natural selection acts on comes from mutation and sexual reproduction. Mutation is the only way new alleles can be created in a population, whereas sexual reproduction creates new combinations of alleles.
- The factors scientists use to study evolutionary relationships include anatomical structures, embryonic development, the fossil record, and the molecular studies of genes.
- In genetics, a population is a group of sexually reproducing and interbreeding organisms that can produce fertile offspring. A population of organisms is the basic unit of evolution.
- A species is defined by most biologists as a population that can interbreed and produce fertile offspring. If two groups cannot interbreed and produce fertile offspring, they are not the same species.
- Different models of speciation and extinction include models in which species gradually evolve over time, with small, continuous, incremental changes and sudden, rapid changes interspersed with longer periods with little change.
- Taxonomy is the system of assigning scientific names to classify living things according to evolutionary relationships.
- The Earth likely formed about 4.5 billion years ago with the rest of our solar system. The first fossil evidence of life on Earth dates from about 3.5 billion years ago.
- Early in the history of life, three domains of living things emerged: the archae, bacteria, and eukaryotes.
- Life on Earth probably began when the chemicals of life gradually developed from spontaneously formed prebiotic compounds, such as amino acids, in an atmosphere that was most likely composed of elements and compounds such as nitrogen, carbon dioxide, methane, ammonia, and water vapor, with virtually no molecular oxygen. The addition of solar radiation and energy from lightning strikes may have sparked the formation of the prebiotic compounds.

[illegible]

- [illegible]
- [illegible]
- [illegible]
- [illegible]
- [illegible]
- [illegible]
- [illegible]
- [illegible]
- [illegible]
- [illegible]

Prokaryotes, Protists, Fungi, and Plants

chapter 12

OVERVIEW

- **Prokaryotes**
- **Protists**
- **Fungi**
- **Plants**
- **Summing it up**

Organisms reflect their diverse evolutionary history. Despite the diversity of organisms, all deal with similar challenges as they live, grow, and reproduce. We have this much in common with mushrooms, pine trees, and amoebas: we must all get energy and biosynthetic building blocks, avoid predation and disease, and reproduce.

PROKARYOTES

Prokaryotes, including the domains bacteria and archae, are small, simple organisms lacking the internal subcellular organelles found in eukaryotes. Prokaryotic cells are, on average, about one tenth the size of eukaryotic cells.

Bacteria

Descriptions of prokaryotic cells often begin with their characteristic shape under the microscope: round (**cocci**), spiral-shaped, or cylinder-shaped (**bacilli**). These characteristic shapes are determined by the cells' rigid walls. Bacterial cell walls are built partly of a mesh of peptide and sugar polymers called **peptidoglycan**. Some antibiotics, such as penicillin, work by inhibiting the formation of the peptidoglycan cell wall.

Bacteria are also characterized by their reaction with a colored stain, called Gram stain. "**Gram positive**" bacteria take in more of this stain and appear darker under a microscope than "**Gram negative**" bacteria. The reason for this difference in color stems from the differing chemical structure of the cell walls of these groups. Gram stain penetrates the cytoplasm through the cell walls. Gram positive bacteria have thicker peptidoglycan layers in their cell walls and so retain more stain, while Gram negative bacteria have much thinner layers of peptidoglycan. Both Gram positive and Gram negative bacteria have plasma membranes, but Gram negative bacteria also have outer membranes containing lipopolysaccharide, a dangerous toxin for those infected by Gram negative bacteria.

Q Are Gram positive bacteria more sensitive to lysozyme than Gram negative bacteria? How about penicillin?

A Gram positive bacteria have their peptidoglycan cell walls exposed to their external environment, accessible to digestion by lysozyme or to inhibition of cell wall synthesis by penicillin or related antibiotics. Gram negative bacteria, on the other hand, have outer membranes outside of their peptidoglycan cell walls, which block access to lysozyme or penicillin. Gram positive bacteria are more sensitive to both lysozyme and penicillin.

Some bacteria are also surrounded by a **capsule**, a layer of sticky sugars that can help attach to surfaces and help bacteria hide from the immune system. Bacteria with capsules are often much more pathogenic (disease-causing) than related bacteria lacking capsules.

Q A vaccine is developed against a bacterial protein found in the outer membrane of a pathogenic Gram negative bacterium. When the vaccine is tested in animals, it does not prevent infection, despite the presence of antibodies in the blood of animals after vaccination and the continued presence of the protein in the bacteria. What is one possible explanation?

A There are a variety of explanations for the failure of a vaccine, but one explanation is that the pathogenic strain is surrounded by a capsule, preventing antibodies from reaching the surface of the bacterial cell, where the protein is, thereby rendering the vaccine ineffective.

Many bacteria have **flagella**, slender projections from the bacterial cell wall that move the cell through its surroundings. Prokaryotic flagella help provide movement, like eukaryotic flagella, but they are smaller and do not work by the same mechanism. Bacterial flagella spin to create a whip-like motion that propels the cell, while eukaryotic flagella move back and forth, bending rapidly due to the action of motor proteins on microtubules in the flagella.

Q Do bacterial flagella move by means of tubulin-associated motor proteins?

A No. This describes eukaryotic flagella, which contain microtubules. The resemblance of bacterial and eukaryotic flagella is strictly superficial. They operate by completely different mechanisms, and the bacterial flagella do not use tubulin.

Bacteria have smaller and simpler genomes than eukaryotes and, as a result, can rapidly replicate, often producing a new generation in an hour or less when the supply of nutrients is optimal. This rapid generation time, and the presence of many bacteria, can result in rapid adaptation of bacteria to a changing environment. Antibiotic resistance develops readily in bacteria, one measure of the great ability of bacteria to adapt. When environmental conditions are less favorable, some bacteria form **endospores**, dehydrated structures containing the genome and a thick protective coat. Endospores can survive many years until they are once again placed in favorable conditions and become metabolically active.

Q Does natural selection operate in bacteria if they reproduce asexually?

A Yes. Natural selection does not require sexual reproduction to occur. Sexual reproduction increases the genetic diversity that natural selection acts on, but even without it, there will still be genetic diversity in bacteria. Bacteria reproduce so rapidly and create such huge numbers of cells that mutation is a much bigger factor in bacteria than in higher eukaryotes. In addition, bacteria can exchange genetic information by transduction, transformation, and conjugation. When a selective pressure is present, some bacteria will survive more than others, and those that survive will pass on genes and traits to their progeny that improve their survival. Look at antibiotic resistance. An antibiotic may kill 99.9% of the bacteria in an infection, but if 0.1% of bacteria have mutations or carry plasmids providing antibiotic resistance and grow back, all of the new population of bacteria will share their resistance.

The rapid evolution of bacteria has allowed them to adapt to a broad range of metabolic conditions and environments. Bacteria and archae have been discovered practically everywhere on earth, including the depths of the oceans, mountain peaks, high in the atmosphere, and buried in solid rock. The metabolic needs of bacteria and other organisms are often described according to their carbon source and their energy source (remember to look at the first part and the second part of each of these names to keep them clear):

- **Chemoheterotrophs:** Organic carbon compounds are consumed both for energy and for carbon, which is used to build new biological compounds. (Humans are chemoheterotrophs.)
- **Photoheterotrophs:** Use light as an energy source, but cannot use CO_2 for a carbon source, relying on other organic compounds for biosynthesis
- **Photoautotrophs:** Use light for energy and carbon dioxide for carbon, like plants
- **Chemoautotrophs:** Use inorganic compounds for energy and use carbon dioxide for biosynthesis

Q A strain of bacteria oxidizes lactose for energy and milk proteins, sugars, and fats for metabolic building blocks. Is this organism a chemoautotroph?

A No. It is a chemoheterotroph, using organic compounds for energy and for building blocks.

Q A bacteria living in rice fields uses light for energy and organic compounds from the surrounding water as a carbon source for biosynthesis. This type of bacteria is which of the following?

(A) Chemoautotroph
(B) Chemoheterotroph
(C) Photoautotroph
(D) Photoheterotroph

A The correct answer is (D). Photoheterotrophs are photosynthetic and use carbon from organic compounds rather than carbon dioxide.

Another way of classifying organisms is based on how they use or tolerate oxygen:

- **Obligate anaerobes**: Can only survive without oxygen present
- **Obligate aerobes**: Can only survive with oxygen present
- **Facultative anaerobes**: Can survive in either condition. If oxygen is present, they will use it; if it is not, they will survive using anaerobic respiration.

Q A yeast species can grow aerobically and perform oxidative phosphorylation when oxygen is present but can also use fermentation to produce energy when oxygen is not present. How is the yeast best described?

A The yeast described is a facultative anaerobe.

Archae

Another group of prokaryotes, the archae, were not recognized as such until recently, but are so distinct from both bacteria and eukaryotes that they now occupy one of the three fundamental domains of living things.

Archae resemble bacteria in some ways and eukaryotes in others. Like bacteria, archae cells are relatively simple, lacking organelles and possessing a circular DNA genome. On the other hand, their form of transcription and translation is more akin to that of eukaryotes than bacteria. Like eukaryotes, archae genes sometimes have introns, and their genes are transcribed by more than one RNA polymerase. Archae cell walls are not made of peptidoglycan, and the lipids in their membranes are distinct from both bacteria and eukaryotes. These fundamental and unique differences have caused biologists to classify archae in their own domain.

One method used to compare archae, eukaryotes, and bacteria is the molecular characterization of the genes in each group, such as the sequence of ribosomal RNA genes. In some cases, it appears as though in the remote past, genes may have moved horizontally among these three domains, complicating the molecular analysis of the evolutionary relationship among the domains.

One of the most notable traits of some archae is their ability to live as "extremophiles" in harsh environmental conditions where nothing else can survive. Archae that are thermophiles (lovers of heat) thrive in hot springs or deep ocean vents that can exceed the boiling point. The DNA polymerases used in PCR have been adapted from such thermophilic archae. Archae membrane lipids are often long, ether-linked lipids, spanning all the way across the cell membrane with polar groups on each end. Lipids with this unusual structure may help explain the ability of archae to live and thrive in such harsh environments.

Archae are not limited to volcanic vents and extremely acidic or salty conditions. Analysis of sea water suggests that even the open ocean may be home to large numbers of archae cells that are difficult to grow in laboratory cultures, playing a previously unappreciated role in the world's ecosystems.

Q Which of the following is not a characteristic of archae?

(A) Small linear genome
(B) Presence of ether-linked membrane lipids
(C) Presence of introns in some members
(D) Absence of organelles
(E) Ability of some members to survive in water under pressure above 100°C

A The correct answer is (A). Archae have a circular DNA genome. The other choices are all characteristics of archae.

PROTISTS

Protists include a wide variety of single-celled eukaryotes, as well as a few odd multicellular organisms. In past taxonomic systems, these organisms were often grouped together in their own kingdom, since biologists did not know where else to put them. Realizing today that these organisms are not closely related to each other in their descent or morphology, scientists no longer group them together.

Protists get nutrition in a variety of ways, including photosynthesis, absorption of food, and ingestion of food. They reproduce by a variety of methods, both sexual and asexual. Some are plant-like (photoautotrophs), others animal-like (ingestive chemoheterotrophs), and others fungus-like (absorptive chemoheterotrophs). Various taxonomists treat the classification of protists differently, with some protists classified with the other groups they resemble. Common examples of protists include:

- **Euglena:** These single-celled organisms with chloroplasts are photosynthetic when the sun is shining but are also able to use heterotrophic nutrition, ingesting food like smaller cells. Euglenas have a flagellum, chloroplasts, and a nucleus.
- **Giardia:** A parasite that causes severe diarrhea when ingested. They have mitochondria that are reduced in size and function, lacking most normal mitochondrial functions.

- **Plasmodium:** The malaria parasite, plasmodium has a complex life cycle involving two hosts—mosquitoes and humans (or other mammals). Malaria is one of the world's most serious health problems, infecting hundreds of millions of people. The parasite's increasing resistance to drugs, the difficulty of vaccine development, and setbacks with mosquito control have hampered treatment. The parasite appears to evade immune responses by constantly changing the proteins it expresses on the cell surface.
- **Paramecium:** Like other ciliates, paramecium cells have large numbers of cilia on their cell surface that are involved in locomotion and gathering food. Paramecium cells are large and complex, with a fairly rigid shape and an oral groove in which food is collected and ingested, forming food vacuoles. A contractile vacuole pumps out water that enters the cells, since they live in fresh water and must maintain osmotic balance. Ciliates have large macronuclei that control most cell functions, as well as small micronuclei. They usually reproduce asexually but have the ability to exchange micronuclei through conjugation.
- **Amoebas:** Amoebas are singled-celled organisms, without a defined rigid shape, that move and capture prey by means of flexible projections called pseudopodia. Prey is ingested by phagocytosis. Amoebas reproduce asexually by binary fission.
- **Diatoms:** Diatoms are algae cells, unicellular photosynthetic phytoplankton, with silica cell walls formed in intricate shapes and patterns. Much of the phytoplankton in the oceans is composed of diatoms.
- **Radiolarians:** Single-celled marine organisms with a glassy, silica-based shell, radiolarians have spiky pseudopodia that extend out in all directions to catch passing prey.
- **Dinoflagellates:** Members of the marine environment, these flagellated organisms are a major component of plankton, and many are photosynthetic phytoplankton. Some dinoflagellates cause red tide, algal blooms that can release toxins that poison other marine life.
- **Red and green algae:** Of all protists, these have the closest relationship to plants. Photosynthetic unicellular protists, they have chloroplasts with two membranes. They have a life cycle with generations alternating between haploid and diploid forms in sexual reproduction.
- **Brown algae and red algae seaweeds:** Although protists are often generalized to be single-celled, brown algae are an exception. The seaweed and kelp along the coasts of many parts of the world are brown algae species. Although resembling plants, with apparently complex tissues, these organisms are not directly related to land plants. They also have complex life cycles.

 Are protists classified in the same group, despite their varied appearance, due to their apparently close relationship, by molecular analysis, in phylogenetic trees?

 No. They are basically grouped together because biologists do not know where else to put them.

FUNGI

Fungi are eukaryotes and absorptive chemoheterotrophs. This group includes molds, yeasts, and mushrooms. They are not flashy, like flowering plants, or mobile, like fun-loving animals, but they play a huge role in ecosystems. Through absorptive nutrition, they digest material outside their body and then absorb the digested material, secreting enzymes into their environment that break down biological macromolecules of dead or living organisms. In digesting dead organisms, fungi help to decompose organic material, returning the material to nutrient cycles. When fungi live on other organisms and draw their nutrition from them, they can live as parasites, or in some cases, establish a symbiotic relationship. Most plants live in symbiosis with fungi called mycorrhizae, which operate in conjunction with the roots of plants. The fungi in these symbiotic relationships help plants extract nutrients from the soil, increasing the surface area used for this purpose beyond the plant's own roots. In return, the fungi get a supply of sugars produced by plants through photosynthesis. Lichens are another symbiotic relationship between fungi and algae.

 Are fungi found on a dead tree acting as parasites?

 No. If the tree was living this might be the case, but since the tree is dead, the fungus is acting as a decomposer, a common role for fungi in ecosystems.

Yeasts are single-celled fungi, but other fungi are multicellular, with bodies consisting of interconnected tube-like filaments called **hyphae**. If a large number of hyphae are present, they form a web-like collection called a **mycelium**. The fungi cell wall is made of chitin.

Fungi can reproduce sexually or asexually. In either case, they produce spores that are dispersed into the environment in large numbers. If spores encounter good growth conditions, they will start to grow into a new fungus. The hyphae and spores of fungi are usually haploid; diploid cells are only found in a small part of the sexual life cycle, followed by meiosis and spore production. Asexual reproduction can occur through production of spores or through pieces of hyphae broken from the rest of a fungus, which can become a new individual. Sexual reproduction produces spores as well, but varies considerably in the details between different fungal groups. Some fungi reproduce only asexually.

Q Asexual reproduction in a mold can allow it to rapidly spread. What advantage can sexual reproduction offer?

(A) The ability to produce haploid spores
(B) Greater dispersion of spores
(C) Avoidance of the energetically costly diploid stage of the life cycle
(D) Increased genetic variation in future generations
(E) Escape from predation

A The correct answer is (D). Sexual reproduction does not allow faster growth but increases genetic variation to contend with environmental variation.

In several cases, fungi have been cultivated for use by man. Yeasts, for example, have played an important role in human history, fermenting the ethanol in beer and wine as well as producing leavened bread. Some cheeses are produced with the help of fungi, and fungi have been the source of key antibiotics, such as penicillin. Fungi are farmed and eaten or collected in the wild, although mistakes in mushroom collecting can be deadly due to the toxins that some mushrooms produce (presumably to fend off predators).

PLANTS

Terrestrial plants evolved from single-celled green algae about 470 million years ago, starting their conquest of the land perhaps from green algae living in shallow water that evolved to tolerate periods of dryness on the shore. Biologists sometimes classify green algae as plants due to this evolutionary relationship. To live on land, plants evolved:

- greater structural support to deal with gravity
- the ability to get necessary nutrients to support photosynthesis
- the ability to reproduce in a nonaquatic environment
- mechanisms to prevent drying, such as a waxy cuticle layer

The major groups of surviving plant phyla are classified based on whether they produce seeds, how they produce seeds, and whether or not the plant has vascular tissue. Some of the key traits of all plants include:

- **Alternation of generations** between haploid **gametophytes** and diploid **sporophytes**
- Embryos that develop from diploid zygotes and remain closely associated with maternal tissues, depending on the female parent for nutrition
- Walled haploid spores produced by sporophytes that grow into multicellular haploid gametophytes. The spore wall helps plants reproduce on land, preventing desiccation of the spore. The gametophytes produce gametes that complete the alternation of generations.

- **Apical meristem**, with growth of plants focused at the tips of roots and stems. It is through the proliferation of cells at the apical meristem that plants grow, elongating roots and stems.

The alternation of generations in all plants involves two different stages of plant development—the **gametophyte** and the **sporophyte**. Haploid gametophytes grow through mitosis and produce haploid gametes. When two haploid gametes join through fertilization to form a zygote, the diploid zygote then grows to form a sporophyte. The sporophyte produces haploid spores through meiosis and the spores grow to form the haploid gametophyte again, back to where things started. In the more primitive nonvascular plants, such as mosses, the haploid gametophyte is larger and more dominant in the plant life cycle, with the sporophyte playing a minor role. In more recently evolved groups, the sporophyte is more important in the life cycle. The plant we see is the sporophyte and the gametophyte is hidden deep inside the tissues of the sporophytes.

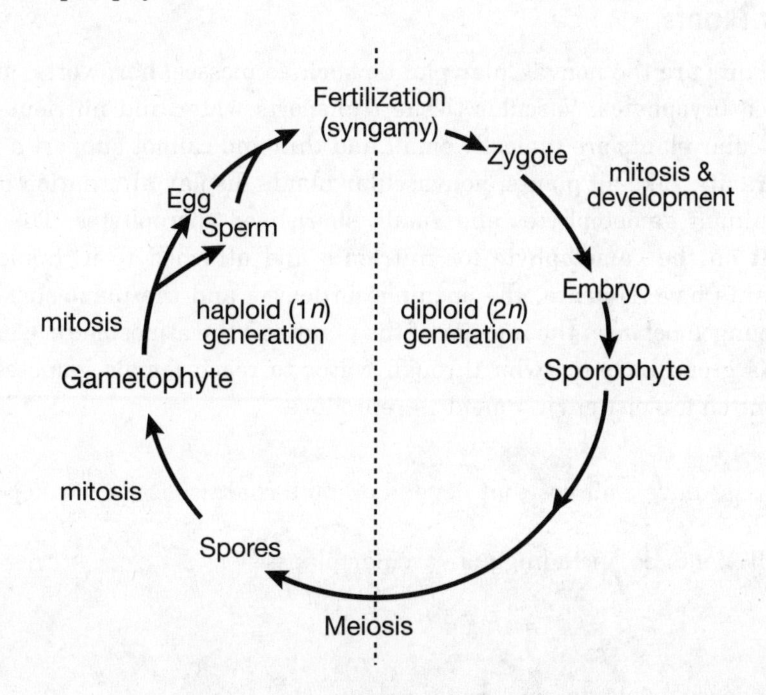

Alternation of Generations in Plants

Q Do sporophytes and gametophytes both perform meiosis?

A Only sporophytes are diploid; they produce haploid spores through meiosis.

While animals and many other organisms have developmental programs that limit growth to a certain size or age, one of the unique traits of plants is that most never stop growing. This trait relates to the growth of plant tissue called meristem, which is found dispersed at many locations in plants, either at the tips of roots and shoots (**apical meristem**) or along the lengths of roots and shoots (**lateral meristem**). The proliferation of cells in apical meristem makes roots and shoots grow longer, while proliferation of cells in lateral meristem makes roots and shoots grow thicker. The apical meristem creates primary growth, while the lateral meristem creates secondary growth.

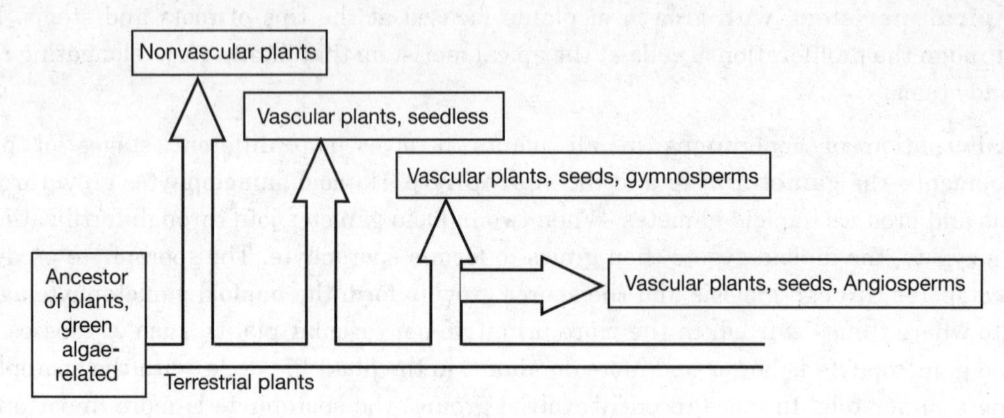

Evolution of the Major Plant Groups

Nonvascular Plants

The simplest plants are the nonvascular plants, such as mosses, hornworts, and liverworts, sometimes called bryophytes. Vascular tissue transports water and nutrients in the plant body, so nonvascular plants are typically small and thin and cannot support a large, upright mass against gravity. Like all plants, nonvascular plants display alternation of generations, with large, dominant gametophytes and small, short-lived sporophytes. The sporophyte is often dependent on the gametophyte for nutrients and attached to it. Some, but not all, nonvascular plants have stomata, the openings in leaves and the plant surface that allow gases to be exchanged between the interior of the plant and the atmosphere. Generally, sperm of plants in this group have to swim through water to reach female gametes, making the plants dependent on a moist environment to reproduce.

Q Do mosses have embryos that depend on maternal tissues to develop?

A Yes, all plants do, including nonvascular plants.

Q Do nonvascular plants have roots?

A No. They would not be able to move nutrients from the roots to the rest of the plant even if they had roots, since they have no vascular tissue.

Q Do bryophytes produce seeds?

A Not even close. Bryophytes are nonvascular plants and produce neither seeds nor vascular tissue.

Vascular Seedless Plants

Although mosses and other nonvascular plants succeeded in colonizing land, their lack of vascular tissue restricted where they could grow. The evolution of vascular tissue in plants like ferns and horsetails allowed these plants to grow taller than mosses—reaching toward, and blocking out, sunlight—and to live in dryer environments. Vascular plants also have stronger cell walls, providing more structural support against gravity. In addition to vascular tissue, these plants also have roots and leaves. Roots allow vascular plants to absorb water and essential minerals from soil, while leaves allow specialization to maximize the collection of light for photosynthesis, producing more nutrients that are transported to the rest of the plant through the vascular system.

The vascular tissue includes **xylem** and **phloem**. Xylem is the tissue that moves water and soil nutrients up from roots to the rest of the plants through **tracheids** or vessel elements. Tracheids, which carry water, are the remains of dead cells, leaving only their tube-like cell walls to carry water upward through the plant from the roots. The phloem carries sugar and other nutrients from leaves to the rest of the plant through **sieve cells** or **sieve-tube members.** These cells lack most organelles, including a nucleus, but are connected to neighboring companion cells that maintain all metabolic activities and support the cells. Sieve-tube member cells are connected end-to-end with a permeable section of cell wall, called the **sieve plate**, further allowing nutrients to pass directly through the interconnected chains of cells.

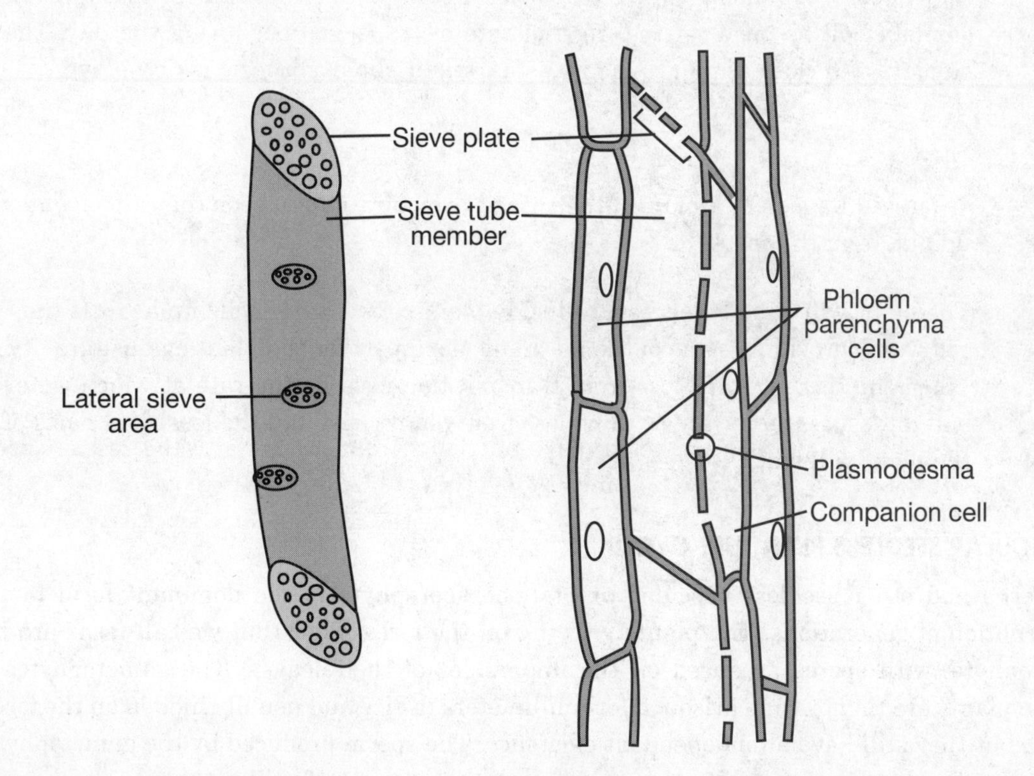

Q Does transport in the xylem or phloem involve dead cells?

A Cells in the phloem are alive, while the tracheids carrying water in the xylem are the cell walls of dead cells.

Water is moved through the vascular tissues of plants by forces both pushing and pulling at it, including osmotic potential and physical pressure. Diffusion can move fluids and solutes between adjacent cells, but movement throughout the plant requires large-scale transport of fluid and solutes through the xylem and phloem.

Here's how it works. Evaporation of water from leaves in, say, a tree reduces the pressure in the higher regions of the plant's xylem. Lower pressure in this part of the vascular system draws water upward. The adhesive nature of water in the narrow tracheids and vessel tube members also helps draw the water pulled by the transpiration of water from leaves up from the roots. Roots draw water in from the soil, pushing water up through the xylem as well, but the upward pressure of transpiration and surface tension are more important.

Q During cool, sunny weather, does more water move upward through xylem in day or nighttime in a C3 plant?

A C3 plants open their stomata during the day, unless the weather is too hot and they must keep them closed to conserve water. In cool, sunny weather, the stomata will be open, increasing the rate of transpiration during the day. The stomata will close at night when it is dark and photosynthesis is not active.

Q What will happen to a plant if the rate of transpiration exceeds the rate of water uptake from the roots?

A The plant will wilt. The pressure inside plant cells pushes cells up against their cell walls, providing some of the mechanical support for the plant against gravity. When the rate of water loss from transpiration exceeds the rate at which water enters tissues, then water will be lost from plant cells, and the less rigid parts of the plant will wilt.

VASCULAR SEEDLESS PLANT LIFE CYCLE

In ferns and other seedless vascular plants, the sporophyte is the dominant form in the alternation of generations. The plants growing on the forest floor that we call ferns are the sporophyte, with spores produced on the undersides of their leaves. The gametophytes of these plants are tiny by comparison, a few millimeters in size and usually hidden on the forest floor, but they still have an independent existence. The sperm produced by the gametophytes of these plants still require water to reach female gametes and fertilize them.

Q What stage of the fern life cycle is created by the fertilized egg?

A The fertilized egg becomes the diploid zygote, which will grow into the diploid sporophyte, which is the fern plant we know and love.

Q Do mosses have tracheid cells?

A No, tracheids are part of vascular tissue. Mosses are nonvascular plants.

Q Do the walled spores of ferns allow them to live in arid environments?

A Spores resist desiccation, but the plants still need some water for sperm to fertilize female gametes.

Seed Plants (Gymnosperms and Angiosperms)

In the colonization of land by plants, vascular tissue was a big advance. Another was the evolution of seeds. Seeds pack the developing plant embryo and nutrients inside a protective coat that can lie dormant and resist drying out until conditions are favorable for growth. Seeds allowed plants to further colonize new areas on land, reducing their dependence on wet environments for reproduction. Since seeds carry a lot more nutrients than spores, these plants could survive into a broader range of environments.

The gametophytes in seed plants are much smaller than those in nonvascular plants and ferns, so small that they do not have an independent existence, developing instead inside parental tissues. Shielded inside the parent, they do not have to contend with the harsh terrestrial environment. Male gametophytes are contained within pollen, whereas female gametophytes are contained within the ovule of the female plant. When pollen is carried from the male to the female, the male gametophytes move through a pollen tube in the ovule to fertilize the female gamete. Once their pollen reaches the female, the male gametophytes only have to pass within the female flower. Male gametes do not have to swim in the open to reach the female gametes in seed plants.

GYMNOSPERMS

Often referred to as evergreens, gymnosperms (which means "naked seeds" in Latin) have seeds that are not enclosed in an ovary. While seedless vascular plants dominated the Carboniferous period, laying down deposits that were later transformed into coal, forests of gymnosperms, including pine trees, other conifers, and gingko trees, gradually took over between 350 to 150 million years ago.

Like all plants, gymnosperms have an alternation of generations between haploid sporophytes and diploid gametophytes. Their sporophyte is very dominant—it is the tree we see—with only a very small gametophyte enclosed within the tissues of the sporophyte.

Q If a redwood tree is a sporophyte, where are the spores it produces?

A The spores are produced in the cones, with male cones producing pollen and female cones containing ovules.

Q The sperm of nonvascular plants and vascular seedless plants are flagellated and motile, while the sperm of many gymnosperm species lack flagella and are nonmotile. How does the sperm reach the egg in such a species?

A The sperm is carried in a pollen grain on the wind to reach the ovule. Once it reaches the ovule, it is carried by the developing pollen tube to reach the egg nucleus within the female gametophyte. In more primitive plants, the sperm must reach eggs by swimming in the open for a short distance, so they must be flagellated.

Conifers produce large volumes of pollen from male cones. The pollen is released to the wind to reach female cones. Each pollen grain develops from a microspore and contains a male gametophyte. The female cone contains megaspores that develop into female gametophytes inside the ovule. When a pollen grain reaches the female ovulate cone, the pollen grain creates a pollen tube, allowing male nuclei to reach the ovule and fertilize the female gamete. The union of male and female haploid gametes creates the diploid zygote that develops inside the ovule. The ovule becomes a seed containing the embryo.

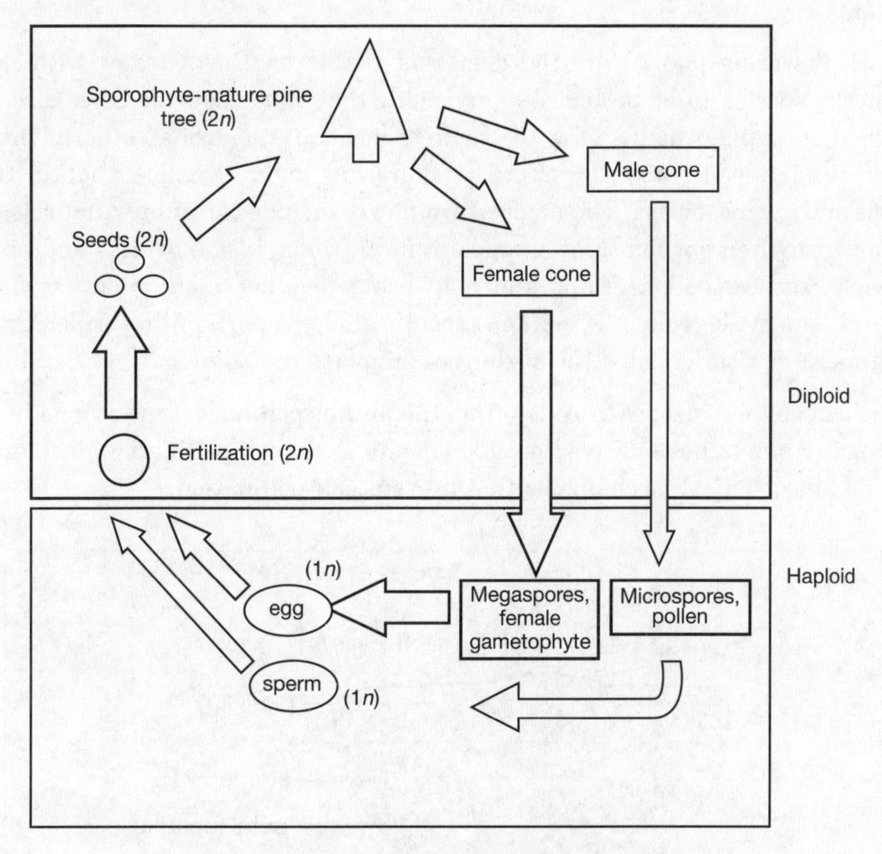

Gymnosperm Life Cycle

One note about fertilization in gymnosperms: it can take a long time. A year or more can pass between pollination until fertilization, and even more time is needed for a seed to mature.

Q Do gymnosperms use insect pollinators?

A No, generally they rely on the wind for pollination.

Q Drier environments would most tend to favor which of the following?

(A) Nonvascular plant

(B) Gymnosperm

(C) Vascular seedless plant

A The correct answer is (B). Drier environments would most favor the gymnosperm, as it has seeds. The nonvascular plant and the seedless vascular plant have sperm that must swim through water to reach the female gamete, requiring a moist environment for reproduction. This step does not occur in gymnosperms and the dry nature of seeds allows them to resist desiccation as well.

ANGIOSPERMS

Angiosperms (flowering plants) are the dominant plants on Earth today, with more than 250,000 known species (and probably many more that have not yet been discovered by Western scientists). Angiosperms have the same basic adaptations of all other plants, as well as vascular tissue, seeds, and flowers. The flower is the innovation that distinguishes angiosperms and gymnosperms. The angiosperm flower includes **stamens** that release pollen from the upper **anther** portion. Flowers also contain ovules inside ovaries and **carpels** (or **pistils),** which receive pollen grains and help bring together male gametes with female gametes inside the ovules and ovaries. The carpel includes a portion that pollen grains stick to, the **stigma**, and a stalk called the **style** that supports the stigma.

Some plants can self-fertilize, with pollen from the anther pollinating the stigma in the same flower, but many plants have evolved mechanisms to avoid self-pollination, including flower structures and biochemical mechanisms that prevent self-pollination.

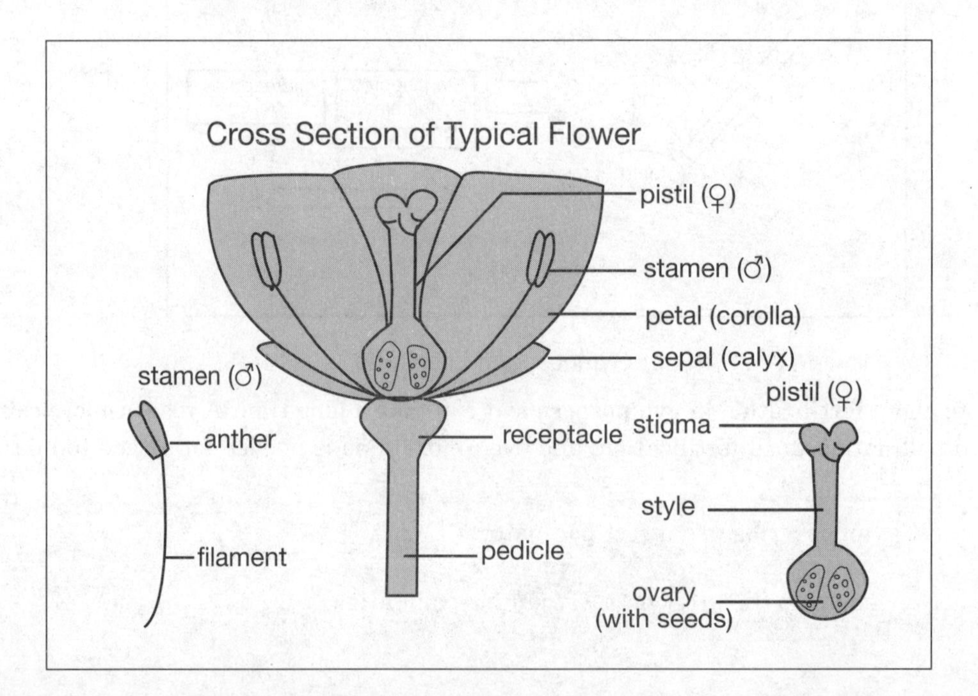

Q Self-pollination would be the easiest, lowest energy route for plants to achieve pollination. For plants to evolve such that they avoid self-pollination, there must be an advantage to pollination between different plants. What is this advantage?

A Self-pollination restricts mating to those alleles that occur in the same plant. In a population of plants with some genetic diversity, there are probably many more possible combinations of alleles that can be achieved by mating with other plants. This genetic diversity has a real adaptive value for the population and species, maintaining and creating new genetic diversity that improves the ability to survive environmental challenges.

In the angiosperm's alternation of generations (between haploid sporophytes and diploid gametophytes), flowers produce both male and female gametophytes. Ovules contain the female gametophytes, including the unfertilized egg. Male gametophytes are produced inside pollen grains, each of which contains two haploid cells. When a pollen grain lands on a stigma, one of the haploid cells goes through mitosis to form two sperm cells, while the other haploid cell forms a tube from the stigma to the ovule. The two sperm cells move through the pollen tube until they reach the female gametophyte. There, one sperm cell fertilizes the egg and the other sperm cell fuses with two nuclei in the female gametophyte to create a triploid ($3n$) cell. This **double fertilization** is one of the key features of all angiosperms. The diploid fertilized egg develops into the plant embryo and the triploid fertilized cell develops into the **endosperm**. The early plant embryo has a simple root and cotyledons, simple embryonic leaves. The endosperm is part of the seed that helps nourish the plant embryo when it starts to germinate and develop.

In addition to sexual reproduction, many plants can readily reproduce asexually, due to the decentralized nature of plants. Asexual reproduction allows plants to locally reproduce rapidly, although plants reproduced this way lack genetic diversity. Shoots or roots from a plant can spread, allowing asexual growth, or pieces of a plant can create new individuals. Asexual reproduction is often exploited in agriculture, particularly for varieties of plants that are bred to be seedless.

Plant Evolution and Diversity

Animals and plants have coevolved in many ways, including the intimate interactions between flowering plants and pollinating insects, the defenses plants create to protect against predation, and the role of animals in dispersing plant seeds. Many flowering plants rely on insects, birds, or other animals to assist in pollination. To accomplish this, the plant evolves to attract the pollinator and improve the ability of the pollinating animal to carry pollen from flower to flower. The brightly colored petals of flowers are specialized leaves that help to attract the animals that pollinate them. Not all angiosperms have brightly colored flowers; angiosperms that are wind-pollinated, like grasses, will often have less obvious flowers. Many plants have evolved with flowers that require specific insects for pollination, and these insects have coevolved to fit that ecological niche.

There are a few ways that plants rely on animals to disperse their seeds. In some plants, the ovary grows in size after fertilization to become a fruit. When a fruit ripens, the tasty, sweet fruit attracts birds or mammals who eat, and then disperse, the seeds. Or the fruit can fall to the ground, rot, and provide nutrients to the seeds. Dispersal of seeds to new locations can help plants exploit broader ecosystems and better survive local changes. Animals can carry the seeds and defecate in a distant location, depositing the seed, along with a fresh supply of fertilizer. Some plants have seeds that evolved to stick to animal fur and thus be carried to more distant locations that would not be accessible if seed deposition was left to wind or gravity.

In addition to relying on animals, in some cases, plants have evolved the means to repel animals that might otherwise eat and destroy the plant. Plants have physical defenses like thorns (acacia, roses, raspberries) and spines (cacti), chemical irritants (poison oak), and chemical defenses inside, such as toxins that keep animals from eating plants (nicotine in tobacco). These close interactions of plants and animals as they coevolve create important parts of ecosystems.

Plant Hormonal Activity

Like animals, plants have a variety of hormones that act through signaling mechanisms, including second messengers like intracellular calcium, which turn transcription of genes on and off and regulate protein kinases. The signals that plants respond to include growing toward light (phototropism) and water, gravity, changes in the seasons, and environmental stress. Plant hormones include:

- **Auxin:** The chemical hormone that causes auxin responses is indoleacetic acid. Auxin can produce many different responses in plants depending on when, where, and how it is applied. Auxin can cause stems or roots to grow faster or cause plants to grow in one direction or another by causing cells on one side of the plant to grow faster.
- **Ethylene:** Ethylene is a volatile gas that plants produce as a hormone during fruit ripening and in response to many different types of plant stress. Commercially, it is used in agriculture to ripen fruit that is picked early for easier transportation.
- **Abscisic acid:** While most other plant hormones increase the rate of growth of specific cells, abscisic acid slows the rate of cell growth. Abscisic acid is present in high concentration in seeds, keeping the plant embryo dormant until conditions are right for germination.
- **Gibberellins:** Plant hormones that causes growth of plants, including growth of fruit, growth of stems, and germination of seeds.
- **Cytokinin:** Plant hormone that is a modified version of adenine and causes cell growth.

Plant responses to hormones often involve multiple hormones working together. It is often hard to isolate the effects of one hormone from the other. Auxins can cause stem elongation at low concentrations, but inhibit stem growth at higher concentrations by causing ethylene to be released. The effects that cytokinins have on cells depend on how much auxin is present.

EXERCISES: PROKARYOTES, PROTISTS, FUNGI, AND PLANTS

1. In which of the following is the gametophyte larger than the sporophyte?

(A) Moss
(B) Fern
(C) Palm tree
(D) Rose
(E) Spruce tree

2. A plant is identified with a mutation causing seeds to germinate prematurely in winter. Which of the following is most likely the nature of the mutation in this plant?

(A) Silent mutation in the receptor for gibberellins
(B) Nonsense mutation in enzyme that produces abscissic acid
(C) Missense mutation in receptor for cytokinin
(D) Deletion of gene encoding enzyme that produces ethylene
(E) Nonsense mutation in receptor for gibberellins

3. If an angiosperm is grown in soil that has been sterilized to prevent the growth of mycorrhizal fungi, the effect on the plant is most likely to be which of the following?

(A) The plant will grow smaller and slower than a plant grown in nonsterilized soil.
(B) The plant will grow taller and bloom earlier than a plant grown in nonsterilized soil.
(C) The plant will be shorter but produce larger fruit than normal.
(D) The plant will be unable to perform photosynthesis.
(E) Transpiration will be greatly increased.

4. One of the traits shared by bacteria and eukaryotes is:

(A) Mitochondria
(B) Krebs cycle
(C) Flagella with microtubules
(D) Introns
(E) PolyA tail addition in mRNA

5. Which of the following protists utilizes photosynthesis for nutrition?

(A) Diatoms
(B) Radiolarian
(C) Paramecium
(D) Amoeba
(E) Plasmodium

6. Which of the following is commonly diploid?

(A) Hyphae of fungi
(B) Gametophyte of gymnosperm
(C) Sporophyte of vascular seedless plant
(D) Sperm cell of a nonvascular plant
(E) Gametophyte of a nonvascular plant

7. In angiosperms, fertilization occurs inside the:

(A) Pistil
(B) Ovule
(C) Stamen
(D) Microspore
(E) Carpel

8. Angiosperm pollen grains contain which of the following?

(A) A single diploid cell
(B) A single haploid cell
(C) Two diploid cells
(D) Two haploid cells
(E) A triploid cell and a diploid cell

ANSWER KEY AND EXPLANATIONS

1. A	3. A	5. A	7. B	8. D
2. B	4. B	6. C		

1. **The correct answer is (A).** Only in the nonvascular plants like mosses is the gametophyte the dominant larger stage in the alternation of generations.
2. **The correct answer is (B).** Abscisic acid accumulates in seeds and keeps plant embryos dormant until conditions are ready for germination. A nonsense mutation would create a premature stop codon in the enzyme producing this hormone, so none of it could be made, and the seed embryos would germinate early.
3. **The correct answer is (A).** Plants gain increased access to soil nutrients through their symbiosis with mycorrhizal fungi. Without the fungi, their growth is impaired.
4. **The correct answer is (B).** Both bacteria and eukaryotes have the Krebs cycle as part of cellular ATP production in aerobes. Bacteria lack mitochondria, flagella with microtubules, introns, and polyA tail addition.
5. **The correct answer is (A).** Diatoms are photosynthetic protists and a significant part of marine phytoplankton.
6. **The correct answer is (C).** Sporophytes of plants are diploid. The sporophyte produces haploid spores, which yield haploid gametophytes and gametes. The union of gametes through fertilization brings the cycle around to the diploid sporophyte again.
7. **The correct answer is (B).** Ovules are inside the ovary. Each ovule contains the female gametophyte that produces the egg that gets fertilized by sperm from pollen.
8. **The correct answer is (D).** These both become sperm cells—one fertilizes the egg to form the plant embryo and the other creates the triploid endosperm.

SUMMING IT UP

- Prokaryotes, including the domains bacteria and archae, are small, simple organisms lacking the internal subcellular organelles found in eukaryotes. Descriptions of prokaryotic cells often begin with their characteristic shape under the microscope: round (**cocci**), spiral-shaped, or cylinder-shaped (**bacilli**). These characteristic shapes are determined by the cells' rigid walls.
- Some bacteria are surrounded by a **capsule**, a layer of sticky sugars that can help attach to surfaces and help bacteria hide from the immune system. Bacteria with capsules are often much more pathogenic (disease-causing) than related bacteria lacking capsules.
- Bacteria have smaller and simpler genomes than eukaryotes and, as a result, can rapidly replicate, often producing a new generation in an hour or less when the supply of nutrients is optimal. This rapid generation time, and the presence of many bacteria, can result in rapid adaptation of bacteria to a changing environment.
- Archae resemble bacteria in some ways and eukaryotes in others. Like bacteria, archae cells are relatively simple, lacking organelles and possessing a circular DNA genome. On the other hand, their transcription and translation have some aspects that are more like those of eukaryotes than bacteria. Like eukaryotes, archae genes sometimes have introns and have their genes transcribed by more than one RNA polymerase. Archae have different cell walls than bacteria—not made of peptidoglycan—and the lipids in their membranes are distinct from both bacteria and eukaryotes.
- Protists include a wide variety of single-celled eukaryotes, as well as a few odd multicellular organisms. Common examples of protists include **euglena, giardia, plasmodium, paramecium, amoebas, diatoms, radiolarians, dinoflagellates, red and green algae,** and **brown algae** and **red algae seaweeds.**
- Fungi are eukaryotes and absorptive chemoheterotrophs. This group includes molds, yeasts, and mushrooms. Using absorptive nutrition, they digest material outside their body and then absorb the digested material, secreting enzymes into their environment that break down biological macromolecules of dead or living organisms.
- To live on land, plants evolved:
  - greater structural support to deal with gravity
  - the ability to get necessary nutrients to support photosynthesis
  - the ability to reproduce in a nonaquatic environment
  - mechanisms to prevent drying, such as a waxy cuticle layer
- The alternation of generations in all plants involves two different stages of plant development—the **gametophyte** and the **sporophyte**. Haploid gametophytes grow through mitosis and produce haploid gametes. When two haploid gametes join through fertilization to form a zygote, the diploid zygote then grows to form a sporophyte. The

sporophyte produces haploid spores through meiosis and the spores grow to form the haploid gametophyte again, back to where things started.

- The simplest plants are the nonvascular plants, such as mosses, hornworts, and liverworts, sometimes called bryophytes. Nonvascular plants are typically small and thin and cannot support a large weight against gravity. Like all plants, nonvascular plants display alternation of generations, with large, dominant gametophytes and small, short-lived sporophytes. Some, but not all, nonvascular plants have stomata, the openings in leaves and the plant surface that allow gases to be exchanged between the interior of the plant and the atmosphere.
- Vascular plants have stronger cell walls, providing more structural support against gravity and helping them grow taller on land. In addition to vascular tissue, these plants also have roots and leaves. Roots allow vascular plants to absorb water and essential minerals from soil, while leaves allow specialization to maximize the collection of light for photosynthesis, producing more nutrients that are transported to the rest of the plant through the vascular system.
- Self-pollination restricts mating to those alleles that occur in the same plant. In a population of plants with some genetic diversity, there are probably many more possible combinations of alleles that can be achieved by mating with other plants. This genetic diversity has a real adaptive value for the population and species, maintaining and creating new sexual diversity that improves the ability to survive environmental challenges.
- Vascular tissue includes xylem, which moves water and nutrients from the roots up to the rest of the plant, and phloem, which moves nutrients from the leaves down to the rest of the plant.
- Vascular seedless plants, like ferns, have a dominant sporophyte and a small, independent gametophyte. Sperm must swim through water to reach female gametes.
- Gymnosperms, like pine trees, have seeds, a large dominant sporophyte, and a gametophyte that is surrounded and enclosed within the sporophyte.
- Angiosperms are the flowering plants. Angiosperm flowers include the stamen, where pollen is produced, and the ovary, where the female gametes inside ovules are produced and fertilized. Angiosperm fertilization includes a double fertilization that creates the endosperm in the angiosperm seed.

Animal Structure and Function

chapter 13

OVERVIEW

- **Body design**
- **Embryonic development: protostomes vs. deuterostomes**
- **Invertebrates**
- **Vertebrates**
- **Mammal structure and function**
- **Summing it up**

Animals often receive more attention from biologists than other forms of life, perhaps because we, as animals ourselves, find this kingdom particularly captivating. As diverse as animals are, however, all animals share some key features. For example, all animals are chemoheterotrophs, which means that they receive energy and carbon from food sources that are ingested. All animals are also multicellular and most reproduce sexually (although there are some exceptions), producing haploid sperm and egg that join during fertilization to create the next diploid generation.

Despite the role of haploid gametes and diploid organisms in animal reproduction, this kingdom of living creatures does not display alternation of generations as members of the plant kingdom do. The haploid stage in animals is limited to gametes and their direct precursor cells. There are also, of course, many distinct characteristics that define each major group of animals.

BODY DESIGN

Animal bodies have a few common features that differentiate them from various other groups. Among these are two types of body symmetry, **bilateral symmetry** and **radial symmetry**. Bilateral symmetry occurs when one side of an animal's body is a mirror image of the other side. The body structure of animals with radial symmetry, on the other hand, is arranged like a wheel.

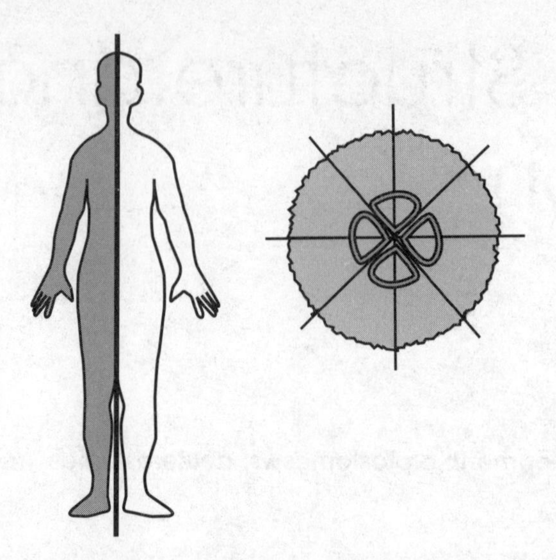

Q Do humans have bilateral or radial symmetry?

A Bilateral. The left and right halves of the human body are more or less a mirror image of one another. This is mainly true of vertebrates as well as of arthropods and many other animal groups.

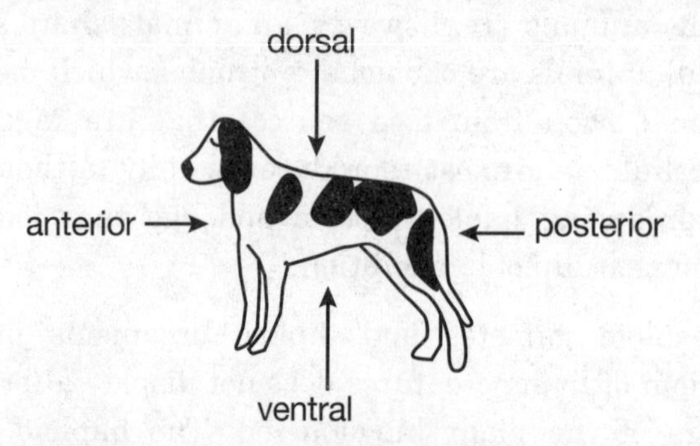

Body plans are also described according to their layout along two main body axes. The head region is often called the anerior, while the opposite end at the rear of the animal is called the posterior. The lower surface of animals is often called ventral, while the back is called dorsal.

Q Is the human spinal cord on the dorsal or ventral side of the body?

A Dorsal.

Another basic feature that differentiates animal phyla is the type of internal cavities they possess for digestive and other bodily functions. Some simple animals do not have a digestive tract, so they either ingest nutrients directly into cells (as sponges do), or they have a digestive sack with only one entrance (as cnidarians and nematodes do). Most other phyla have a complete digestive tract, including a gastrointestinal tract and an alimentary canal. In a digestive tract, nutrients enter via one opening (the mouth) and pass in one direction through the system; waste is excreted via an opening at the other end (the anus).

In some organisms, such as annelids and vertebrates, another body cavity called the **coelom** surrounds the digestive tract and other organs. The outer body wall and the inner organs are lined with tissue derived from mesoderm-like muscle containing the liquid-filled coelom cavity. Having a coelom allows an earthworm to contract and relax the bands of muscle around this cavity, using the liquid in the coelom as if it were a skeletal structure. For vertebrates, the coelom includes the space around the organs of the chest and abdomen, allowing the organs to move and function. In some animals, such as nematodes, a **pseudocoelom**, in which only one side of the internal cavity is lined with muscle from mesoderm, is present instead. **Acoelomate** animals (those without a coelom), like flatworms, have solid tissue filling the space from the digestive tract to the external surface.

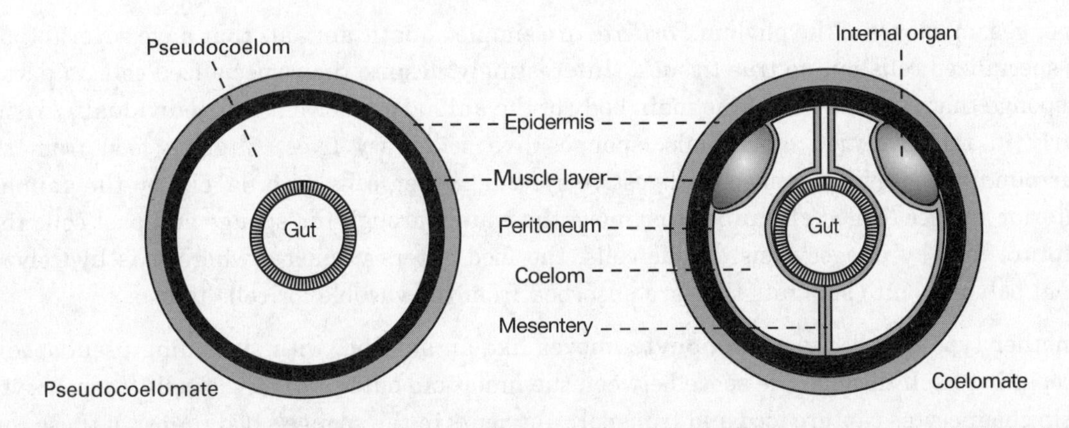

Q Can animals with a pseudocoelom have a complete digestive tract?

A Yes. Nematodes, for example, have a complete digestive tract with a mouth and separate anus, but they also have a pseudocoelom.

EMBRYONIC DEVELOPMENT: PROTOSTOMES VS. DEUTEROSTOMES

One of the basic divisions among animals that possess a coelom is determined by how the embryo develops. There are two groups:

1. **Protostomes** include annelids, arthropods, and mollusks. In early embryonic development, the fertilized zygote of a protosome undergoes **spiral cleavage.** Protostome cells rapidly become differentiated into specific cell types. The **blastopore** (the single opening in the early embryo that leads into the internal cavity), formed during gastrulation, eventually becomes the digestive tract. In the coelom, cells form a "pocket" that branches off from cells lining the gut.

2. **Deuterostomes** include echinoderms and chordates. Deuterostomes undergo **radial cleavage** and remain totipotent (that is, embryonic cells retain the potential to differentiate into any type of tissue) for a longer time during embryonic development than protostomes. The blastopore of a deuterostome eventually becomes the anus rather than the digestive tract. In the coelom of a blastopore, the mesoderm proliferates from cells lining the inner cavity.

For the next several pages, we'll take a quick survey of the key phyla in the animal kingdom. For purposes of brevity and clarity, the more obscure animal phyla that include just a few species won't be covered here and are not likely to appear on your exam.

The phyla are presented in rough order of evolutionary descent and relative complexity, starting with sponges and ending with big-eyed chordates, including vertebrates.

INVERTEBRATES

Porifera

Sponges, members of the phylum *Porifera*, are simple aquatic animals that have several types of specialized cells but no true tissues. (Interestingly, despite their specialized cells, a part of a sponge that is broken off of the main body of the animal can form a new individual.) Arising early in the evolution of animals, sponges live sedentary lives, filtering food from the surrounding water. They have no digestive system; rather, cells with flagella on the animal's interior surface (called **choanocytes**) move the water through the sponge and past cells that capture food by phagocytosis. Inside cells, the food enters vacuoles, where it is hydrolyzed from polymers into subunits that are absorbed from the vacuole for cells to use.

Another type of cell, the **amoebocyte**, moves like an amoeba, with streaming pseudopodia, propelling itself through the space between the inner and outer surfaces of cells. Amoebocytes help choanocytes capture food and transport nutrients in the sponge's body. Most of these cells are in contact with the water outside the sponge, and they move through the pores and channels that run through the sponge to draw gases from and eliminate waste directly into the aquatic environment. Sponges have no circulatory, respiratory, nervous, or excretory systems.

Q Do sponges have a coelom or a pseudocoelom?

A Sponges have neither, because they do not have an internal body cavity between the different cell layers.

Q Do sponges move water by contracting muscular tissue around the internal water channels?

A No. Sponges have no muscle cells or any other true tissues. They move water using the flagella of choanocytes.

Cnidaria

Cnidarians (in the past known as coelenterates) include jellyfish, sea anemones, corals, and hydra. They can live either as swimming **medusa,** such as jellyfish, whose tentacles point downward around the mouth, or as stationary **polyps,** such as anemones, with the mouth and tentacles facing upward. Cnidaria are distinguished by the following characteristics:

- Radial symmetry
- Digestive gastrovascular cavity (gut) with a single opening
- Stinging cells called **nematocysts** for protection and capturing prey
- Tentacles arranged around the mouth opening to the digestive sac
- A simple, decentralized nerve net that helps coordinate motion
- Two cell layers (endoderm and ectoderm), with a jelly-like substance called **mesoglea** between them

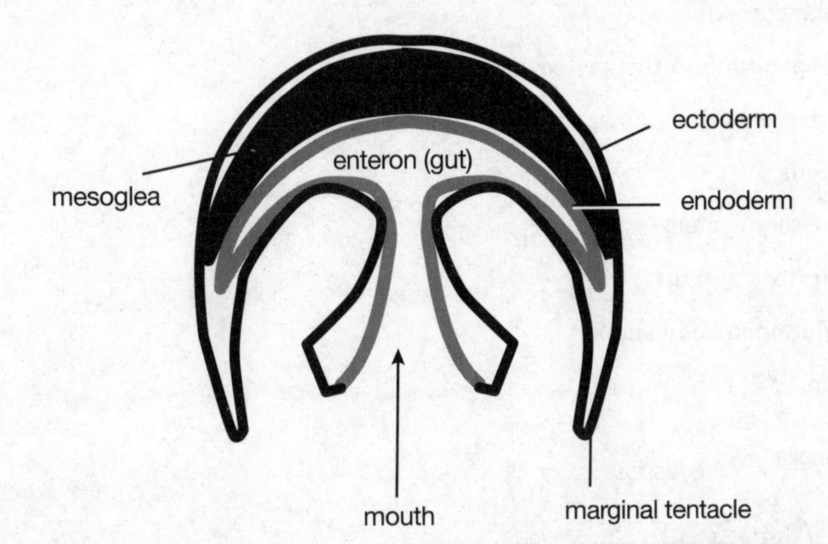

Cnidarians do not have muscle tissue. Instead, the cytoskeleton of their cells contracts and creates motion that is coordinated by the cnidarians' nerve net. They do not have a centralized nervous system or a head-like structure. Cnidarians reproduce sexually by releasing eggs and sperm, but many also reproduce asexually—small cnidarians will bud from older, larger individuals. Still other cnidarians alternate between asexual and sexual reproduction: they asexually reproduce via polyps that bud from individuals and become free-swimming medusa, and they sexually reproduce medusa to produce new polyps. An interesting example: A group of sea anemones clustered together on a rocky shore may actually be clones that have budded asexually from one another. These clones may even battle other sea anemone clones for space on the rocky surface.

Q Where are nematocysts located in cnidarians?

A In most cnidarians, they are located on the tentacles.

Platyhelminthes

Phylum *Platyhelminthes* consists of flatworms. It includes both free-living species, such as planarians, and parasitic species, such as liver flukes. Flatworms are acoelomate and have a gastrovascular cavity with a single opening for feeding. The small, flat shape and branching gastrovascular cavity inside the body of Platyhelminthes animals means that all cells are just a short distance from their aquatic surroundings. This allows for a direct exchange of gases.

Planarians are flatworms known for their regenerative ability. They have the following characteristics:

- Bilateral symmetry
- A single opening into the gastrovascular cavity
- Ventral nerve cords
- Muscle cells
- No circulatory system
- No respiratory system
- A thin, flattened body shape
- No coelom

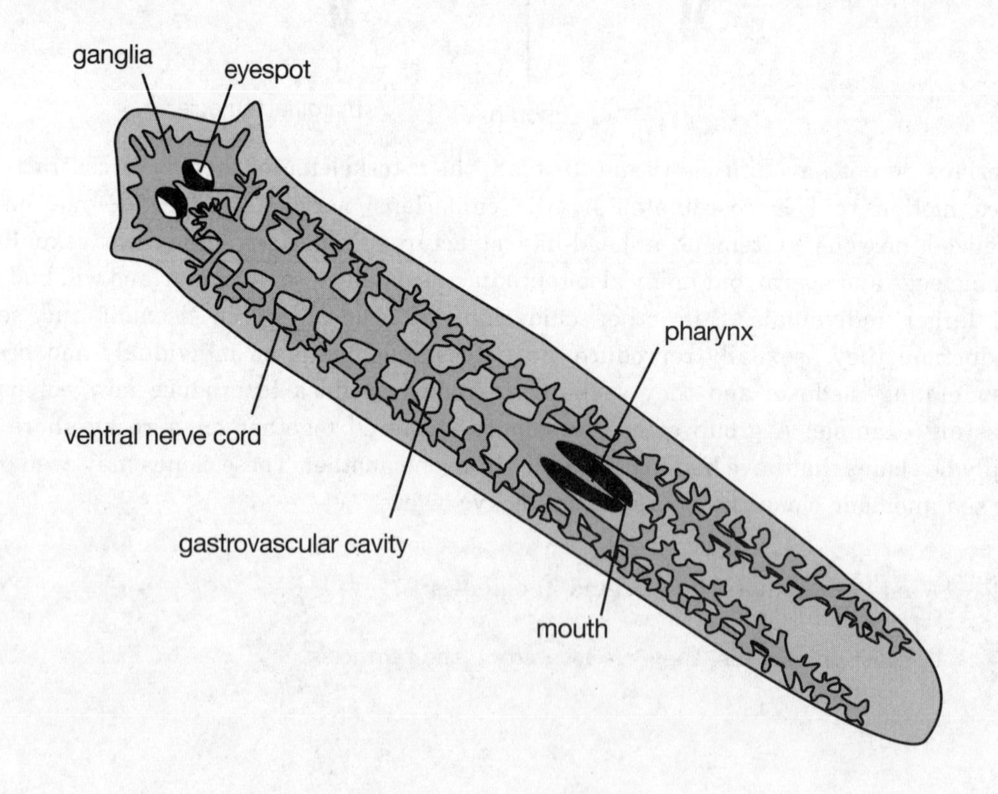

Q Do flatworms use oxygen? If so, how do cells receive it?

A Yes, they do use oxygen, and they get it by diffusion across their wet exterior surface and through the gastrovascular cavity.

Q What feature is shared by most free-living members of the Phyla *Cnidaria* and *Platyhelminthes*?

(A) Bilateral symmetry

(B) Gastrovascular cavity with a single opening

(C) Specialized excretory organs

(D) Anterior ganglia centralizing some nervous responses

(E) Muscle cells

A The correct answer is (B). Both phyla have a gastrovascular cavity with a single opening. Flatworms have bilateral symmetry, but cnidarians have radial symmetry, so (A) is incorrect. Neither has specialized excretory organs, so (C) is incorrect. Flatworms often have anterior ganglia in the head region centralizing some sensory responses, but cnidarians have a decentralized nerve net and they have no head region, so (D) is incorrect. Flatworms have muscle cells, but cnidarians do not, so (E) is incorrect.

Flatworm reproduction is predominantly sexual and individuals are hermaphroditic (having both male and female features). Parasitic species often have special adaptations for living and reproducing inside host organisms. Intestinal parasites often have degenerated digestive systems, relying on the host for digestion, and devote a great deal of their body mass to reproductive systems.

Nematodes

Although they are small and not obvious to the naked eye, nematodes are abundant in numbers and species, with billons of nematodes per acre of rich soil. Nematodes are nonsegmented roundworms with a complete alimentary tract, a separate mouth and anus, and a pseudocoelom. Nematode locomotion involves contraction of muscles running the length of its body. These muscles use the fluid-filled pseudocoelom as a hydrostatic skeleton. The writhing and wriggling motion is not a very effective way of moving in liquid, but in soil or against a solid substrate, it performs well. Because of their small size, nematodes do not require a circulatory or respiratory system, although some parasitic specimens have been discovered that measured several meters long.

Mollusca

The phylum *Mollusca* includes such varied animals as octopi, snails, clams, mussels, and slugs. These animals may look very different from one another, but they share some basic anatomical features and evolutionary history, including the following:

- Three germ layers during development
- Bilateral symmetry (may not always be obvious because of torsion of the body during development)
- Coelom (greatly reduced in some adults)
- External shell (except for squid, octopi, and slugs)
- Mantle, which secretes the shell and covers the visceral mass
- Visceral mass, which includes most of the internal organs
- Muscular foot, used for attachment and movement
- Rasping radula, used to scrape food
- Complete digestive tract, including mouth, stomach, intestine, and anus
- Swimming larval stage called a **trocophore**
- Gills for gas exchange (except for terrestrial species), often in the mantle cavity
- Open circulatory system (except for cephalopods)

Different classes of mollusks differ from the basic body layout in various ways. The **cephalopods** (squid, octopi, and nautilus) often lack a shell, but they have tentacles and use a modified foot to move water and propel the body. As predators, cephalopods have a well-developed brain and sensory apparatus, including an eye that resembles the vertebrate eye in many ways but which evolved independently. Cephalopods also have a closed circulatory system, perhaps to provide more efficient nutrient transport and respiratory gas exchange for this more active group.

Gastropods (snails and slugs) have body torsion during development, which twists and reduces organs on one side of the body. **Bivalves**, including clams, oysters, scallops, and mussels, have shells with two halves and feed by filtering water through gills and moving water through the mantle cavity with siphons.

Q Is the octopus eye an example of a homologous structure or of an anlogous structure?

A The complex eye of cephalopods greatly resembles the structure of the vertebrate eye, but the two evolved independently. This is an example of an analogous structure.

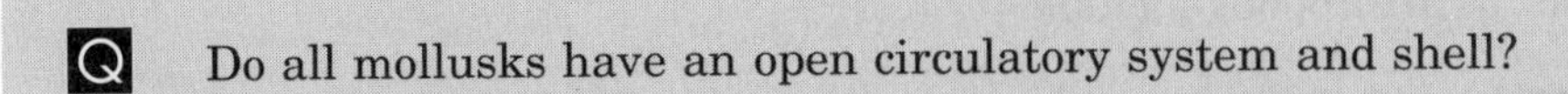

Q Do all mollusks have an open circulatory system and shell?

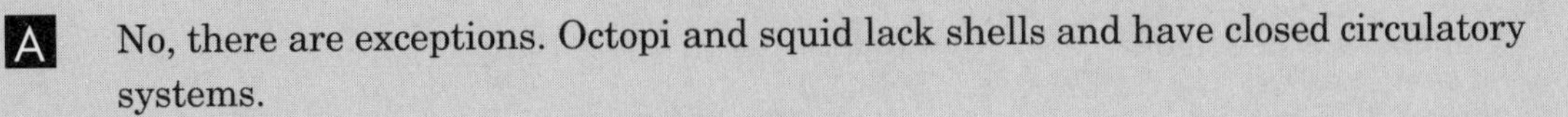

A No, there are exceptions. Octopi and squid lack shells and have closed circulatory systems.

Annelida

The phylum *Annelida* includes segmented worms such as earthworms, marine worms (called polychaetes), and external parasites (leeches). Annelids have a coelom along the length of their bodies and the body and coelom are divided into discrete segments. Rings of muscle surround each segment, and there are muscles oriented to the worm's length. By alternating contraction of these two types of muscles, the earthworm produces its characteristic motion, a sequential lengthening and shortening of segments to push forward. Animals that use this combination of muscle acting against a fluid-filled coelom have a **hydrostatic skeleton**. Annelids have simple excretory organs called metanephridia in each segment. They also have the following features in common:

- Specialized regions of the digestive tract (mouth, pharynx, esophagus, crop, intestine)
- Ventral nerve cords
- Ganglia in anterior region, collections of nerves near the pharynx
- Closed circulatory system with muscular, heart-like segments
- Chaetae on their external surface, bristles that aid forward motion by preventing backward motion
- Moist skin perfused with blood vessels for gas exchange

Q Does the volume of annelid segments change during annelid movement?

A The volume of each segment does not change, just its shape.

Arthropoda

Arthropods inhabit almost every corner of the earth. This phylum includes animals such as dust mites, ants, and crustaceans. The evolutionary success of arthropods is no doubt attributable to their body plan. The ancestral arthropod had a hard exoskeleton with a segmented body and one pair of appendages attached to each segment. Over time, many arthropods have evolved to lose some appendages or perform different functions with them (including reproduction, defense, movement, and feeding).

The arthropod exoskeleton protects against predation, but it also restricts growth. Consequently, as arthropods grow, they must periodically molt, shedding the old exoskeleton and producing a new, larger one. As you might expect, they are unusually vulnerable during molting periods.

Common characteristics of arthropods include:

- Bilateral symmetry
- Chitinous exoskeleton
- Segmented body
- Jointed appendages
- Coelom
- Open circulatory system with dorsal heart
- Hemolymph, which carries oxygen and circulates throughout tissues
- Ventral nervous system with ganglia in head region
- Extensive specialization of head with sensory organs

The body structure of terrestrial arthropods imposes a size limit beyond which their weight would crush the body or prevent movement. The amount of oxygen in the atmosphere also restricts their size, since tracheoles provide only a limited amount of air. Water allows arthropods like lobsters to grow relatively large, but, overall, the evolution of a monstrously large arthropod, such as those seen in some science fiction films, is impossible.

Here's a review of the major subphyla in the *Arthropoda* phylum.

CHELICERATES

This subphylum includes spiders, mites, and scorpions. They have specialized appendages called chelicerae for grasping food, and they have lungs for respiration. Arachnids are terrestrial and have eight legs but no antennae.

MYRIAPODS

This subphylum includes millipedes and centipedes. Many millipedes live on the floors of wooded areas and help break down plant material; centipedes, on the other hand, are predators. Members of this group have a pair of antennae.

CRUSTACEANS

Crustaceans include crabs, barnacles, lobsters, and shrimp. These are predominantly aquatic species, although some are terrestrial. Aside from the species that humans consume as food, isopods and copepods such as krill are abundant and important in marine ecosystems. Most crustaceans have a head, thorax, and abdomen region; most have separate sexes.

HEXAPODS

This subphyla is comprised of insects and three small orders of insect-like animals with six thoracic legs. Insects have a head, thorax, and abdomen, and many species are winged. They also have **Malphigian tubules** for excretion and tracheal tubes for respiration. Metamorphosis during insect development is common, with larval insects going through either incomplete metamorphosis, in which larvae resemble small versions of the adult insect (but are not sexually mature and lack wings), or complete metamorphosis, in which larvae undergo a more dramatic transformation from a caterpillar-like larval stage to a winged, sexually mature adult stage.

Q Which of the following has a closed circulatory system?

(A) Flatworm
(B) Earthworm
(C) Clam
(D) Honey bee
(E) Nematode

A Of these animals, only the earthworm has a closed circulatory system. Flatworms and nematodes do not have circulatory systems; clams and insects have open circulatory systems.

Echinodermata

Echinoderms include sea stars, sea cucumbers, sand dollars, and urchins. They are strictly marine organisms, with radial symmetry in adults that is usually organized in multiples of five. Like chordates, they are deuterostomes, evolved from ancestors with bilateral symmetry, and they have larvae with bilateral symmetry.

A **water vascular system** in echinoderms features interconnected canals that run into extensions of the body and tube feet that can extend and retract to help with locomotion and food gathering. Echinoderms have a complete digestive tract and a coelom. Sea stars, urchins, and sand dollars may appear to have exoskeletons, but echinoderms have an endoskeleton of calcium and magnesium that contains spines and plates. Some of these project away from the center of the body to protect against predation. Echinodermata reproduction involves external fertilization, with release of eggs and sperm by separate males and females.

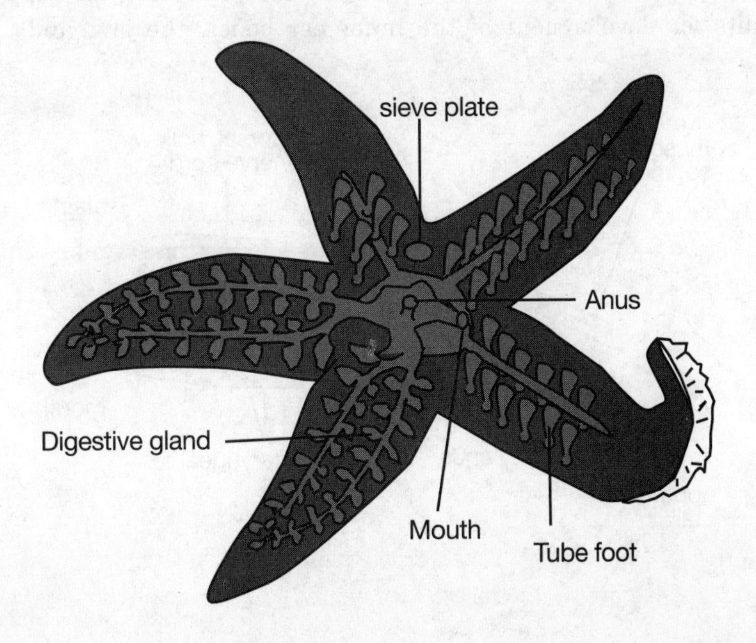

Chordata

While we tend to think of chordates as advanced, they are not evolutionary newcomers. In fact, evidence shows that chordates have been on Earth for more than 500 million years. As with echinoderms, chordates are deuterostomes that have a coelom, and these two groups are more closely related than other animal phyla.

Chordates display bilateral symmetry, a closed circulatory system, and advanced respiratory, excretory, and nervous systems. A few key adaptations distinguish the chordates from other phyla, and while these are not all present in adult organisms, they are present in at least some stages of chordate development for all species, including vertebrates.

- **Notochord**. Every chordate has a stiff but flexible skeletal rod called a notochord, which developed from mesoderm on the dorsal side of the animal. The notochord provides support for muscles. In most vertebrates, it is present in embryos but is partially replaced in adults by the bones of the vertebrae.
- **Dorsal nerve cord**. This is a hollow region between the notochord and the surface of the organism. During early embryonic development, the nerve cord develops from the ectoderm infolding into the neural tube. The nerve cord becomes the brain and spinal cord in vertebrates.
- **Muscular tail**. Unlike other phyla, chordates have a muscular tail that extends past the end of the digestive tract. In humans and some other species, the tail is present in the embryo but disappears during later development.
- **Pharyngeal slits**. Along the pharynx of a chordate, just behind the mouth, are rows of slits interspersed with supports. In non-vertebrate chordates such as tunicates, these slits are used for filter feeding, in which water is passed from the mouth to the outside without going through the rest of the digestive tract. In vertebrates, the slits are present in embryos but are altered in the adult for other functions. In fish, the slits develop into gills, and in terrestrial vertebrates the slits do not open, using the supporting structure between slits for development of the inner ear bones, the jaw, and other structures instead.

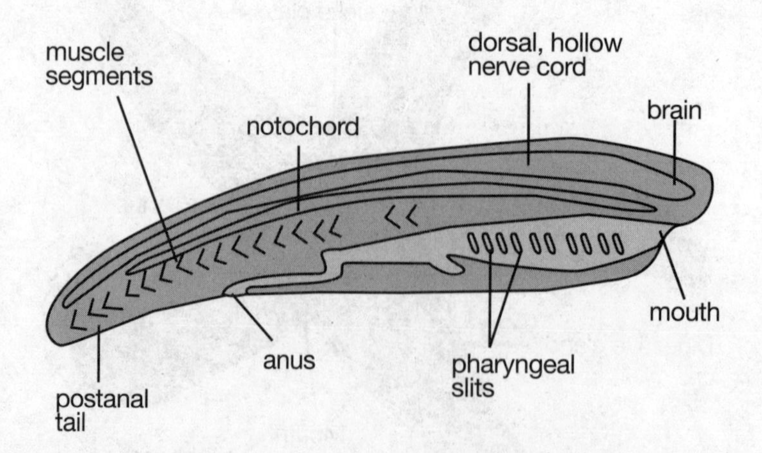

Q Animals in which of the following phyla are characterized by a dorsal nerve cord?

(A) Annelids
(B) Arthropods
(C) Platyhelminthes
(D) Chordates
(E) Cnidarians

A Chordates are the only animals of the phyla listed here that have a dorsal nerve cord.

Some primitive chordates, including tunicates (sea squirts), do not much resemble vertebrates in the adult form but have the characteristics described for other chordates when in the larval stage: a notochord and muscular tail for swimming. Once the tunicate larvae find a spot to settle down and attach, they develop into filter-feeding adults and lose the notochord and tail.

VERTEBRATES

The evolution of a head with a skull, backbone, and jaws lead to the vertebrates. A few species, such as the hagfish (which has a skull but no vertebrae) are still in existence to document the intermediate evolutionary steps leading to the vertebrates. These traits were first fully present in fishes.

Fishes

Cartilaginous fish (**chondrichthyes**), such as sharks and rays, are vertebrates that have all of the innovations already mentioned, as well as advanced skeletons and vertebral systems. Their bones, however, are composed largely of cartilage and are not hardened extensively with calcium mineralization.

Vertebrates with bones hardened by calcium phosphate deposits are called **osteichthyes**. These animals replace cartilage with bone as they develop—a trait seen in other vertebrates, such as mammals. Bony fish also have a swim bladder, which they use to regulate buoyancy, gills for respiration, and a lateral line sensory organ. Water moves through the gills of these animals while they are swimming, and it is "pumped" over the gills with muscles and the movement of the operculum flap that covers the gills.

Fish have fins that are supported by flexible rays or bony lobes. Today, lobe-finned fishes are found only in a few species, such as the coelacanth (often called a "living fossil") and lungfishes. Like lungfishes today, early lobe-finned fishes lived in shallow water and gulped air into primitive lungs, using their fins for movement across land. Over geologic time, these adaptations evolved to reduce dependence on water for tetrapods—animals with four limbs.

Amphibians

Amphibians, including frogs, toads, and salamanders, are tetrapods with varying degrees of dependence on water. All have simple lungs, essentially air sacs, that fill some of their respiratory needs. Additional respiration occurs through moist skin. Amphibians reproduce through external fertilization in aquatic or damp environments, laying eggs that lack a hard external shell. Frog development includes transformation from a swimming tadpole stage with gills and a tail to a sexually mature stage in which the tail and gills degenerate and four limbs develop to support movement on land.

Reptiles

Unlike amphibians' eggs, the amniotic eggs of reptiles have thicker shells that allow gas exchange but slow water loss. This allows eggs to be laid on land. As a consequence, reptile eggs must also be fertilized internally, within the female reproductive tract, before the shell hardens. The membranes of the amniotic egg around the embryo occur in reptiles, birds, and mammals in some form, including the amnion (the fluid cushion around the embryo), the yolk sac (which supplies nutrients), allantois (waste disposal), and chorion (gas exchange).

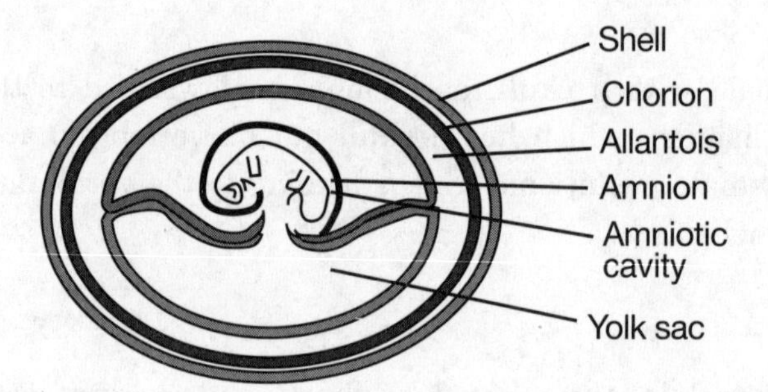

Reptiles also have thicker, less water-permeable skin than amphibians—another adaptation to life on land. While it slows water loss, thicker skin also prevents reptiles from exchanging gases across their skin and thus requires that the animal have more efficient lungs than those of amphibians. Reptiles are commonly described as "cold-blooded," but this term is inaccurate. True cold-blooded animals cannot regulate their body temperature; reptiles do so in a variety of ways. They do not generate a constant internal temperature through metabolic energy, however, as mammals do.

Q Which of the following is NOT a reptilian trait that allows animals to tolerate drier terrestrial climates than amphibians?

(A) thickened skin

(B) eggs with thick shells

(C) internal fertilization

(D) loss of gills during development

(E) amniotic egg

A The correct answer is (D). Reptiles do not lose gills during development. The other answer choices are present in reptiles but not amphibians.

Q Would equivalently sized mammals or reptiles in similar environments need to consume more calories in their food?

A Endotherms like mammals need to consume a lot more energy for their body weight, since they use a lot of it to maintain their body temperature through internally generated heat.

Birds

Birds are believed to have descended from dinosaurs. Their feathers are hollow and filled with air, facilitating flight. The feathers help insulate the body and shape the wing to provide lift and forward motion. The bones of birds' forelimbs are modified for wings. Unlike mammalian lungs, the avian respiratory system is designed with air sacs in addition to lungs, so that the lungs are constantly filled with fresh air. This efficient respiratory system, along with a four-chambered heart and separate pulmonary and systemic circulations, supports the intense metabolic requirements necessary for flight.

Birds have harder, calcified egg shells compared with the eggs of reptiles, although the same extra-embryonic membranes are present in both types of eggs.

Q Birds are distinct from reptiles in which of the following ways?

(A) They undergo internal fertilization.

(B) They do not have teeth.

(C) Amnion is present in the egg.

(D) Bones contain calcium phosphate.

(E) Pharyngeal slits alter during development into other head structures.

A The correct answer is (B). Birds lack teeth; reptiles have them. Both groups have internal fertilization, eggs with the extra-embryonic membranes and amnion, calcification of bones, and adaptation of pharyngeal slits into other structures.

Mammals

Mammals have hair and mammary glands that produce milk for feeding their young. They are endothermic, active and have body hair and fat for insulation. Like birds, they have a four-chambered heart that provides efficient circulation of oxygenated blood to tissues.

Early mammals probably laid eggs similar to those of reptiles; a few mammalian species, called **monotremes**, still do so. Another group of mammals, called **marsupials**, support infant development briefly inside the mother, but the young are born very early in development and require a significant period of further growth in the mother's pouch before they are able to become independent. Marsupials thrive primarily in Australia, which has been geographically isolated from other continents and lacks many placental mammals. A few marsupials also survive in North and South America.

Placental mammals (including humans) represent the great majority of surviving mammalian species. They experience much longer placental development than marsupials do and have undergone several other mammalian adaptations. Among the placental mammals is the primate group, which led to the evolution of hominids and ultimately humans.

All primates have opposable thumbs and binocular vision, perhaps to assist in grasping and climbing. The species that exists today that is believed to be most closely related to humans is the chimpanzee. Fossil records and molecular clues suggest the last common ancestor of humans and chimpanzees existed from 5 to 7 million years ago. Fossils of several subsequent hominid species have been discovered as having existed in the period leading up to the emergence of *H. sapiens* in Africa, about 150,000 to 200,000 years ago.

From nematodes to vertebrates, all animals must eat, escape predation and disease, grow, and reproduce to create new generations, which then meet all of the same challenges all over again. While these challenges are similar from one generation to the next, the solutions, which have evolved as the animals themselves have evolved, are myriad.

MAMMAL STRUCTURE AND FUNCTION

The Digestive System and Metabolism

Animals need energy for almost everything they do. Some animals need most of their energy to fuel locomotion—but movement is not the only use of energy. Biosynthesis, generating body heat, and fueling basic molecular activities also require energy, which comes from chemical fuel—food. Animals are **chemoheterotrophs**, which break down food into components and absorb the molecules into tissues and cells. There, some of the chemical components are oxidized to drive ATP production, and additional molecules are used as building blocks for biosynthesis.

Herbivores eat only plants; **carnivores** eat other animals; **omnivores** eat a broad variety of plants and animals. Food substances include the following important nutrients:

- Sugars/Carbohydrates. These include monosaccharides (glucose, fructose), disaccharides (sucrose, lactose), and polysaccharides (glycogen, starch, cellulose)
- Proteins (essential amino acids are particularly important)

- Fats (triglycerides, cholesterol)
- Nucleic acids
- Water
- Minerals. These include calcium (for neuronal and intracellular signaling and bone building), phosphorous (for bone building and making nucleic acids), copper (enzyme cofactor), iron (for oxygen transport by hemoglobin and electron transport in metabolism), zinc (an enzyme cofactor), and iodine (for thyroid hormone synthesis). For humans, the following are also essential: vitamin C, vitamin B_1 or thiamine, vitamin B_3 or niacin, vitamin B_6, pantothenic acid, folic acid, biotin, vitamin A, vitamin D, and vitamin K.

Q A deficiency in which vitamin or mineral is associated with anemia?

A Iron, which is part of hemoglobin.

Q A deficiency in which vitamin or mineral is associated with night blindness?

A Vitamin A, which is involved in vision.

"Essential" nutrients are those that an animal cannot synthesize. As animals have evolved to receive nutrients from consuming other organisms, they have in many cases lost the ability to synthesize some of the components necessary for biosynthesis, such as some of the amino and fatty acids. For example, of the twenty amino acids in proteins, humans can synthesize twelve, so they must take in the remaining eight essential amino acids through diet.

Q Would animals with greater metabolic needs evolve a gastrovascular cavity or a complete digestive tract for nutrition?

A A complete digestive tract is more efficient at extracting energy from food than a gastrovascular cavity. Only animals with low metabolic energy needs, such as cnidarians, have a gastrovascular cavity.

In humans (and other vertebrates0, the digestive process begins in the mouth. Chewing food breaks it into smaller pieces with greater surface area so that digestive enzymes can more easily break it down further. Saliva secreted by glands in the mouth help moisten and lubricate food as it passes through the digestive tract. Hydrolysis of macromolecules also begins in the mouth with the enzyme salivary amylase, which breaks down starch into smaller sugars.

The type of food an animal eats is closely related to the design of its digestive tract. Humans eat a variety of foods and have a variety of teeth, including sharp, pointed incisors, cutting teeth in the front of the mouth, and blunt molars at the sides and back of the mouth for grinding and crushing. Carnivores, such as dogs and cats, have sharp, fanged teeth for catching prey; herbivores have teeth designed to break down plant material. Birds, which don't have teeth, have a muscular crop and a gizzard that help grind food into smaller pieces.

Swallowed food is pushed by the tongue and mouth muscles toward the **esophagus**, and moves past the **pharynx**, which connects to both lungs and the esophagus. To keep food or liquid from entering the lungs during swallowing, a flap called the **epiglottis** closes access to the trachea. Once food is in the esophagus, a muscular, wavelike action called **peristalsis** moves it toward the stomach. Peristalsis occurs in different regions of the gastrointestinal (GI) tract and involves controlled contraction of bands of muscle.

Q Does peristalsis in the esophagus require conscious muscular control?

A No. Swallowing is initiated as a conscious action, but the movement of food through the pharynx and esophagus is reflexive and occurs without conscious control.

The liquid in the stomach (**chyme**) is very acidic because of the secretion of hydrochloric acid by parietal cells. Chyme helps break up food particles and molecules and kills most bacteria found in food. Additional cells secrete digestive enzymes that digest proteins. Pepsinogen is secreted as an inactive precursor form of **pepsin;** the pepsinogen is activated by acid in the stomach to form pepsin. Enzymes such as proteases are often secreted in inactive forms called zymogens that are activated in the gastrointestinal tract. Releasing proteases in an initially inactive form helps protect the cells that secrete the enzymes from sustaining damage.

Other cells in the stomach lining secrete mucus, protecting the walls of the stomach from damage by acid and enzymes. The human stomach has muscular rings of muscle, called **sphincters**, that control the movement of food in and out of the stomach. To allow food to move from the stomach to the small intestine, the pyloric sphincter relaxes.

Q Which of the following is not secreted in the stomach: hydrochloric acid, mucus, or pepsin?

A Pepsin is not secreted in the stomach. Pepsinogen is secreted and then activated in the lumen of the stomach to form pepsin.

As food moves into the small intestine, several additional secretions are added from the **pancreas** as the food passes into the duodenum, the small intestine region closest to the stomach. The pancreas functions as an endocrine gland, releasing insulin, glucagon, and other hormones into the bloodstream, and also as an exocrine gland, secreting proteases and other material through the pancreatic duct into the lumen of the intestine. The exocrine secretions of the pancreas that are released into the GI tract include the following:

- **Trypsin:** A protease acting in the small intestine to cleave proteins next to positively charged amino acids
- **Chymotrypsin:** A pancreatic protease acting next to hydrophobic amino acids in proteins

- **Bicarbonate:** A basic solution, neutralizing the acidity of stomach secretions
- **Amylases:** Enzymes that hydrolyze sugars into smaller units for absorption
- **Lipase:** An enzyme that hydrolyzes triglycerides into glycerol and free fatty acids
- **Nuclease:** Breaks down nucleic acids into individual nucleotide subunits for absorption

Other secretions released into the small intestine are bile salts, which are produced in the liver and stored in the gall bladder. Bile salts are charged, detergent-like derivatives of cholesterol that help to solubilize fats in the small intestine and break them into smaller particles that pancreatic lipase can access more easily. The gall bladder releases bile salts into the intestine in response to hormones signaling that a fatty meal has been eaten. Without bile salts, fats are not digested and absorbed as efficiently.

Q How will fat digestion be affected if a person's gall bladder is removed but the duct from the liver to the small intestine remains in place and functional?

A Bile salts will be released at a constant pace from the liver into the small intestine rather than being stored in the gall bladder and released in response to hormones. The ability to digest fatty meals will be impaired, although not eliminated.

The small intestine also expresses a variety of enzymes on the inner surface of the intestine facing the lumen. These enzymes are mostly bound to the surface of the cells, further breaking down proteins, sugars, and nucleic acids into their subunit monomers for ready absorption. This layer of the intestine is sometimes called the "brush border."

The hydrolysis of food molecules prepares them for absorption. The human intestine is highly specialized for absorbing food; it has an extensive interior surface area and a rich supply of blood. The cellular layer facing the lumen, the intestine epithelium, is folded with **villi** to create a greater expanse of epithelium. Each cell in the villi has even smaller projections on its surface, called **microvilli**. Every villus has a supply of arteries, veins, and lymphatic vessels called lacteals. Once nutrients are absorbed by cells in the intestinal epithelium, they are passed on to blood vessels or lacteals for transport to other parts of the body.

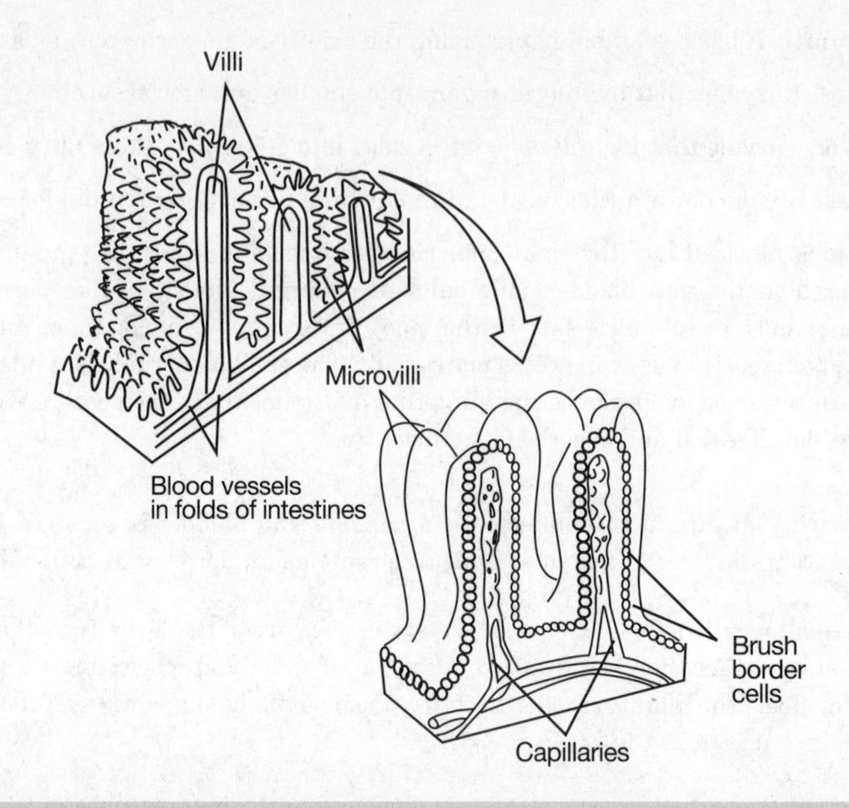

Q The presence of microvilli in the epithelium of the small intestine does which of the following: increases secretion of trypsin, stimulates intestinal peristalsis, increases the surface area for nutrient absorption, neutralizes the acid from stomach contents, or increases the supply of blood to the intestine?

A Microvilli increase the surface area for nutrient absorption. The small intestine is highly specialized for this purpose, with a great deal of surface area in villi and microvilli.

Most nutrients are absorbed after digestion as small units that are transferred to capillaries in the villi and then transported throughout the body. Glycogen, for example, is hydrolyzed by salivary amylase, pancreatic amylase, and enzymes on the brush border, until the sugars become monosaccharides such as glucose, fructose, and galactose. These sugars are transported into the intestinal epithelial cells and then to the blood. The blood carries the sugars throughout the body to tissues, where glucose is used to generate energy in glycolysis or is stored in glycogen.

Protein digestion follows a similar procedure. Proteins are hydrolyzed into smaller pieces beginning with pepsin in the stomach, followed by trypsin and chymotrypsin in the small intestine. Additional proteases in the brush border of the intestinal epithelial cells complete the process, so that individual amino acids can be absorbed by the epithelium and transported in the blood throughout the body to support biosynthesis.

Q Would an absence of proteases in the intestinal brush border affect the efficiency of amino acid absorption from digested proteins?

A Yes. Absorption requires breaking up proteins into the smallest possible pieces, amino acids. Bigger pieces of polypeptides aren't well absorbed and reduce the percentage of amino acids released and absorbed. The brush border proteases complete the process begun by secreted proteases like pepsin and trypsin.

Fat digestion follows a somewhat different path. Fats are mostly water-insoluble triglycerides that clump together when mixed with water. Bile salts produced in the liver and released by the gall bladder break up fats into smaller globules. Globules with greater surface area help pancreatic lipase hydrolyze triglycerides into free fatty acids and glycerol, which are absorbed by the intestinal epithelial cells. These do not directly enter blood, however. Instead, these epithelial cells produce triglycerides again and assemble the triglycerides with proteins and cholesterol into lipoprotein particles called **chylomicrons**. Chylomicrons are transported into small lymphatic vessels called lacteals that drain into larger lymphatic vessels, and they eventually drain into veins leading to the heart. In the circulation, triglycerides are hydrolyzed again by a lipase expressed in the surface of cells, releasing free fatty acids that cells use.

Q Blood that has passed through the small intestine moves into the hepatic portal vein and through the liver. The liver helps regulate the blood level of nutrients such as glucose. Would disruption of circulation through the hepatic portal vein disrupt levels of glucose or triglycerides in blood to a greater extent?

A Other nutrients such as glucose, absorbed in the small intestine, are transported directly to the liver in the hepatic portal circulation, where they are processed. One consequence of the unique pathway of fat absorption is that fats are not transported directly to the liver. They first bypass the liver by going into the lymphatic system. Glucose would be affected more than triglycerides if the hepatic portal vein were disrupted.

Most digestion occurs in the first segment of the small intestine, and the absorption of nutrients is mostly completed in the rest of the small intestine. However, the process is not yet complete. Nutrients pass from the small intestine into the **large intestine.** Much shorter than the small intestine, the large intestine absorbs the water remaining in its contents until only indigestible solid material remains to be excreted as feces. The large intestine harbors symbiotic bacteria that produce vitamins, such as K, required by humans.

The digestive process is regulated by nervous and hormonal signals. Local signals within the part of the nervous system "attached" to the GI system help integrate its responses. Moreover, the sympathetic inhibits digestion whereas the parasympathetic stimulates digestion.

Hormones that regulate digestive function include the following:

- **CCK:** Released from the small intestine in response to proteins and fats, this peptide hormone acts on the stomach to reduce secretion and motility, slowing the movement of food into the intestine. It also stimulates the secretory functions of the pancreas, gall bladder, and liver.

- **Ghrelin:** Released from the stomach in response to periods of fasting, ghrelin is a peptide hormone that stimulates hunger.
- **Gastrin:** A peptide hormone released by the stomach in response to proteins in the stomach, gastrin increases stomach secretion and motility as well as motility in the intestine.
- **Secretin:** Released by the small intestine in response to acid, secretin inhibits stomach acid secretion and motility and stimulates bicarbonate production.

Q A lack of secretin production would do which of the following: increase the acidity of the contents of the small intestine, inhibit CCK release, stimulate acid production by the stomach, increase the rate of chyme movement through the pyloric sphincter, or stimulate protein digestion in the intestine?

A It would increase the acidity of the contents of the small intestine. Secretin is released by increased acidity in the small intestine, stimulating bicarbonate production to neutralize the acid. In the absence of secretin, the contents of the small intestine would be more acidic after a meal.

Q Is CCK secreted into the lumen of the gastrointestinal system to exert its effects?

A Some probably is released into the lumen, but most of its actions as a hormone require transport in the blood to act on other, more distant, tissues.

The Respiratory System

Human respiration begins in the mouth and nose. Typically, we breathe through the nose while at rest, which helps to warm and moisten air before it reaches the membranes of the lungs. The nose and linings of human airways also help remove contaminants before they reach the more delicate membranes where gas exchange occurs. The **larynx** and **epiglottis** direct air into the trachea rather than the esophagus. Once air enters the trachea, it passes first through two bronchi, one leading into each lobe of the lung, and then inside each lung into increasingly smaller bronchioles. Ultimately, air reaches the **alveoli**—tiny grape-like sacs at the ends of the airways.

Almost all gas exchange occurs in the alveoli. They are highly perfused with deoxygenated blood that flows through capillaries and is delivered by the pulmonary arteries. The alveoli contain a layer of thin, flat cells under outside capillaries, and a thin layer of moisture coating these cells allows gas exchange to occur. Oxygen diffuses down a gradient from atmosphere to blood, and carbon dioxide diffuses in the other direction.

Q Do birds' lungs have alveoli?

A No. Birds require one-way movement of air through the lungs. Rather than alveoli, they have minute tubes through which air passes.

The human lungs fill the chest space and are surrounded by the ribs and a muscular lower membrane called the diaphragm. The lungs are also enclosed in the **pleura** that form a tight seal between the organ and the surrounding space inside the chest. With each breath, the diaphragm contracts and flattens, thereby increasing the volume of the chest. Muscles around the chest connecting the ribs also contribute, particularly during exercise. As the volume of the chest increases, the membranes around the lungs are drawn out, increasing the volume of the lungs. The tissues of the lungs are highly elastic, and as the volume of the lungs increases, air is drawn into the lungs. When the diaphragm relaxes, the chest volume decreases again, increasing pressure around the lungs, decreasing their volume and causing air to be expelled.

Q In a healthy person at rest, which requires muscular contraction: inhalation or exhalation?

A Inhalation requires the diaphragm to contract; exhalation is a passive process at rest, with tissues drawn inward as the diaphragm relaxes.

Human respiration is described and measured using the following terms:

- **Lung capacity:** Total volume of the lungs (about 6 liters)
- **Vital capacity:** The maximum breathable volume (about 5 liters)
- **Residual volume:** The amount of air left in the lungs after the strongest possible exhalation (about 1 liter)
- **Tidal volume:** The amount of air exchanged in the lungs in a normal breath (about 0.5 liter). Not all of the air in the lungs is exchanged with each breath.

Q Which is larger, the vital capacity or the tidal volume?

A The vital capacity is larger. Tidal volume is the volume of a normal breath.

Breathing is controlled by the pons and medulla oblongata in the brain. We can voluntarily control breathing to an extent, but, at some point, the autonomic systems take over. The rate and depth of breathing match metabolic needs. The more CO_2 in blood, the lower the pH level. If the pH of blood and cerebrospinal fluid becomes more acidic, breathing rate increases in an attempt to remove carbon dioxide from the blood. The breathing centers of the body normally respond more readily to the amount of carbon dioxide than oxygen in the blood.

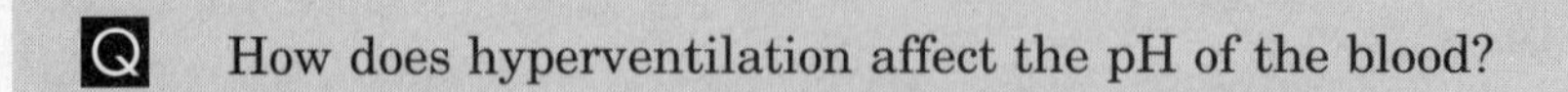
Q How does hyperventilation affect the pH of the blood?

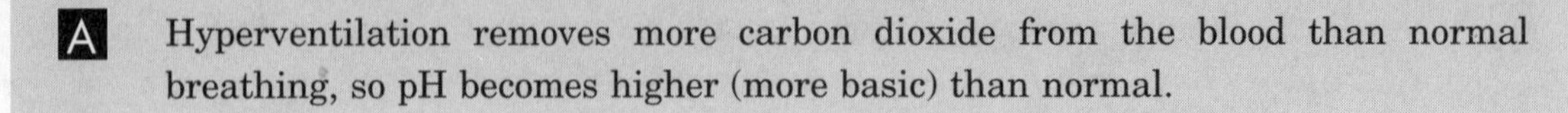
A Hyperventilation removes more carbon dioxide from the blood than normal breathing, so pH becomes higher (more basic) than normal.

HEMOGLOBIN

Oxygen does not dissolve in blood to be transported from lungs to tissues. Rather, it binds to the carrier protein **hemoglobin**, located in red blood cells. Hemoglobin proteins have a quaternary protein structure with four subunits in each functional unit. These interact to create cooperative oxygen binding. The absence of cooperativity in oxygen binding produces an oxygen saturation plot more like that of myoglobin. Myoglobin also binds oxygen but acts as a monomer, displaying no cooperativity.

Q Does the cooperative binding of oxygen to hemoglobin allow the active transport of oxygen in lungs against a pressure gradient?

A No. Respiratory gases diffuse down a pressure (or concentration) gradient. Cooperative binding helps hemoglobin to bind oxygen in places where it is abundant (lungs) and quickly release oxygen where the partial pressure of oxygen is low (in active tissues).

Fetal hemoglobin has a higher affinity for oxygen than maternal hemoglobin, ensuring that oxygen flows from the maternal to the fetal circulation through the placenta. Metabolic factors like pH and carbon dioxide change hemoglobin affinity, causing more oxygen to be released from blood into tissues when it is most needed, in response to increased metabolic activity. The sensitivity of hemoglobin to pH is called the Bohr shift.

Q Under more acidic conditions, is the affinity of hemoglobin for oxygen increased or decreased?

A The Bohr shift makes the affinity of hemoglobin for oxygen lower when the pH falls, as occurs in tissues during exercise.

Carbon dioxide is generally not transported as a free dissolved gas in the blood. Some carbon dioxide is carried by hemoglobin in red blood cells. The enzyme carbonic anhydrase in red blood cells also converts CO_2 into carbonic acid, which then forms bicarbonate ions. As carbon dioxide is removed in the lungs, the shift in equilibrium also shifts the bicarbonate and carbonic acid back to CO_2 to be exhaled.

The Circulatory System

For large organisms, a circulatory system is essential for allowing cells to transport material internally and externally. The larger the organism and the more active its metabolism, the more efficient its circulatory system must be. The vessels of circulatory systems carry fluids through the system, and a pump (or pumps) moves the fluid through the vessels. Organisms in various animal phyla have evolved to have either an **open circulatory system** or **closed circulatory system**. In closed systems, the circulatory fluid, the blood, is always contained with the vessels and the heart, relying on transport and diffusion of material across the lining of the vessels to reach tissues. In open systems, vessels open into the rest of the body, mixing the circulatory fluid with the extracellular fluids throughout the body.

Q Which provides more efficient circulation: an open or closed circulatory system?

A A closed system, which is a kind of "one-way street" for circulatory fluids.

In organisms with closed circulatory systems, the circulatory fluid is called blood, and the blood vessels are subdivided into arteries, which carry blood from the heart to the rest of the body, and veins, which return blood from the tissues to the heart. As arteries move away from the heart, they branch into smaller and smaller arteries and arterioles, until they reach the smallest vessels, the capillaries. It is in the capillaries where most of the exchange of material between tissues and the blood takes place. From the capillaries, vessels gather together again into progressively larger veins and venules until they reach the heart.

Q Which vessels overall have more blood moving through them in the body at any given moment: capillaries or arteries?

A The amount of blood flowing through capillaries and arteries must be the same, since any blood going through arteries ends up going through capillaries.

All vertebrates have closed circulatory systems with a chambered heart. The vertebrate heart includes **atria**, which receive blood returning to the heart from the veins, and **ventricles**, which pump blood out from the heart back to the tissues. The further the circulation is from the heart, the lower the pressure becomes. The arteries carrying blood from the heart to capillaries are under pressure from the heart, while the capillaries are under lower pressure, and veins under the lowest pressure. Different classes of vertebrates differ in the layout of their circulatory system and the structure of their heart.

Q In which portion of the cardiovascular system is most carbon dioxide absorbed from tissues: the pulmonary artery, venules, capillaries, arterioles, or aorta?

A The exchange of oxygen, carbon dioxide, and other substances between blood and tissues occurs in the capillaries.

Fish have a two-chambered heart, so their single ventricle must pump blood through two successive sets of capillaries in the gills and the rest of the body. Amphibians have a three-chambered heart which is somewhat more efficient.

The evolution of the vertebrate heart continued in birds and mammals, with both groups independently evolving a four-chambered heart that fully divides the circulation of deoxygenated blood to the lungs and oxygenated blood to the rest of the tissues. Both birds and mammals have two atria, and two ventricles divided into two separate circulations. The division of the heart and circulation in this way provides for more efficient circulation and more efficient transport of oxygen to tissues, supporting their high activity level and high rate of metabolism as endotherms.

BLOOD

Human blood is a mixture of cells and liquid with about 45% cells. Most blood cells are red blood cells (**erythrocytes**) that carry oxygen. Others are white blood cells (**lymphocytes** and **leukocytes**), which are part of the immune system, and **platelets**, which are involved in blood clotting. Human erythrocytes are highly specialized for oxygen transport. They lack a nucleus and mitochondria and are packed with hemoglobin for transporting oxygen.

The fluid in which blood cells are suspended is called **plasma** and is made mostly of water with various salts, proteins, and other dissolved substances. The plasma maintains the necessary osmotic balance and pH required by bodily tissues; it also carries nutrients such as glucose, respiratory gases (oxygen and carbon dioxide), metabolic waste products (urea), and proteins (albumin, antibodies, clotting proteins, hormones).

One of the functions of blood is to help close wounds through which more blood might be lost. A wound reveals factors in its exposed edges that trigger platelets to release clotting factors. This in turn triggers a cascade of proteases and culminates in the proteolytic activation of thrombin, which cleaves fibrinogen into fibrin. Platelets adhere to one another in the wound, and activated fibrin fibers adhere in the wound site as well to form a clot. Once the wound heals, the clot dissolves. A lack of an essential clotting factor can lead to **hemophilia**. Too much clotting may cause a heart attack or a certain type of stroke.

THE HEART

All mammals, including humans, have a four-chambered heart with separate circulations for pumping blood through the lungs and pumping blood to the rest of the body. This physiological adaptation enables humans and other warm-blooded (**endothermic**) mammals to have a higher rate of metabolism than reptiles, fish, or amphibians.

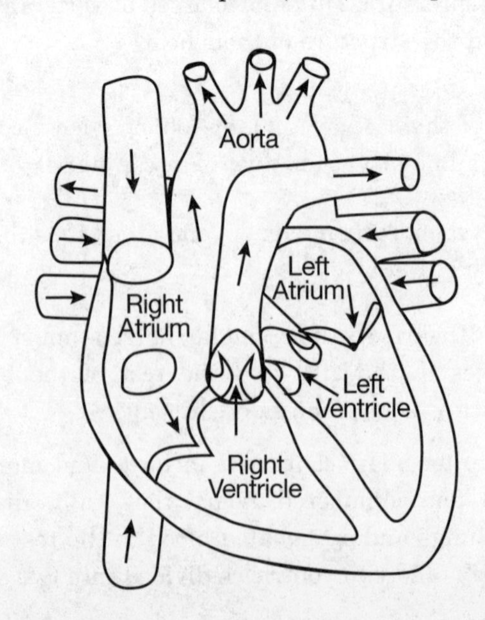

The human heart is composed of muscle tissue, with two atria that receive blood and pump it into two ventricles, which then pump the blood out of the heart. The **aorta** is the biggest artery in the body, carrying blood from the left ventricle to the rest of the circulatory system. The **pulmonary artery** carries blood from the right ventricle to the lungs (blood is returned to the left atria by the pulmonary vein). The ventricles have much thicker walls and contract more powerfully than the atria.

Each heart beat, each cardiac cycle, includes a complete contraction of the atria and the ventricles. This cycle is divided into the **systole**, which is when contraction occurs, and the **diastole**, the period between contractions. Valves in the heart open and close to keep blood from flowing backward. **Semilunar** valves are located where the arteries carry blood away from the ventricles. They open when the ventricles contract and close between contractions to keep blood from flowing backward. The **atrioventricular** valves in each side of the heart divide the atria and ventricles and keep blood from flowing backward into the atria when the ventricles contract.

Q Are there any arteries in the body that carry deoxygenated blood?

A The pulmonary artery carries deoxygenated blood to the lungs.

Q Are the semilunar valves open or closed during atrial contraction?

A The atria contract to push blood into the ventricles. The semilunar valves lie at the exit of the ventricles into the major arteries. For the ventricles to fill, the semilunar valves must be closed.

Blood flows from the heart directly into a bed of capillaries and then back to the heart. In a couple of locations in the body, called portal circulations, blood flows through two beds of capillaries before returning to the heart. One of these portal circulations occurs between the intestines and the liver and is called the hepatic portal circulation. The capillaries that perfuse the small intestine and help with nutrient absorption do not collect blood into veins that go directly to the heart. Instead, these vessels move blood to the liver so that it enters another set of capillaries. The second portal circulation in humans occurs between the hypothalamus and the pituitary gland.

The heart pumps constantly, matching the needs of tissues with the required supply of blood and respiratory gases. The cardiac muscle is contractile tissue with interconnected cells that spread action potentials directly and rapidly throughout the heart. Each contraction is triggered by an action potential that moves in a wave, first through the cardiac muscle tissue of the atria, and then, after a delay, through both ventricles. This action potential initiates in the heart and is triggered spontaneously in a special **pacemaker** region of the heart known as the **sinoatrial node**, in the right atrium.

Q An action potential will be detected in which part of the heart first in each cardiac cycle?

A The right atrium, where the sinoatrial node lies. Each heart beat is initiated with an action potential, starting at the sinoatrial node and spreading to other areas of the heart.

Although the heart spontaneously initiates each cardiac cycle, it is also influenced by nervous and hormonal signals to beat more quickly or slowly. Without input from other factors, the heart's internal pacemaker runs much more quickly than is actually observed in the resting human heart. The pacemaker is inhibited by the parasympathetic nervous system, which slows the heart rate. Similarly, stimulation of the sympathetic nervous system, the so-called "fight or flight response," speeds up the pacemaker and increases the heart rate. The hormones epinephrine and norepinephrine (also called adrenaline and noradrenaline) are released by the adrenal glands in response to sympathetic stimulation, and they also increase the heart rate.

Q What effect on heart rate would occur if the neurotransmitter used by the parasympathetic system is blocked?

A The parasympathetic nervous system slows the heart rate, so blocking its signal would increase the heart rate.

BLOOD PRESSURE

The arteries have strong muscular walls made of smooth, contractile muscle and elastic, connective tissue that allows them to stretch and contract. With each contraction of the heart, the pulse of pressure passes through arteries, pushing against the muscular walls, which push back to retain the shape of the artery. Veins, on the other hand, have much less smooth muscle and are under much less pressure after blood passes through capillaries. Blood flow through veins relies on valves that keep blood from flowing backward and on the movement of skeletal muscle around the veins to push the blood forward.

Q How does a lack of physical activity affect blood flow through veins in the lower extremities?

A Veins have little pressure to push blood forward, so they rely in part on the action of skeletal muscles around the veins to push blood forward. Without this, gravity will cause blood to accumulate in the lower extremities.

The movement of blood through capillaries is controlled by valves in each tissue that locally regulate the flow of blood. At any given moment, different tissues have different needs, depending on what the body is doing. For example, following a meal, the flow of blood increases to the gastrointestinal tract, but during exercise the flow of blood is greatest to the muscles. The overall blood pressure in a person is the combination of how hard the heart is pumping blood through the system and how much resistance to the blood flow exists in the arteries. The more arterioles relax, the less resistance exists to the flow of blood into peripheral tissues and the lower a person's overall blood pressure will be.

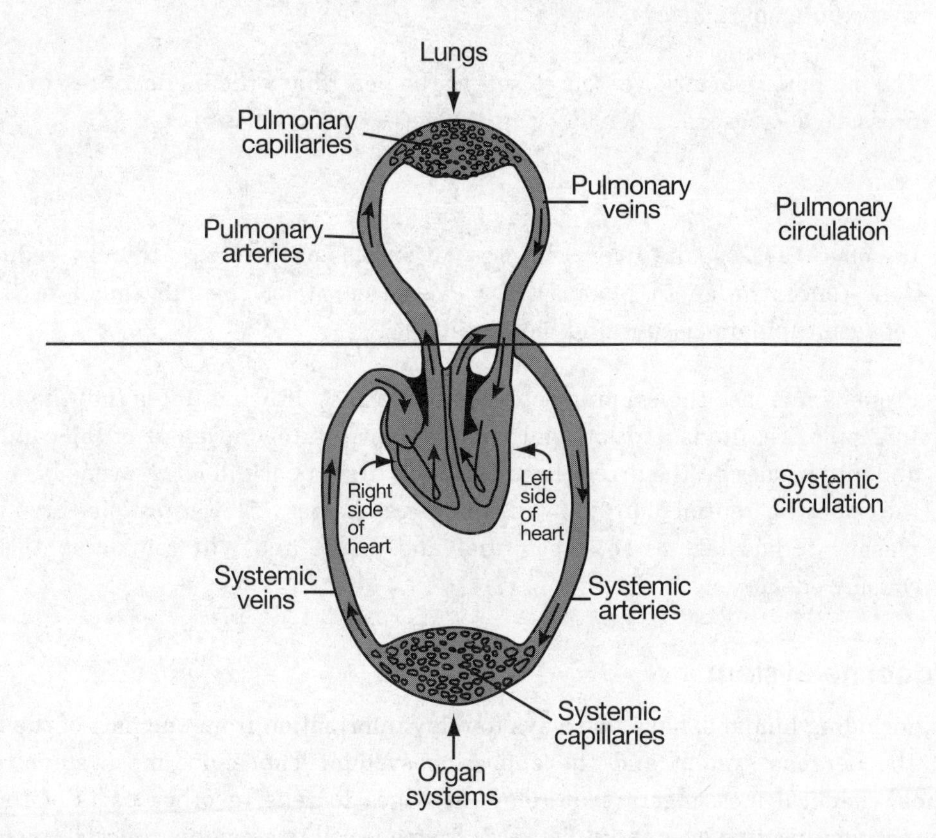

Ultimately, the purpose of the circulatory system is to supply tissues with essential ingredients and remove waste. All of this occurs in the capillaries. Essentially, the heart, the arteries, and the veins all serve the capillaries. While the arteries and veins have thick walls, the capillaries have extremely thin walls that are just one cell thick. This allows close contact between the blood, the cells in the tissues, and the extracellular fluids around the cells. Oxygen and carbon dioxide diffuse across this layer of cells in a concentration gradient between the blood and tissues. The layer of cells in capillaries also allows molecules of a low enough molecular weight to pass between cells into the extracellular fluids. In normal, healthy tissue, proteins, cells, and other large molecules are retained in the blood. At the arterial end of capillaries, enough pressure still remains to drive the movement of fluids out of the capillaries and into the tissues. At the venous end, the pressure is reduced and fluids are drawn back into the blood by osmotic pressure, created in part by the proteins left behind inside the capillaries.

Q In which portion of the cardiovascular system is the pressure greatest: the arterioles, the capillaries of the leg, the lung capillaries, the superior vena cava, or the pulmonary artery?

A The pulmonary artery. Vessels closest to the heart have the highest pressure. The pressure decreases farther away, in arteries, capillaries, and veins.

Q In some diseases, the liver produces fewer plasma proteins, thereby reducing their concentration in plasma. How does this affect the movement of fluids between capillary tissues and blood?

A Proteins increase the osmotic potential of plasma, drawing fluids into the blood. In capillaries, fluids and low-molecular-weight solutes move out of the capillary and into tissues on the arteriole end and are drawn back in on the venous end by the osmotic potential of proteins inside the capillaries. If fewer proteins are in the plasma, it has less osmotic potential and more fluid will remain in tissues, causing edema.

The Endocrine System

Animals, including humans, have two ways to relay information from one part of the body to another: the nervous system and the endocrine system. The endocrine system releases **hormones**, chemical messengers that relay messages to cells in other parts of the body, sometimes communicating over short distances from one cell to a neighboring cell (**paracrine** signaling) and at other times signaling longer distances through blood or other extracellular fluid. Some chemical messengers are even received by the same cell releasing the hormone (**autocrine** signaling). Once hormones arrive at their destination at the cells of the target tissue, they bind to specific receptors to exert a response.

The distinction between endocrine and nervous signals is not always clear. The endocrine and nervous systems often coordinate signaling, so hormones of the endocrine system alter nervous system signals and nervous system impulses control hormone release in some cases. The nervous system uses chemical messengers that act locally at synapses (neurotransmitters) or more distantly (neurohormones). In a few cases, chemical messengers such as adrenaline (epinephrine) are released by both endocrine and neuronal cells.

Q If a neuron releases a chemical messenger at synapses with smooth muscle in an endocrine gland and this chemical messenger diffuses across the synaptic cleft to bind to a receptor in the target cell, is this chemical messenger acting as a hormone?

A Hormones are chemical messengers, but in this particular case the chemical messenger is acting as a neurotransmitter (more on this in the section about the nervous system).

HORMONES

Hormone messenger types include proteins, steroid hormones, and other small-molecule hormones that are charged, usually based on modified amino acids. Steroid hormones include testosterone, progesterone, estrogen, and cortisol, all of which have structures based on cholesterol and are fairly hydrophobic. When these hormones circulate in the plasma, they diffuse through cell membranes to bind to their receptors in the cytoplasm of target cells. Steroid hormone receptors act as transcription factors, binding to specific DNA sequences in the promoters and enhancers of genes to turn hormone-responsive genes on and off. Other small, hydrophobic hormones, such as thyroid hormone and retinoic acid, also diffuse through membranes to bind to intracellular receptors in this class.

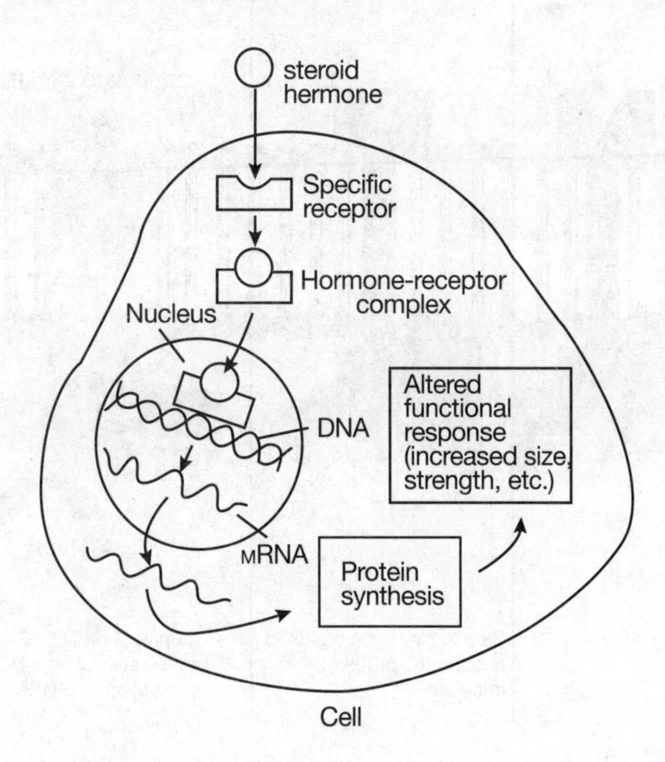

Q If antibodies that recognize the rabbit estrogen receptor are injected in a rabbit, will the antibodies elicit an immune response against the rabbit's own uterus cells?

A Probably not. The estrogen receptors in the rabbit will be in the cytoplasm of cells, hidden from the antibody.

Other hormones include polypeptides, small charged molecules that are too large and too hydrophilic to diffuse across membranes. Examples of peptide hormones include parathyroid hormone (PTH), vasopressin (or anti-diuretic hormone, ADH), and oxytocin. The receptors for these hormones are on the surface of cells as transmembrane proteins in the plasma membrane. Many such hormones act through **G-protein coupled receptors (GPCRs)** that activate signal transduction which initiates second messenger production.

For example, one class of receptors stimulates the enzyme adenylate cyclase, increasing production of the second messenger cyclic AMP (cAMP). One action of cAMP is to bind to protein kinase A (PKA), causing it to phosphorylate other proteins and regulate their activity.

G-protein activation of adenylate cyclase

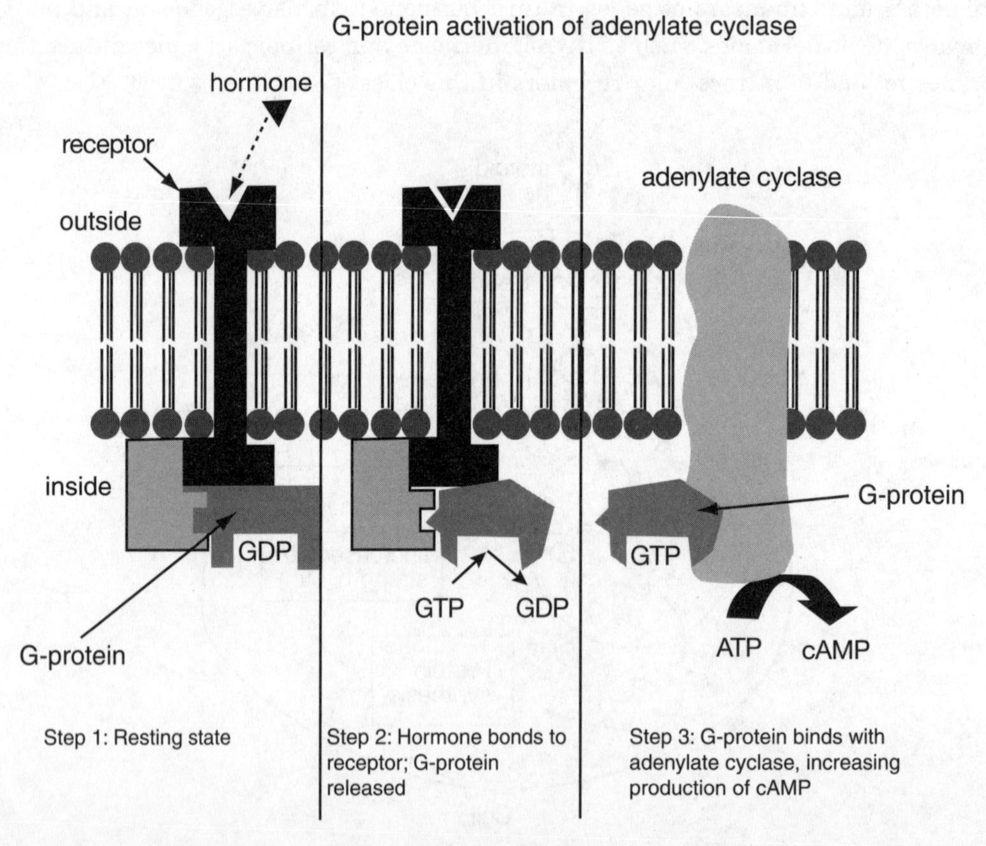

Step 1: Resting state

Step 2: Hormone bonds to receptor; G-protein released

Step 3: G-protein binds with adenylate cyclase, increasing production of cAMP

Another common signaling pathway used by GPCRs causes the release of intracellular calcium from the endoplasmic reticulum, modulating the activity of various enzymes. In addition to affecting the activity of protein kinases and other targets in the cytoplasm, cAMP and calcium also activate genes, causing more lasting changes to cells that persist after the hormone disappears from blood.

Q Once a hormone activates a GPCR and causes the release of calcium into the cytoplasm, how is that hormone's signal switched off?

A Hormone activity is closely regulated, and there are methods of turning off a hormone's signal as readily as it is turned on. One way is to remove the hormone from circulation in the blood by having the liver remove it. Enzymes could be used to degrade it, or the receptor can be switched off, which often happens after prolonged activation of a hormone receptor. Another way is to remove the second messenger from inside cells. The calcium released from the endoplasmic reticulum can also be rapidly sequestered again, ready for the next time hormone activation of its receptor occurs.

GROWTH FACTORS

Proteins called growth factors act on a wide variety of cells to cause them to proliferate or differentiate. For example, epidermal growth factor acts on cell surface receptors in many tissues to cause proliferation. Erythropoietin is a growth factor produced by the kidney that acts on cells in the bone marrow to produce more red blood cells. Growth hormone acts on cells throughout the body during youth to stimulate growth and control metabolic functions of many tissues. The receptors for growth factors are often linked directly or indirectly to protein kinases, which modulate downstream signaling cascades in cells.

To fully understand hormones, one must understand how specific areas of the brain function. The **hypothalamus** is a small region located in the middle lower region of the brain that closely links the nervous and endocrine systems. The actions of the hypothalamus are closely connected to the pituitary gland located just below it. The pituitary gland consists of the **anterior pituitary** and the **posterior pituitary**. The latter contains extensions of neurons from the hypothalamus, so that hormones produced in the cell bodies of the hypothalamus are transported via axons into the posterior pituitary, where they are released.

The posterior pituitary secretes two peptide hormones: vasopressin and oxytocin. The anterior pituitary works differently, however. Cells in the hypothalamus release hormones that are carried in blood to the pituitary through local portal circulation, where they act on the pituitary to regulate its own endocrine secretion of hormones. The hormones released by the hypothalamus to act on the anterior pituitary are called **releasing hormones**; as the name suggests, they cause the pituitary to release hormones that act on tissues throughout the body, including other endocrine glands.

Q If the portal circulation between the hypothalamus and anterior pituitary were damaged, how would the endocrine system be affected?

A The anterior pituitary directly receives blood from the hypothalamus that has a high concentration of important releasing hormones. If portal circulation is damaged, the anterior pituitary cannot efficiently receive hormones from the hypothalamus, thus hindering the normal regulation of a wide array of hormones.

OTHER GLANDS OF THE ENDOCRINE SYSTEM

The endocrine system includes a variety of glands and tissues that release hormone messengers to communicate with other cells and tissues.

- Adrenal glands (medulla) release epinephrine
- Adrenal glands (cortex) release mineralocorticoids and glucocorticoids
- Pancreas releases insulin and glucagon
- Thyroid glands release thyroid hormone in response to TSH
- Parathyroid glands release PTH
- Pituitary gland secretes hormones regulating homeostasis, including trophic hormones that stimulate other endocrine glands.
- Testes and ovaries release sex steroids

Adrenal Glands

The adrenal glands, located above the kidneys, are made up of an adrenal cortex (the center of the glands) and an adrenal medulla (the outside of the glands). The adrenal medulla releases adrenaline (epinephrine) and noradrenaline (norepinephrine) in response to stimulation by the sympathetic nervous system. These hormones act in concert with the system for stress or danger responses to mobilize energy reserves and prepare the body for action. The receptors for these hormones are GPCRs located in a variety of tissues throughout the body.

How does glycogen synthase in skeletal muscle respond to epinephrine?

Its activity is reduced. Epinephrine stimulates a fight-or-flight response, mobilizing the body for quick action. In these conditions, skeletal muscle must begin mobilizing its energy reserves from glycogen and glucose in the blood, rather than storing energy as new glycogen.

The adrenal cortex produces glucocorticoids, also called corticosteroids, which are released in response to the pituitary stress hormone ACTH. Glucocorticoids act on intracellular receptors to modulate gene transcription. In acute stress, they affect energy metabolism by placing more glucose in the bloodstream. Cortisol and related derivatives such as hydrocortisone have strong anti-inflammatory properties, which make them useful drugs. However, with prolonged use, they also suppress the immune system and create unfavorable changes in energy metabolism. The adrenal cortex also produces mineralocorticoids such as aldosterone, which helps the kidneys retain sodium.

The Pancreas

Insulin and glucagon are both produced in the pancreas and are important modulators of energy metabolism. They work in concert to stabilize glucose levels in the blood, but each exerts the opposite action from the other. High and sustained blood glucose levels can damage tissues and proteins; low blood glucose may deprive the brain and other organs of the supply they need for metabolic purposes.

Insulin is secreted by beta cells in the pancreas—small islands of endocrine cells surrounded by pancreatic tissues that secrete material into the digestive tract. It is released after meals, when more glucose is present in the blood than is needed for immediate energy production. The released insulin acts on tissues throughout the body—especially liver cells and skeletal muscle, which store the glucose in the form of glycogen—to uptake the glucose. Glucagon is released when glucose levels fall; it stimulates the liver to uptake some stored glycogen and use it to release glucose into the bloodstream. The liver also performs gluconeogenesis in these conditions, synthesizing glucose from other metabolic intermediates. A lack of insulin secretion leads to Type I diabetes, in which the body requires insulin injections to survive. Type II diabetes involves a decreased responsiveness to the body's own insulin, although normal or even unusually high levels are present. It is often associated with obesity.

Q Is the glucagon level elevated after a meal?

A After a meal, insulin is secreted and glucagon is not.

Q Is glucagon secreted during fasting?

A After a prolonged period without food, insulin levels drop and glucagon secretion increases, stimulating energy reserves to maintain the proper glucose level in the blood.

Thyroid

Thyroid hormone is synthesized by modifying the amino acid tyrosine. There are two forms: one with three atoms of iodine (T_3) and another, called thyroxine, with four atoms of iodine (T_4). Thyroid hormone is involved in human development and in regulating metabolic rate and heart rate. A releasing hormone from the hypothalamus called TRH causes the pituitary to secrete thyroid-stimulating hormone, TSH. TSH travels through the bloodstream to the thyroid, stimulating thyroid hormone to be released. High thyroid hormone levels produce feedback inhibition of TRH and TSH release, keeping the overall thyroid hormone level in a narrow range in healthy individuals. Those who have too much thyroid hormone have hyperthyroidism, with accompanying altered heart function and increased metabolic rate; those who have too little thyroid hormone have hypothyroidism, with accompanying low metabolic rate.

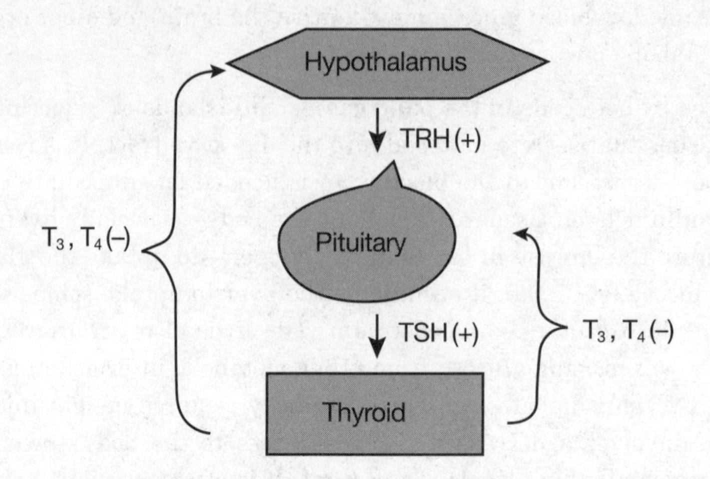

(–) = inhibition; (+) = stimulation

Q How is the secretion of TSH affected when thyroid hormone levels fall?

A Thyroid hormone secretion is regulated by feedback control. When thyroid hormone levels fall, production of TSH and TRH is stimulated.

Parathyroid

Parathyroid hormone (PTH) regulates the body's extracellular calcium concentration and bone metabolism. The body's calcium levels are closely controlled in a narrow range; changes outside of this range can produce uncontrolled muscle contraction. If the calcium level of blood plasma falls, PTH is released by the parathyroid gland to cause bone resorption and the release of calcium into the plasma. PTH also reduces the loss of calcium through urine excretion and increases vitamin D production, which in turn helps to raise calcium levels in serum and increase intestinal absorption.

Q Cancers of the parathyroid glands can result in unusually high secretion of parathyroid hormone. How does this affect the level of calcium in blood?

A Too much PTH will cause excessive bone resorption and elevated calcium in blood.

Q A lack of exposure to sunlight and a lack of fortified dairy products would have which of the following effects on the human body?

(A) Bones become more dense than normal.

(B) Bones become longer.

(C) Intestinal calcium absorption is reduced.

(D) Increased amounts of fat are deposited in adipose tissues.

(E) Parathyroid hormone production is reduced.

A The correct answer is (C). Vitamin D is produced in the skin upon exposure to sunlight; it is also commonly added to milk as a supplement. Without sunlight exposure and supplemental vitamin D, the body can be deficient in this vitamin. Because the body's main response to vitamin D is to increase intestinal absorption of calcium, this absorption will be low when vitamin D levels are low.

Pituitary Gland: Prolactin, Oxytocin, and Vasopressin

Prolactin is a peptide hormone produced in large quantities in the anterior pituitary during pregnancy. Its main response is to prepare breast tissue for milk production (lactation) after birth. Lactation stimulates further prolactin release and supports ongoing milk production.

Oxytocin, another peptide hormone, is secreted by the posterior pituitary and is thought to produce a wide range of bodily responses, including stimulating the uterus during labor and birth, stimulating the release of milk during nursing, and acting in concert with prolactin to support milk production.

A third peptide hormone, vasopressin, is produced by the posterior pituitary and helps raise blood pressure and reduce fluid loss by increasing the concentration of urine in the kidney. With dehydration or blood loss, these responses can help preserve blood flow to essential tissues.

The Nervous System

The vertebrate nervous system is divided at the highest level into a **central nervous system** (CNS), including the brain and spinal cord, and a **peripheral nervous system**. One of the distinguishing features of chordates, including vertebrates, is the dorsal nerve cord, which in vertebrates becomes the CNS. The **cerebrum** occupies most of the human brain and carries out higher mental functions. The surface of the cerebrum is convoluted, which increases surface area and thus increases the number of neurons and neuronal connections in the human brain. The so-called **grey matter** of the brain surface contains neuronal cell bodies and rich interconnections with dendrites. The **white matter** in the center of the brain is rich in myelinated axons. The cerebral cortex receives information from sight, sound, touch, taste, and other sensory systems and controls voluntary motion of skeletal muscle. Information from each sense and from body parts is directed to sensory areas of the brain for processing and then to associative areas for relating different types of information. Specific motor cortex regions provide motor control of different parts of the body. A small region of the frontal lobe called **Broca's area** provides motor control of speech, and another region called **Wernicke's**

area is required for understanding but not producing speech. The cerebral cortex is responsible for higher mental functions and abstract thought such as math and music.

The cerebrum is divided into left and right hemispheres connected by the **corpus callosum**. The left hemisphere of the cerebrum is thought to be the portion that focuses on logic, math, and fine motor control; the right hemisphere is thought to be concerned with context, emotions, music, and object relationships.

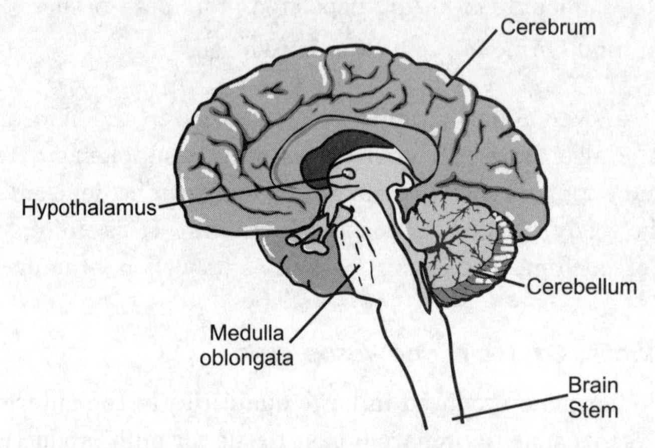

Cross section of Human brain

NEURONS

The basic unit of the nervous system is the **neuron**, the nerve cell that carries electrochemical signals called action potentials through the body to create responses in target tissues. The vertebrate neuron has a central cell body that includes the nucleus, most of the cytoplasm and organelles, smaller branches extending from the cell body called **dendrites**, and a long projection called the **axon**. Dendrites receive signals from other neurons and carry them toward the cell body; and axons carry signals away from the cell body to other cells. The axons of many neurons are wrapped in a fatty insulating sheath called **myelin** that helps the cells transmit electrochemical signals.

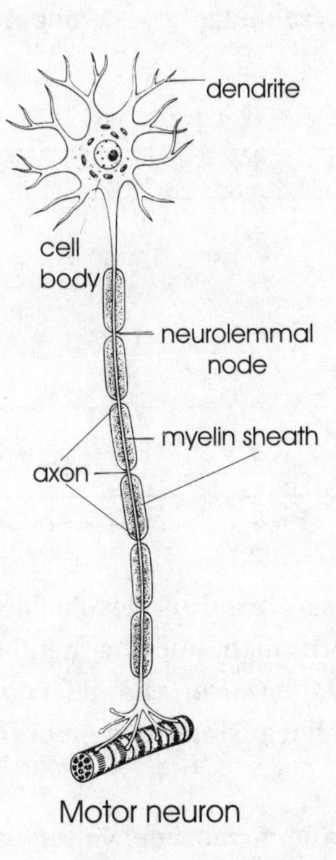

Motor neuron

Action potential is an electrochemical wave that moves in one direction through neurons, from the cell body to its axon, terminating in synapses with other neurons, glands, or muscle. To have action potential, cells must first have a resting potential created by pumping ions across the plasma membrane. This gives the neuron's cytoplasm a negative charge. The resting potential is negative if the interior of the membrane has more negative charges than on the exterior of the membrane. Because cells are filled with charged molecules (proteins, nucleic acids, and small solutes), the resting potential maintains the osmotic balance of cells with the extracellular environment.

Q Does the movement of sodium ions from a region of low concentration to one of high concentration have a negative or positive change in free energy (ΔG)?

A Moving something against a concentration gradient, or more precisely, against an electrochemical gradient, takes energy. Since the reaction will only occur spontaneously in the other direction, going from low to high concentration has a positive ΔG and requires energy input from another source (such as ATP hydrolysis) to make it move in this direction.

The action potential changes the neuron's membrane potential from the resting potential and moves that change along the neuron's membrane as a wave of depolarization. The voltage-gated ion channels in neuronal membranes make this possible when they open to allow sodium ions to enter. This is what gives neurons their uniquely excitable membranes. Although potassium channels allow potassium to move across the cell's membrane to establish the resting potential, the neuron's membrane is fairly impermeable to sodium ions. If the voltage-gated sodium channels open and sodium ions enter, they will change the

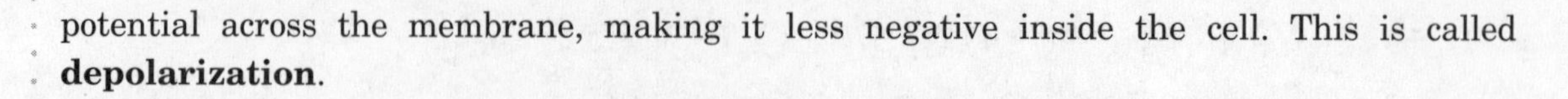

potential across the membrane, making it less negative inside the cell. This is called **depolarization**.

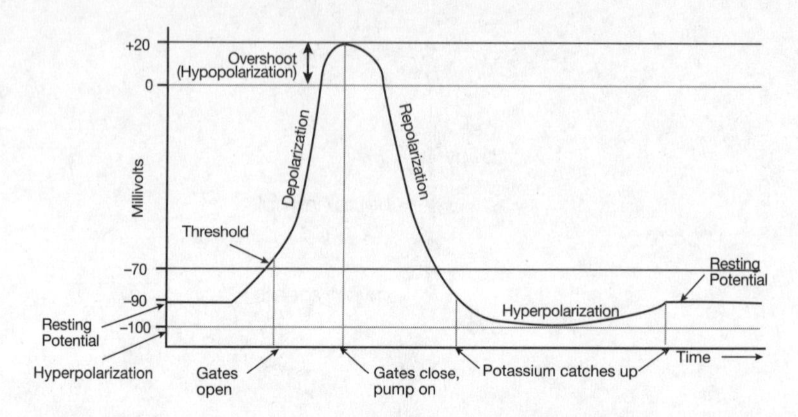

If the membrane potential reaches a **threshold** level of depolarization, then the voltage-gated channels open, sodium ions pass through, and the membrane potential may momentarily be positive. Within a few milliseconds, however, channels open in this section of membrane and potassium ions move out of the cell, causing the membrane potential to return to its original state.

At the same time, in the neighboring membrane, voltage-gated sodium channels also respond to changes in membrane potential by opening, thereby causing another patch of membrane to depolarize. The action potential moves along a neuron in this way, with a wave of depolarization and repolarization caused by voltage-gated ion channels. In myelinated neurons, action potentials jump from node to node between patches of myelin, moving more rapidly along the axon than in the absence of myelin. This type of movement of the action potential is called **saltatory** (jumping) **conduction**.

Action potentials are often described as all or none. This means that, once the threshold depolarization is reached and the cell responds with an action potential, the size of the action potential is always the same, no matter what the original stimuli were.

Q The main effect of axon myelination is to:

(A) conserve axonal heat

(B) create larger action potentials

(C) allow axons to transmit action potentials in two different directions

(D) allow action potentials to travel more quickly

(E) eliminate the need for axons to generate a resting potential

A The correct answer is (D). Myelination of axons allows saltatory conductance, in which the action potential jumps forward from node to node between myelinated regions and moves much more quickly along the axon as a result. Action potentials are always the same size and always move in one direction. Axons still need to generate a resting potential, however, to generate an action potential.

SYNAPSES AND NEUROTRANSMITTERS

When the action potential reaches the end of an axon, it must cross either an electrical or a chemical synapse between one neuron and another. In an **electrical synapse**, the cytoplasm of two cells is connected by a gap junction so that an action potential can be directly transmitted between the cells. Most of what we'll discuss here, however, concerns **chemical synapses**, in which a small gap (the **synaptic cleft**) separates the axon of the neuron from the target cell and chemical messengers called neurotransmitters carry the signal across this gap. When the action potential reaches a chemical synapse at the end of an axon, synaptic vesicles filled with neurotransmitters fuse with the cell membrane to dump neurotransmitter into the synaptic cleft. The neurotransmitter molecules diffuse a short distance across the cleft to bind on receptors on the target cell and create a response.

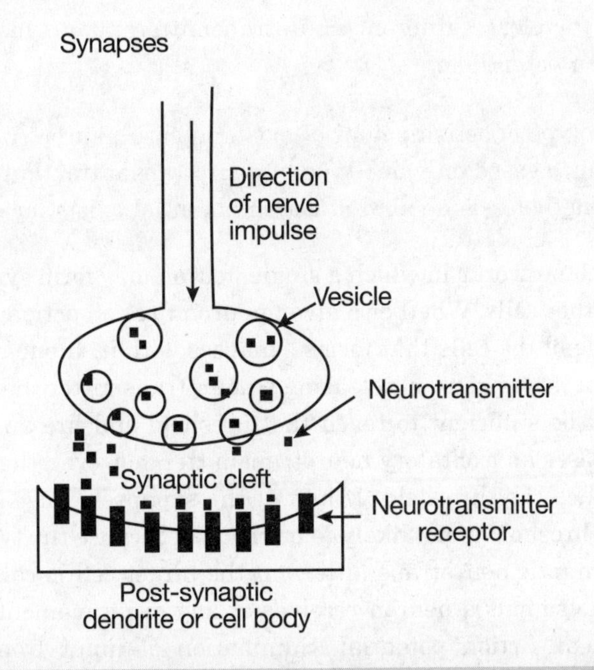

Q Are neurotransmitters involved in transmission of action potentials across an electrical synapse?

A Electrical synapses do not have a synaptic cleft. Two cells with an electrical synapse are joined by gap junctions that carry an action potential directly between cells.

The receptors for neurotransmitters on the target cell are often ligand-gated ion channels that allow ions across the membrane to alter the membrane potential of the target cell. Other neurotransmitters act through G-protein coupled receptors that do not directly open ion channels, but stimulate second messenger pathways that alter ion channel function. If neurotransmitters make the membrane of the target cell more negative than the resting potential, an **inhibitory postsynaptic potential** (IPSP) exists, moving the membrane potential further away from the threshold for firing an action potential. Alternatively, if neurotransmitters and their receptors make the postsynaptic membrane potential less negative by moving closer to the action potential threshold, an **excitatory postsynaptic potential** exists.

Q How is the postsynaptic membrane potential affected if a neurotransmitter opens a ligand-gated chloride channel and allows chloride ions to enter the cell?

A The interior of the membrane will become more negative than normal. This creates an inhibitory postsynaptic potential, moving the membrane potential farther away from the threshold for an action potential.

Q Does a neuron release different neurotransmitters at a synapse with different types of action potentials?

A No. Only one type of action potential exists in each neuron. It was once believed that a neuron released only one type of neurotransmitter, but this appears to be incorrect. Nonetheless, one kind of action potential exists for each neuron.

Neurons are part of rich networks in which a single neuron may form synapses with hundreds or even thousands of other cells. Whether a given neuron fires an action potential or not is the result of input from one of the cells that forms synapses with it. If one neuron fires an action potential that reaches a target neuron, releasing a neurotransmitter that causes an EPSP, the depolarization will not be sufficient to reach the threshold and fire an action potential. If a group of synapses releases an excitatory neurotransmitter, however, the depolarization of the combined EPSPs reaches the threshold. If half of the synapses cause EPSPs and the other half cause IPSPs, the threshold is not likely to be reached because the two cancel one another. The total impact of synaptic neurotransmitters in the target cell is called **summation**. The summation of all of the signals a neuron receives at any given moment determines whether that neuron will fire an action potential. Summation of input from chemical synapses provides great flexibility in neuronal response—flexibility that would be absent if connections were all hard-wired with electrical synapses.

Once neurotransmitters complete their work at the synapse, the signal needs to be switched off. A signal that lasts too long or never turns off may cause significant problems, such as uncontrollable muscle contraction. This can occur when the neurotransmitters diffuse from the synaptic cleft into the surrounding extracellular fluids. Since the distances and quantities of material involved are minute, this occasionally occurs. In most cases, however, enzymes at the synapse rapidly degrade the neurotransmitter, and transporters reuptake and recycle the material into the presynaptic cell for repackaging in new synaptic vesicles.

Blocking these activities can increase the action of neurotransmitters and cause the neurotransmitter to accumulate in the synapse. For example, acetylcholine, which occurs at the neuromuscular junction where motor neurons connect to skeletal muscle, terminates the signal of the neurotransmitter by the activity of the enzyme **acetylcholinesterase**, which degrades acetylcholine in the synaptic cleft. Chemicals that irreversibly inhibit insect acetylcholinesterase are used as insecticides; compounds that do the same to human acetylcholinesterase have been used as chemical warfare agents. Other cholinesterase inhibitors that act in the brain instead of the neuromuscular junction are used to treat Alzheimer's disease.

Q If an irreversible inhibitor of acetylcholinesterase that binds covalently at the enzyme active site is examined in the lab with purified enzyme, can the inhibition be overcome with increased substrate concentration?

A No. An irreversible inhibitor is insurmountable—the inhibitor reduces the number of active sites and is bound covalently so it cannot be competed off of the active site with higher substrate concentrations.

Most neurotransmitters are amino acids or are synthesized from amino acids and are packaged into synaptic vesicles that reside near the presynaptic membrane, waiting for an action potential to trigger their release into the synaptic cleft. Amino acid neurotransmitters include **aspartate**, **glycine**, and **glutamate** and an additional amino acid that does not occur in proteins, **gamma-amino butyric acid (GABA)**. GABA and glycine are important inhibitory neurotransmitters in humans, allowing chloride ions to enter cells, while aspartate and glutamate are excitatory, both acting on the NMDA receptor that is believed to reinforce synapses over time in memory and learning.

Q Is the glycine used as a neurotransmitter chemically different from the glycine found in proteins?

A No. Amino acid neurotransmitters (glycine, glutamate, aspartate) are the same amino acids used in proteins.

Acetylcholine is an important neurotransmitter in multiple ways. First, it is used by motor neurons at synapses with skeletal muscle and plays an essential role in voluntary motion. The release of acetylcholine at the neuromuscular junction opens ligand-gated sodium channels, triggering the action potential in the muscle cell that leads to muscle contraction. It is also an important neurotransmitter in the sympathetic and, particularly, in the parasympathetic autonomic nervous system. Acetylcholine released by the parasympathetic nervous system alters the activities of many of the body's organs. It slows the heart rate and increases the activity of some digestive glands and digestive smooth muscle tissues. Acetylcholine acts through several receptors, including nicotinic receptors (ion channels) and muscarinic receptors (GPCRs).

Q Would the actions of acetylcholine through a nicotinic receptor be faster or slower than actions through a GPCR receptor?

A Ion channel responses are faster, but they are briefer.

Q If acetylcholine regulates so many different systems as part of the nervous system, how does it avoid activating all of them at once upon release?

A Acetylcholine is a neurotransmitter, not a hormone. All of its actions are local, meaning they are restricted to synapses. It does not activate all of these systems at one time because its actions are always restricted to the synapses where it is released and to the receptors at those synapses. Drugs that bind to acetylcholine receptors can circulate in the blood to act more broadly in the body, but the variety of receptors involved in acetylcholine makes it possible to target specific receptors and functions.

A broad class of neurotransmitters derived from amino acids includes serotonin, dopamine, epinephrine, and norepinephrine, often called collectively the **biogenic amines** or **monoamines**. They are all relatively small polar molecules that act through cell surface receptors. They are all important in the CNS, involved in normal behavior and sometimes involved in disease. Their receptors are the targets of many pharmaceutical treatments and some illegal drugs. These neurotransmitters act through a variety of receptors in the CNS and in the periphery. They produce such varied responses that it's not possible to describe everything they do here. However, here are some of their more important functions:

- **Dopamine:** Acts on the brain's areas of reward, motivation, cognition, and movement control. A precursor of dopamine, L-Dopa, is used to improve the movements of Parkinson's patients.
- **Serotonin:** Apparently involved in regulating emotions, such as euphoria, depression, anger, and anxiety. Drugs targeting the serotonin system are available to treat migraine headaches, clinical depression, and nausea, depending on which receptors or transporters are targeted.
- **Norepinephrine:** Acts as a stress hormone when released by nerves or the adrenal glands in the periphery but acts as a neurotransmitter in the brain, stimulating wakefulness.

The vertebrate neuron is associated with several other cell types in the nervous system that help neurons work without conducting nerve impulses themselves. These are called **glia**. **Schwann cells** create the myelin that surrounds the axons of some neurons, aiding in transmitting action potentials. Another type of glial cell, **astrocytes**, support nervous system structure and how neurons connect with one another. Glial cells create the environment in which neurons work.

THE AUTONOMIC NERVOUS SYSTEM

The peripheral nervous system communicates with the CNS to regulate many body functions, including motor control through the **somatic nervous system** and involuntary actions controlled through the **autonomic nervous system**, which includes the sympathetic and parasympathetic nervous systems. The tissues targeted by these systems are smooth muscle, cardiac muscle, and glandular tissues, including the heart, lungs, arteries, digestive organs, reproductive organs, and many endocrine glands. Refer to the chart below for more specific information on how these systems affect the rest of the body.

EFFECTS OF SYMPATHETIC AND PARASYMPATHETIC NERVOUS SYSTEMS

	Sympathetic Nervous System	**Parasympathetic Nervous System**
Heart	increases heart rate	slows heart rate
Lungs	dilates bronchi, airways	constricts airways
Stomach and intestines	decreases secretion and digestion rate	increases secretion and digestion rate
Pupils of the eye	dilates pupils	constricts pupils
Pancreas	decreases secretion	increases secretion
Liver	decreases glycogen synthesis	increases glycogen synthesis

Q The drug atropine blocks the effects of the parasympathetic nervous system on the eye. How does this affect the pupil size?

A The parasympathetic nervous system constricts the pupil, so blocking that system will cause the pupil to dilate.

THE BRAIN: LEARNING AND MEMORY

One of the key traits of mammals is their ability to learn and to remember. Molecular mechanisms involved in these traits appear to include changes at existing synapses, which can occur quickly, and the formation of new synapses, which are reinforced over time. Short-term memory may involve changes in existing synapses. Learning more complex behavior may require developing new synapses over a longer period.

In a form of simple learning called **habituation**, animals grow accustomed to a specific stimulus after they experience it repeatedly; eventually the same stimulus causes less of a response. For example, the first time you poke a snail's eyestalk, it withdraws immediately; if you do it several more times, the snail will habituate and ignore the stimulus. The opposite of habituation is **sensitization**, in which a repeated stimulus causes a stronger response over time.

In a more complex form of learning called **associative learning**, one stimulus is associated with another one. Associative learning is often used by scientists or animal trainers to reinforce or prevent a specific behavior. For instance, if you give a dog an unpleasant electrical shock every time it leaves the backyard, it will associate leaving the yard with getting that shock and will eventually stop trying to leave. This is referred to as **classical conditioning**.

Here's a review of how parts of the nervous system and brain are involved in memory and learning.

- **Limbic system.** Involved in generating emotions. The limbic system includes structures in and near the interior of the cortex around the brainstem, including the thalamus, hypothalamus, hippocampus, and amygdala. Damage to any of these regions may disrupt the interactions among emotion, perception, and action.

- **Thalamus.** A relay station of the brain that helps move information between the cortex and lower parts of the CNS and periphery.
- **Hippocampus.** Essential for forming new memories from experience. The hippocampus may also play a role in spatial memory.
- **Hypothalamus**. A key part of the endocrine system, the hypothalamus regulates pituitary function. It is also important as a nervous system regulator of body temperature, thirst, hunger, pleasure, and sexual activity.
- **Amygdala**. The amygdala is responsible for memory formation, particularly those that involve a strong emotional component.
- **Brainstem**. This includes the **medulla oblongata** and **pons** above the spinal cord. The brainstem controls several autonomic functions, such as breathing, digestion, and heart rate. It also integrates sensory information, coordinates some of the broader aspects of body movement, and contains the reticular activating system that regulates sleep, arousal, and attention.
- **Cerebellum.** The cerebellum coordinates fine motor movement. Damage to this area of the brain does not prevent motion but may cause awkwardness and uncoordination.
- **Choroid Plexus**. This area is where cerebrospinal fluid is produced.

Q Damage to a brain region that impairs the regulation of breathing probably involves which of the following: the amygdala, the cerebrum, the cerebellum, the medulla oblongata, or the hypothalamus?

A The medulla oblongata in the brainstem helps regulate essential autonomic functions, including breathing.

Sensory Systems

The nervous system receives information from many sources about what is going on inside and outside the body. This includes data about light, sound, touch, smell, taste, body position, osmolarity of body fluids, pressure, heat, cold, pain, and more. All of this information comes via action potentials initiated by sensory receptors, each of which is uniquely sensitive to specific stimuli.

POSITION

Mechanoreceptors are sensory cells that detect physical stimuli and their effects on tissues. Often these cells detect how physical forces change the shape of a cell membrane and its permeability to ions. One type of mechanoreceptor detects the relative position of a mammal's body and monitors the effects of gravity to help orient an animal as it moves.

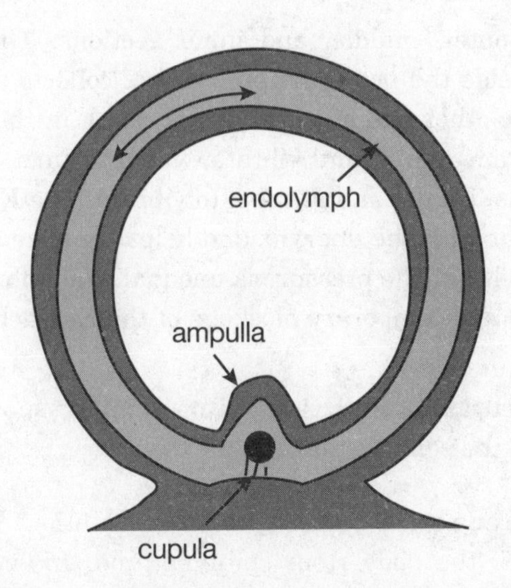

In humans, the **semicircular canals** of the inner ear sense the body's position compared to gravity. These canals are situated on three different planes; each is a curved tube filled with fluid and connected on both ends to a fluid-filled chamber. Movement of the head or body causes the fluid to shift in the canals. In each canal, hair cells protruding into the fluid, in a structure called a cupula, detect movement by bending, thus triggering action potentials that travel to the CNS. Because each canal has a different orientation, the direction of movement affects each canal differently. This allows the brain to integrate the information and sense the overall movement and position of the head through three-dimensional space.

HEARING

Sound is comprised of pressure waves that are propagated through air or water. The magnitude of the pressure change constitutes the volume of a sound (the wave's amplitude). The distance between waves is the wavelength, or frequency. Sound with a high frequency has shorter waves (more waves per second); sound with low frequency has longer waves (fewer waves per second).

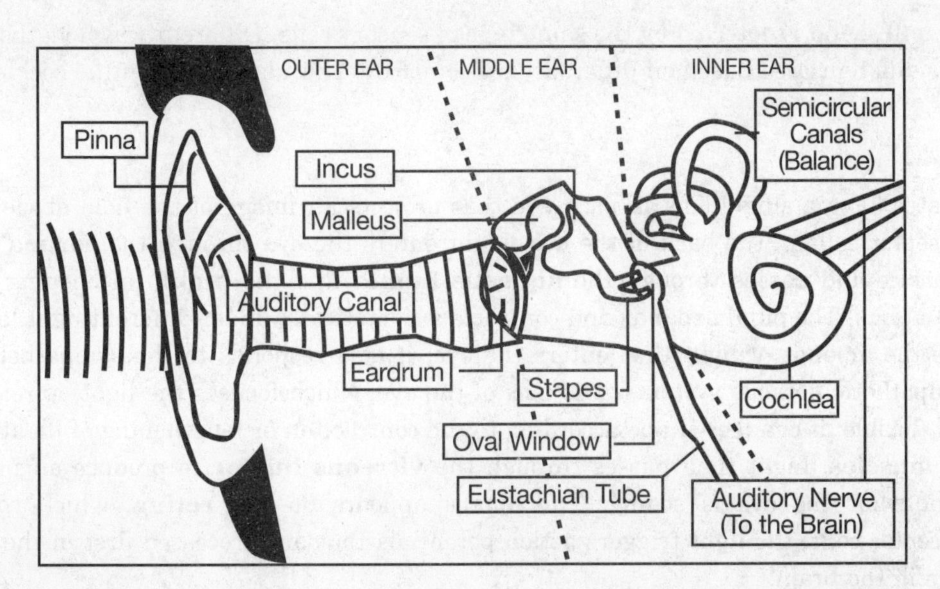

The human ear includes outer, middle, and inner sections. The outer ear includes the cartilaginous structure outside the body (the pinna) that collects and directs sounds toward the **auditory canal**. At the other end of the auditory canal, in the middle ear, sound waves make the tympanic membrane (ear drum) vibrate, which in turn causes three tiny bones of the middle ear, the **malleus, incus,** and **stapes**, to vibrate. The **Eustachian tube** connects the interior of the middle ear with the pharynx and helps equalize air pressure on either side of the ear drum. You can feel that the pressure is unequal when, for example, you have a cold or are flying in an airplane and temporary blockage of the Eustachian tube occurs.

Q Where does the sensation of dizziness from spinning motion occur: the cochlea, the middle ear, or the semicircular canals?

A The semicircular canals are responsible for orientation. The fluid in the canals keeps moving after the body stops spinning, and this creates the sensation of dizziness.

In the inner ear, sounds are converted into action potentials. Here, vibrations from the malleus, incus, and stapes are converted to pressure waves in the fluid-filled **cochlea**. Sound waves travel through the cochlea's **vestibular canal**, above sensory hair cells inside the cochlear duct and atop the **basilar membrane** that vibrates with the sound waves. The basilar membrane vibrates differently according to different frequencies of sound. High-pitched sound waves travel the shortest distance in the cochlea and make the basilar membrane vibrate closest to the point where they enter the cochlea; low-pitched sounds travel the farthest from the point of entry. Louder sounds make the membrane vibrate more (if the sounds are very loud, the excessive vibration can damage the inner ear). In the region that vibrates, hair cells bend back and forth and fire action potentials to the brain through the auditory nerves.

Q Are different receptors required to detect sounds at different wavelengths?

A All sound is detected by the same type of sensory cells. Different wavelengths are differentiated based on different position of the stimulated cells in the cochlea.

VISION

Vertebrates have a single lens in each eye that projects an image of the field of view onto photoreceptor cells in the back of the eye. Light enters the eye through the **cornea** on the front surface and passes through the **aqueous humor** and the **pupil**, an opening in the pigmented **iris**. The pupil expands and contracts automatically under different light levels to regulate the amount of light that enters the eye; it also responds to the sympathetic and parasympathetic nervous systems. The lens of the eye, which focuses the light, is relatively soft and flexible; it changes shape according to the contraction or lengthening of its attached **ciliary muscles**. Light then passes through the **vitreous humor** to produce an inverted image of what the eye is seeing. This image appears on the **retina**, which contains photoreceptor cells; the light triggers action potentials that are processed first in the retina and then in the brain.

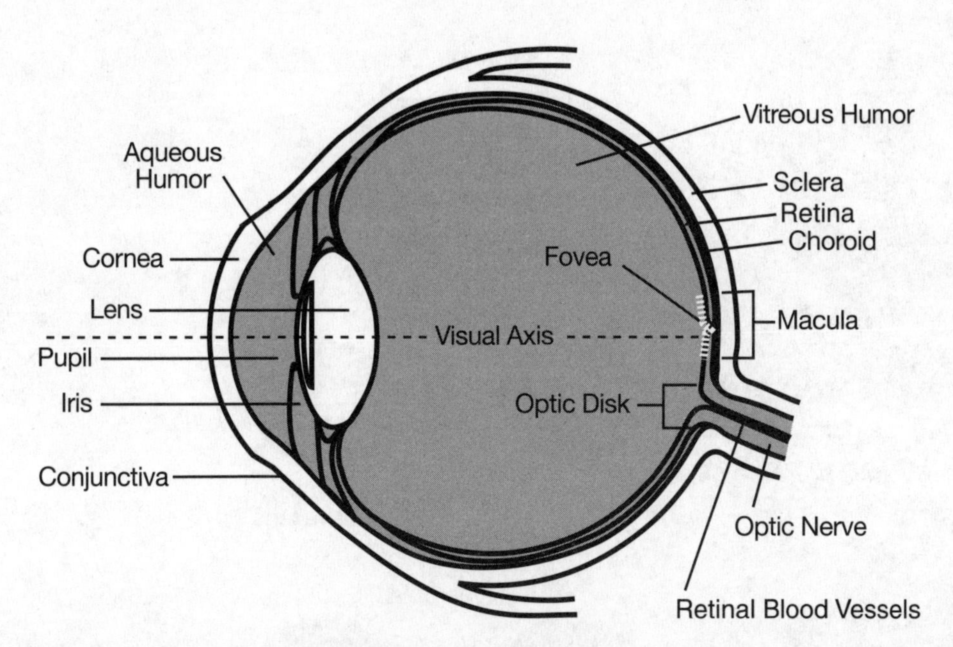

The eye has two types of photoreceptor cells called **rods** and **cones**. Cones are responsible for color vision; rods are more sensitive to light than cones but do not distinguish colors. The cones are focused around the **fovea** and **macula** where vision is sharpest. Three types of cone cells detect red, blue, and green. Farther away from the fovea, rods predominate.

Q Which are more responsible for night vision: rods or cones?

A Rods are more sensitive to light than cones and are primarily responsible for vision in dim conditions.

Inside the photoreceptor cells, light detection begins with proteins that act as visual pigments when paired with the co-factor **retinal**, which is derived from vitamin A. The protein components of the visual pigments are all G-protein coupled receptors with slightly different amino acid sequences that provide different sensitivity to light in different photoreceptor cells. The visual pigment in rods in **rhodopsin**; **photopsins** are the visual pigments in red, blue, and green cone cells.

Q Would a mutation in a photopsin gene affect all cone cells?

A No. Each type of cone cell expresses a different visual pigment photopsin protein encoded by a different gene. A mutation in one of these genes does not affect the other two photopsin genes in other cone cells.

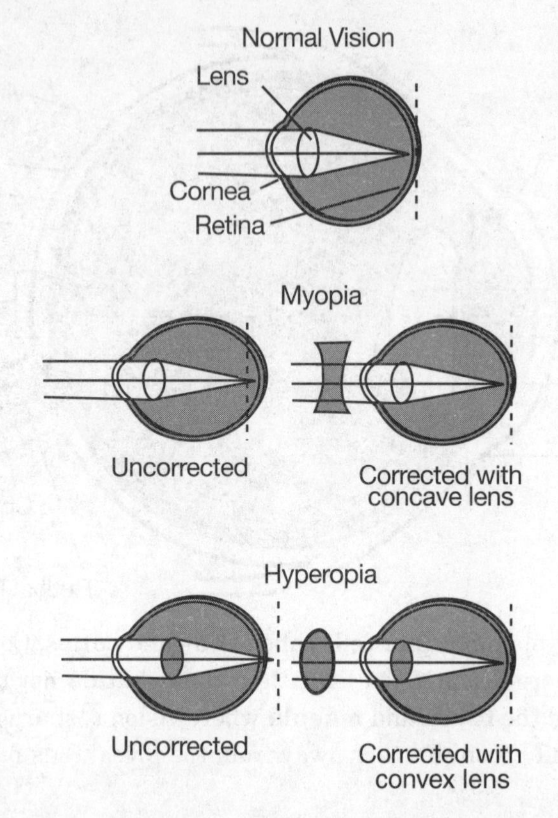

Changing the eyeball's shape focuses an image in the eye. In people who have less than perfect vision, the eye does not fully focus the image. Those with nearsightedness or **myopia** have an elongated eyeball that causes the focus of the eye to fall short of the retina. Those with farsightedness or **hyperopia** have a shortened eyeball shape that causes the focus of the eye to fall beyond the retina. Another type of vision problem, **presbyopia,** occurs with increasing age. The lens of the eye gradually becomes less flexible and is unable to adjust sufficiently for near vision. In the case of myopia or hyperopia, corrective lenses (eyeglasses or contact lenses) return the focus to the retina for better acuity. In the case of presbyopia, reading glasses (which are magnifying lenses) or a specific type of eye surgery can provide better near vision.

Q Do corrective lenses for a hyperopic person move the focal point closer or farther from the lens of the eye?

A Corrective lenses for hyperopia move the focal point closer to the lens, so that the image focuses on the retina.

Q Does the human eye change shape when focusing on objects at various distances?

A No. Only the lens of the eye changes shape.

TASTE AND SMELL

Taste and smell involve chemoreceptors—proteins that have evolved to detect and respond to specific chemical stimuli. Taste receptors in the human mouth and tongue in the form of taste buds can detect dissolved chemicals that produce saltiness, sweetness, bitterness, sourness, and what has recently been called "umami," a meaty taste associated with glutamate or monosodium glutamate.

Most of the taste sensations humans experience are derived from the olfactory receptors (which explains why food may taste bland when one has a cold or other sinus problem that temporarily blocks the sense of smell). Chemicals responsible for aroma diffuse through the air and dissolve in the thin layer of mucus on the nasal epithelium. Humans can distinguish thousands of smells, but our sense of smell is not as acute as that of some other mammals: There appear to be about 400 functional olfactory receptors in the human nose but as many as 1,000 in some other mammals.

Q Is there a smell receptor for saltiness of food?

A Probably not. Saltiness is given by the charged ionic sodium ions in solution. Sodium ions are not volatile and most will not reach the olfactory epithelium. Most smells come from fairly nonpolar substances.

The Excretory System

To remove urea and maintain osmotic balance, organisms in many groups have evolved organs in which a dilute solute is filtered, from which material is selectively reabsorbed as needed, and through which nitrogenous wastes are excreted. In annelids, for instance, **metanephridia** in each segment draw fluid from the coelom into the tubule, from which material is reabsorbed by epithelial cells. Whatever is left, such as nitrogenous waste, is excreted from the body. Insects use Malphigian tubules that function similarly.

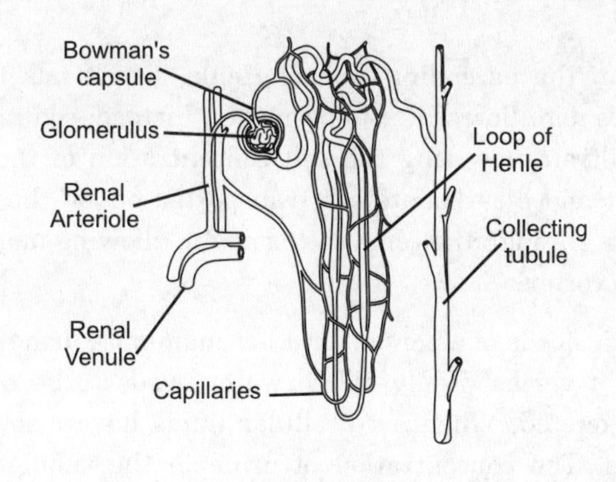

Nephron

The vertebrate kidney works in a related way. The kidney contains great numbers of small tubules called **nephrons**. Blood flows into small capillaries under pressure in each nephron; these form a ball called the **glomerulus**. The capillaries of the glomerulus surround the end of a tubule, the **Bowman's capsule**. Blood is filtered in the glomerulus and Bowman's capsule. Cells and high-molecular-weight material such as proteins remain in the blood, but smaller molecules, including water, salts, amino acids, and glucose, are filtered through Bowman's capsule. This creates the initial urinary filtrate. A large volume of blood is filtered each day in the human body. Most of the water and many solutes are reabsorbed to prevent loss, and waste material is left behind or actively secreted into the filtrate as it moves through the **proximal tubule**, the **loop of Henle**, the **distal tubule**, and the **collecting duct**.

Q Where does filtration of blood take place in the nephron?

A In the glomerulus and Bowman's capsule.

The material filtered from blood moves through the proximal tubule, where some of the solutes in the filtrate are reabsorbed by transporters in the epithelial cells lining the tubule to prevent their loss through the urine. The reabsorbed material includes glucose, sodium chloride, bicarbonate ions, amino acids, and potassium. Other compounds, including hydrogen ions, are actively secreted into the filtrate. One of the ways in which the body regulates blood pH and extracellular fluids is through removal of H^+ in the kidneys.

The filtrate next reaches the loop of Henle. The extracellular fluids of the cortex (the inner region of the kidney) have a very high solute concentration, much higher than that of the medulla (the outer kidney region), and are high in sodium, chloride, and urea. The first part of the loop of Henle, the descending loop, is permeable to water. Since the cortex is very salty, water is drawn by osmosis out of the filtrate and into the renal cortex, decreasing the volume of the filtrate and making it saltier.

Q Which has a higher concentration of solutes: the medulla or the cortex of the human kidney?

A The cortex.

The filtrate continues in the ascending loop of Henle, which has thicker walls than the descending loop and does not allow water to permeate. Instead, sodium chloride diffuses and is pumped out of the filtrate, lowering the salt concentration of the filtrate. In the distal tubule, more water, salt and bicarbonate are transported out of the filtrate. The collecting duct carries filtrate down through the renal cortex again, allowing more water to diffuse from the filtrate and into the cortex.

Depending on the osmotic needs of a person (or other mammal), urine may be excreted with a solute concentration that varies widely. When water needs to be conserved, for example, concentrated urine is excreted. When extracellular fluids have a low solute concentration, dilute urine is excreted. The concentration of urine in the tubules requires the osmotic gradient established from the renal medulla to the cortex. This gradient is constantly maintained via sodium being pumped out of the ascending loop of Henle, even as urinary filtrate continues moving through the tubules and the final urine is produced. The blood vessels around the tubules in the nephron have a countercurrent exchange system to maintain this gradient. Blood flows into each nephron in an afferent arteriole that branches into the capillaries of the glomerulus. It then moves into capillaries in a network around the rest of the nephron. Capillaries around the loop of Henle, called the vasa recta, have blood flowing in the direction opposite that of the filtrate in the tubules. This system helps maintain the osmotic gradient as material is exchanged among the filtrate, extracellular fluids, and blood in the vasa recta.

The concentration and volume of urine excreted is closely controlled to maintain the osmolarity and volume of blood and extracellular fluids. If the solute concentration of extracellular fluids increases and volume decreases, the sensation of thirst is triggered. Another factor regulating urine concentration is the peptide hormone vasopressin, also called **antidiuretic hormone** (**ADH**), which makes the collecting duct walls more permeable to water so that more will be reabsorbed from the filtrate. The end result is more concentrated urine and less water loss through urine.

Q How might the fluid intake be affected in a person lacking a normal gene for vasopressin?

A Without vasopressin, a person will not form concentrated urine and may lose a great deal of water via the urine. He or she will need to consume large quantities of fluids to compensate or risk death from dehydration.

Another factor that regulates urine formation is the hormone **angiotensin**. Rather than responding to the concentration of blood solutes, this system responds to the volume of extracellular fluids and blood pressure. If fluid volume drops, such as with injury and blood loss, then blood pressure falls. Cells monitor blood pressure near the glomerulus and release an enzyme called renin when it drops. This in turn cleaves a peptide in the blood to create hormone angiotensin II. This peptide constricts the arteries, triggers the release of aldosterone, and causes greater water and salt reabsorption in the kidney.

Q Does angiotensin create the opposite effect of vasopressin?

A No. Although angiotensin responds more readily to blood pressure and vasopressin responds more readily to the concentration of solutes, the hormones complement one another. Both retain water and solutes and maintain extracellular fluid.

BODY TEMPERATURE REGULATION

The internal temperature of animals is a key component of homeostasis. Some animals, including most fish, reptiles, and invertebrates, are **ectotherms**, meaning that they have an internal temperature near the external ambient temperature. Others, like birds and mammals, are **endotherms**, meaning that they maintain tight and constant internal control of body temperature. Endotherms use a great deal of energy to maintain a constant internal temperature, but they are also able to remain active in environments that range widely in temperature.

Q In a cold habitat, assuming similar body size, which would consume more food: an endotherm or an ectotherm?

A An endotherm requires more energy to maintain its body temperature, so it must eat more.

Whatever their internal temperature, the heat that animals generate is derived from metabolism. Producing ATP from glucose through glycolysis and oxidative phosphorylation, for example, produces heat as a byproduct. Hydrolyzing ATP to use its energy to power biosynthesis, movement, or active transport also releases heat. The heat either accumulates in the body or is lost to the surrounding environment. Endotherms have an internal set point for body temperature that regulates heat production and heat loss to maintain a constant internal temperature.

Q During a fever, which alters the set point: the immune response or infectious organisms?

A The immune system secretes substances that alter the set point and cause fever. The infection triggers the immune system to do so, but the immune system changes the set point.

If a body's core internal temperature falls, the body will increase heat production. Some mammals generate heat over a short term by shivering, which is a series of rapid muscle contractions that generate metabolic heat.

Another mechanism for heat generation in cold conditions is called **non-shivering thermogenesis**. This occurs in a tissue called brown fat. Brown fat cells are a type of adipose rich in mitochondria. The mitochondria consume nutrients, feeding metabolic substrates into the Krebs cycle to produce the mitochondrial pH gradient. These mitochondria also allow the protons to diffuse across the mitochondrial membrane without producing ATP. This rapid metabolism without ATP is not wasted. Remember that energy is never created nor destroyed; it merely changes forms until, ultimately, it becomes heat.

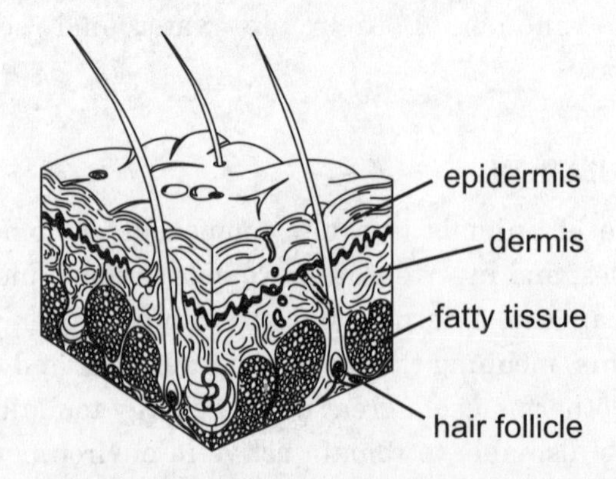

Cross section of Human skin

Animals, including humans, must also find ways to eliminate excess heat when the body's internal temperature rises and the external environment is too hot. Blood flow just under the skin is regulated to help transfer heat from inside the body to the environment. Arterioles in skin respond to nerves to vasodilate or vasoconstrict as needed. Because water evaporation has a high molar heat level, it can absorb heat and help cool the body. For mammals that sweat, the sweat layer cools the skin as it evaporates, helping to remove heat from the body.

The Skeletal and Muscular Systems

All chordates have an endoskeleton. In animals such as sharks and rays, it is made of flexible cartilage; in other vertebrates—including humans—it is mostly made of calcified bone. The human skeleton is composed of an **axial skeleton**, which guards the CNS, organs of the chest, the skull, spine (vertebrae), and ribs, and an **appendicular skeleton**, which is made up of the appendages—arms, legs, hands, and feet.

Bones are joined together by three main types of joints:

1. Discs of cartilage between vertebrae allow for spinal flexibility but protect the nerves inside the spine.
2. Fused joints of the skull provide solid protection for the brain.
3. Ball-and-socket joints, such as in the femur (pelvis) and humerus (shoulder) allow a broad range of rotation for great flexibility.

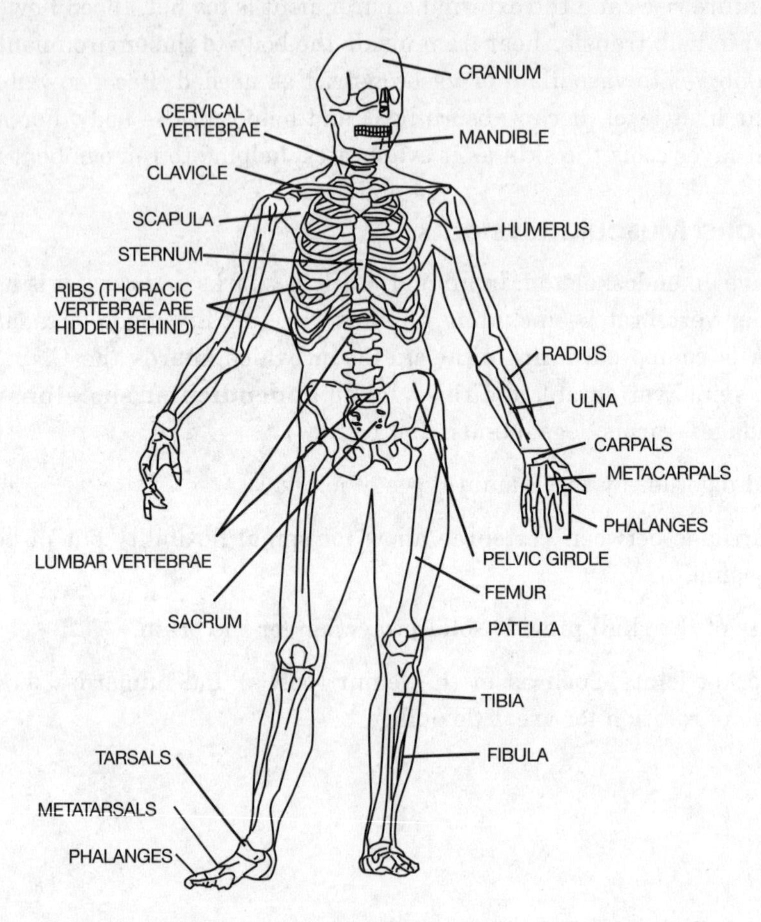

Motion of the vertebrate endoskeleton is always driven by skeletal muscle, cartilage, and the tendons that join bones together. When muscles contract, they bring the attachment sites of muscle and bone closer together. Movement in the opposite direction occurs when the muscle on one side relaxes and the other contracts.

Muscle contraction occurs when two types of protein filaments slide past one another inside muscle cells. These are the **thick filaments,** made of the protein **myosin,** and **thin filaments,** made of **actin**. They are tightly organized in rows inside the cell, forming bands where the filaments overlap, giving muscle a banded appearance under the microscope. Myosin proteins have a head that latches onto actin filaments and hydrolyzes ATP to drag the myosin and actin filaments past one another. When the muscle cell is not contracting, proteins called **tropomyosin** and **troponin** bind to actin filaments and block myosin, preventing contraction. When a motor neuron reaches the neuromuscular junction, it releases the neurotransmitter acetylcholine and triggers an action potential in the muscle cell. Once started, this action potential propagates throughout the muscle cell along the membrane, including infoldings called **transverse tubules**. The action potential causes the release of calcium from the **sarcoplasmic reticulum**, and the released calcium causes the tropomyosin and troponin to move and allow myosin to bind to actin.

Q Which of the following best describe the role of troponin and tropomyosin in skeletal muscle contraction?

(A) They propagate action potentials in muscle cells.
(B) They act as motor proteins to move myosin and actin filaments past one another.
(C) They hydrolyze ATP in response to calcium.
(D) They permit myosin heads to bind actin in the presence of calcium.
(E) They trigger acetylcholine release.

A The correct answer is (D). When there is no action potential and no calcium released in the interior of the muscle cell, troponin and tropomyosin bind actin and prevent myosin from binding to actin. When calcium is present, they move and allow myosin to bind actin, triggering the sliding of muscle filaments past each other.

The extent of depolarization of the muscle cell membrane is the same with each action potential. However, each muscle contains not one skeletal muscle cell but many cells bundled together. The greater the number of cells contracting, the greater the strength of the overall muscle contraction. Also, a motor neuron can deliver more frequent action potentials to a muscle cell. If the action potentials happen quickly enough, the muscle cell does not relax between action potentials and stays contracted. As the action potentials accumulate, the strength of contraction increases. At its limit, it will reach a state of tetanus, the maximal strength of contraction.

Q Motor neurons cause stronger skeletal muscle contraction by doing which of the following?

(A) Transmitting larger action potentials
(B) Releasing more acetylcholine at the neuromuscular junction with each action potential
(C) Causing a stronger response in the muscle cell to each action potential
(D) Using electrical rather than chemical synapses
(E) Firing more rapid action potentials

A The correct answer is (E). Action potentials always do the same thing. They are always the same size and virtually always release the same amount of neurotransmitter. The muscle cell does not respond differently to one action potential over another. Motor neurons always communicate with skeletal muscle through chemical synapses.

There are three types of muscle cells:

1. skeletal
2. cardiac
3. smooth

The same basic mechanism of actin and myosin filaments sliding past one another is involved in contraction of all three types. Skeletal muscle cells are large and have multiple nuclei; the other types of muscle cells have single nuclei. Skeletal muscle is involved in voluntary motion when stimulated by motor neurons. Cardiac muscle causes contraction of the heart, and smooth muscle responds to hormones or the autonomic nervous system to cause involuntary contraction of smooth muscle in the walls of internal organs such as the gastrointestinal tract and the arteries.

	Actin-myosin	Striated	Multinucleated	Cytoplasm of cells connected
Skeletal	yes	yes	yes	no
Cardiac	yes	yes	no	yes
Smooth	yes	no	no	yes

Q What type of muscle cell does not appear striated under a microscope, has a single nucleus in each cell, and has cells connected by gap functions to form electrical synapses between cells?

A Smooth muscle.

The Reproductive System

Species survival requires reproduction. Failure to reproduce is a sure road to low fitness, and alleles that hinder reproduction will not survive long in a population. The site of fertilization and development is an important factor. Many aquatic species use external fertilization, in which large numbers of eggs and sperm are produced and released into the water. Internal fertilization in the female reproductive tract permits terrestrial species to live in drier environments. Eggs laid by species that fertilize externally are soft and highly permeable to water. On the other hand, species that fertilize internally produce harder, more water-impermeable eggs, or they have internal embryonic development.

Q In organisms with amniotic eggs, is fertilization internal or external?

A An amniotic egg has a shell when it is laid and cannot be fertilized externally. It must be fertilized internally.

In humans, sexual reproduction involves the union of male and female gametes in fertilization. In women, ova, also called eggs, are produced in the ovaries. Women in their reproductive years usually have one ovum mature and ovulate each month, in concert with the menstrual cycle. When ovulation occurs, the egg travels from the ovary to the nearby **fallopian tubes** and from there, propelled by cilia, toward the **uterus**. The **vagina** is the passage from the **cervix** at the end of the uterus to the female genitalia, which include the **clitoris** and the labia. The menstrual cycle begins during puberty and continues until approximately age 50, when menopause occurs. At menopause, menstruation ceases and estrogen and progesterone levels fall.

Q What is the structure by which ova move from the ovaries to the uterus?

A Oviducts in most animals; fallopian tubes in humans.

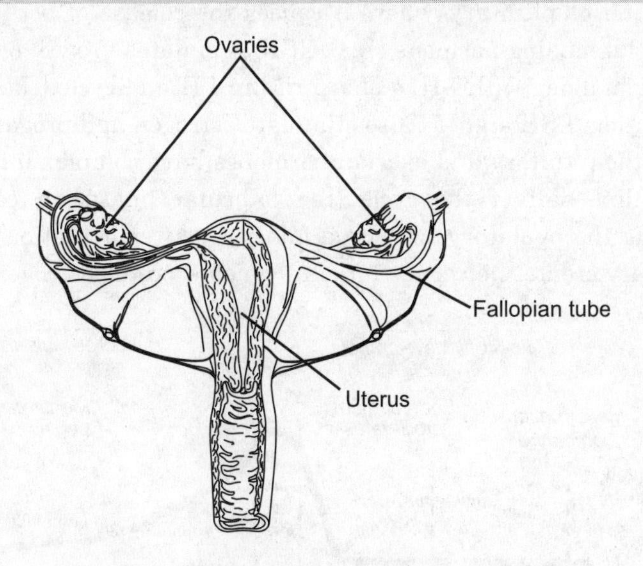

Female reproductive system
(front view)

The meiotic formation of gametes in the ovaries is called oogenesis. Diploid stem cells that produce the female gametes, **oogonia**, produce primary oocytes that enter meiosis but pause at meiotic prophase I. These cells can remain paused at this stage for many years, throughout childhood and into puberty, until one or more ova are stimulated during each menstrual cycle to develop further. When this happens a polar body that has almost no cytoplasm is released. The resulting maturing secondary oocyte can be ovulated while still frozen at meiosis II. In humans, meiosis is not complete until after fertilization.

Q How many female gametes are produced by meiosis from each primary oocyte?

A One. Oogenesis does not divide cytoplasm equally among cells, producing only one ovum.

Q If fertilization of an ovum does not occur, when does the ovum complete meiosis II?

A It does not. Meiosis II of an ovum is not completed (in humans at least) until after fertilization.

Ova do not develop on their own but in the context of the menstrual cycle, as pituitary and ovary hormone levels vary. GnRH secreted from the hypothalamus travels through the portal circulation to the anterior pituitary, where it causes the release of both follicle-stimulating hormone (FSH) and luteinizing hormone (LH). FSH stimulates growth of the ovarian follicle and oocyte maturation, along with LH; a sharp rise in LH mid-cycle (the ovulatory surge) is responsible for ovulation. FSH and LH also stimulate estrogen and progesterone secretion by the ovary, although the pituitary and ovarian hormones have complex interactions. Estrogen levels rise in the first half of the cycle (the follicular phase) in concert with follicle development, causing the ovulatory LH surge midcycle. After ovulation, the follicular cells remaining in the ovary create the corpus luteum, which secretes estrogen and progesterone.

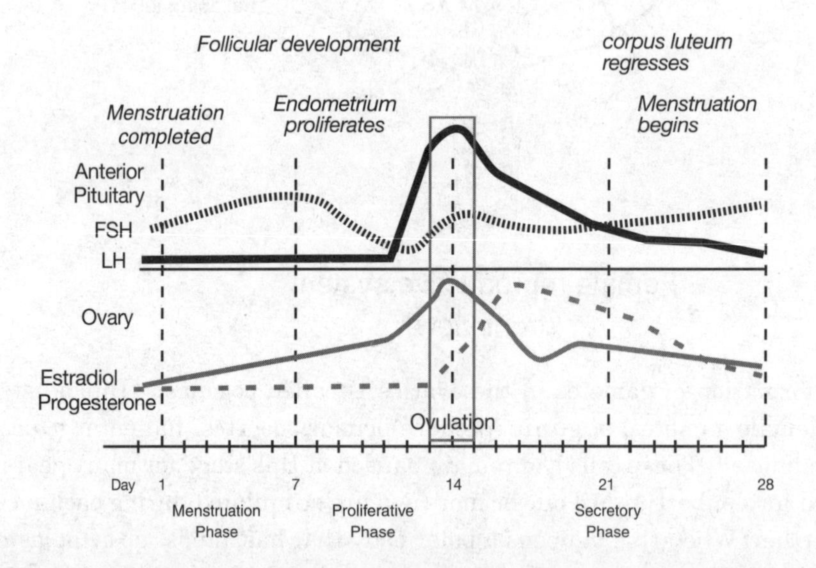

Q Where is estrogen produced during the part of the menstrual cycle leading up to ovulation?

A In follicles in the ovary.

As estrogen increases in the latter part of the follicular phase, the endometrial lining of the uterus thickens in preparation for a potential embryo. After ovulation, the progesterone and estrogen secreted together by the corpus luteum increase endometrial thickening, and the endometrium becomes more vascularized and secretes material that will support an embryo in the event of fertilization. If no fertilization occurs, the corpus luteum in the ovary degenerates and estrogen and progesterone levels drop rapidly. Without hormonal support, the blood flow to the upper layer of endometrial tissues stops, and the tissue is lost through menstruation. As menstruation occurs, FSH and LH are already setting the stage for the next cycle, causing another batch of follicles to begin maturing.

Q Which best correlates with ovulation: secretion of FSH, LH, or progesterone?

A An LH surge mid-cycle is responsible for and correlates with ovulation.

Q Where is progesterone mostly produced during the secretory phase of the menstrual cycle?

A In the corpus luteum.

If fertilization does occur, an embryo becomes implanted in the endometrium of the uterus and continues to develop. The developing embryo secretes a hormone called human chorionic gonadotropin (hCG) that keeps the corpus luteum from degenerating and allows it to continue producing estrogen and progesterone to maintain the endometrium. The hormone hCG is not normally produced by women except during pregnancy; for this reason, it is a fairly reliable marker for pregnancy.

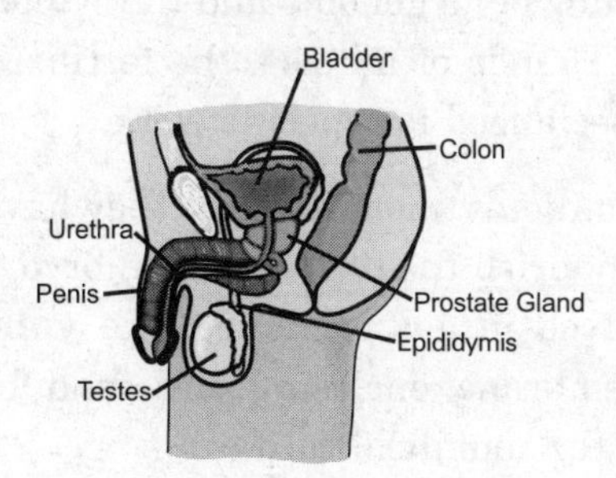

The male human reproductive system centers on the male gamete or sperm. The process by which meiosis produces mature sperm is known as **spermatogenesis**. Sperm production starts in the **seminiferous tubules** of the testes, where diploid precursor cells constantly produce primary spermatocytes that enter meiosis. After completing meiosis, each primary spermatocyte has produced four sperm cells that can move through the male reproductive tract. The testes are enclosed in the scrotum, which permits sperm development at lower-than-normal body temperature. Before ejaculation, sperm is mixed with secretions from the prostate gland, the seminal vesicles, and **bulbourethral glands** to produce semen. As the sperm mature, they travel through the **epididymis** until they are propelled through the vas deferens and the urethra of the penis during ejaculation. Sperm are usually not fully motile until they are activated in the female reproductive tract.

FERTILIZATION AND DEVELOPMENT

Spermatozoa are much smaller than ova (egg cells) and contribute almost no cytoplasm to the zygote; most of the sperm cell is occupied by the nucleus. In mammals, the sperm also contribute a centriole that will form the centrosome for the first cell division after fertilization. The flagella of sperm allow them to "swim" toward the egg.

In mammals, fertilization takes places in the female reproductive tract. The ovum is surrounded by follicular cells and the **zona pellucida,** made of polysaccharides and proteins from extracellular matrix. Once a sperm locates an ovum, the acrosomal process at the head of the sperm releases hydrolytic enzymes that digest the protective layers surrounding the egg cell plasma membrane. As the membrane of the sperm cell fuses with the membrane of the egg, it triggers additional changes in the zona pellucida that prevent fertilization by additional sperm.

The mammalian early embryo before gastrulation (called the blastocyst) is not simply an empty ball of cells. It includes an **inner cell mass** and an external layer called the **trophoblast**. The inner cell mass differentiates into the embryo; the trophoblast carries out implantation of the embryo in the endometrium of the mother and helps to form the extra-embryonic membranes and the placenta that protect and nourish the developing embryo throughout pregnancy.

Q Does the human embryo develop from the inner cell mass or the trophoblast?

A The inner cell mass.

The human fertilized egg contains all the information needed to become an adult organism, including the genetic programming of its genome and the cytoplasmic determinants contained within the egg. Through many rounds of mitosis, the fertilized human egg ultimately gives rise to trillions of cells, each specialized for specific work.

Birds and mammals are both amniotes, meaning that they have extra-embryonic membranes that enclose, protect, and help nourish the developing embryo. These membranes include the **chorion**, the outer layer involved in gas exchange; the **yolk** or placenta, which contains nutrients; the **amnion**, a membrane enclosing the fluid bathing the embryo; and the **allantois**, which forms part of the umbilical cord.

After gastrulation occurs in vertebrate development, the organs and details of the body plan develop. All chordates have a **notochord** that differentiates from the mesoderm along the dorsal side of the embryo. The notochord sends a signal to the neighboring ectoderm cells along the back of the embryo to fold inward and differentiate into the **neural tube**—the start of the CNS. The influence of cells on the differentiation of neighboring cells in this manner is called **induction**. To the sides of the notochord, more mesoderm differentiates in strips called **somites**. These develop into the bones of the vertebral column.

Q Does the notochord differentiate into nerve tissue?

A No. It induces neighboring ectoderm cells to differentiate into nerve tissue.

At the cleavage stage, every cell in the embryo is exactly the same and has the potential to differentiate into any type of adult cell. This ability is called totipotency. In some species, if a four-cell embryo is divided into individual cells, each cell can continue developing and differentiating into a complete adult organism. Cells in protostome embryos become committed to specific fates earlier than cells in deuterostomes, which remain totipotent for a longer time.

Stem cells in embryos often remain totipotent at early stages. As development continues, many cells commit to specific developmental fates. Cells that retain the ability to differentiate into multiple different types are termed **pluripotent**. Even in adult organisms, pluripotent stem cells continue to give rise to multiple different cell types. Stem cells allow some tissues to heal and regenerate, and hematopoietic stem cells in mammalian bone marrow constantly give rise throughout life to all of the cells found in blood. Tissues such as that of the CNS that were long thought to be devoid of new cell formation in adults have since been found to harbor special populations of stem cells that continue proliferating in adults.

Many of the molecular signals involved in cellular differentiation and development also involve the regulation of gene expression, such as interacting cascades of proteins that bind to gene promoters to turn transcription on and off. Specific-specific transcriptional regulatory factors cause cells to differentiate into muscle or liver, for example, and the development of animals along the axis of their bodies is determined by transcriptional regulation as well. A series of conserved genes called homeotic genes controls development along the body axis in animals ranging from fruit flies to humans.

The Immune System

Our bodies harbor a rich "soup" of nutrients, but at the same time we are surrounded by a sea of microbes, including bacteria, viruses, and fungi. Many of these cause disease when they manage to colonize in the body. The vertebrate immune system keeps this from happening in two ways: through **innate immunity** and **adaptive immunity**.

INNATE IMMUNITY

Innate immunity is the series of barriers that does not target any specific microbe but offer general protection against infection. The first barrier to infection is the skin, a relatively thick (for a microbe) layer of dry cells. Intact skin provides an effective barrier to microbe entry. Sweat also has antimicrobial properties. For example, it contains the enzyme lysozyme that degrades the peptidoglycan cell wall of Gram-positive bacteria. Lysozyme in tears helps protect the eye against infection as well.

The moist linings of tracts in the gastrointestinal, reproductive, and respiratory systems provide other possible routes of entry for infection. Mucus lining these membranes helps trap potential pathogens, and cilia on the epithelial layer of the respiratory pathways move trapped pathogens up and out of the body. Microbes entering the GI tract must contend with the harsh acidity of the stomach.

Q When cilia move microbes upward in the respiratory tract, where do they go?

A The microbes are either swallowed or they are expelled by coughing.

Microbes that manage to enter the bloodstream or extracellular fluids of tissues encounter additional components of the innate immune system. Their entry triggers the activation of mast cells throughout the body, which release histamine and other molecules in an inflammatory response. Inflammatory signaling molecules cause local blood vessels to dilate and become more permeable, increasing blood flow to the region and drawing protein and cells to the site to help fight infection. Inflammatory signaling molecules released at the site of infection may include chemokines, prostaglandins, interferons, and cytokines such as interleukin-1 and TNF. In addition to increasing blood flow at the site of infection, these signaling molecules attract macrophages, neutrophils, and eosinophils that phagocytose pathogens. Pathogens internalized by phagocytosis go to lysosomes, where they are degraded by hydrolytic enzymes.

A group of proteins making up what is called the complement system in the blood can also kill microbes directly. The complement proteins form a proteolytic cascade triggered by the presence of pathogens. This leads to the creation of pores in the cell surface of invading cells and their subsequent lysis.

Q Which of the following is not a component of innate immunity: the skin, complement proteins in blood, macrophages, mucus in the respiratory tract, or B cells?

A B cells are part of the adaptive immune system rather than the innate immune system.

ADAPTIVE IMMUNITY

The components of innate immunity are always present and are essentially the same for all people. But many pathogens trigger an immune response specific to themselves, which increases with time and repeated exposure. These responses are called adaptive or acquired immunity. Adaptive immune responses involve two main types of lymphocytes, **B cells** and **T cells**. When activated, B cells produce antibodies, part of the **humoral response**, clearing foreign agents, such as bacteria, ranging free in the extracellular fluids. T cells provide **cell-mediated immunity**, recognizing and killing virus-infected cells (cytotoxic T cells) and coordinating the responses of other immune cells (helper T cells).

Q Do B cells respond to infectious agents found inside other cells?

A No. They respond to foreign antigens found in extracellular fluids. Viruses must be present in extracellular fluids at some point in their life cycles, however, and they can elicit a B cell response when they are present.

All blood cells, including lymphocytes, originate from proliferation and differentiation of stems cells in the bone marrow. Different growth factors cause differentiation into different types of blood cells. T cells mature in the thymus; B cells in vertebrates differentiate in the bone marrow. As T cells and B cells mature, they express on their surfaces a protein receptor that may recognize potential pathogens. B cells express immunoglobulins; T cells express T cell receptors. These cell surface proteins can bind specific molecules called antigens on the surface of pathogens.

Q Chemotherapy to treat cancer might cause a suppressed immune response for which of the following reasons?

(A) It kills previously differentiated B cells.
(B) It destroys erythrocytes.
(C) It kills rapidly dividing cells in the bone marrow.
(D) Macrophages are blocked from phagocytosis of microbes.
(E) B cells proliferate.

A The correct answer is (C). Chemotherapy blocks replenishment of immune cells from stem cells in the bone marrow. With fewer immune cells present, the patient's immune response is suppressed.

Not all B cells or T cells express the same protein receptor. B cells express a wide range of immunoglobulin types as they mature, using pieces of genes stitched together by recombination. The mature immunoglobulins that B cells produce have four polypeptide sections that join in each antibody molecule: two heavy, longer chains and two lighter, smaller chains. The heavy and light chains are held together by disulfide bonds to create constant regions that vary little from antibody to antibody and variable regions that bind to antigen. To assemble such antibodies, the immunoglobulin genes in mature B cells are put together using several versions of each domain. Each B cell pieces together its own version of an immunoglobulin gene that it will express when it is mature.

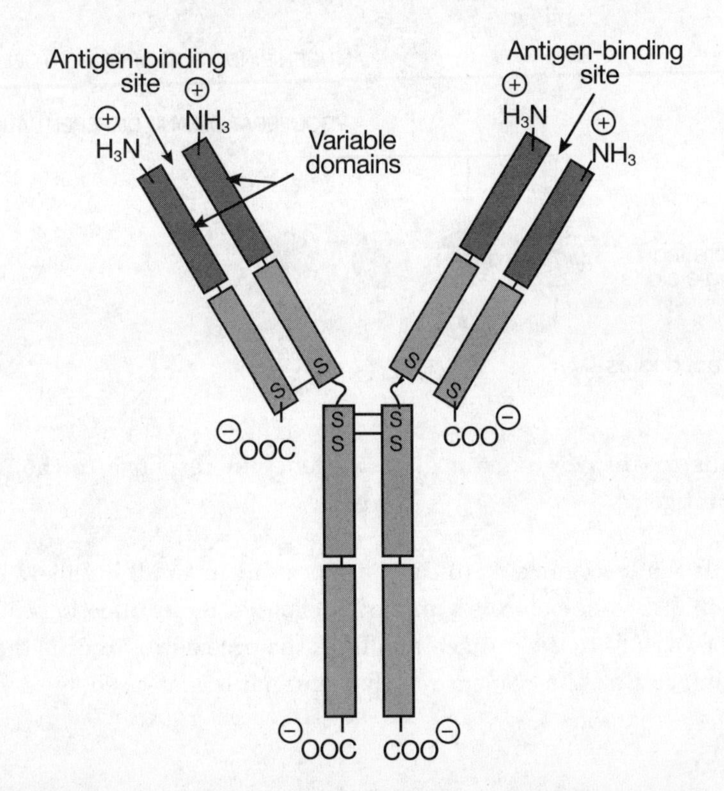

If the immunoglobulin genes of B cells are randomly assembled, what prevents some B cells from recognizing proteins on the body's own cells rather than infectious agents? Some of them do, initially, but in a healthy person, B cells that recognize the body's own molecules are eventually removed to prevent the immune system from harming rather than protecting the body. If this system fails, this "self-tolerance" fails and an autoimmune disorder will develop. Multiple sclerosis is an example of an immune response against the body's own myelin, which coats neurons.

With such a broad range of B cells, how do we respond to specific antigens when bacteria or viruses invade? The key is the selective amplification of specific clones of B cells from the preexisting diverse population. Each infecting agent has a variety of proteins and sugars present on its surface for immunoglobulin to bind to. If the particular immunoglobulin on a B cell binds to antigen on a bacterium or other foreign substance, it will cause that B cell to proliferate. The new B cells will express the same version of the immunoglobulin.

Some of the B cells differentiate into **plasma cells**, short-lived cells that secrete immunoglobulins into the blood. Others form **memory cells** that circulate in the blood for long periods and express the same immunoglobulin receptors on their cell surfaces.

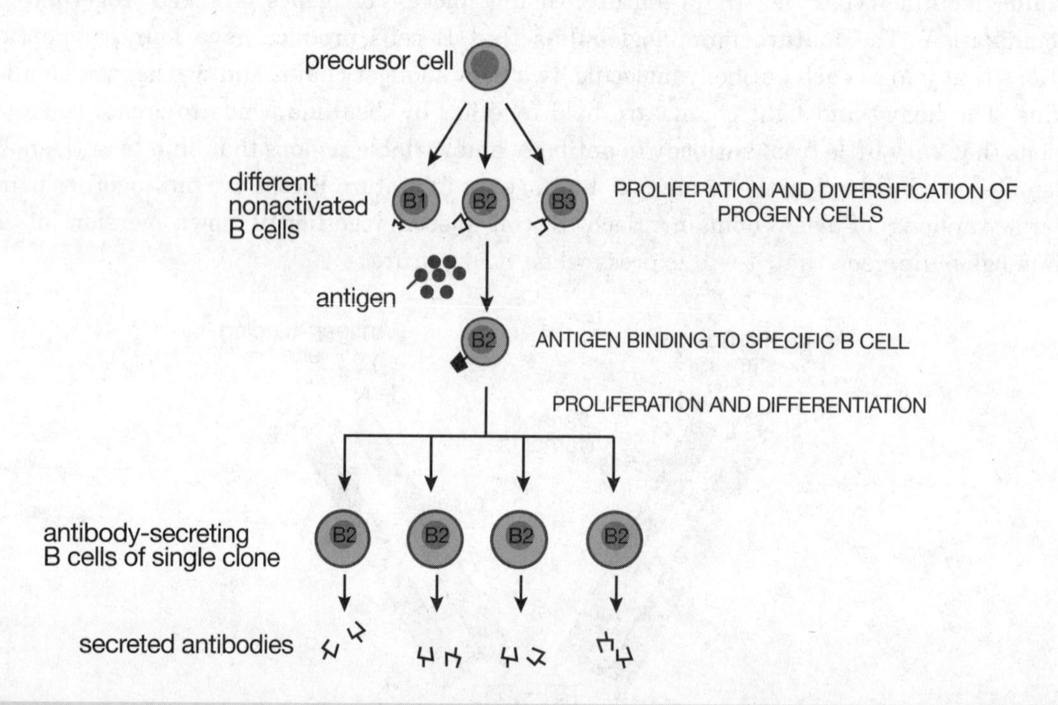

Q Do B cells create new immunoglobulin genes in response to the presence of a specific antigen?

A No. B cells that recognize an antigen are present in small numbers in the overall population of B cells before an animal is exposed to an infectious agent or other antigen. Exposure to the antigen amplifies the preexisting B cells that respond to an antigen, so that the response to that antigen is increased.

Q If the immunoglobulin genes are examined in a population of B cells, will the genes be the same in all of these cells?

A No. Different cells contain different versions of the gene assembled randomly by recombination.

There are five different types of immunoglobulins with slightly different structures and different functions. Each class of antibody has a different constant region on the end:

1. **IgD** is a cell surface antigen receptor on B cells before activation with antigen.
2. **IgG** is the most common form of antibody, secreted by B cells into plasma.
3. **IgM** joins together in groups of five.
4. **IgE** is believed to be involved in allergy response.
5. **IgA** is secreted in tears, mucus, and other bodily fluids.

When antibodies bind to antigens on pathogens like bacteria, they can help remove the bacteria in several ways:

- Binding of the antibody to the surface of the antigen can keep the pathogen from binding to other substances, blocking the proteins or sugars on its surface.
- The antibody, having two antigen binding sites in each antibody molecule, can cross-link antigens in large complexes, making them insoluble.
- Antibodies sticking to the surface of a pathogen can alert phagocytotic cells, such as macrophages, to destroy the infectious cells.
- Antibodies can activate the complement system, triggering the formation of a pore complex that can lyse and kill the invading cell.

Like B cells, T cells also express a special form of antigen receptor on their surface, called the **T cell receptor**. T cell receptors have a constant region and a variable region that recognizes antigen. They do not recognize free antigen in blood. They bind antigen only if pieces of the antigen are presented to T cells bound to protein complexes called **major histocompatibility complex (MHC) I** and **MHC II** on the surface of other cells.

Each MHC protein picks up pieces of antigen from inside the cells where they are produced. MHC I and II are found on different cells. Almost all cells of the body express MHC I. Any cell infected with a virus binds pieces of the virus to MHC I and provokes a T cell response. MHC I binds pieces of cellular proteins as well. When cytotoxic T cells encounter a viral antigen in MHC I and the viral antigen is recognized by the T cell receptor of that T cell, then the T cell will be activated to kill the infected cell, killing the viruses along with the cell.

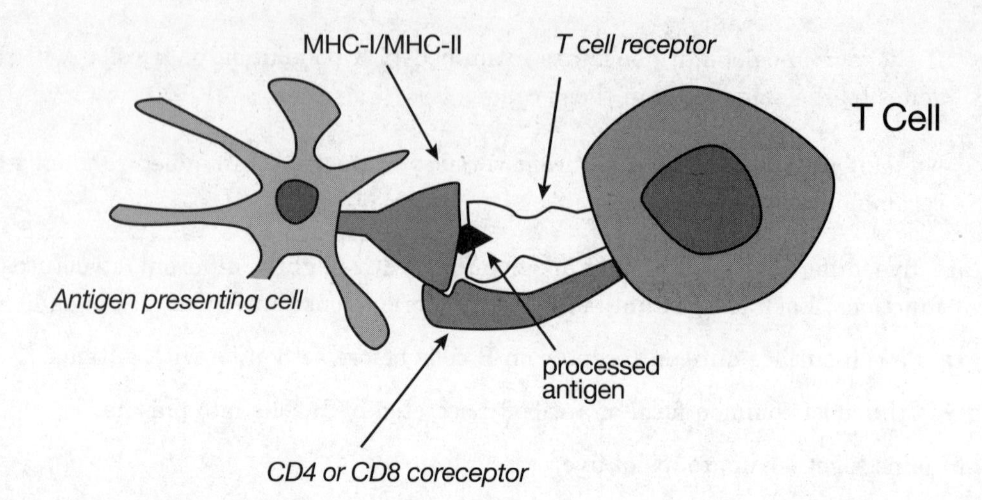

Unlike MHC I proteins, MHC II proteins occur in specific types of immune cells called **antigen presenting cells**. These include macrophages, dendritic cells, and B cells. When these cells encounter a pathogen such as a bacterium, they internalize it through phagocytosis and express the pathogen antigens in MHC II. Then helper T cells with a T cell receptor that recognizes the antigen can stimulate an immune response.

Helper T cells do not themselves kill pathogens or infected cells, but they play an essential role in immune response by helping to activate cytotoxic T cells and B cells. B cells cannot respond on their own to a pathogen encountered in blood, even if the pathogen is a perfect match for the B cell immunoglobulin receptor. They need helper T cells that recognize the same pathogen to communicate with and activate them. What can happen to the body without the vital helper T cells is easily evident in the immunosuppression caused by human immunodeficiency virus (HIV), the virus that causes AIDS. HIV selectively infects and kills helper T cells.

Q Are pieces of a cell's own proteins also displayed in the MHC on the cell's surface?

A Yes. Cells are constantly displaying a range of their own peptides on the cell surface in MHC, but the immune system ignores the cell's own peptides because T cells responding against these are eliminated. Only if a foreign substance is found in the MHC is the immune system triggered.

Since the immune system is prevented from responding to self antigens, most of the cell's own proteins in MHC will not create a T cell response. In some cases, however, cells will express unusual proteins that are bound to MHC I on the cell surface, and T cells will respond against these self antigens. A classic example is with cancer cells. These cells can alter the proteins they express, and the new proteins—although they are self—are enough to stimulate a T cell response if they are in MHC I. In this way, the immune system can help prevent cancer as well as inhibit infections. The genes encoding MHC genes can vary greatly from person to person, and they are found in several different alleles. If cells with an MHC I allele are injected into a person with a different MHC I allele, they can provoke an immune response.

This response against different MHC alleles is what causes organ recipients to reject donated organs following transplant.

Do the T cell receptor genes undergo recombination as T cells mature?

Yes. Just as the immunoglobulin genes undergo recombination to create the immunoglobulins each mature B cell expresses, the T cell receptor genes undergo recombination to create the broad array of antigen specificities in the "naïve" population of T cells before exposure to antigen.

Vaccines

Memory B cells circulate for years after the first occurrence of an infection. To protct against another bout, they can quickly initiate a strong immune response against the original antigen. People who have had chicken pox, for instance, seldom contract the disease a second time, because their memory B cells and similarly long-lived T cells remain in the body to block the same disease from recurring. Vaccines exploit this aspect of the body's immune system for protection against disease.

Vaccines are commonly made from:

- live but weakened versions of a virus that cannot cause disease
- killed infectious organisms
- recombinant proteins

All of these types of vaccines can provide a primary immune response that will amplify the B cells and T cells that respond to the pathogen.

EXERCISES: ANIMAL STRUCTURE AND FUNCTION

1. Which of the following would be most likely to prevent a person from hearing sounds of a specific pitch?

(A) Blockage of the auditory canal
(B) Puncture of the tympanic membrane
(C) Severed auditory nerve
(D) Damage to hair cells along a particular region of the basilar membrane
(E) Fusion of the stapes and incus in the inner ear

2. The neurotoxin tetrodotoxin, found in the puffer fish, can cause a slight numbness when eaten or even death if a large dose is consumed. It binds to voltage-gated sodium channels in axons. Which of the following would best describe the impact of tetrodotoxin on neuronal function?

(A) It prevents movement of action potentials down axons.
(B) It blocks postsynaptic neurotransmitter receptors.
(C) It increases the size of inhibitory postsynaptic potentials.
(D) It causes excessive quantities of neurotransmitter to accumulate in synapses.
(E) It prevents summation in postsynaptic neurons.

3. Which of the following is most likely involved in the production of the large number of olfactory genes of mammalian genomes?

(A) Nonhomologous recombination
(B) Gene duplication through recombination errors, followed by mutation and natural selection
(C) Horizontal gene transfer from other species
(D) Convergent evolution of receptors previously involved in other functions
(E) Defective proofreading function in DNA polymerase

4. Which of the following is NOT a response of the sympathetic nervous system?

(A) Decreased glycogen synthesis in liver
(B) Increased heart rate
(C) Constricted bronchi
(D) Decreased secretion of bicarbonate by pancreas
(E) Release of epinephrine from adrenal glands

5. In an animal that can reproduce sexually or asexually, when might conditions favor asexual reproduction?

(A) When weather conditions change dramatically
(B) When animals are exposed to a new disease organism
(C) When conditions are constant and well-suited for rapid growth
(D) When non-random mating occurs
(E) When Hardy-Weinberg equilibrium has been established

6. Which of the following traits do tunicates and sharks share during development?

(A) Notochord
(B) Hollow dorsal nerve cord
(C) Muscular tail posterior to the anus
(D) Pharyngeal slits
(E) All of the above

7. Which of the following does NOT occur during the menstrual cycle?
   (A) A surge in LH causes ovulation.
   (B) The corpus luteum produces progesterone.
   (C) GnRH stimulates LH and FSH release from the anterior pituitary.
   (D) Progesterone supports follicular development.
   (E) In the absence of fertilization, degeneration of the corpus luteum precedes menstruation.

8. Which of the following is most likely to be affected by damage to the hippocampus?
   (A) Ability to understand spoken words
   (B) Coordination of fine motor movements in fingers
   (C) Regulation of body temperature to a specific set point
   (D) Consolidation of new memories
   (E) Processing of visual information to recognize faces

9. What is a key difference between insulin and glucagon?
   (A) Insulin binds to cell surface receptors; glucagon binds to receptors in the cytoplasm.
   (B) Insulin is a peptide hormone; glucagon is not.
   (C) Glucagon acts on the liver; insulin does not.
   (D) Insulin is produced by the pancreas; glucagon is not.
   (E) Insulin stimulates glycogen production in skeletal muscle; glucagon does not.

10. Which of the following features is NOT found in cnidarians?
   (A) Free-swimming medusa stage in many species
   (B) Nerve cells
   (C) Excretory system
   (D) Sexual reproduction
   (E) Stinging cells called nematocysts

ANSWER KEY AND EXPLANATIONS

1. D	3. B	5. C	7. D	9. E
2. A	4. C	6. E	8. D	10. C

1. **The correct answer is (D).** Sound causes vibration in the basilar membrane and activates hair cells. Sounds of different wavelengths move the basilar membrane at different distances from the point where sound enters the cochlea. The other choices would have more global effects on hearing.

2. **The correct answer is (A).** Voltage-gated sodium channels are essential for the movement of action potentials down axons, and tetrodotoxin would block this.

3. **The correct answer is (B).** Gene duplication followed by mutation is responsible for the creation of large gene families such as the olfactory receptors. Errors in recombination are one likely mechanism involved in gene duplication events.

4. **The correct answer is (C).** The sympathetic nervous system causes bronchi to dilate, not constrict.

5. **The correct answer is (C).** Asexual reproduction may have evolved in some species to allow rapid proliferation in mild conditions. Sexual reproduction would be more optimal in the face of environmental challenges.

6. **The correct answer is (E).** Answer choices (A) through (D) list the four key traits of chordates, and both tunicates and sharks are chordates. These traits may not all be present or obvious in every adult tunicate or shark, but they are all present at least during development.

7. **The correct answer is (D).** Progesterone does not support or cause follicular development.

8. **The correct answer is (D).** The hippocampus is involved in forming new memories. The cerebrum handles speech comprehension and processing sensory information. The cerebellum is involved in fine motor coordination, and the hypothalamus regulates body temperature.

9. **The correct answer is (E).** Both hormones are peptides that bind to cell surface receptors, they are both are produced in the pancreas (although in different cells), and they both act on the liver.

10. **The correct answer is (C).** Cnidarians have all of these features except for an excretory system. Because they have two cell layers and most cells are in close contact with the external, aquatic environment, wastes and gases are exchanged directly with the environment.

SUMMING IT UP

- Animal bodies can be differentiated from various other groups of organisms by body symmetry, type of internal cavities for digestive and other bodily functions, and embryo development.
- In protostomes (including annelids, arthropods, and mollusks), the fertilized zygote undergoes spiral cleavage and cells rapidly differentiate. In deuterostomes (including echinoderms and chordates), the fertilized zygote undergoes radial cleavage and remains totipotent for a longer time during embryonic development.
- Members of the phylum Porifera, such as sponges, are simple aquatic animals that have several types of specialized cells but no true tissues. Most live stationary lives, filter food from the surrounding water, and have no digestive system.
- Cnidarians live either as swimming medusa such as jellyfish or as stationary polyps such as anemones. They have radial symmetry, a digestive gastrovascular cavity (gut) with a single opening, stinging cells, tentacles arranged around the mouth opening, a decentralized nerve net, and two cell layers (endoderm and ectoderm), with mesoglea between them.
- Platyhelminthes are acoelomate and have a gastrovascular cavity with a single opening for feeding. They have bilateral symmetry, a single opening into the gastrovascular cavity, ventral nerve cords, muscle cells, no circulatory or respiratory system, and no coelom.
- Nematodes are nonsegmented roundworms with a complete alimentary tract, a separate mouth and anus, and a pseudocoelom. Nematode locomotion involves contraction of muscles running the length of the body.
- Mollusks, including octopi, snails, clams, mussels, and slugs, have bilateral symmetry, a coelom, an external shell (in most cases), a mantle that secretes the shell, a muscular foot for attachment and movement, a radula for scraping food, a digestive tract, gills for gas exchange (except for terrestrial species), and, in most cases, an open circulatory system.
- Annelids include earthworms, polychaetes, and external parasites. They have a segmented coelom and body, specialized regions of the digestive tract, ventral nerve cords, ganglia in anterior region and collections of nerves near the pharynx, and a closed circulatory system with muscular, heart-like segments.
- Arthropods are among the largest phyla. The exoskeleton protects against predation but restricts growth, so arthropods must molt periodically. Common characteristics include bilateral symmetry, segmented body and jointed appendages, a coelom, an open circulatory system with dorsal heart, hemolymph, and a ventral nervous system with ganglia in the head region.
- Echinoderms are strictly marine organisms, with radial symmetry in adults usually organized in multiples of five. They are deuterostomes, evolved from ancestors with bilateral symmetry, and they have larvae with bilateral symmetry.

- As with echinoderms, chordates are deuterostomes that have a coelom, and these two groups are more closely related than other animal phyla.
- Chordates display bilateral symmetry, a closed circulatory system, and advanced respiratory, excretory, and nervous systems. They are distinguished from other phyla for the most part by a notochord, a dorsal nerve cord, a muscular tail, and pharyngeal slits. In humans, all of these characteristics are either modified or replaced during late development or adulthood.
- The development of a head with a skull, backbones, jaws, and bones led to the evolution of vertebrates. These traits were first fully present in fishes. Vertebrates include fishes, amphibians, reptiles, birds, and mammals.

Ecosystems

chapter 14

OVERVIEW

- Nonbiological aspects of environments
- Biomes
- Behavior of animals in their environment
- Chemical cycles
- Impact of humans on the biosphere
- The future of the biosphere
- Summing it up

Together, all of Earth's creatures form, and share, the biosphere. As we humans have only one planet to live on, the events of the biosphere and the pressure humanity places on it are of particular interest to us all.

NONBIOLOGICAL ASPECTS OF ENVIRONMENTS

The environment living things inhabit includes both living and nonliving (or abiotic) components. Life on Earth has adapted to a huge variety of environmental conditions, and the community of organisms living in each region reflects the conditions there.

Sunlight is the basis of almost all energy produced and used by living things on Earth, so varying sunlight exposure and temperature affect living things greatly. Different parts of the globe are exposed to different levels of sunlight, making the Earth generally warmer near the equator and colder at higher latitudes, like the poles. Proximity to coastal regions often greatly moderates temperatures on land, whereas inland regions experience greater extremes of temperature. Ocean currents redistribute heat on a global scale and nearby ocean currents can also affect terrestrial temperatures. Water is another essential ingredient for living things that varies from almost none in desert regions to aquatic environments. The combination of varying sunlight, precipitation, terrain, and temperature create the climate of each biogeographic region and shape ecosystems.

Q Ocean currents greatly affect ecosystems along the Pacific and Atlantic coasts of the United States. Is this a biotic or abiotic component of ecosystems?

A Ocean currents are not living, so they are an abiotic component of ecosystems.

Another key to understanding each ecosystem is the structure of interactions between the variety of organisms that live together in a community—the living component of each ecosystem. The variety of organisms living in a region is often described as biological diversity. The relationships between organisms are largely based on who eats whom, starting at the bottom with the **primary producers**. The primary producers are the autotrophs that obtain energy from the abiotic environment. Usually, this energy is sunlight absorbed by photosynthetic organisms and converted into chemical energy. Photosynthesis by primary producers provides all of the energy that most ecosystems rely on, transferring the energy in a food web between species. In addition to the primary producers in an ecosystem, the **primary consumers** are herbivores that eat the plants, algae, or other photosynthetic primary producers. **Secondary** and **tertiary consumers** eat other animals, acting as predators of the herbivores and other carnivores. **Decomposers** return the nutrients of the other organisms back to the Earth, to enter the food web again.

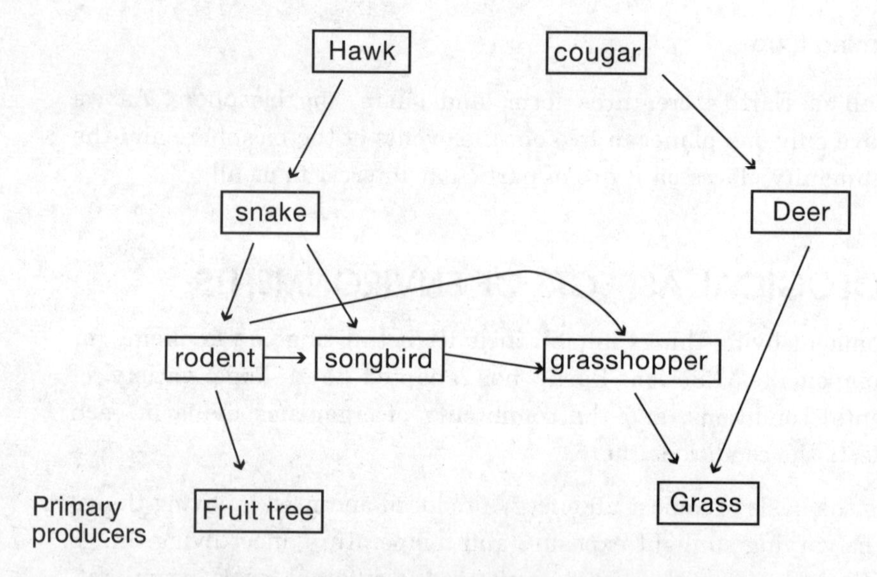

Example of a food web (highly simplified)

Q In a deep ocean vent, is sunlight the source of energy for primary producers?

A No, this is one exception to the general rule that photosynthesis provides energy for primary producers. The primary producers in this ecosystem get their energy from inorganic chemicals released from the vent.

Biologists often work to measure the amount of energy flowing through biological systems to see how all of the pieces fit together. The amount of energy flowing into the system always equals the amount of energy flowing out, one way or another. The amount of sunlight absorbed by plants through photosynthesis is the **gross primary productivity**, while the amount of energy stored by plants after using some for their own metabolic needs is the **net primary productivity**. The productivity of aquatic or terrestrial ecosystems is limited by the amount of light they receive and nutrients that are available for photosynthesizers. Aquatic biomes with the greatest primary productivity include coastal waters, coral reefs, and wetlands, all of which have abundant dissolved nutrients and sunlight. The open ocean often has low gross primary productivity (per square kilometer) due to low levels of dissolved nutrients near the surface and limited penetration of light through water into deeper levels. In regions where cold, nutrient-rich water from deeper layers wells up into the photic zone (the region into which light penetrates), gross primary productivity can be much higher. Among terrestrial ecosystems, the greatest gross primary productivity is seen in regions with the right combination of sunlight, water, and temperatures similar to those in rain forests.

Q Is the gross primary productivity greater or less than the net primary productivity?

A Gross productivity is greater, since net productivity subtracts the energy plants use for their own respiration, making ATP for their metabolic needs.

Q Can net productivity be measured by the amount of biomass of producers that are present in an ecosystem?

A No. Some plants, like trees, may accumulate more biomass but have less net productivity because less of their energy and biomass is consumed by herbivores.

Each level in the food web is often described as a **trophic level**. As one level consumes organisms from another level, it is transferring energy between levels, but the process is never 100% efficient. When one organism eats another one, it loses some of the energy as metabolic heat, and more is lost through excretion. Only a relatively small percentage of energy (about 20% or less) is actually transferred and incorporated as biomass at a higher trophic level.

Q In a terrestrial ecosystem, is the energy content generally higher in primary producers or primary consumers?

A Primary producers generally have greater energy content in the ecosystem because the transfer of energy between trophic levels is always limited.

Biomes are often characterized by the types of species that are abundant or help maintain the structure of the community. The **dominant species** in a community is that with the greatest number of individuals, while the **keystone species** might not be the most abundant but has an unusually large impact on other species. Examples of dominant species are pine or spruce trees growing in temperate coniferous forests, where those trees represent a large percentage of the plant life and biomass in that biome and the trees themselves shape the biome for other organisms living there. Keystone species are sometimes predators that both help control populations of herbivores and increase the diversity of the overall population. In the absence of predators, the herbivore population explodes and plant life can be adversely affected. An example of a keystone species at work is the impact of sea otters on the coastal kelp forest ecosystem. Sea otters prey on large numbers of sea urchins, keeping their population in check. When sea otter populations crash, sea urchins proliferate, decimating the kelp forest and the diversity of other species that rely on the kelp forest.

Q In a grassland biome, are grasses, large herbivores, or tertiary consumers the dominant species?

A The dominant species has the most individuals and/or the most biomass present. In grasslands, the grasses are dominant.

BIOMES

The biomes of the biosphere are categorized in a variety of ways but commonly include aquatic biomes and terrestrial biomes.

Aquatic Biomes

Lakes and Rivers: Freshwater biomes have varying oxygen and nutrient levels. Small rivers and lakes will generally have more mixing and more oxygen. In some cases, the more nutrients present, the less oxygen there is, especially with low mixing. Fertilizer runoff from cropland can cause algal and bacterial blooms, depriving other organisms of oxygen. Lakes with a high level of nutrients are called **eutrophic**, while those with low levels of nutrients are **oligotrophic**.

Wetlands: Wetlands have either fresh water along shallow lakes and rivers or varying salt concentration in river estuaries and coastal wetlands. Wetlands generally have very high productivity and can be low in terms of oxygen content in the water. They have great biological diversity, with many aquatic organisms relying on coastal wetlands and estuaries for breeding and as part of their food web.

Ocean, Intertidal: Intertidal coastal regions are subject to wave and tidal movement of ocean water, and organisms have adapted to periods of exposure to air. Wave action results in a high level of water oxygenation and mixing and nutrients from land create a high nutrient level.

Ocean, Coral Reefs: Coral reefs found in tropical equatorial or sub-equatorial waters generally are exposed to abundant sunlight and temperatures between 20°C and 30°C. Corals are the dominant organisms in this biome. These cnidarians form the reefs—from their calcium carbonate shells left behind over many generations. Algae that live symbiotically with the coral provide photosynthetic productivity and are sensitive to increases in water temperature, with global warming threatening coral reef survival. Coral reefs have very high biological diversity.

Open Ocean (Pelagic): The oceans cover 70% of the Earth's surface, most of which is open ocean, away from coastal regions. The nutrient and oxygen levels in deep ocean waters are affected by their latitude, wind, and ocean currents. Oxygen levels in surface waters are usually high, while nutrient levels for phytoplankton are low. Photosynthesis is restricted to surface layers by the low penetration of light into water. Wind and ocean currents tend to mix water, increasing the nutrient level of surface waters and the phytoplankton photosynthetic productivity. Tropical open oceans often have a layer of sun-warmed water on the surface that does not mix with the deeper water, reducing the nutrient content and productivity of surface water.

Deep Ocean (Benthic): This zone receives minimal sunlight and is cold. Oxygen is present. Nutrients come from the remains, debris, and detritus of organisms in upper layers of water.

Q Which aquatic biome is best described as having a warm surface layer, low nutrient levels, and low photosynthetic productivity?

(A) Intertidal
(B) Benthic, abyssal
(C) Pelagic, tropical
(D) Estuaries
(E) Coral reef

A The correct answer is (C). The open oceans in tropical regions often have warm surface layers that do not mix with deeper water, making the surface relatively low in nutrients and phytoplankton.

Q The photic zone of the oceans is the region in which:

(A) Nutrients are most concentrated
(B) Enough sunlight can penetrate in order to conduct photosynthesis
(C) Bioluminescent organisms predominate
(D) Upwellings of nutrients occur
(E) A steep temperature gradient occurs

A The correct answer is (B). The photic zone has light. A big clue is the root of photic, *photo-*, which relates to all things with light.

Terrestrial Biomes

Tundra and Polar Regions: The polar regions have very low photosynthetic productivity and are frozen year-round. Life is still present, from microscopic organisms to large mammals. Life in both the Arctic and Antarctica depends on ocean life. The fringes of Antarctica are inhabited by birds that survive by eating marine organisms. The Arctic is home to birds and mammals living on ice or around the frozen ocean. The tundra covers large areas of North America and Eurasia that lie near the Arctic. The tundra's brief summer gives rise to small, quick growing plants, like grasses and mosses, but no trees or large plants inhabit this region. Migratory mammals and birds inhabit the region in the summer, with some species persisting through the long, cold winter.

Coniferous Forest (Boreal, Taiga): South of the tundra in Asia and North America, coniferous forests, including pines, firs, and spruce trees, cover broad swaths of land. Winters are still generally cold, and summer is generally short but can be very warm and intense with growth and insect life. There are more year-round inhabitants in the boreal forests than the tundra, and the ground is not frozen year-round with permafrost. Large mammalian herbivores (moose) are common, as are carnivores (wolves) and omnivores (bears); all have adapted to the cold winters. Amphibians and reptiles are not common and most birds migrate through the region, moving south in the winter. Although much of these forests still exist, intense logging is a pressing threat.

Temperate Deciduous Forest: These occur at middle latitudes, mainly in the Northern Hemisphere. Rainfall is significant, sun is abundant, and a great variety of animals inhabit these forests year-round, even with some snow in the winter. The temperate forests have more biological diversity, higher primary productivity, and more complex food webs. They have more complex structure with multiple layers, from a high canopy of the taller trees down through smaller trees and undergrowth on the ground. Much of the eastern and southeastern U.S. was once covered in temperate forest, which has been removed with the development of cities and agriculture.

Tropical Forest: Tropical forests near the equator have abundant light and rain year-round, the highest terrestrial primary productivity, and very high biological diversity. Tropical rainforests may contain half or more of the Earth's species, although the true number is unknown. Covering large areas of South and Central America, Asia, and Africa, this biome is being disrupted by development at a rapid pace, with about half of its area already gone.

Grasslands: Grasslands predominate in inland regions of continents, with cold winters, hot summers, and broad expanses dominated by the grasses. Large herds of mammalian grazers were common in these areas, although today these herbivores are highly reduced in most areas. Native grasses have deep roots and grow as perennials, preserving and enhancing soil nutrient levels and a rich soil community of organisms; these ecosystems have largely been replaced or changed by grazing livestock and agriculture.

Chapparal: Small specific regions around the globe have a combination of dry summers, bright sun, and winter rain in coastal areas called chaparral. Plants are mostly shrubs, grasses, and herbs that have adapted to dry summers. Biological diversity of all vertebrates is large, increased perhaps by the variety of microclimates and terrain. The Mediterranean was largely surrounded by chaparral prior to modern urbanization and development.

Savanna: This is mostly dry grassland, with a few scattered trees. Fire and drought adaptations are found in these species. Large mammalian grazers like wildebeests are common, but seasonal drought and growing seasons cause large migrations in some cases.

Desert: These are areas of very little rain, and what rain does occur is unpredictable, leading plants to evolve around water conservation. Animal inhabitants are often nocturnal to avoid the heat of the day.

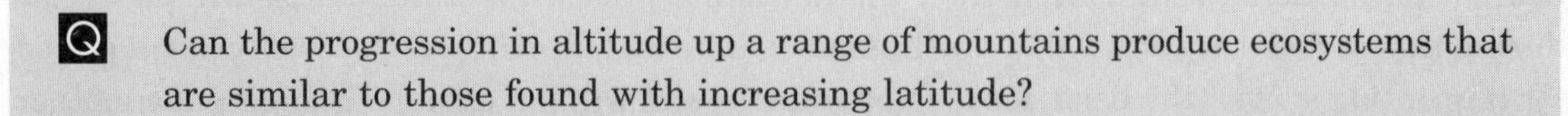

Q Can the progression in altitude up a range of mountains produce ecosystems that are similar to those found with increasing latitude?

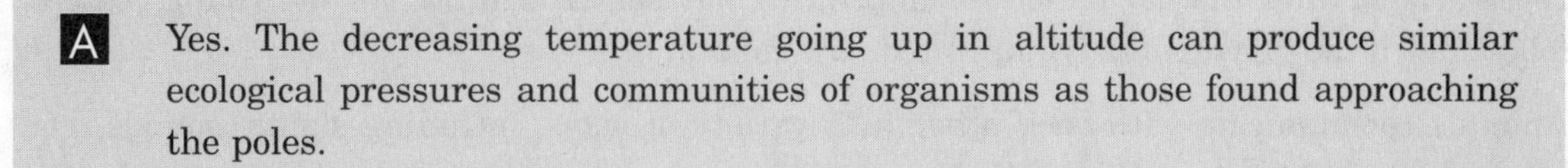

A Yes. The decreasing temperature going up in altitude can produce similar ecological pressures and communities of organisms as those found approaching the poles.

BEHAVIOR OF ANIMALS IN THEIR ENVIRONMENT

Animals exhibit a variety of behaviors related to feeding, mating, avoiding predators, and taking care of their young. These behaviors affect how they relate to others of the same species and how they interact with the rest of their ecosystem. Some behaviors have a strong genetic component and appear, at least in part, to be preprogrammed genetically as part of nervous system development. Movement toward certain positive stimuli or away from noxious stimuli is one example of this. Other examples are **chemotaxis**, movement resulting from chemical stimuli, and **phototaxis**, movement toward light. Even single-cell organisms often display the ability to move toward stimuli, like food molecules, and away from a drop of acid placed in their environment. Planaria avoid light, snails withdraw their eyestalks, and moths are drawn to a porch light not because of learning, but as inborn behaviors automatically activated by their environment. Migratory behavior is common in many animals and further proof of an inborn genetic component.

Q If young animals display the same migratory behavior as their parents, does this mean the behavior is genetic?

A Not necessarily. It might have been learned from the parents. To see if it is genetic, it is necessary to do experiments to make sure it was not learned.

A more complex type of behavior, called a **fixed action pattern,** involves a sequence of behaviors that are not learned but are triggered by a **sign stimulus** or **releaser** from the environment. The sign stimulus can often be reproduced with a highly simplified version of the normal stimulus. For example, young kelp gulls peck at a red spot on their mother's beak to get the mother to feed them. If almost any object with a red dot is presented to the chick, it will peck at the dot in the same way it pecks at its mother's beak.

Q In the example of the kelp gulls, what is the releaser?

A In this case, the pecking of the young bird is the fixed action pattern and the releaser is the red dot.

Another commonly studied type of behavior is called **imprinting**. Imprinting is a learned behavior that can only be learned during a specific developmental period. A common example is the way young geese learn to relate to their mother. This information is imprinted in the birds during a key period soon after they hatch. They will identify with the animal they are nearest to, including a human, if their mother is not present. If a human is with the geese during this period, they will follow the human as if he is their mother. Once the critical period for imprinting is over, the behavior is fixed in place and will not change. Bringing the mother goose around after the key period for imprinting has passed will not get the young geese to follow her if they have already imprinted on a human or a dog.

Animals communicate with each other in a variety of ways, including sights, sounds, and smells. The information being communicated can mark territory, warn of danger, inform others of food, or find mates. Chemical smells can be potent means of communication, both in water and on land, and still elicit a response even after great dilution. Chemical messengers that are released by one individual to cause a response in other members of the same species are called **pheromones**. Female moths release chemical attractant pheromones that can attract male moths from miles away. Visual signals work best in bright light and require a direct line of sight, so they are more common in animals that are active during the day. Birds that are active during the day and use visual cues for communication may use bright colors to communicate. Sounds can convey information around the clock, do not require line of sight, and can travel great distances if they are loud enough. In a dense forest, sound may be a more useful form of communication than visual communication.

Q Do pheromones affect individuals of a different species?

A No. By definition, pheromones are signals between individuals of the same species.

The genetic component that underlies many behaviors (which are phenotypic by definition) subjects them to natural selection. The behaviors that are inherited and display variation in a population will select for alleles that create behaviors providing the highest fitness. When looking for food, animals have to balance the amount of energy expended in looking for food, or foraging, with the amount of energy they receive from the food. Animals cannot calculate this, but natural selection automatically selects for foraging behaviors that provide the greatest benefit, while avoiding predation and excessive energy loss.

Q In some species, relatives help to feed the young, despite the apparent energetic cost and potentially increased risk of predation. How can this be explained in evolutionary terms?

A This is an example of kin selection. Altruism like this may seem to run counter to natural selection, but natural selection does not always occur through an individual's own descendents. By increasing the survival of related individuals, they are increasing their own fitness, since they share alleles with their relatives.

Mate selection and sexual behavior are also subject to natural selection in many species, like selecting for specific traits and behaviors in mates. The bright colors of male birds, the antlers of deer, and the competition between males may select healthier individuals through selective mating.

Interactions Between Species

One way species can interact is **predation**, in which one organism consumes another. Needless to say, this is not a win-win situation. The predator comes out on top in this interaction, gaining nutrition. The prey? Well, he contributes to the ecosystem. Any interaction involving eating another organism is a form of predation, even if it involves an herbivore and a plant. Sometimes the word predation is used more specifically to refer to interactions in which one organism is consumed and killed as a result of an interaction.

The interaction between predator and prey can become an evolutionary arms race, with the predator evolving more effective means to catch prey and the prey evolving ever better means of escaping. Increasing speed of both parties is one evolutionary strategy, like the cheetah and gazelle, or increasing armor in the prey and means of defeating the armor in the predator, like the shell of a mussel and the arms and vascular tube feet of the starfish that pries it open. An additional prey strategy is camouflage, to evolve coloring or shape that matches the surrounding, making it hard for predators to find the prey. Prey sometimes evolve chemical defenses, like toxins that make predators ill, and predators in turn either avoid the prey or evolve means of detoxifying or defeating the prey's defenses. Some insects and frogs advertise their toxins with bright colors, warning off potential predators that either instinctively avoid them or quickly associate the colors with an unpleasant eating experience. In **Batesian mimicry**, other nontoxic organisms often mimic the coloration of toxic ones, to gain the benefit of avoiding predation. Another form of mimicry, called **Mullerian mimicry,** involves multiple species that are either toxic or venomous and resemble each other to gain stronger avoidance by predators.

Q If a toxic and a nontoxic butterfly resemble each other, is this an example of Batesian mimicry, camouflage, or Mullerian mimicry?

A Batesian mimicry, in which chances that the nontoxic butterfly will be eaten are less because predators associate its appearance with the effects of the toxic butterfly.

Parasitism is another interaction between species that is advantageous for one member of the interaction and negative for the other. Parasites gain nutrition from other organisms, the hosts, causing them harm by damaging tissues and diverting nutrition. Usually, the parasite is much smaller than the host and can be found on the exterior (ectoparasites) or interior (endoparasites) of the host. Parasites include many microbes, flatworms (flukes and tapeworms), segmented worms (leeches), and arthropods (ticks, mosquitoes and fleas). Parasites such as mycobacterium can even live inside other cells. Every group of living things is prone to parasitism, including microbes; even parasites can have parasites!

Q Is the interaction of a bacteriophage virus with a bacterial cell a form of parasitism, even if viruses are not living on their own?

A Viruses are obligate intracellular parasites. They can't do anything on their own, but their ability to divert cellular resources to make new viruses earns them the parasite title.

A **mutualistic** relationship between organisms is a true win-win for both parties involved. Everybody wins with mutualism. For example, cleaner fish are small fish that eat parasites from the mouths, gills, and bodies of larger fish. The cleaner fish get food and the larger fish gets its parasites removed. Most plants live symbiotically with mycorrhizal fungi on their root system, which helps plants absorb nutrients from soils. Plants in turn provide nutrients, like sugars from photosynthesis, back to the fungi. Algae and fungi live together symbiotically in lichens. Algae and coral also live together symbiotically.

Commensalism is an interaction in which one organism benefits and the other is not affected at all, either positively or negatively. In practice, most interactions probably affect both parties in some way, even if only slightly, so truly commensal interactions may not occur in nature. Birds that follow large herbivores, waiting for insects to be disturbed, suggests a commensal relationship, in which the birds get food and the herbivore is unaffected. It's hard to know, though, if the herbivores are truly unaffected or if we just can't measure how they are affected by the presence of the bird.

Q Pollinating insects expend a great deal of energy in moving from flower to flower, carrying pollen with them as they feed on the flowers' nectar. What type of interaction is this?

(A) Commensalism
(B) Parasitism
(C) Mutualism
(D) Predation
(E) Competition

A The correct answer is (C). This is a win-win interaction in which the insect gets food and the flowers get pollinated.

Each species interacts in a unique way with its environment, using a unique set of resources. No two species fit into their ecosystem in exactly the same way. This is the result of evolution and the pressure of competition between species that have overlapping and competing interactions with their environment. If two species do compete for resources, one competes more effectively than the other and evolves to reduce competition or one of the species loses the competition and disappears.

The unique interaction of each species with the rest of its environment is called its **ecological niche**. Each organism has a unique niche, describing everything about how it fits in with the rest of the ecosystem around it, including what it eats, when it eats, how it eats, where it lives, etc. No two species can occupy and compete for the same niche (not for long anyway), but they can coexist if they use their resources differently. For example, if there are two species of squirrel that both live in walnut trees and eat the same nuts, they might still be able to coexist if one species eats only nuts found on the top of the tree and the other only eats nuts found on the bottom of the tree. They may also overlap their resource use, perhaps with one species selected to be smaller, and eating smaller nuts, while the other species evolves to increase in size and eat the large walnuts.

Q Is extinction of one species the inevitable result of two species competing for the same ecological niche?

A No. One or both species can evolve to reduce the overlap in niches.

A population's size is determined by how many organisms die and how many new organisms are born (ignoring immigration for now). Different organisms display different strategies for maintaining their population size. Some organisms (like clams) maintain their population by producing very large numbers of young that receive little or no care and have a relatively high mortality rate in early life. Other organisms (like primates) produce few young, invest a great deal of care in them, and have a high survival rate of their young.

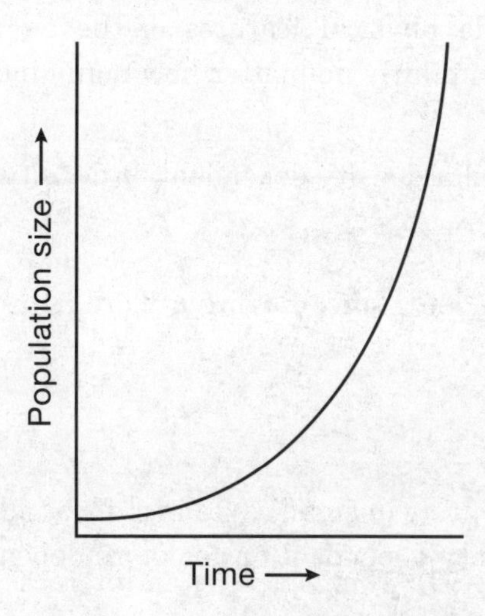

In conditions where resources are not a limiting factor and other normal limitations on the number of individuals (like predation and disease) do not come into play, the number of individuals in a population can increase exponentially. For example, some bacteria can divide to reproduce asexually, by binary fission, every 20 minutes, under optimal growth conditions. One bacterium can produce 6.8×10^{10} bacteria in this way in just 12 hours, weighing more than the Earth before long. Clearly, this cannot happen, for in the real world, exponential growth quickly runs up against factors that limit it, including predation, lack of food, and disease. The bacteria will run out of nutrients a long time before they take over the planet.

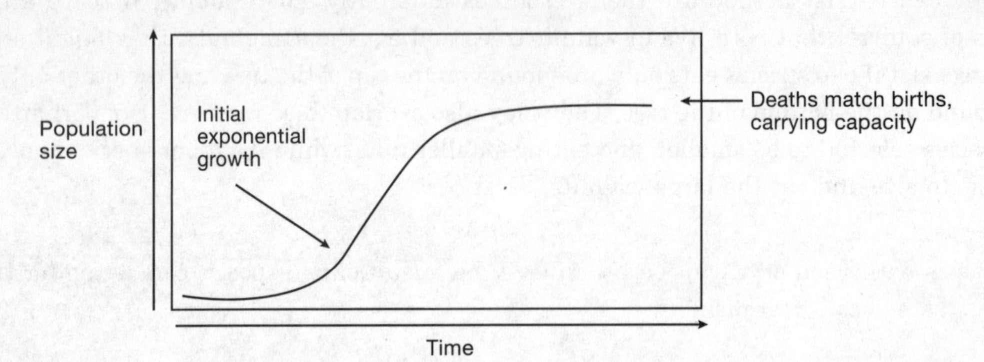

If other factors counterbalancing proliferation in a species are factored in, the **logistic model** is created, in which populations may display rapid exponential growth at first. But as the population grows and resources become limited, population growth slows and the size of the population levels off. The size of the population at this point is called the **carrying capacity**, in which the number of deaths equals the number of births. Like exponential growth, the logistic model is a theoretical model used to compare against the behavior of real populations. Factors that can limit a population's growth are either **density dependent** or **density independent**. Density dependent factors include the spread of disease in a population, which will be more likely if the population is denser. Factors that can place density independent limits on a population include physical features of the environment or weather-related factors, affecting populations similarly no matter how many individuals are present.

Q Is competition for food a density dependent or density independent factor affecting a population?

A This is density dependent, since having a more dense population increases the competition for food.

Q A species of sea birds nests in small caves in cliffs beside the ocean. Is competition for nesting sites density dependent or density independent?

A There are a finite number of nesting sites. The competition for the sites will depend on how many birds are present. Since the number of sites does not change, this scenario is density dependent.

Although ecosystems are sometimes depicted as static and unchanging, natural systems are constantly affected by a variety of factors that disturb them in one way or another. Factors like drought, floods, fires, volcanoes, diseases, and humans keep ecosystems in a state of flux. After a disturbance, particularly a large one, the community of organisms in the disturbed region's composition is altered. The changes in the community after a disturbance, over time, are called **succession**. After a volcanic eruption, succession may start with just the hardiest organisms that can grow on bare rock, like lichen. Once soil has started to accumulate, mosses, small grasses, and insects may follow. Then, over the years, trees and larger organisms will again populate the region. Trees can go through succession as well, starting with smaller trees that need a great deal of light, followed by taller conifers that block out light from the understory, and then becoming the dominant species in later succession. Some plants have evolved to fill the niche as colonizers that do particularly well after a disturbance. Some plants have seeds that even require a disturbance, like fire, in order to reproduce effectively.

CHEMICAL CYCLES

Water, nitrogen, and carbon are three of the most essential and basic chemical components of life on Earth. These chemicals and elements are always moving from one form to another through the living and abiotic components of ecosystems, often on a global scale.

Nitrogen Cycle

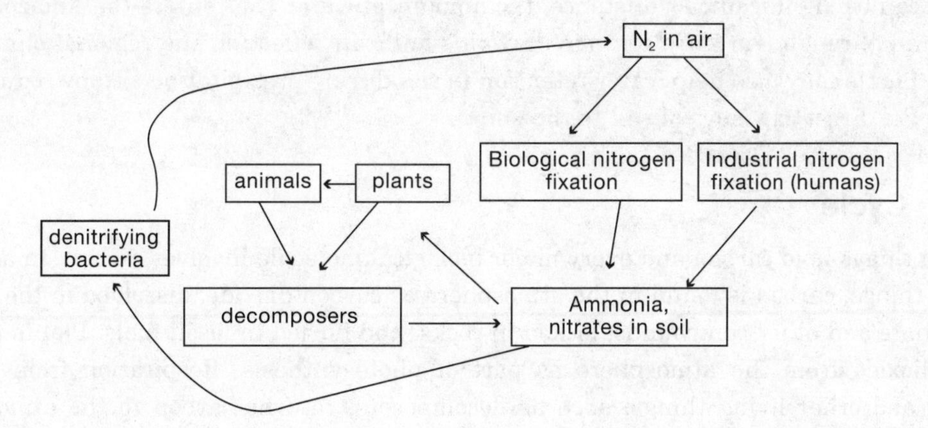

Most of the atmosphere (about 78%) is made up of nitrogen, but most of it cannot be used by living things directly since it is in the form of molecular nitrogen, N_2. Living organisms need nitrogen in proteins and nucleic acids, as well as other compounds they synthesize. Nitrogen-fixing bacteria in soil convert atmospheric molecular nitrogen into other inorganic forms that plants or other bacteria can use, such as ammonia and nitrates. Nitrifying bacteria in soil help the process along, converting ammonia to nitrates. These forms of nitrogen in soil allow plants to grow and can either be eaten by animals or decompose to return nitrogen to the soil. Some nitrogen is also released back to the atmosphere by denitrifying bacteria.

Q Can animals use inorganic forms of nitrogen, such as nitrates, directly for the biosynthesis of amino acids?

A No. They get all of the nitrogen they use for biosynthesis from plants in the form of organic compounds.

Water Cycle

Water plays a unique role in living systems. The Earth is unique in having water in all three forms: liquid, solid (ice), and gas (water vapor). The water cycle includes the movement of water between these forms and between the land and the major reservoir of water on Earth, the oceans. The oceans hold about 97% of Earth's water at any moment, with the remainder in fresh water, ice caps, and water vapor. These other forms of water may not be a big percentage of the water on Earth, but they are essential for the survival of terrestrial life. Terrestrial life depends on rain and other forms of precipitation to bring water to land. Rain comes from condensation of water vapor in the atmosphere to form clouds. The water vapor comes from evaporation from the oceans and land and from transpiration from plants. Plant transpiration is the biggest biological component of the water cycle.

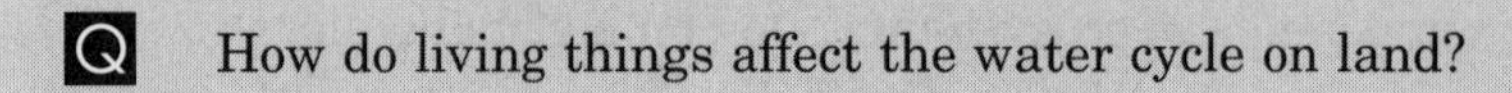

Q How do living things affect the water cycle on land?

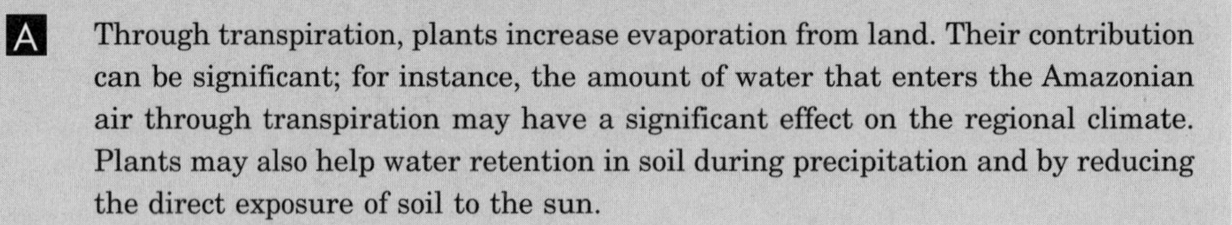

A Through transpiration, plants increase evaporation from land. Their contribution can be significant; for instance, the amount of water that enters the Amazonian air through transpiration may have a significant effect on the regional climate. Plants may also help water retention in soil during precipitation and by reducing the direct exposure of soil to the sun.

Carbon Cycle

All living things hold carbon and every major biological molecule involves carbon. In addition to living things, carbon is found in the atmosphere as carbon dioxide, dissolved in the oceans as carbonate and other compounds, locked in rocks, and buried in fossil fuels. Plants remove carbon dioxide from the atmosphere as part of photosynthesis. Respiration from plants, animals, and other living things, such as decomposers, returns carbon to the atmosphere again as carbon dioxide. The amount of carbon dioxide that is removed from the atmosphere by plants roughly equals the amount returned by respiration. The burning of fossil fuels by humans has greatly increased the amount of carbon dioxide entering the atmosphere, changing the carbon cycle and acting as a greenhouse gas to cause climate change. In turn, climate change may further alter the carbon cycle as the oceans, polar regions, and other biomes release carbon dioxide from other sources, like permafrost, in response to warming.

Carbon Cycle

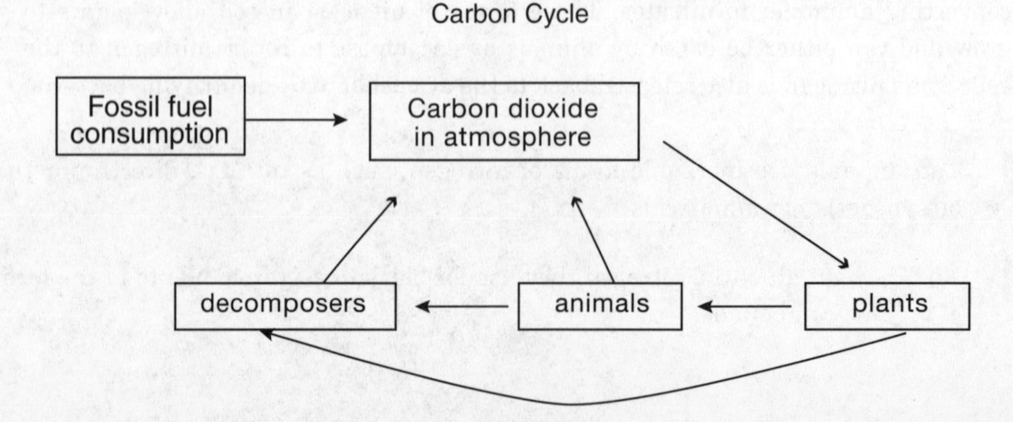

IMPACT OF HUMANS ON THE BIOSPHERE

The number of humans on Earth has increased dramatically (in fact, nearly exponentially) in the last century—from 2 billion in 1930 to about 3 billion in 2002, bringing steadily rising urban growth, industrial activity, and agriculture. All of this has left few, if any, parts of the globe unchanged. The rate of change is so rapid that few fully appreciate the impact of humans on the biosphere. A short list of the human impact on ecosystems includes:

- **Acid Rain:** Burning fossil fuels, particularly coal, often produce sulfur and nitrogen-containing air pollutants that are as acidic as pH 4 when mixed with water in rain clouds. These pollutants can drift large distances in the air, affecting others even far away. On the ground, acid rain removes nutrients from soil, damages plants, and acidifies lakes and streams, harming aquatic organisms. These are problems that have been seen in recent decades on a wide scale in the U.S. and Canada and Europe. Legislation in some countries now restricts sulfur emission, but many parts of the world continue emitting sulfur pollution without effective regulations.

- **Ozone Hole:** Were it not for a thin layer of ozone high in the Earth's atmosphere, our planet's surface would be bombarded by much higher levels of dangerous UV irradiation. Once ozone levels were first observed to be falling in the 1970s, the problem was traced to chlorofluorocarbons (CFCs), chemicals used as refrigerants in air conditioners, refrigerators, and aerosol sprays. If left unchecked, ozone depletion may lead to an increased cancer rate in humans and unknown harmful effects on plants and animals, particularly near the poles where depletion is greatest. The use of the most ozone-depleting CFCs has since been banned and atmospheric levels are decreasing. However, recovering former ozone levels may take decades.

- **Nitrogen Pollution:** Modern agriculture uses huge quantities of chemically produced nitrates to supplement biological nitrogen fixation and increase crop production. The level of human nitrogen fixation has surpassed biological nitrogen fixation, releasing vast quantities of nitrates into ecosystems. From agricultural land, nitrates often wash into rivers and lakes, stimulating massive algae growth through **eutrophication**. When algae and other plant material fall into deeper water and decompose, bacterial action can remove oxygen from the water, making life impossible for other organisms, vertebrate or invertebrate. For instance, the agricultural nitrates that enter the Mississippi River and drain into the Gulf of Mexico now routinely create large "dead zones" due to oxygen depletion in this manner.

- **Climate Change:** The Earth's climate is moderated by greenhouse gases, such as carbon dioxide, in its atmosphere. These gases absorb infrared heat energy and prevent it from escaping into space, helping make Earth habitable. Since the Industrial Revolution, man has been burning fossil fuels and releasing large quantities of carbon dioxide back into the atmosphere, changing the carbon cycle. This trend accelerated in the 20th century, increasing atmospheric carbon dioxide from a preindustrial level of about 280 ppm (parts per million) to about 380 ppm today. This trend correlates with increasing global temperatures, a trend that is expected to continue in the next century with a further 2–5°C increase in global average temperature. In addition to melting poles and rising sea levels (and the impact of this on humans), biomes around the world will be greatly affected. The distribution of biomes would move rapidly northward in many cases, and distribution of rainfall may change in unpredictable ways. Ecosystems that are already fragmented and under stress from human activity may not survive these rapid and large changes in climate.

- **Biodiversity Loss:** The evolution of life has created a rich diversity of species in all Earth's biomes and ecosystems, species that live with each other in an intricate interwoven web. Extinction is not new to our planet, but man is now causing one of the largest mass extinctions in Earth's history. We are destroying habitats, polluting land and water, changing the climate, and hunting or over-fishing a variety of species to extinction. The world's amphibians have suffered dramatic losses, perhaps due to an imported fungus spreading through their populations, as well as habitat loss, climate change, and pollutant exposure. A third or more of amphibian species may already be extinct. The loss of species is a loss of biological innovation accumulated in 3.5 billion years of evolution that we have hardly explored. Many key medicines have come from plants (aspirin, penicillin, taxol) and today we are losing species of tropical plants before they are even characterized. With each species loss, the web of interactions that hold ecosystems together may be undone, and the essential services ecosystems perform, providing soil, air, and water, may be lost.

THE FUTURE OF THE BIOSPHERE

With aggressive action, dramatic losses of the Earth's habitats and species may be avoided. We are increasingly finding that protecting the biosphere is necessary for our own survival. Living organisms provide a great range of services that we have traditionally taken for granted, assuming they were virtually limitless. We emit pollutants into the oceans and air, assuming that the seemingly endless skies and seas would dilute the pollutants and take the problems away. As it turns out, as vast as they are, they are not limitless and the world's organisms and ecosystems are being gravely harmed by our actions. The living systems of Earth, from tropical forests to marine phytoplankton, produce the oxygen, soil, and water we rely on for survival. Climate change and other human impacts on the biosphere may irreversibly harm the planet and humans if left unchecked. It would be wise to avoid this risk and to take care of the rest of Earth's biosphere.

EXERCISES: ECOSYSTEMS

1. The main source of biological nitrogen fixation is:

(A) Respiration of animals
(B) Photosynthesis of plants
(C) Actions of bacteria producing ammonia and nitrates from N_2
(D) Action of decomposers in soil
(E) Photo-oxidation of smog

2. The impact of chemical fertilizers on a deep mountain lake ecosystem can best be associated with:

(A) Enhanced primary productivity
(B) Enhanced oxygen content at lake bottom
(C) Elevated temperature of surface water
(D) Reduced decomposition
(E) Loss of photosynthetic activity

3. Habitat fragmentation into small sections of land might affect populations in what way?

(A) Hardy-Weinberg equilibrium will be established more rapidly.
(B) The populations will be subjected to natural selection to enhance genetic variation.
(C) The rate of migration will increase as habitat fragmentation increases.
(D) Smaller isolated populations divided among several small pieces have less viability than one larger habitat with the same overall area.
(E) It will increase the biodiversity of each fragment.

4. Aphids live on a tomato plant, feeding on plant nutrients that are withdrawn with sucking mouthparts. Ants tend to the aphids, keeping away aphid predators and the aphids provide the ants with a sugary substance called honeydew. What is the relationship of the aphid with the plant?

(A) Commensalism
(B) Predation
(C) Mutualism
(D) Symbiosis
(E) Parasitism

5. Which of the following might take place at point Q in the chart below?

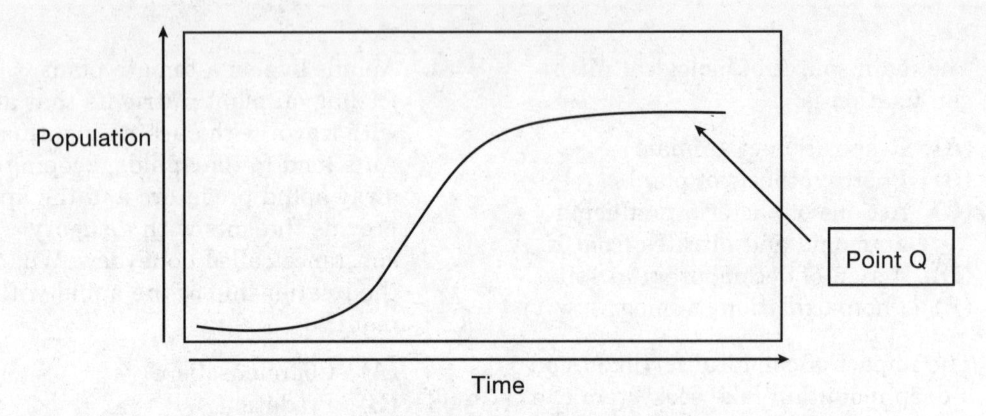

(A) Density independent factors increase the death rate.
(B) Density dependent factors increase the birth rate.
(C) The rate of deaths exceeds the rate of births.
(D) Density dependent factors decrease the birth rate.
(E) Hardy-Weinberg equilibrium is established.

6. Which of the following is most likely to occur if two species compete for the same food resources in a region?

(A) Allopatric speciation
(B) Directional selection
(C) Succession
(D) Habituation
(E) A stable but reduced carrying capacity for both species

7. Male stickleback fish will aggressively attack other males who are red on the bottom. Display of almost any object that is red on the bottom will stimulate an aggressive response from these fish, while a male that lacks red does not stimulate any response. Even males that are raised in isolation without exposure to other fish will display this behavior. What type of behavior does this describe?

(A) Imprinting
(B) Habituation
(C) Fixed action pattern
(D) Associative learning
(E) Long term potentiation

8. The oceans affect terrestrial biomes in all of the following ways except:

(A) Absorbing carbon dioxide from the atmosphere with photosynthesis
(B) Serving as a source for atmospheric water vapor that reaches terrestrial biomes through precipitation
(C) Delivery of nitrates to support terrestrial plant growth
(D) Redistribution of heat between equatorial and polar regions
(E) Providing sources of nutrition for vertebrates in coastal ecosystems

ANSWER KEY AND EXPLANATIONS

1. C	3. D	5. D	7. C	8. C
2. A	4. E	6. B		

1. **The correct answer is (C).** Biological nitrogen fixation is carried out by bacteria, many living symbiotically with plants.
2. **The correct answer is (A).** Chemical fertilizers add large amounts of nitrogen and phosphorus to aquatic ecosystems, stimulating the growth of algae and aquatic plants, primary producers.
3. **The correct answer is (D).** Smaller populations that are isolated will have reduced genetic variation compared to a single larger population and will have reduced viability as a result.
4. **The correct answer is (E).** The aphids feed on the plant but do not kill it. The aphid is benefited while the plant is harmed by the loss of nutrients. The part about the ants is included just to confuse things and does not affect the answer.
5. **The correct answer is (D).** This plot is the logistic model, with population growth leveling off and stopping as the population reaches the carrying capacity. At this point, the population size is constant—the rate of deaths is the same as the rate of births. These factors will be density dependent, since they require the population to reach a certain size, and they will either increase the death rate or increase the birth rate. The rate of deaths does not exceed the rate of births. If it did, then the population would shrink (answer choice C is incorrect). The problem has nothing to do with Hardy-Weinberg (answer choice E is incorrect).
6. **The correct answer is (B).** When two species compete for the same resources, they are never going to be exactly the same in their ability to compete. Either selection will create a difference between the species, reducing the competition, or one of the species will go extinct in this region (answer choice E is incorrect). The other choices are irrelevant.
7. **The correct answer is (C).** A fixed action pattern is a programmed behavior that is released by a specific stimulus, even in a highly simplified state.
8. **The correct answer is (C).** Nitrates generally enter the terrestrial nitrogen cycle from nitrogen fixation by bacteria in soil. The oceans do all of the other items through physical and biological processes.

SUMMING IT UP

- Sunlight is the basis of almost all energy produced and used by living things on Earth, so varying sunlight exposure and temperature affect living things greatly. Water is another essential ingredient for living things that varies from almost none in desert regions to aquatic environments. The combination of varying sunlight, precipitation, terrain, and temperature create the climate of each biogeographic region and shape ecosystems.
- The relationships between organisms are largely based on who eats whom, starting at the bottom with the **primary producers**. The primary producers are the autotrophs that obtain energy from the abiotic environment. In addition to the primary producers in an ecosystem, the **primary consumers** are herbivores that eat the plants, algae, or other photosynthetic primary producers. **Secondary** and **tertiary consumers** eat other animals, acting as predators of the herbivores and other carnivores. **Decomposers** return the nutrients of the other organisms back to the Earth, to enter the food web again.
- The amount of energy flowing into the system always equals the amount of energy flowing out, one way or another. The amount of sunlight absorbed by plants through photosynthesis is the **gross primary productivity**, while the amount of energy stored by plants after using some for their own metabolic needs is the **net primary productivity**. The productivity of aquatic or terrestrial ecosystems is limited by the amount of light they receive and nutrients that are available for photosynthesizers.
- Each level in the food web is often described as a **trophic level**. As one level consumes organisms from another level, it is transferring energy between levels, but the process is never 100% efficient. When one organism eats another one, it loses some of the energy as metabolic heat, and more is lost through excretion. Only a relatively small percentage of energy (about 20% or less) is actually transferred and incorporated as biomass at a higher trophic level.
- The biomes of the biosphere are categorized in a variety of ways, but commonly include aquatic biomes (Lakes and Rivers; Wetlands; Ocean, Intertidal; Ocean, Coral Reefs; Ocean, Open Ocean [Pelagic]; Ocean, Deep Ocean [Benthic]) and terrestrial biomes (Tundra and Polar Regions, Coniferous Forest [Boreal, Taiga], Temperate Deciduous Forest, Tropical Forest, Grasslands, Chapparal, Savanna, Desert).
- Animals exhibit a variety of behaviors related to feeding, mating, avoiding predators, and taking care of their young. These behaviors affect how they relate to others of the same species and how they interact with the rest of their ecosystem. Some behaviors have a strong genetic component and appear, at least in part, to be preprogrammed genetically as part of nervous system development. **Chemotaxis** is movement stimulated by chemical stimuli and **phototaxis** is movement toward light. A more complex type of behavior, called a **fixed action pattern,** involves a sequence of behaviors that are not learned, rather they are triggered by a **sign stimulus** or **releaser** from the environment. Another commonly studied type of behavior is called **imprinting**. Imprinting is a learned behavior that can only be learned during a specific developmental period.

- One way species can interact is **predation**, in which one organism consumes another. In **Batesian mimicry**, nontoxic organisms often mimic the coloration of toxic ones to gain the benefit of avoiding predation. Another form of mimicry, called **Mullerian mimicry,** involves multiple species that are either toxic or venomous and resemble each other to gain stronger avoidance by predators.
- **Parasitism** is another interaction between species that is advantageous for one member of the interaction and negative for the other. Parasites gain nutrition from other organisms, the hosts, causing them harm by damaging tissues and diverting nutrition.
- A **mutualistic**, or **symbiotic,** relationship between organisms is a true win-win for both parties involved.
- **Commensalism** is an interaction in which one organism benefits and the other is not affected at all, either positively or negatively. In practice, most interactions probably affect both parties in some way, even if only slightly, so truly commensal interactions may not occur in nature.
- Although ecosystems are sometimes depicted as static and unchanging, natural systems are constantly affected by a variety of factors that disturb them in one way or another. Factors like drought, floods, fires, volcanoes, diseases, and humans keep ecosystems in a state of flux. After a disturbance, particularly a large one, the community of organisms in the disturbed region's composition is altered. The changes in the community after a disturbance, over time, are called **succession**.
- Water, nitrogen, and carbon are three of the most essential and basic chemical components of life on Earth. These chemicals and elements are always moving from one form to another through the living and abiotic components of ecosystems, often on a global scale.
- A short list of the human impact on ecosystems includes acid rain, ozone hole, nitrogen pollution, climate change, and biodiversity loss.

PART IV

TWO PRACTICE TESTS

PRACTICE TEST 2

PRACTICE TEST 3

answer sheet

ANSWER SHEET PRACTICE TEST 2

SECTION I

1. Ⓐ Ⓑ Ⓒ Ⓓ Ⓔ
2. Ⓐ Ⓑ Ⓒ Ⓓ Ⓔ
3. Ⓐ Ⓑ Ⓒ Ⓓ Ⓔ
4. Ⓐ Ⓑ Ⓒ Ⓓ Ⓔ
5. Ⓐ Ⓑ Ⓒ Ⓓ Ⓔ
6. Ⓐ Ⓑ Ⓒ Ⓓ Ⓔ
7. Ⓐ Ⓑ Ⓒ Ⓓ Ⓔ
8. Ⓐ Ⓑ Ⓒ Ⓓ Ⓔ
9. Ⓐ Ⓑ Ⓒ Ⓓ Ⓔ
10. Ⓐ Ⓑ Ⓒ Ⓓ Ⓔ
11. Ⓐ Ⓑ Ⓒ Ⓓ Ⓔ
12. Ⓐ Ⓑ Ⓒ Ⓓ Ⓔ
13. Ⓐ Ⓑ Ⓒ Ⓓ Ⓔ
14. Ⓐ Ⓑ Ⓒ Ⓓ Ⓔ
15. Ⓐ Ⓑ Ⓒ Ⓓ Ⓔ
16. Ⓐ Ⓑ Ⓒ Ⓓ Ⓔ
17. Ⓐ Ⓑ Ⓒ Ⓓ Ⓔ
18. Ⓐ Ⓑ Ⓒ Ⓓ Ⓔ
19. Ⓐ Ⓑ Ⓒ Ⓓ Ⓔ
20. Ⓐ Ⓑ Ⓒ Ⓓ Ⓔ
21. Ⓐ Ⓑ Ⓒ Ⓓ Ⓔ
22. Ⓐ Ⓑ Ⓒ Ⓓ Ⓔ
23. Ⓐ Ⓑ Ⓒ Ⓓ Ⓔ
24. Ⓐ Ⓑ Ⓒ Ⓓ Ⓔ
25. Ⓐ Ⓑ Ⓒ Ⓓ Ⓔ
26. Ⓐ Ⓑ Ⓒ Ⓓ Ⓔ
27. Ⓐ Ⓑ Ⓒ Ⓓ Ⓔ
28. Ⓐ Ⓑ Ⓒ Ⓓ Ⓔ
29. Ⓐ Ⓑ Ⓒ Ⓓ Ⓔ
30. Ⓐ Ⓑ Ⓒ Ⓓ Ⓔ
31. Ⓐ Ⓑ Ⓒ Ⓓ Ⓔ
32. Ⓐ Ⓑ Ⓒ Ⓓ Ⓔ
33. Ⓐ Ⓑ Ⓒ Ⓓ Ⓔ
34. Ⓐ Ⓑ Ⓒ Ⓓ Ⓔ
35. Ⓐ Ⓑ Ⓒ Ⓓ Ⓔ
36. Ⓐ Ⓑ Ⓒ Ⓓ Ⓔ
37. Ⓐ Ⓑ Ⓒ Ⓓ Ⓔ
38. Ⓐ Ⓑ Ⓒ Ⓓ Ⓔ
39. Ⓐ Ⓑ Ⓒ Ⓓ Ⓔ
40. Ⓐ Ⓑ Ⓒ Ⓓ Ⓔ
41. Ⓐ Ⓑ Ⓒ Ⓓ Ⓔ
42. Ⓐ Ⓑ Ⓒ Ⓓ Ⓔ
43. Ⓐ Ⓑ Ⓒ Ⓓ Ⓔ
44. Ⓐ Ⓑ Ⓒ Ⓓ Ⓔ
45. Ⓐ Ⓑ Ⓒ Ⓓ Ⓔ
46. Ⓐ Ⓑ Ⓒ Ⓓ Ⓔ
47. Ⓐ Ⓑ Ⓒ Ⓓ Ⓔ
48. Ⓐ Ⓑ Ⓒ Ⓓ Ⓔ
49. Ⓐ Ⓑ Ⓒ Ⓓ Ⓔ
50. Ⓐ Ⓑ Ⓒ Ⓓ Ⓔ
51. Ⓐ Ⓑ Ⓒ Ⓓ Ⓔ
52. Ⓐ Ⓑ Ⓒ Ⓓ Ⓔ
53. Ⓐ Ⓑ Ⓒ Ⓓ Ⓔ
54. Ⓐ Ⓑ Ⓒ Ⓓ Ⓔ
55. Ⓐ Ⓑ Ⓒ Ⓓ Ⓔ
56. Ⓐ Ⓑ Ⓒ Ⓓ Ⓔ
57. Ⓐ Ⓑ Ⓒ Ⓓ Ⓔ
58. Ⓐ Ⓑ Ⓒ Ⓓ Ⓔ
59. Ⓐ Ⓑ Ⓒ Ⓓ Ⓔ
60. Ⓐ Ⓑ Ⓒ Ⓓ Ⓔ
61. Ⓐ Ⓑ Ⓒ Ⓓ Ⓔ
62. Ⓐ Ⓑ Ⓒ Ⓓ Ⓔ
63. Ⓐ Ⓑ Ⓒ Ⓓ Ⓔ
64. Ⓐ Ⓑ Ⓒ Ⓓ Ⓔ
65. Ⓐ Ⓑ Ⓒ Ⓓ Ⓔ
66. Ⓐ Ⓑ Ⓒ Ⓓ Ⓔ
67. Ⓐ Ⓑ Ⓒ Ⓓ Ⓔ
68. Ⓐ Ⓑ Ⓒ Ⓓ Ⓔ
69. Ⓐ Ⓑ Ⓒ Ⓓ Ⓔ
70. Ⓐ Ⓑ Ⓒ Ⓓ Ⓔ
71. Ⓐ Ⓑ Ⓒ Ⓓ Ⓔ
72. Ⓐ Ⓑ Ⓒ Ⓓ Ⓔ
73. Ⓐ Ⓑ Ⓒ Ⓓ Ⓔ
74. Ⓐ Ⓑ Ⓒ Ⓓ Ⓔ
75. Ⓐ Ⓑ Ⓒ Ⓓ Ⓔ
76. Ⓐ Ⓑ Ⓒ Ⓓ Ⓔ
77. Ⓐ Ⓑ Ⓒ Ⓓ Ⓔ
78. Ⓐ Ⓑ Ⓒ Ⓓ Ⓔ
79. Ⓐ Ⓑ Ⓒ Ⓓ Ⓔ
80. Ⓐ Ⓑ Ⓒ Ⓓ Ⓔ
81. Ⓐ Ⓑ Ⓒ Ⓓ Ⓔ
82. Ⓐ Ⓑ Ⓒ Ⓓ Ⓔ
83. Ⓐ Ⓑ Ⓒ Ⓓ Ⓔ
84. Ⓐ Ⓑ Ⓒ Ⓓ Ⓔ
85. Ⓐ Ⓑ Ⓒ Ⓓ Ⓔ
86. Ⓐ Ⓑ Ⓒ Ⓓ Ⓔ
87. Ⓐ Ⓑ Ⓒ Ⓓ Ⓔ
88. Ⓐ Ⓑ Ⓒ Ⓓ Ⓔ
89. Ⓐ Ⓑ Ⓒ Ⓓ Ⓔ
90. Ⓐ Ⓑ Ⓒ Ⓓ Ⓔ
91. Ⓐ Ⓑ Ⓒ Ⓓ Ⓔ
92. Ⓐ Ⓑ Ⓒ Ⓓ Ⓔ
93. Ⓐ Ⓑ Ⓒ Ⓓ Ⓔ
94. Ⓐ Ⓑ Ⓒ Ⓓ Ⓔ
95. Ⓐ Ⓑ Ⓒ Ⓓ Ⓔ
96. Ⓐ Ⓑ Ⓒ Ⓓ Ⓔ
97. Ⓐ Ⓑ Ⓒ Ⓓ Ⓔ
98. Ⓐ Ⓑ Ⓒ Ⓓ Ⓔ
99. Ⓐ Ⓑ Ⓒ Ⓓ Ⓔ
100. Ⓐ Ⓑ Ⓒ Ⓓ Ⓔ

SECTION II

Essay Question 1

answer sheet

Essay Question 2

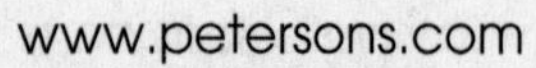

answer sheet

Essay Question 3

answer sheet

Essay Question 4

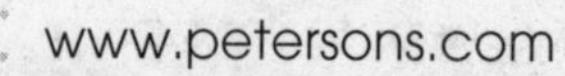

answer sheet

Practice Test 2

SECTION I

100 QUESTIONS • 80 MINUTES

Directions: Each of the questions or incomplete statements below is followed by five suggested answers or completions. Select the one that is best in each case and mark the corresponding circle on the answer sheet.

1. Organisms in which of the following groups have a coelom, closed circulatory system, notochord, and a muscular tail posterior to the anus?

(A) Mollusk
(B) Arthropod
(C) Annelid
(D) Chordate
(E) Echinoderm

2. How many valence electrons does carbon have?

(A) 0
(B) 1
(C) 2
(D) 4
(E) 8

3. Which of the following amino acids would have an overall negative charge at pH 7?

(A) Lysine
(B) Glycine
(C) Glutamate
(D) Alanine
(E) Tyrosine

4. If the concentration of hydroxide ions (OH^-) in a solution of water is 0.01 M, what is the pH of the solution?

(A) 1.99
(B) 2.00
(C) 9.99
(D) 10.00
(E) 12.00

5. An enzyme that hydrolyzes DNA has optimal activity at 37°C. If the enzyme is heated for 15 minutes to 80°C at neutral pH, and then cooled again to 37°C, no enzyme activity is detected. The reason for the lack of enzyme activity after cooling is

   (A) the structure of the enzyme's active site was disrupted
   (B) the primary structure of the enzyme was denatured
   (C) DNA is denatured at high temperatures like 80°C
   (D) the enzyme requires a vitamin-based cofactor that was absent
   (E) the enzyme does not function well at low temperatures due to reduced kinetic activity of substrate molecules

6. Which of the following best describes the path taken by newly synthesized liver transmembrane proteins acting as ion channels in the plasma membrane?

   (A) ER to Golgi to lysosome to plasma membrane
   (B) Cytoplasm to ER to Golgi to plasma membrane
   (C) ER to Golgi to plasma membrane
   (D) Golgi to ER to lysosome to plasma membrane
   (E) Nucleus to Golgi to ER to plasma membrane

7. Which of the following do animal cells have that plant cells lack?

   (A) Mitochondria
   (B) Chloroplasts
   (C) Centrioles
   (D) Rough endoplasmic reticulum
   (E) Plasma membrane

8. The following chemical reaction has a negative ΔG as written:

   $A + B \leftrightarrows C + D$

   Which of the following statements best describes this reaction?

   (A) The reaction will spontaneously move forward toward the production of C and D.
   (B) ATP hydrolysis will be required for this reaction to occur.
   (C) A and B will quickly be converted to C and D in solution.
   (D) The reaction is at equilibrium.
   (E) The reaction is endergonic and will absorb heat from its surroundings.

9. When a paramecium consumes yeast in a sealed tube and digests the yeast cells, which of the following can be said of the overall system, including the paramecium, yeast cells, and their culture medium?

   (A) Overall energy decreases and entropy stays constant
   (B) Overall energy increases and entropy is reduced
   (C) Overall energy stays the same and entropy increases
   (D) Overall energy decreases and entropy increases
   (E) Overall energy stays the same and entropy stays the same

10. In cellular respiration in mitochondria, what is the role of NADH?

   (A) Serve as the final electron acceptor
   (B) Oxidize ATP to synthesize high energy phosphate bonds
   (C) Reduce pyruvate to form acetyl-CoA
   (D) Transfer energy to the electron transport chain to drive the creation of a pH gradient
   (E) Carry electrons across the inner mitochondrial membrane

11. In which of the following would photorespiration be the greatest?

(A) C3 plants on a hot day
(B) C3 plants on a cold day
(C) CAM plants on a cold day
(D) C4 plants on a hot day
(E) Photosynthetic bacteria

12. At what stage of the mitotic cell cycle does the nuclear envelope break down?

(A) Interphase
(B) Prophase
(C) Anaphase
(D) Telophase
(E) S phase

13. In a cross between two heterozygotes for a gene, both parents carry a copy of a recessive allele and of a dominant allele. The gene resides on an autosomal chromosome. What percentage of offspring will express the recessive phenotype?

(A) 0%
(B) 25%
(C) 50%
(D) 75%
(E) 100%

14. The role of DNA polymerase I in DNA replication is which of the following?

(A) Make DNA single stranded so it is available as a template for DNA synthesis
(B) Extend DNA synthesis from RNA primers
(C) Relief of torsional stress
(D) DNA synthesis on the leading strand
(E) Replace RNA primer with DNA

15. In capillaries, which of the following are able to move from the blood into the extracellular fluid in noninflamed tissues?

(A) Immunoglobulins
(B) Red blood cells
(C) Glucose
(D) Platelets
(E) Serum albumin protein

16. When antigen binds to antigen receptor on a B cell, it causes which of the following to occur most directly?

(A) Proliferation of this B cell and differentiation into plasma B cells and memory B cells
(B) Recombination of B cell immunoglobulin genes to assemble V, D, and J segments
(C) Interaction of the B cell receptor with an MHC complex
(D) Fragmentation of the B cell genome and apoptosis
(E) Destruction of cells infected by a virus expressing this antigen

17. The movement of oxygen out of blood occurs in

(A) bronchioles
(B) alveoli
(C) veinules
(D) capillaries
(E) arterioles

18. A hormone released by the posterior pituitary is

(A) oxytocin
(B) prolactin
(C) epinephrine
(D) growth hormone
(E) gonadotropin releasing hormone

19. A comparison of the evolutionary relationship among bacteria, eukaryotes, and archae would best be focused on

(A) comparison of mitochondrial DNA sequences
(B) comparison of introns
(C) comparison of enzymes involved in glycolysis
(D) comparison of ribosomal RNA sequences
(E) comparison of histone genes

20. Membrane depolarization during an action potential along an axon is associated with which of the following?

(A) Ligand-gated sodium channels
(B) Voltage-gated potassium channels
(C) The sodium-potassium pump
(D) Voltage-gated sodium channels
(E) Potassium leak channels

21. The sympathetic nervous system stimulates which of the following to occur?

(A) Release of parathyroid hormone
(B) Secretion of epinephrine by the adrenal medulla
(C) Skeletal muscle contraction
(D) Constriction of bronchioles
(E) Slowing of heart rate

22. Bacterial conjugation involves which of the following?

(A) Uptake of naked DNA by bacteria
(B) Transfer of bacterial genes packaged in viruses
(C) Transfer of plasmid with F factor from one bacterium to another
(D) Replication of plasmid with F factor and transfer of copy to an F- bacterium
(E) Sexual reproduction in bacteria between male and female strains

23. How is ATP produced in photosynthesis?

(A) Excitation of electrons in photosystem I creates phosphodiester bonds.
(B) A pH gradient created by an electron transport chain drives ATP synthesis.
(C) The final electron acceptor of Photosystem II transfers a phosphate to ADP.
(D) NADPH transfers a phosphate to ADP.
(E) In cyclic electron flow, energy from photosystem I is diverted through an electron transport chain that transfers high energy electrons to ATP.

24. In epigenetic inheritance

(A) mutations in the genome accumulate with reduced DNA repair activity
(B) DNA replication errors cause codons to be repeated in genes
(C) chromatin structure affects inheritance of traits from mother to offspring
(D) chromosomes do not distribute normally during meiosis
(E) the expression of one gene is affected by a second gene

25. The lac operon in bacteria includes three genes that are expressed in the presence of lactose. Which of the following best describes the mechanism regulating the expression of these genes?

(A) A transcriptional activator binds lactose to recruit RNA polymerase to the promoters of the genes at three different locations in the bacterial genome.
(B) The genes are transcribed in all conditions, but translation of the genes only occurs in the presence of lactose when the messenger RNA is exported from the nucleus.
(C) Lactose binds to RNA polymerase, increasing its transcriptional activity and resulting in increased production of the proteins involved in lactose metabolism.
(D) RNA stability is increased by polyA addition in the presence of lactose.
(E) A transcriptional repressor blocks transcription in the absence of lactose. In the presence of lactose, the repressor does not bind the promoter and all three genes are transcribed in one RNA.

26. The difference between grey matter and white matter in the human central nervous system is that

(A) grey matter is better perfused with capillaries
(B) grey matter does not perform oxidative phosphorylation
(C) grey matter contains cells that are being replaced by neuronal stem cells
(D) grey matter contains little or no myelin
(E) grey matter produces much faster action potentials

27. The effect of mycorrhizal fungi in their interaction with plants is

(A) to perform nitrogen fixation for plants
(B) to increase the availability of soil nutrients for the plant
(C) to provide photosynthetic pigments
(D) to increase the availability of carbon dioxide for the plant
(E) to provide glucose to the plant to support plant respiration

28. Which of the following statements regarding DNA and RNA is false?

(A) DNA and RNA both contain nucleotides joined by phosphodiester bonds.
(B) DNA contains deoxyribose while RNA contains ribose.
(C) DNA and RNA are both synthesized in a 5' to 3' direction.
(D) DNA and RNA are both synthesized using an existing DNA template.
(E) DNA contains uracil while RNA contains thymine.

29. If a plant cell with an intact cell wall is exposed to a hypertonic solution, what will happen to the plant cell?

(A) The plant cell will shrink away from the cell wall.
(B) Salts will be transported across the plasma membrane to equilibrate with the external environment.
(C) The plant cell will swell, pressing up against the cell wall and creating pressure against the cell wall.
(D) The plant cell will burst the cell wall.
(E) Cell walls do not allow the exchange of salts or water in and out of plant cells so there will be no effect.

30. The function of complement proteins found in blood is which of the following?

(A) To cleave fibrinogen during clot formation
(B) To carry hormones in the blood, transporting them to target tissues
(C) To prevent the immune system from reacting against self antigens
(D) To form pores in the membrane of infectious cells
(E) To provide missing fragments of protease enzymes secreted from the pancreas, activating the proteases.

31. The HIV retrovirus includes all of the following in its life cycle except

(A) integration of viral DNA in the genome of host cell
(B) copying of viral RNA into double-stranded DNA
(C) viral ribosomes produce reverse transcriptase
(D) budding of virus from cell with membrane envelope
(E) replication of viral RNA genome

32. In translation, each codon in mRNA is matched with the corresponding amino acid in the genetic code by which mechanism?

(A) Ribosomal RNA recognizes each codon in mRNA and initiates transfer of the correct amino acid to the polypeptide chain.
(B) tRNA molecules form matching base pairs with each codon to align the correct amino acid.
(C) mRNA molecules fold into a tertiary structure, which indicates the position of the amino acid in the polypeptide structure.
(D) Ribosomes align mRNA molecules with each codon in a gene to minimize errors in transcription.
(E) Amino-acyl tRNA synthetase enzymes bind to each codon in mRNA to ensure the accuracy of translation.

QUESTIONS 33–35 REFER TO THE FOLLOWING FIGURE.

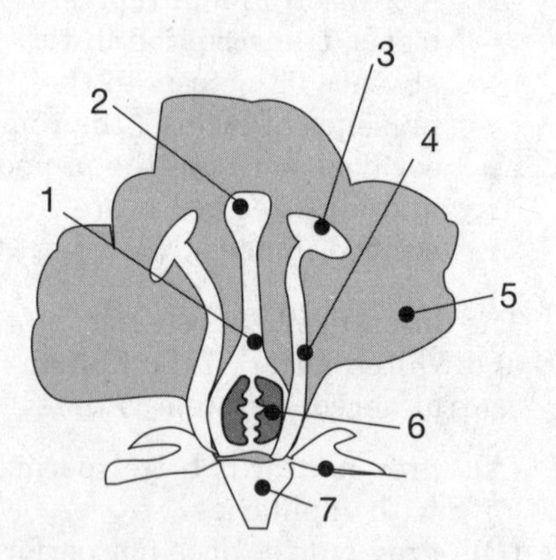

33. In the flower structure, where are male gametophytes produced?

(A) #1
(B) #2
(C) #3
(D) #4
(E) #6

34. Which of the structures in this figure is the stigma?

(A) #1
(B) #2
(C) #3
(D) #4
(E) #6

35. Where does fertilization take place?

(A) #1
(B) #2
(C) #3
(D) #4
(E) #6

36. Which of the following is not a mechanism used by humans to adapt to cold climates in winter?

(A) Release of thyroid hormone to increase the metabolic rate
(B) Reduction of the temperature set point to consume less metabolic energy
(C) Constriction of arterioles leading to skin
(D) Shivering
(E) Wearing clothing that conserves heat

37. The large intestine plays the greatest role in which of the following?

(A) Water reabsorption
(B) Fat absorption
(C) Acid neutralization
(D) Protein hydrolysis
(E) Lipid absorption

38. The initial stages of spermatogenesis in humans occur in which of the following?

(A) Vas deferens
(B) Urethra
(C) Epididymus
(D) Seminiferous tubules
(E) Spermatogonium

39. How does the energetic yield of fermentation compare to oxidative phosphorylation?

(A) Fermentation does not occur in the absence of oxidative phosphorylation so they cannot be compared.
(B) Oxidative phosphorylation always produces the same energetic yield whether fermentation occurs or not.
(C) Fermentation occurs more rapidly than oxidative phosphorylation so it has a higher yield.
(D) Fermentation has a lower energy yield and does not produce any high energy electron carriers to support the creation of a mitochondrial pH gradient.
(E) When oxygen is present, fermentation produces a higher energy yield than oxidative phosphorylation.

40. The exoskeleton of aquatic arthropods allows all of the following functions except

(A) flexibility of joints in appendages
(B) transport of respiratory gases across chitin
(C) protection of tissues from predation
(D) internal attachment sites for muscle
(E) support of body against gravity

41. The main product of the Calvin cycle in C4 plants is which of the following?

(A) Organic acids
(B) Oxygen
(C) ATP
(D) Carbohydrates
(E) NADPH

$$\text{pyruvate} + NAD^+ + \text{Coenzyme A} \leftrightarrows \text{Acetyl-CoA} + NADH + H^+ + CO_2$$

42. In the above forward reaction, which molecule is the reducing agent?

(A) Pyruvate
(B) NAD^+
(C) Coenzyme A
(D) Acetyl-CoA
(E) Carbon dioxide

43. A nondividing cell from somatic tissues in a buffalo contains sixty chromosomes. A spermatogonium cell from a buffalo in G2 phase of the cell cycle contains how many chromosomes?

(A) 15
(B) 30
(C) 60
(D) 120
(E) 240

44. A flower species with six chromosomes has a gene controlling flower color on chromosome 1 and a gene controlling leaf size on chromosome 3. There are two alleles for the flower color gene: a recessive allele for white, and a dominant allele for blue. There are also two alleles for leaf size: a dominant allele for long and a recessive allele for short. If two plants that are heterozygous for both alleles are crossed, what fraction of their offspring will have white flowers and long leaves?

(A) $\frac{1}{16}$
(B) $\frac{1}{8}$
(C) $\frac{1}{4}$
(D) $\frac{1}{2}$
(E) $\frac{9}{16}$

QUESTIONS 45–46 REFER TO THIS FIGURE:

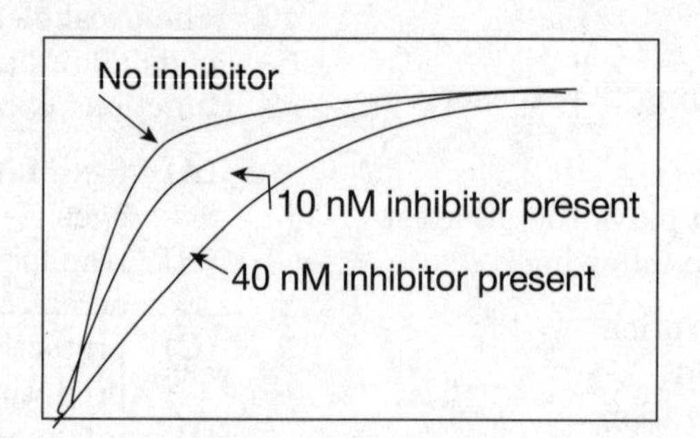

45. What type of inhibitor is tested in this experiment?

(A) Noncompetitive
(B) Un-competitive
(C) Competitive
(D) Irreversible
(E) Allosteric

46. Which of the following best describes the interaction of the inhibitor with the enzyme?

(A) The inhibitor covalently modifies polar amino acid side chains exposed on the enzymes's surface.
(B) The inhibitor disrupts the secondary structure of the enzyme protein.
(C) The inhibitor binds at the enzyme's active site.
(D) The inhibitor is reversible but does not interact with enzyme at the active site.
(E) The enzyme displays cooperative binding of substrate that is disrupted by the inhibitor.

47. Compared to gram negative bacteria, gram positive bacteria generally have which of the following traits?

(A) More prone to releasing large quantities of endotoxin, lipopolysaccharide into the blood of infected animals
(B) Much greater amount of DNA in a large linear genome, increasing their sensitivity to staining
(C) Reduced sensitivity to antibiotic related to penicillin
(D) Greater content of peptidoglycan in their cell wall
(E) Reduced reliance on glycolysis for energy production

48. The stomata are located where in plants?

(A) The liquid inside chloroplasts
(B) Within the thylakoid membrane
(C) In the epithelial layer of cells in leaves
(D) Connecting sieve-tube members in the phloem
(E) In the inner chloroplast membrane

49. The human fetus is able to acquire sufficient oxygen through the placenta because of which of the following?

(A) Active transport of oxygen against a pressure gradient
(B) Chemical synpases between the maternal and fetal circulation
(C) Movement of red blood cells from the maternal to the fetal circulatory system
(D) Supplementation of oxygen supply from gases dissolved in amniotic fluid
(E) Expression of fetal hemoglobin genes that have higher affinity for oxygen than maternal hemoglobin

50. In the metabolism of fructose to generate energy, how many acetyl-CoA molecules are produced for every fructose that enters catabolic pathways?

(A) 0
(B) 1
(C) 2
(D) 3
(E) 4

51. A key difference between the digestion of triglycerides and proteins is that

(A) after absorption, triglycerides are transported in the lymph
(B) the pancreas is not involved in the digestion of triglycerides
(C) the majority of hydrolysis of proteins occurs in the small intestine while lipids are hydrolyzed mainly in the stomach
(D) triglycerides are very hydrophobic so they are not subject to hydrolysis before absorption
(E) triglycerides are absorbed directly in the intestine without hydrolysis first

52. When a ribosome reaches a stop codon in an mRNA, which of the following occurs?

(A) A tRNA binds to the stop codon, transferring methionine to the C-terminus.
(B) A tRNA binds that does not have an amino acid bound, hydrolyzing the peptide from the neighboring tRNA on the ribosome.
(C) A release factor binds to the stop codon, causing translation termination and release of the peptide and mRNA from the ribosome.
(D) The ribosome cleaves the mRNA, releasing it from the ribosome.
(E) A polyA tail is added to the mRNA.

```
  H     O            H     O
   \   //             \   //
     C                  C
     |                  |
OH---C---H          H---C---OH
     |                  |
 H---C---OH         H---C---OH
     |                  |
 H---C---OH         H---C---OH
     |                  |
     H                  H
```

53. The two carbohydrates depicted have what relationship to each other?

(A) Structural isomers
(B) Diastereomers
(C) Anomers
(D) Enantiomers
(E) None of the above

54. A wooden handle from a possible campsite of prehistoric humans is subjected to determination of its C^{14} content. The half life of C^{14} is about 5,700 years. The sample contains 12% as much C^{14} as modern samples of wood. Assuming that atmospheric C^{14} carbon has been constant over the time period examined, what is the approximate age of the wooden handle?

(A) 600 years
(B) 2,800 years
(C) 5,700 years
(D) 11,000 years
(E) 17,000 years

55. Which of the following traits arose earliest in chordate evolution?

(A) Amniotic egg
(B) Placental nutrition of embryos
(C) Hollow dorsal nerve cord
(D) Lungs for terrestrial respiration
(E) Three-chambered heart with closed circulatory system

56. The functions of human chorionic gonadotropin include

(A) prevention of regression of the corpus luteum after fertilization
(B) stimulation of breast development
(C) triggers ovulation
(D) stimulation of regression of the endometrium during menstruation
(E) release of hormone from hypothalamus for gonadotropins

57. One of the differences between myelinated neurons and non-myelinated neurons is

(A) only non-myelinated neurons are found outside of the vertebrate brain
(B) non-myelinated neurons use acetylcholine as a neurotransmitter while myelinated neurons use other neurotransmitters
(C) action potentials travel more rapidly in myelinated axons
(D) myelinated neurons propagate action potentials using voltage-gated potassium channels instead of sodium channels
(E) myelinated neurons form electrical synapses with other cells

58. Some seeds need to soak in water if germination is to occur, mimicking the effects of spring rain. What substance is removed by the water?

(A) Gibberelin
(B) Starch
(C) Cotyledon
(D) Cytokinin
(E) Abscisic acid

59. In the catabolism of fatty acids, where does carbon from fatty acids with an even number of carbons enter other metabolic pathways?

(A) Fatty acids are converted into glucose, which is metabolized through glycolysis and oxidative phosphorylation.
(B) Fatty acids are oxidized and split to produce acetyl-CoA that enters the Krebs cycle.
(C) Fatty acids produce triglycerides that are broken into glycerol and monosaccharides.
(D) Fatty acids produce succinate that enters the Krebs cycle.
(E) Fatty acid metabolites enter the pentose phosphate pathway.

60. Which of the following is produced from ectoderm in chordates?

(A) Heart
(B) Intestinal epithelium
(C) Bone
(D) Arterial smooth muscle
(E) Cerebellum

61. In the polymerase chain reaction, the highest temperatures used in each cycle of the reaction are used to accomplish which of the following?

(A) Annealing of primers to single stranded DNA
(B) Breaking hydrogen bonds between strands in double-stranded DNA
(C) Stimulate activity of DNA polymerase
(D) Allow DNA polymerase to replicate DNA without starting from a primer
(E) Inhibition of lagging strand synthesis

62. Where does blood from the lungs return to the heart?

(A) Right atrium
(B) Left atrium
(C) Right ventricle
(D) Left ventricle
(E) Aorta

63. In a deep sea ocean vent community, which of the following is the primary producer?

(A) Photoautotrophic cyanobacteria
(B) Chemoautotrophic prokaryotes
(C) Chemoautotrophic zooplankton
(D) Scavengers involved in decomposition of detritus from surface water
(E) Chemoheterotrophic nematodes

64. Cellulose is largely indigestible by humans because

(A) it is a large polysaccharide so it cannot be absorbed
(B) humans do not have bacteria in their digestive tract
(C) humans do not produce enzymes that can hydrolyze beta-glycosidic linkages between glucose monomers in a polysaccharide
(D) it contains phosphorous-based cross linkages between carbohydrate groups
(E) humans do not eat foods containing cellulose

65. The different responses of animals to testosterone and estrogen are due to

(A) the vastly different physical nature of these two hormones
(B) the binding of testosterone primarily to receptors on the cell surface while estrogen binds mainly to receptors located in the cellular interior
(C) estrogen regulates transcription while testosterone regulates translation
(D) binding to different receptors that regulate transcription of different genes in different tissues
(E) estrogen triggers production of the second messenger cAMP while testosterone triggers intracellular calcium release

66. Disulfide bonds involve which of the following?

(A) Methionine
(B) Cysteine
(C) Glutathione
(D) Immunglobulins
(E) Alpha helices

67. In an organism from phylum porifera, nutrition occurs in which of the following manners?

(A) Digestive enzymes are secreted into a gastrovascular cavity from which nutrients are absorbed.
(B) Digestive enzymes are secreted externally, and hydrolyzed molecules are then absorbed.
(C) Food particles are internalized by cells and hydrolyzed in the cellular interior.
(D) Food moves through a complete digestive tract from mouth to anus, with material hydrolyzed and absorbed primarily in an intestine.
(E) Photosynthesis is the primary source of nutrition.

68. In non-cyclic electron flow, which of the following is the ultimate destination of photo-excited electrons?

(A) ATP
(B) NADPH
(C) Oxygen (O_2)
(D) Cytochrome c
(E) Glucose

QUESTIONS 69–70 REFER TO THE FOLLOWING FIGURE:

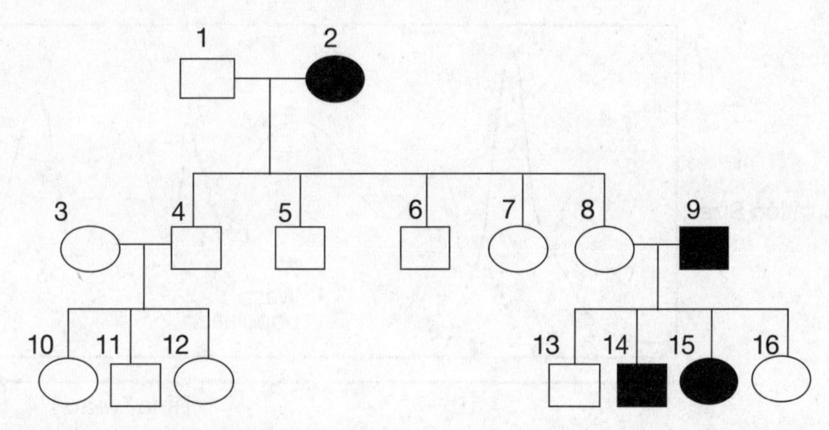

69. This pedigree reflects a trait causing pointed toe-nails in affected individuals. There are two alleles in the group of individuals shown in this pedigree, as determined by genetic analysis. What is the mechanism for the expression of the allele associated with this trait?

(A) Recessive autosomal
(B) Maternal inheritance
(C) Incomplete dominance
(D) Co-dominance
(E) Dominant sex-linked

70. What is the probable phenotype of individual 8 in the pedigree?

(A) Homozygous recessive
(B) Heterozygous
(C) Homozygous dominant
(D) Polygenic
(E) Not possible to determine

71. Which of the following is most active at an acidic pH?

(A) Pancreatic lipase
(B) Pepsin
(C) Trypsin
(D) Chymotrypsin
(E) Amylase

72. Blood pressure is greatest in which of the following?

(A) Systolic in the aorta
(B) Diastolic in the pulmonary artery
(C) Systolic in peripheral arterioles
(D) Diastolic in the superior vena cava
(E) Capillaries in the lower extremities

73. A region of marine upwelling along a continental shelf may affect an ecosystem in which of the following ways?

(A) Reduced oxygen content of water
(B) Increased acidity of water
(C) Reduced biodiversity
(D) Increased gross primary productivity
(E) Reduced zooplankton activity

QUESTIONS 74–75 REFER TO THE FOLLOWING FIGURE:

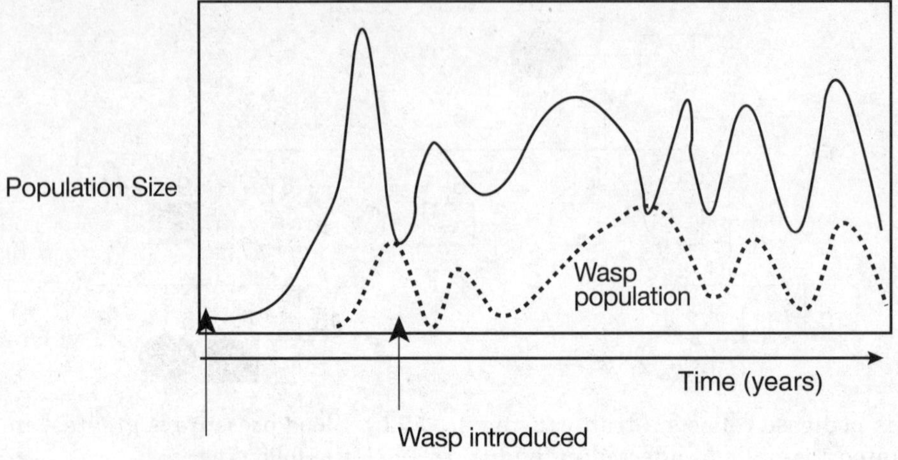

74. On an island, a caterpillar (species #1) was introduced by farmers to control a weed population. However, the caterpillar population began to eat crop plants as well as weeds. In a later year a wasp species from another region was also introduced to the island. The size of each population was tracked over time, producing the data in this graph. Which of the following statements is best supported by these data?

(A) The wasp drives directional selection of the caterpillar population.
(B) The caterpillar population follows the logistic model for population growth.
(C) The climate has a big impact on the carrying capacity of the environment for the caterpillar population.
(D) The wasp and caterpillar populations follow similar cycles, but it is not possible to state the cause of this based on the data given.
(E) Wasps provide effective control of the caterpillar population.

75. To test the relationship of the wasps and caterpillar populations to each other, a biologist studied the two and found that the wasp did not kill or prey on the caterpillar species #1 directly. The caterpillar #1 species eats a weed plant that is abundant on the island, and when the caterpillar population increases, more of the weeds are eaten, clearing space for another grassy plant that is eaten by caterpillar species #2. This second caterpillar species is the prey of the wasp. To test the relationship of the species to each other, the biologist enclosed a small space of land with a net and excluded wasps but left the caterpillar species inside the net. The population of caterpillar species #1 inside the net fluctuated in the same manner as outside the net. Which of the following best describes the relationship of the wasps and caterpillar species #1?

(A) Parasitic
(B) Predation
(C) Mutualism
(D) Commensalism
(E) Competition

76. Which of the following reflects the net reaction for photosynthesis in CAM plants?

(A) $C_2H_4O_2 + 4CO_2 + light \leftrightarrows C_6H_{12}O_6 + O_2$
(B) $C_2H_4O_2 + CO_2 + H_2O + light \leftrightarrows C_6H_{12}O_6 + O_2$
(C) $6\ H_2O + 6\ CO_2 + light \leftrightarrows C_6H_{12}O_6 + 6\ O_2$
(D) $3\ H_2O + 6\ CO_2 + light \leftrightarrows C_6H_{12}O_6 + 3\ O_2$
(E) $4\ H_2O + 4\ CO_2 + light \leftrightarrows 4\ C_6H_{12}O_6 + 4\ O_2$

77. Oxidative phosphorylation of glucose produces water. Where does the oxygen in the water come from?

(A) Glucose
(B) Atmospheric molecular oxygen
(C) Carbon dioxide
(D) NADH
(E) Glycolysis

78. The ureter does which of the following?

(A) Carries urine from bladder to exterior
(B) Carries urine from kidney to bladder
(C) Sperm maturation
(D) Carries blood into the medulla of the kidney
(E) Carries initial urinary filtrate in the nephron

79. A pH of 5 indicates that the hydrogen ion concentration is which of the following?

(A) 5 micromolar
(B) 5 millimolar
(C) 5×10^{-2} molar
(D) 1×10^{-5} molar
(E) 5×10^{-5} molar

80. Rod cells do which of the following in the eye?

(A) Provide pigment in the iris
(B) Visual acuity for reading using the fovea
(C) Accurate detection of color in dimly lit conditions
(D) Detection of color in bright conditions
(E) Sensitive detection of light in a variety of light intensities

81. A buffer in blood does which of the following?

(A) Prevents large pH changes in different metabolic conditions
(B) Carries carbon dioxide
(C) Converts lactate into pyruvate
(D) Provides part of the innate immune defense against infection
(E) Prevents blood loss from a wound

82. The lymphatic system does which of the following?

(A) Red blood cell maturation
(B) Carries hormones from pancreas
(C) Returns extracellular fluids to blood
(D) Produces antibodies
(E) Carries digested carbohydrates from the intestine to the blood

83. Strenuous exertion in muscle can sometimes lead to an accumulation of lactate. Where does the lactate come from?

(A) Fatty acid beta oxidation
(B) Fermentation of glucose in anaerobic conditions
(C) Gluconeogenesis
(D) Breakdown of glycogen
(E) Krebs cycle

84. Chymotrypsin does which of the following?

(A) Cleavage of proteins next to hydrophobic amino acids
(B) Cleavage of proteins next to positively charged amino acids
(C) Cleavage of disulfide bridges
(D) Cleavage of clotting factors to close wounds
(E) Dissociation of proteins from fats

85. Which of the following is generally most closely associated with higher species diversity in terrestrial ecosystems?

(A) Colder climates
(B) Smaller habitat sizes
(C) Higher primary productivity
(D) Removal of keystone species
(E) Disruption of an ecosystem by an invasive alien species

86. Fungi contribute to the carbon cycle in which of the following ways?

(A) Releasing CO_2 through absorption and respiration
(B) Absorbing CO_2 from decomposing organisms
(C) Increased transpiration
(D) Sequestering carbon into biologically unavailable forms
(E) Producing carbohydrates from CO_2

87. A pseudogene can be identified as

(A) encoding a reverse transcriptase
(B) similar to other genes in a cluster, but with premature stop codon
(C) transposition from one place to another in the genome
(D) modification of DNA with methylation
(E) using a different genetic code

88. In a population of butterflies with a range of sizes, those in the middle size range in wing span are preyed on by sparrows, while larger and smaller butterflies are not. The effect on the butterfly population is

(A) allopatric speciation
(B) disruptive selection
(C) directional selection
(D) divergent evolution
(E) punctuated equilibrium

89. The toxin cyanide binds to cytochrome C oxidase in the mitochondrial electron transport chain in the inner mitochondrial membrane, preventing the transport of high energy electrons through the system. This protein is the last carrier in the electron transport system. Which of the following would be observed in cells treated with cyanide and provided with glucose and oxygen?

(A) Increased oxygen consumption
(B) Increased pH gradient
(C) Increased carbon dioxide production
(D) A complete loss of ATP production
(E) The Krebs cycle will stop producing NADH

90. Insect antennae are involved in which of the following?

(A) Echolocation
(B) Chemosensation
(C) Photoreceptors
(D) Feeding
(E) Balance

91. Which of the following is different between an atom of C^{13} and an atom of C^{14}?

(A) Protons
(B) Electrons
(C) Neutrons
(D) Number of valence electrons
(E) Atomic number

QUESTION 92 REFERS TO THE FOLLOWING REACTIONS:

#1 $X + C \leftrightarrows Y + B$ $\Delta G = 5$ kcal/mol

#2 $ATP + H_20 \leftrightarrows ADP + P_i$ $\Delta G = -7.3$ kcal/mol

#3 $X + C + ATP + H_20 \leftrightarrows Y + B + ADP + P_i$

92. For the reaction #1 shown, the ΔG is 5 kcal/mol

For the coupled overall reaction described, the ΔG is −2.5. The effect of coupling this reaction to ATP hydrolysis is to

(A) change the equilibrium
(B) reduce the activation energy
(C) eliminate the need for an enzyme
(D) release more energy from X
(E) produce more Y than B

93. The reason enzymes are selective for specific substrates is

(A) substrates bind to many different locations on each enzyme
(B) the shape of the substrate must fit precisely in the enzyme's active site
(C) having the wrong substrate bind will denature an enzyme
(D) the substrate determines the primary structure of enzymes
(E) enzymes are very sensitive to changes in temperature and pH

94. Bacterial ribosomes are distinct from eukaryotic ribosomes because

(A) bacterial ribosomes do not bind to DNA
(B) bacterial ribosomes cannot perform splicing
(C) bacterial ribosomes are made of two subunits
(D) bacterial ribosomes are assembled in the nucleus
(E) bacterial ribosomes can translate RNA while it is still being transcribed

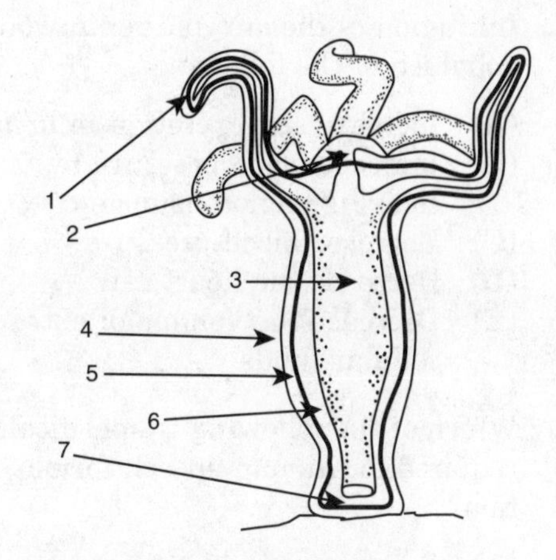

95. Item #3 is called the

(A) mesoderm
(B) blastocoel
(C) coelom
(D) gastrovascular cavity
(E) protostome

96. A triploid banana does not produce seeds because

(A) meiosis cannot produce gametes
(B) mitosis does not occur normally
(C) triploid fruits are prone to errors in DNA replication
(D) post-zygotic barriers to fertilization prevent seed formation
(E) it lacks genes required for seed formation

97. Ethylene does which of the following in plants?

(A) Cell elongation
(B) Plant response to light
(C) Protection against insects
(D) Fruit growth
(E) Induces fruit ripening

98. The human diaphragm is made of

(A) skeletal muscle
(B) smooth muscle
(C) cartilage
(D) mesenteric membrane
(E) respiratory tissue

99. Inhibition of the enzyme renin would do which of the following?

(A) Increase salt excretion in urine
(B) Increase blood pressure by cleaving angiotensinogen
(C) Decrease blood pressure
(D) Decrease the heart rate
(E) Decrease the volume of extracellular fluids

100. Which of the following is specifically required for human speech formation?

(A) Broca's area
(B) Amygdala
(C) Limbic system
(D) Reticular formation
(E) The creation of speech is too complex to associate with one brain region.

SECTION II

4 QUESTIONS • 90 MINUTES

Four mandatory questions will be asked. In general, there will be two on molecules and cells, one on genetics and evolution, and one on organisms and populations. Some questions may cover more than one of these areas. One or more of the four questions may be designed to test analytical and reasoning skills. Laboratory experiences may be reflected in these questions.

Essay Question 1

	Population 1			Population 2		
	Allele H Frequency	**Allele h Frequency**	**Pop 1 Size**	**Allele H Frequency**	**Allele h Frequency**	**Pop 2 Size**
Year 1	.7	.3	5000			
Year 2	.63	.37	5100			
Year 3	.6	.4	5000			
Year 4	.65	.35	4900			
Year 5	.7	.3	5000			
Year 6	.7	.3	4900			
Year 7	.7	.3	4800			
Year 8	.68	.32	4500	.4	.6	100
Year 9	.7	.3	4800	.3	.7	250
Year 10	.7	.3	4500	.25	.75	300
Year 20	.7	.3	4500	0	1	300

A researcher is studying the genetics of a diploid population of newts living in an oligotrophic freshwater lake ecosystem. The first time he examines the population (Population 1 in the table above), he finds two different alleles of an enzyme involved in the production of bile acids. The alleles are H and h, where H is a dominant allele and h is a recessive allele associated with reduced bile acid production in homozygotes. Heterozygotes have a phenotype that is the same as homozygous dominant individuals. He examines the population yearly and finds the allele frequencies and population sizes shown in the table. In Years 6 and 7, the water level of the lake falls, splitting the lake in two, with Population 1 in a larger section, and Population 2 in a smaller section. The two lakes remain separated from each other in subsequent years.

(A) How many newts in Population 1 will have normal bile acid production phenotype in Year 7, if the population displays random mating and a sufficiently large population size to avoid random drift?

(B) Which of the following provides the best explanation for the difference in allele frequency between Populations 1 and 2 in Year 8? Why?
- Random drift
- Mutation
- Non-random mating

(C) In Year 20, how can it be determined whether speciation has occurred?

Essay Question 2

Angiosperms have adapted to a broad array of environments, including dry terrestrial environments.

(A) Describe three angiosperm structural adaptations that increase fitness in a dry climate. For each of these adaptations, describe the structural change and how the adaptation improves fitness for the plant.

(B) For either CAM or C4 plants, describe the alternative mechanism for carbon fixation and how this mechanism provides a selective advantage.

Essay Question 3

Describe the skeletons found in chordates, arthropods, and annelids. Describe the elements that provide strength and support, how each element enables movement, and the limitations the skeleton might impose on animals.

Essay Question 4

A single base pair mutation is created in an essential bacteriophage gene, creating a stop codon in the fifth codon of the viral capsid gene encoding a protein that is 100 amino acids. This virus is called Mut5. When the Mut5 bacteriophage genome was packaged into viruses and used to infect bacteria, no infectious viral particles were produced in the bacterial media. The wild-type bacteriophage genome, however, produced 1 x 10^8 viruses/ml of bacterial media when infecting the same bacterial cells. The bacterial strain used in the first experiment was then irradiated; the surviving bacteria were grown in media, and this new strain was tested for its ability to support growth of the wild-type virus. With the wild-type bacteriophage, the same number of viruses were produced in the wild-type and the irradiated bacteria. In the mutant virus, while no viruses were produced in the wild-type bacteria, 2 x 10^7 viruses/ml were produced by infection in media containing the strain derived from irradiated bacteria.

Infectious Viruses Produced by Bacterial Infection

	Wild-Type virus	Mut5 Virus
Wild-type bacteria	1×10^8/ml	0 /ml
Irradiated bacteria	1×10^8/ml	2×10^7/ml

(A) Discuss a mechanism by which the strain of irradiated bacteria can support growth of the Mut5 bacteriophage, while the wild type bacteria do not support growth of this virus.

(B) How might translation in the irradiated bacterial strain be affected more broadly in the absence of bacteriophage?

(C) If wild-type bacteria are infected with both wild-type virus and Mut5 virus at the same time, will any infectious viruses be produced that can reinfect the same wild-type bacterial cells? Will viruses be produced containing the Mut5 genome as well? Explain why or why not.

ANSWER KEY AND EXPLANATIONS

Section I

1. D	21. B	41. D	61. B	81. A
2. D	22. D	42. A	62. B	82. C
3. C	23. B	43. D	63. B	83. B
4. E	24. C	44. A	64. C	84. A
5. A	25. E	45. C	65. D	85. C
6. C	26. D	46. C	66. B	86. A
7. C	27. B	47. D	67. C	87. B
8. A	28. E	48. C	68. B	88. B
9. C	29. A	49. E	69. A	89. E
10. D	30. D	50. C	70. B	90. B
11. A	31. C	51. A	71. B	91. C
12. B	32. B	52. C	72. A	92. A
13. B	33. C	53. B	73. D	93. B
14. E	34. B	54. E	74. D	94. E
15. C	35. E	55. C	75. D	95. D
16. A	36. B	56. A	76. C	96. A
17. D	37. A	57. C	77. B	97. E
18. A	38. D	58. E	78. B	98. A
19. D	39. D	59. B	79. D	99. C
20. D	40. B	60. E	80. E	100. A

1. **The correct answer is (D).** Chordates have all of these characteristics. None of the other groups has a notochord or a muscular tail posterior to the anus. Annelids and a few mollusks have a closed circulatory system, but most have an open system. All of these groups have a coelom.

2. **The correct answer is (D).** Carbon has four valence electrons, allowing it to form four covalent bonds with other atoms, and giving it the great versatility of bonding that makes the chemistry of carbon so central to biology.

3. **The correct answer is (C).** All amino acids have an amine group on one side and a carboxylic acid group on the other. The amine is basic so it will be protonated at pH 7 (which is pretty much physiological pH), giving it a positive charge, and the carboxylic acid is acidic, so it will be deprotonated at this pH, giving it a negative charge. The difference between amino acids is the functional group on their side chain. For an amino acid to have an overall net negative charge, the side chain must be acidic and negatively charged. The only amino acid in this group with a carboxylic acid side chain is glutamate. Each glutamate molecule will have one positive charge and two negative charges at this pH, giving the molecule a net charge of -1. Lysine has a basic side chain and will have a net charge of +1, and the other choices are neither acid or basic, so their net charge is 0.

4. **The correct answer is (E).** The pH of a solution = -log[H^+] and the concentration of hydrogen ions related to the concentration of hydroxide ions:

$$[H^+] \times [H^-] = 10^{-14}$$

From this it is possible to figure out the answer without a calculator. If the concentration of hydroxide ions is 0.01, this is 10^{-2} M, meaning that the concentration of hydrogen ions is 10^{-12}M, and the product of the two is 10^{-14}. The pH = -log[10^{-12}], which is 12.

5. **The correct answer is (A).** Heating an enzyme imparts kinetic energy to the molecule in several ways, causing vibration and movement of the non-covalent bonds that hold the enzyme in its properly folded structure. If the kinetic energy is great enough, these bonds break, and the enzyme unfolds, becoming denatured, and causing the carefully folded active site to unfold as well. Without the active site, the enzyme has no activity. The primary structure would not be disrupted, because covalent bonds in proteins would probably require harsher treatment to be disrupted (choice (B) is wrong). The question says that the enzyme was heated, not the DNA (choice (C) is wrong). There is no indication that a cofactor is involved in any way (choice (D) is wrong). The enzyme was just returned to the original temperature that was tested, so it should have the same activity as it did the first time it was tested (choice (E) is wrong).

6. **The correct answer is (C).** Proteins that reside in the plasma membrane are synthesized in the secretory pathway, starting with protein synthesis by ribosomes on the rough ER. In the ER the transmembrane protein will be threaded through the membrane, and it will retain this orientation as it moves in membrane vesicles through the secretory pathway. After the ER, the protein is transported by vesicles to the Golgi, and then to the cell surface where the transmembrane protein is delivered to the plasma membrane.

7. **The correct answer is (C).** Plants cells do have mitochondria to produce ATP for their own consumption. They also have ER, a plasma membrane, and most of the other features that animal cells have, as well as a couple that are unique to plants, like chloroplasts and the cell wall, but the question is not about these.

8. **The correct answer is (A).** ΔG describes the thermodynamics of a reaction, but not the kinetics. With a negative ΔG, the reaction is favored to spontaneously move forward. The word "spontaneous" is used in a special way here, not to mean fast, but just to say that forward movement is energetically favored to occur. This rules out choices (B) and (C). The reaction cannot be at equilibrium; if this was the case the ΔG would be 0.

9. **The correct answer is (C).** This problem relates to the first and second law of thermodynamics. The first law says that energy of a system is always conserved, that energy is never lost or destroyed, that it can only change from one form of energy to another. The energy contained in the compounds of the yeast cells will be converted by the paramecium into other forms of energy, such as movement and heat, but the energy is not destroyed so the overall energy content of the system stays the same. Entropy of a system, on the other hand, is always increasing, as stated by the second law.

10. **The correct answer is (D).** NADH is a high energy electron carrier that is reduced in glycolysis and the Krebs cycle. NADH in turn reduces the intermediates in the electron transport chain, transferring electrons to the chain. Electron transport, in turn, pumps protons out of the mitochondria to create a pH gradient that drives ATP synthesis. NADH is not the final electron acceptor—this is oxygen (choice (A) is wrong). NADH does not act directly either to make ATP (choice (B) is wrong), or to carry electrons (choice (E) is wrong). Pyruvate is oxidized to make acetyl-CoA (choice (C) is wrong).

11. **The correct answer is (A).** Photorespiration is an unproductive side reaction from normal photosynthesis that produces no ATP. It occurs in C3 plants when their stomata are closed on hot sunny days to prevent excessive water loss. Oxygen accumulates inside the leaf, and carbon dioxide gets scarce, so Rubisco uses oxygen as a substrate instead of CO_2 and performs photorespiration. CAM and C4 plants have evolved ways of avoiding photorespiration, fixing carbon by other mechanisms.

12. **The correct answer is (B).** Prophase includes the breakdown of the nuclear envelope, chromosome condensation, and initial formation of the spindle in preparation for cell division. Interphase is the period between mitotic cell divisions (choice (A) is wrong). Anaphase is the period in which chromosomes are separated into the two poles of the dividing cell (choice (C) is wrong). Telophase is when the nuclear envelopes reform and the chromosomes decondense, reversing the events of prophase in preparation for cytokinesis.

13. **The correct answer is (B).** This can be solved with a quick Punnett Square or an equation based on the probability. The probability of a parent's gamete being the recessive allele is $\frac{1}{2}$. The probability of two recessive alleles coming together in a zygote is therefore $\frac{1}{2} \times \frac{1}{2} = \frac{1}{4}$.

14. **The correct answer is (E).** First, primase makes an RNA primer, then DNA polymerase III extends the primer. DNA polymerase I replaces the RNA primer with DNA. DNA does need to be single-stranded for DNA replication to occur, but this is not the job of DNA polymerase I (choice (A) is wrong). DNA polymerase III extends from the RNA primer, not DNA polymerase I (choice (B) is wrong). Other enzymes like helicase take care of torsional stress associated with unwinding DNA and opening up double-stranded DNA for replication (choice (C) is wrong). Almost all of the DNA synthesis on the leading strand is carried out by DNA polymerase III and DNA polymerase I does very little since there is almost no need for RNA primers and there are no Okazaki fragments on the leading strand (choice (D) is wrong).

15. **The correct answer is (C).** The wall of capillaries is thin, one cell thick, allowing the ready movement of oxygen, carbon dioxide, and small solutes like glucose between the blood and extracellular fluid. Larger objects like proteins (choices (A) and (E) are wrong), cell fragments (choice (D) is wrong), or red blood cells (choice (B) is wrong) are too big to pass through the capillary wall into the extracellular fluid.

16. **The correct answer is (A).** The adaptive immune response of B cells and T cells involves selection and amplification of specific clones of cells that respond to antigen out of a population of cells with fairly random antigen affinities. Recombination of immunoglobulin genes during B cell maturation creates the population of cells with

random antigen affinities. Recombination does not occur in response to antigen but prior to antigen exposure (choice (B) is wrong). B cells do not recognize antigen bound to MHC—they bind to antigen in solution in plasma (choice (C) is wrong). B cells proliferate and differentiate in response to antigen; they do not enter apoptosis (choice (D) is wrong). B cells mediate the humoral immune response, not the cell-mediate immune response (choice (E) is wrong).

17. **The correct answer is (D).** All exchange of materials between blood and tissues happens in capillaries. Everything else is pretty much plumbing.

18. **The correct answer is (A).** Prolactin and growth hormone are from the anterior pituitary, epinephrine from the adrenal medulla, and GnRH is produced in the hypothalamus.

19. **The correct answer is (D).** To provide a useful comparison, something needs to be present in all three groups. Bacteria and archae don't have mitochondria (choice (A) is wrong), and bacteria don't have introns. Histones are only present in eukaryotes (choice (E) is wrong). The enzymes involved in glycolysis would be present, but would be conserved less than ribosomal RNA probably, since their function may not be as basic as that of ribosomal RNA.

20. **The correct answer is (D).** As the membrane depolarizes in one section of membrane with the passing of the action potential, the voltage-gated sodium channels in the neighboring stretch of membrane open in response to the action potential depolarization, moving the action potential farther down the membrane of the axon. Ligand-gated sodium channels may be involved in the transmission across the synaptic cleft, but not movement of the action potential along the axon (choice (A) is wrong). Voltage-gated potassium channels cause repolarization, allowing potassium ions out, in the opposite direction as the movement of sodiums into the cell during depolarization. The sodium-potassium pump maintains the resting potential (choice (C) is wrong). Potassium leak channels allow some potassium to leak out of cells to maintain the resting potential, and they also help to create the spontaneous depolarization of the pacemaker region in the heart.

21. **The correct answer is (B).** Epinephrine from adrenals works together with the sympathetic nervous system to cause a variety of "fight or flight" responses throughout the body. The release of parathyroid hormone (PTH) occurs in response to serum calcium levels (choice (A) is wrong). The autonomic nervous system (including sympathetic nerves) does not cause skeletal muscle to contract (choice (C) is wrong). Bronchioles dilate and the heart rate increases in response to the sympathetic system (choices (D) and (E) are wrong).

22. **The correct answer is (D).** Conjugation is the transfer of a bacterial F factor from one bacterium that has it to another that does not. It occurs from one bacterium to another, not with naked DNA (choice (A) is wrong). The uptake of naked DNA by bacteria is called transformation. Transduction is the transfer of bacterial genes packaged into viruses (choice (B) is wrong). Choice (C) is wrong because conjugation always copies the plasmid before it is transferred, so the donor cell does not lose the

plasmid. Sexual reproduction does not occur in bacteria, and conjugation is not sexual reproduction (choice (E) is wrong).

23. **The correct answer is (B).** As in mitochondria, chloroplasts use the production of a pH gradient across a membrane to power ATP production.

24. **The correct answer is (C).** Epigenetic inheritance occurs by a mechanism other than changes in the nucleotide sequence of the genome. Mutations in the genome are not epigenetic, no matter what mechanism creates the mutation. Choice (E) sounds like epistasis, but this is entirely distinct from epigenesis.

25. **The correct answer is (E).** Bacteria often have genes involved in a related function arranged next to each other in the genome and regulated together in the same transcriptional unit. The genes in an operon like the lac operon are all transcribed in one RNA by RNA polymerase. When lactose is not present, the lac repressor protein binds to the operator region in front of the promoter in DNA and prevents RNA polymerase from transcribing the genes. When lactose is present, the lac repressor releases its hold on DNA, and RNA polymerase starts transcribing the genes. The genes in an operon are next to each and transcribed on the same RNA (choice (A) is wrong). The genes are only transcribed and translated when lactose is present (choice (B) is wrong). Lactose binds to the lac repressor, not RNA polymerase (choice (C) is wrong). Bacteria do not have a nucleus or polyA addition (choices (B) and (D) are wrong).

26. **The correct answer is (D).** Grey matter is cell bodies and dendrites, while white matter is myelinated axons.

27. **The correct answer is (B).** Most plants live symbiotically with mycorrhizal fungi in and around their root system. The fungi greatly increase the nutrient gathering surface area of the root system with fine hyphae, improving plant health and increasing plant growth. The fungi do not fix nitrogen however (choice (A) is wrong), supply glucose, or produce photosynthetic pigments. Nor do they provide carbon dioxide—plants get this from the atmosphere.

28. **The correct answer is (E).** DNA and RNA have many similarities, as well as some differences. They are both built of polynucleotides assembled with phosphodiester bonds between each nucleotide, and they are both assembled in a 5' to 3' direction on an existing template of DNA. DNA does contain deoxyribose and thymine, not uracil.

29. **The correct answer is (A).** A hypertonic solution has a higher concentration of solutes than is present inside the cell. By osmosis, water will flow through the plasma membrane and through the cell wall, making the cell shrink. When it shrinks, the cell wall will not change in size and shape, so the cell will shrink away from the cell wall. The cell will not import salts to produce the same osmotic concentration inside the cell (choice (B) is wrong). It will not swell—this is what would happen in a hypotonic solution, in which water would flow into the cell. Even if it did swell, the cell would not burst the cell wall, but just press up against it. Cells walls need to allow material in and out of cells for the cells to survive (choice (E) is wrong).

30. **The correct answer is (D).** The complement system is a part of innate immunity. In response to bacteria, a cascade of proteases in the complement system in plasma is activated, resulting in the assembly of a protein pore in the bacterial membrane. This pore allows water and ions to flow freely through the membrane, making the cell lyse. The complement cascade is not part of clotting (choice (A) is wrong) and does not carry hormones or prevent responses to self antigens (choice (C) is wrong).

31. **The correct answer is (C).** In the normal viral life cycle, virus must first recognize and gain entry into host cells, then its reverse transcriptase enzyme copies the RNA genome into DNA, which is inserted in the genome of the host cell. There, viral genes are transcribed and translated, and new copies of the viral RNA genome are produced. Once new viruses are assembled, they can bud from the infected cell and acquire a viral envelope. Viruses do not, however, have their own ribosomes.

32. **The correct answer is (B).** tRNA molecules bind to each codon in mRNA to translate the codon and add the correct amino acid dictated by the genetic code. Ribosomal RNA plays an important role in translation, but it does not bind or recognize codons in mRNA directly (choices (A) and (D) are wrong). mRNA molecules merely contain the message; they do not indicate the amino acids encoded with any higher order structure (choice (C) is wrong). Amino-acyl tRNA synthetase enzymes attach the correct amino acid to each tRNA and are important for the accuracy of translation, but they do not perform this as described here (choice (E) is wrong).

33. **The correct answer is (C).** Male gametophytes are pollen grains, which are produced at the anthers at the tip of the stamens.

34. **The correct answer is (B).** The stigma, located on top of the style and ovary, is where pollen grains land and begin the growth of the pollen tube through the style to the ovary.

35. **The correct answer is (E).** The female gametophytes in the ovary produce the female gametes in the ovule. The pollen tubes grow into each ovule, delivering sperm to the egg, to perform the double fertilization that is a hallmark of angiosperms.

36. **The correct answer is (B).** Humans are endotherms that use a lot of metabolic energy to maintain a constant core body temperature. We do not have a period of torpor like some species. Thyroid hormone can increase the metabolic rate in cold weather, and arterioles to the skin will constrict, reducing heat loss through the skin. If the arterioles in skin constrict, the fat deposits beneath the skin will also help to insulate and reduce heat loss. Shivering can occur to increase heat generation by skeletal muscles if the body temperature is low.

37. **The correct answer is (A).** The large intestine does not absorb much in the way of nutrients and is not the site of hydrolysis of molecules that are to be digested, but it does help conserve water by reabsorbing remaining water prior to defecation. Also, vitamin K produced by bacteria in the large intestine is absorbed and contributes to clotting.

38. The correct answer is (D). Spermatogenesis occurs primarily in the seminiferous tubules. From there, sperm pass through the rest of the ducts. A spermatogonium is one of the precursor cells that conduct spermatogenesis.

39. The correct answer is (D). Fermentation only yields about 2 ATP per each glucose, whereas oxidative phosphorylation, with the complete oxidation of glucose to carbon dioxide, releases 36–38 ATP molecules per glucose consumed. Fermentation does not produce any NADH or $FADH_2$ and does not support the creation of the pH gradient that drives ATP production. Fermentation does occur in the absence of oxidative phosphorylation—in fact that is what fermentation is all about, keeping energy production going even in the absence of oxygen, even if the energy production is much less efficient. Choice (B) is not correct—it does not compare the energy production of the two pathways. Choice (C) is not correct—the question does not compare the rates of things, just the efficiency of energy production.

40. The correct answer (B). The exoskeleton does a lot of the same things that the vertebrate endoskeleton does. They both allow for movement with flexibility of appendages, protect tissues and organs, support the body, and provide attachment sites for muscles. Neither skeletal system is involved in respiratory gas movement. Respiratory gases are not exchanged with the environment across the exoskeleton, but through gills.

41. The correct answer is (D). The light reactions of photosynthesis make ATP and NADPH from the energy of light and photoexcited electrons, and the Calvin cycle uses ATP, carbon dioxide, and NADPH to make sugars.

42. The correct answer is (A). Pyruvate is oxidized in this reaction, making acetyl-CoA and carbon dioxide. This means that the pyruvate is acting as a reducing agent to reduce NAD^+ to NADH.

43. The correct answer is (D). A regular nondividing cell is diploid and, in this species, has sixty chromosomes. Spermatogonium cells do not themselves go through meiosis, and continually replenish their numbers through mitotic cell division. A spermatogonium cell in G2 phase has gone through S phase, so it has twice the normal amount of DNA and twice the normal diploid number of chromosomes, or 120 chromosomes in this case.

44. The correct answer is (A). The two traits are on different chromosomes, and there will be no linkage, so you can look at each trait independently first. For each trait, crossing heterozygotes has a $\frac{1}{2} \times \frac{1}{2} = \frac{1}{4}$ probability of producing a homozygous recessive offspring. The odds of having both traits turn out to be homozygous recessive is $\frac{1}{4} \times \frac{1}{4} = \frac{1}{16}$.

45. The correct answer is (C). The inhibitor does not change the V_{max}, although it takes more substrate to get there. This is the hallmark of competitive enzyme inhibition. Inhibitor and substrate are competing for the same active sites. Adding more substrate competes inhibitor out of the active sites, making competitive inhibitors surmountable. Noncompetitive, irreversible, or allosteric inhibitors reduce the V_{max}, which does not occur in this case.

46. The correct answer is (C). Competitive inhibitors inhibit the binding of substrate to the active site. They do not covalently modify proteins (choice (A) is wrong) or denature enzymes (choice (B) is wrong). There is no information suggesting cooperative binding is involved in this example (choice (E) is wrong).

47. The correct answer is (D). Gram positive bacteria have one cell membrane surrounded by a thick peptidoglycan cell wall. Gram negative bacteria have a thinner peptidoglycan layer sandwiched between the plasma membrane and an outer membrane. This outer membrane in Gram negative bacteria (not Gram positive) contains lipopolysaccharides that provoke a strong immune response and can cause septic shock during infections (choice (A) is wrong). Both bacterial genomes are going to be relatively small circular genomes, nothing like eukaryotic genomes, and DNA is not involved in Gram staining (choice (B) is wrong). Gram positive bacteria are more sensitive to penicillin, not less (choice (C) is wrong). The metabolic needs of gram positive and gram negative bacteria may vary, but they will all use glycolysis.

48. The correct answer is (C). The stomata are the microscopic leaf openings that allow oxygen out of leaves and allow carbon dioxide to enter leaves to support photosynthesis. The openings are on the epithelial surface layer of leaves, usually on the bottom of the leaf.

49. The correct answer is (E). The fetus gets oxygen by passive diffusion of oxygen down a gradient in the placenta from the maternal to the fetal circulation. Oxygen is never transported by active transport (choice (A) is wrong). Chemical synapses are part of the nervous system, not the circulatory system, and they are not present in the placenta (choice (B) is wrong). Red blood cells do not pass between maternal and fetal circulation. The fetus expresses its own red blood cells (choice (C) is wrong). The amniotic fluid is not involved in fetal respiration (choice (D) is wrong).

50. The correct answer is (C). Fructose is a hexose, a six-carbon sugar, and a structural isomer of glucose. For fructose to be metabolized, it is phosphorylated and enters glycolysis. Like glucose, each molecule of fructose is split by glycolysis to produce two pyruvates, and each pyruvate creates one acetyl-CoA. Each molecule of fructose results in two molecules of acetyl-CoA.

51. The correct answer is (A). Triglycerides are hydrolyzed in the small intestine, with pancreatic lipase hydrolyzing them into free fatty acids and glycerol (choices (B), (D), and (E) are wrong). In the cells lining the intestinal villi, the triglycerides are reassembled and secreted as lipid particles called chylomicrons into lymphatic vessels. These lymphatic vessels move lymph through the body to eventually pass into blood in veins in the shoulder area. Proteins are also hydrolyzed mainly in the small intestine and also depend on pancreatic secretion of enzymes for a large part of their hyrolysis (choice (C) is wrong). They are absorbed in the small intestine, but amino acids absorbed in the intestine are transferred to blood and transported to the liver through the hepatic portal circulation, where the liver immediately gets to work on absorbing and regulating the flow of material from the gastrointestinal tract into the systemic circulation.

52. **The correct answer is (C).** When a ribosome reaches a stop codon, a protein called a release factor binds to the stop codon on the ribosome, causing the polypeptide to be released and the ribosome assembly with the mRNA to fall apart.

53. **The correct answer is (B).** Diastereomers are stereoisomers that are not mirror images. These two molecules differ at the stereochemistry of one carbon, and are clearly stereoisomers, but not mirror images of each other. Structural isomers would have the same molecular formula, but different connectivity of atoms, which is not the case here (choice (A) is wrong). Anomers are the different ring forms of sugars that form in solution (choice (C) is wrong). Enantiomers are mirror image stereoisomers (choice (D) is wrong).

54. **The correct answer is (E).** If there is 12% of the original C^{14} left, this means that the sample has been through three half-lives of decay (100%/2 /2 /2 = 12%). Three half lives of C^{14} is 3 × 5,700 years, which is about 17,000 years.

55. **The correct answer is (C).** A hollow dorsal nerve cord arose with the chordates and all chordates have it. All of the other traits appeared later in various subgroups of the vertebrates. Placental nutrition came later in mammals (choice (B) is wrong) and the amniotic egg first appeared with the reptiles living on land (choice (A) is wrong). Amphibians first had lungs for respiration on land and a three-chambered heart (choices (D) and (E) are wrong).

56. **The correct answer is (A).** Human chorionic gonadotropin (hCG) is produced by the fertilized embryo very soon after fertilization and acts on the corpus luteum to prevent its regression. The corpus luteum continues producing progesterone, which maintains the uterine endometrium to support the developing embryo. Estrogen plays more of a role in breast development (choice (B) is wrong) and LH triggers ovulation (choice (C) is wrong). It is the loss of progesterone that triggers menstruation, not the presence of hCG (choice (D) is wrong). GnRH is the releasing hormone for FSH and LH, not hCG (choice (E) is wrong).

57. **The correct answer is (C).** Myelinated neurons have distinct regions of myelination along their axons, interspersed with regions (nodes of Ranvier) with the axonal plasma membrane exposed to the extracellular environment. The regions covered with myelin are insulated and contain few voltage-gated sodium channels that propagate action potentials. The exposed nodes of Ranvier have densely clustered voltage-gated sodium channels. Action potentials in myelinated neurons do not move in a smooth wave of depolarization, but leap from node to node in saltatory conduction. Saltatory conduction is faster and more efficient than transmitting action potentials in un-myelinated axons.

58. **The correct answer is (E).** Abscisic acid causes dormancy of tissues, including seed dormancy. Seed embryos need to lie dormant through the winter until conditions are ready for germination and growth of the vulnerable embryo. When water washes away the abscisic acid, the embryo can begin growing again.

59. **The correct answer is (B).** Fatty acids are metabolized through beta-oxidation, with repeated rounds of oxidation and chain shortening creating acetyl-CoA that enters the Krebs cycle.

60. **The correct answer is (E).** The central nervous system is induced to form from infolding of the ectoderm to form the neural tube during embryogenesis. The heart, bone, and smooth muscle are from mesoderm and the intestinal epithelium from endoderm.

61. **The correct answer is (B).** Heating DNA is required to make DNA single-stranded. Then, the reaction is cooled somewhat so short single-stranded DNA primers can hybridize on the single-stranded DNA, and DNA polymerase can extend them. The annealing step requires cooler temperatures than melting the double helix (choice (A) is wrong). The DNA polymerase used can tolerate high heat, but this is not the highest temperature of the reaction (choice (C) is wrong). No DNA polymerases work without a primer, including those used in PCR (choice (D) is wrong). There is no lagging strand in PCR (choice (E) is wrong).

62. **The correct answer is (B).** The heart has two sides: the right side pumps blood to the lungs and the left side pumps blood to the rest of the body, the systemic circulation. The blood goes to the lungs from the right ventricle, through the pulmonary artery, and then the oxygenated blood is returned to the heart in the left atrium, which sends it to the left ventricles and through the aorta to the rest of the body.

63. **The correct answer is (B).** In deep ocean vents, ecosystems are based around primary producers that extract energy from inorganic chemicals released in the superheated waters from the vent. The organisms that use these chemicals for energy are chemoautotrophs, including bacteria and archae. There is not enough light for photosynthesis in this region (choice (A) is wrong). Zooplankton are primary consumers, not producers (choice (C) is wrong). Scavengers and any chemoheterotroph are not primary producers (choices (D) and (E) are wrong).

64. **The correct answer is (C).** Cellulose is a polysaccharide made of large numbers of glucose monomers linked by beta-glycosidic linkages. Humans eat many foods from plants containing cellulose, but we cannot digest the cellulose because we do not produce enzymes that can hydrolyze the beta-glycosidic bonds between glucose groups. It is a large polysaccharide, but so are starch and glycogen and we have no problem digesting these because they contain alpha instead of beta-linkages between glucose groups and we produce enzymes that hydrolyze these linkages and break down these polysaccharides into sugar monomers that are absorbed (choice (A) is wrong). Humans do have bacteria in their digestive tract, and lots of them, but we just don't have the same bacteria that ruminants have that are capable of breaking down cellulose (choice (B) is wrong). There are no phosphorous based cross-links in cellulose (choice (D) is wrong).

65. **The correct answer is (D).** Testosterone and estrogen are closely related steroid hormones that are very similar in structure and in mechanism. They are both relatively small (compared to peptide hormones) and hydrophobic and diffuse through the plasma membrane to bind to steroid hormone receptors inside cells. These

receptors act as ligand-activated transcription factors that bind to promoters and enhancers to activate transcription of hormone-responsive genes. The testosterone receptor and estrogen receptors bind different sets of genes and are expressed in different tissues, providing different responses to the two hormones.

66. **The correct answer is (B).** Disulfide bonds between two cysteines in a protein help to anchor the folded structure of the protein, stabilizing it against denaturation.

67. **The correct answer is (C).** Poriferans are sponges, which are filter feeders without specialized digestive organs or tissues. Individual cells internalize small food particles and digest them intracellularly. Cnidarians secrete some enzymes into their digestive sac, their gastrovascular cavity (choice (A) is wrong), and fungi secrete enzymes to break down material that can be absorbed from outside the body (choice (B) is wrong). Many phyla, including annelids, nematodes, arthropods, mollusks, and chordates have a complete digestive tract (choice (D) is wrong). Poriferans do not perform photosynthesis (choice (E) is wrong).

68. **The correct answer is (B).** Electrons are first excited in photosystem II, then move through an electron transport chain down to the photosystem I reaction center. Then electrons are again excited by light, and move down through another electron transport chain to NADPH reductase, where they are used to reduce $NADP^+$ to NADPH

69. **The correct answer is (A).** The trait must be recessive because none of the children of 1 and 2 expresses the trait, but grandchildren do. If it was sex-linked, it would be expressed predominantly in male children, but there is no evidence of this, so it is probably autosomal. If the trait displayed maternal inheritance, meaning it was determined by a gene in the mitochondrial genome, then all of the children of an affected mother would get the trait and affected males would never transmit the trait, but this is not observed. None of the children of 2 has the trait. There is nothing suggesting an intermediate phenotype that would involve incomplete dominance or co-dominance.

70. **The correct answer is (B).** The trait must be recessive because none of the children of 1 and 2 expresses the trait, but grandchildren do. Since the trait is recessive, individuals expressing it must be homozygous for the recessive allele. This means that number 2 is homozygous recessive, and all of her children must have one recessive copy from her, including number 8.

71. **The correct answer is (B).** Pepsin is a protease secreted and activated in the stomach. It is most active at acidic pH, and is often used as an example of the effect of pH on enzyme optima.

72. **The correct answer is (A).** The systolic pressure is the pressure wave with each ventricular contraction pumping blood out through the arteries. The pressure is greatest closest to the heart in the large arteries, with their thick muscular walls contributing to the pressure, and the aorta is the largest artery in the systemic circulation. Smaller arteries, arterioles, capillaries, and veins have increasingly lower pressure. The diastolic pressure is the low point between ventricular contractions, lower than the systolic pressure.

73. **The correct answer is (D).** Upwelling brings nutrients from deeper water near the surface where photosynthesis takes place, increasing the growth of phytoplankton. Surface water in oceans generally has oxygen, and the activity of photosynthesizers will not reduce this. Increased productivity would in general increase biodiversity. If there are more phytoplankton, for example, then there is more for zooplankton to eat, and if there are more zooplankton, then other consumers will also increase.

74. **The correct answer is (D).** It looks like the wasp population varies in size with a pattern similar to the caterpillar population but trailing in time. It is easy to assume that the wasp must be a predator of the caterpillar, but this is not stated anywhere and there are no data given to say that it must be true. The data do not regard any particular traits in the population, so there is nothing to say that choice (A) is true. The caterpillar population fluctuates a lot in size, so it's hard to say it really follows any model (choice (B) is wrong). Something affects the caterpillar population, but we don't know if this is the climate or something else (choice (C) is wrong). We don't know if the wasp is a predator (choice (E) is wrong).

75. **The correct answer is (D).** It seems that the presence of caterpillar species #1 helps the wasps, but the wasps do not affect the caterpillar population. This is the definition of a commensal relationship.

76. **The correct answer is (C).** Photosynthesis produces glucose from water, carbon dioxide, and sunlight. Oxygen is also produced. The reaction must reflect this, and must be balanced. Choices (A) and (B) have a carbohydrate as the starting point on the right side of the reaction so these choices are wrong. Choices (D) and (E) are not balanced properly, with the same number of all atoms on both sides of the equation. The fact that the photosynthesis occurs in CAM plants does not change the net result of photosynthesis. CAM and C4 plants use alternative carbon fixation routes, but the net result of photosynthesis is still the same.

77. **The correct answer is (B).** The water comes from reduction of oxygen, where molecular oxygen is the final electron acceptor of the electron transport chain.

78. **The correct answer is (B).** Urine forms in the kidney, then goes to the bladder in the ureter, and goes from the bladder to be eliminated through the urethra.

79. **The correct answer is (D).** pH is the negative log of concentration. To convert from pH to scientific notation, just make the pH negative, and then make it an exponent of 10. That means that a pH value of 5 is the same is 10^{-5} molar hydrogen ions.

80. **The correct answer is (E).** Cone cells are photoreceptor cells that detect colors, and rod cells are for sensitive detection of light across a broad range of intensities, including dim conditions, but they don't detect colors.

81. **The correct answer is (A).** Buffers are compounds that act as weak acids and bases, reacting with protons in solution to prevent large changes in pH. Large changes in blood pH are dangerous, so blood pH is buffered by the presence of proteins and bicarbonate/carbonic acid.

82. **The correct answer is (C).** When blood passes through capillaries, fluids that pass out of the capillaries into the tissues mostly return back into the blood, but some of the

fluid remains in the tissues and passes into the lymphatic system. Fluids move through lymphatic vessels, driven by surrounding tissues and the action of skeletal muscle. Eventually the lymphatic vessels return the fluid to the blood by emptying into veins near the shoulders.

83. **The correct answer is (B).** During strenuous exercise, muscle uses oxygen faster than the cardiovascular system can deliver it. To keep generating the energy needed, muscle can resort to fermentation, using glycolysis to produce pyruvate and ATP, and then regenerating NAD^+ by producing lactate. The lactate can be converted back to pyruvate in another tissue.

84. **The correct answer is (A).** Chymotrypsin is a digestive protease released by the pancreas into the small intestine. It is synthesized as a zymogen, an inactive form of the enzyme called chymotrypsinogen, and then activated in the small intestine through cleavage by trypsin. Its enzymatic specificity is to cleave peptide bonds adjacent to hydrophobic amino acid side chains.

85. **The correct answer is (C).** Tropical ecosystems with abundant rain and sunlight tend to have the greatest species diversity while ecosystems at higher latitudes have lower species diversity. Fragmented small habitats, or even smaller islands, tend to have lower species diversity than larger habitats or larger islands (choice (B) is wrong). Disrupting an ecosystem by removing a keystone species or introducing an invasive alien species tends to reduce species diversity (choices (D) and (E) are wrong).

86. **The correct answer is (A).** Fungi are absorptive chemoheterotrophs. They perform aerobic respiration, digesting food sources and releasing carbon dioxide. They do not absorb carbon dioxide or perform photosynthesis. Transpiration is part of the water cycle, not the carbon cycle.

87. **The correct answer is (B).** Pseudogenes were duplicated from other genes in the evolutionary past and then underwent a mutation that created a gene that could no longer produce a functional gene product. Pseudogenes are not the same as repetitive elements like transposons or retrotransposons (choices (A) and (C) are wrong). There is nothing relating pseudogenes to methylation: this is a more general modification of DNA (choice (D) is wrong). Organisms do not switch genetic codes in different parts of the genome (choice (E) is wrong).

88. **The correct answer is (B).** The predation will select against the middle range of sizes, exerting selective pressure favoring butterflies at either extreme of size.

89. **The correct answer is (E).** The Krebs cycle and electron transport chain are interconnected in their activity. If cyanide is present and the electron transport chain is blocked from the movement of electrons through the chain, then NADH cannot transfer high energy electrons to the system. If NADH (and $FADH_2$) cannot transfer electrons to the system, then they will accumulate and NAD^+ will run out, causing the Krebs cycle to stop producing more NADH. The cells cannot use oxygen because the electron transport chain is blocked and cannot transfer electrons to oxygen (choice (A) is wrong). The pH gradient will not increase because electron transport will halt (choice (B) is wrong). Carbon dioxide is from the Krebs cycle, which will stop, so choice (C) is wrong.

ATP production from oxidative phosphorylation will be blocked, but the cells can perform fermentation to keep producing at least a small amount of ATP.

90. The correct answer is (B). Insects generally have chemoreceptors, such as pheromone receptors, on their antennae.

91. The correct answer is (C). Isotopes differ in the number of neutrons, not protons or electrons. The atomic number is the number of protons each element has.

92. The correct answer is (A). Changing the overall ΔG of the coupled reaction, turning it from positive to negative, changes the equilibrium of the coupled reaction, driving the production of Y and B forward. We don't know anything about the activation energy for the coupled or individual reactions—the free energy changes don't tell us anything about activation energy barriers (choice (B) is wrong). Most reactions in cells involve enzymes, whether ATP is involved or not. Coupling the reaction to ATP hydrolysis makes the reaction energetically favorable but does not predict the speed of the reaction on its own (choice (C) is wrong). We don't know if this reaction is involved in energy production (choice (D) is wrong), and the relative amount of Y and B should not change, since the reaction indicates a 1:1 stoichiometric balance in either case (choice (E) is wrong).

93. The correct answer is (B). The recognition of substrate requires that the substrate fit into the enzyme's active site. The active site is a precise 3D arrangement of amino acids that bind to the substrate(s) and help to stabilize the transition state intermediate to reduce the activation energy. Substrates don't bind just anywhere on an enzyme—they bind at a specific site that has evolved for that purpose (choice (A) is wrong).

94. The correct answer is (E). Bacteria perform transcription and translation in the same compartment and no processing of RNA is required before it is translated. Both have 2 subunits (C is wrong), and bacteria do not have a nucleus (D is wrong).

95. The correct answer is (D). This cavity has one opening to the exterior, so it is not a complete digestive tract. Cnidarians lack mesoderm (choice (A) is wrong), and this is not a blastula (choice (B) is wrong). Cnidarians are acoelomate (choice (C) is wrong).

96. The correct answer is (A). Wild bananas are diploid, with seeds, but are difficult to eat. Seedless bananas were bred to be triploid. With a triploid plant, reductive cell division does not work, so meiosis cannot produce gametes. If a plant is diploid, then it can go through meiosis to produce haploid gametes, but if it is triploid, it cannot put half of its complement evenly in a gamete. Mitosis and DNA replication must occur normally in these plants—they grow and produce fruit like a normal plant (choices (B) and (C) are wrong). The barrier is prezygotic, at the level of gamete formation (choice (D) is wrong). The plant once produced seeds and probably still has the genes for seed formation in its genome (choice (E) is wrong).

97. The correct answer is (E). Ethylene causes fruit ripening.

98. The correct answer is (A). The diaphragm is composed of skeletal muscle and is stimulated with each breath to contract, opening the chest cavity and expanding the lungs, causing air to be drawn in.

99. The correct answer (C). Renin is released by cells in the kidney when blood pressure is low. Renin is protease that cleaves a protein in plasma to make the hormone angiotensin. Angiotensin acts on arterioles, causing them to constrict, raising blood pressure. Blocking the formation of angiotensin is one way to fight high blood pressure.

100. The correct answer is (A). People with lesions to Broca's area have a specific defect in the formation of speech, although they are otherwise normal.

Section II

SUGGESTIONS FOR ESSAY QUESTION 1

(A) Let p = the allele frequency of H, and q = the allele frequency of h. The abundance of genotypes and phenotypes can be calculated based on allele frequencies. The normal phenotype will be observed in homozygous dominant genotypes (HH) and in heterozygotes (Hh).

HH frequency = p^2 = 0.72 = 0.49
Hh frequency = 2pq = 2 × 0.7 × 0.3 = 0.42

Add these together to get the overall abundance of the wild type phenotype = 91% wild type phenotype. To find the number of animals with this phenotype, multiply this by the population size:

= 0.91 × 4800 = 4368 animals

(B) When the population was split in two, the smaller group went through a population bottleneck, receiving an allele distribution different from the larger population. With small groups sampled from a larger group, random differences come into play, and the smaller the sub-group becomes, the larger the element of randomness. Mutation will introduce a rare allele but will not change the allele frequencies of a population in one generation. Non-random mating cannot change a population this quickly—and it would not be expected to, since Population 1 had the existing allele frequencies at a fairly stable ratio for several years.

(C) Speciation would involve reproductive isolation. To see whether the two populations have become two species, find out whether reproductive isolation has occurred. Mix males and females from the two groups in a natural setting and determine whether they interbreed and whether their young are viable. If they are separate species, they will not interbreed.

SUGGESTIONS FOR ESSAY QUESTION 2

(A) Some possible responses:

- Altered shape of leaves to reduce water loss. Leaves that are smaller display less surface area for water loss. Angiosperms that have reduced surface area and use stems for photosynthesis, such as cacti, display a similar adaptation.
- Development of the female gametophyte enclosed within the tissues of the sporophyte. Sperm cells do not have to swim through water in the open from male to female. This allows angiosperms to live in drier environments than ferns or mosses.
- Stomata close at night to conserve water. Most water loss from plants occurs through the stomata. Closing when there is no need to be open helps reduce water loss. Loss of water through stomata is also reduced when the stomata are on the underside of leaves and when they are located in pits, where air cannot "drag" water out of the leaf.
- A waxy surface coating on leaves and other surfaces. The leaves of many types of plants are covered with a waxy substance, which seals in moisture.
- Deep roots and water storage systems. Many plants living in dry environments have evolved very deep roots to tap into ground water. They may also store

water for use during drier periods, and they may enter a partially dormant state until water is sufficiently abundant.

(B) CAM plants fix carbon in organic acids and open stomata at night to avoid water loss. During the day, the Calvin cycle continues to use carbon from the stored organic acids, while the stomata remain closed to conserve water. This adaptation helps CAM plants such as cacti survive in very hot, dry conditions. Under these conditions, C3 plants close their stomata during the day, accumulating oxygen in the leaf and running low on carbon dioxide. This causes an increase in photorespiration and a loss of photosynthetic efficiency.

C4 plants fix carbon in a different cell than those that perform the Calvin cycle, so this helps to avoid photorespiration. Carbon is fixed first from the atmosphere in mesophyll cells, creating the four-carbon oxaloacetate. Oxaloacetate is transported to bundle sheath cells performing photosynthesis, thereby releasing carbon dioxide for the Calvin cycle. The enzyme used by C4 plants for carbon fixation—PEP carboxylase instead of rubisco—is not prone to photorespiration, so C4 plants can avoid reduced efficiency on sunny days when stomata are closed or partially closed, oxygen starts to accumulate, and carbon dioxide concentrations are reduced inside the leaf.

SUGGESTIONS FOR ESSAY QUESTION 3

(A) **Endoskeleton.** Found in chordates. Provides support for the body and anchor points for muscles. Skeletal muscle contracts to bring anchor points together; bones are attached at joints toward a smaller angle. An example is the human bicep muscle, which is attached to the bones of the forearm and upper arm. Contracting this muscle brings the forearm closer to the upper arm. The chordate endoskeleton is either made of flexible cartilaginous tissue or ossified bone comprised of collagen and mineralized calcium phosphate. Vertebrate bone is generally not solid; it is permeated with a mesh of fibers that provides strength without unnecessary increases in weight that might hinder movement. In the vertebrate endoskeleton, growth at the end plates of long bones occurs until mature size is reached.

(B) **Exoskeleton.** Found in arthropods. Arthropods have a chitinous exoskeleton and jointed appendages. The exoskeleton allows muscle attachment to the interior of the skeleton so appendages move at joints. The skeleton also provides external armor for protection against physical injury and predators; however, it limits growth. Arthropods outgrow exoskeletons and must periodically molt, shedding their existing exoskeleton and producing a new, larger one. During molting, they are vulnerable to predation. The size of arthropods is limited in part by the exoskeleton. The weight of the exoskeleton, along with factors such as the mechanism of respiration, appear to limit the size of arthropods compared to that of chordates.

(C) **Hydrostatic skeleton.** Found in annelids. Annelids have segmented bodies and a coelom, which is divided into sealed chambers. The chambers are surrounded by bands of longitudinal and circular muscles that contract sequentially to change the shape of each segment. Since the coelom is fluid-filled and water is incompressible, the action of the muscle presses against the fluid, changing the shape of the segment and the annelid through coordinated action of the nervous system. The hydrostatic skeleton does not provide a stiff support against gravity, so it is found in small and aquatic animals.

SUGGESTONS FOR ESSAY QUESTION 4

(A) The strain of irradiated bacteria must have been changed in a way that suppresses the effects of the stop codon inserted in the Mut5 viral genome. The only way to do this is with a mutated tRNA that suppresses the stop codon, inserting an amino acid at this position into the viral protein as it is translated, so that translation can produce the full-length protein required for a mature infectious virus to replicate and infect new cells. This is similar to the suppressors of stop codons like amber, which were discovered early in the history of molecular biology.

(B) A mutation in a tRNA gene that allows translation through a stop codon in the viral gene will also allow translation through stop codons in bacterial genes. There are three stop codons, and every bacterial gene uses one of them to stop translation at a specific point, thus limiting the size of the proteins produced. Assuming that the three stop codons are used roughly equally by bacterial genes, a significant number of bacterial proteins in the irradiated strain will be longer than their counterparts in the wild-type bacteria, since translation will continue past the normal stop codon.

(C) If a bacterial cell is infected with both viruses and if the very short Mut5 protein (100 amino acids) does not block assembly of wild-type viral capsid proteins into viruses, then there is no reason why both genomes would not be packaged into viruses. Those viruses containing the wild-type bacteriophage genome will be infectious.

ANSWER SHEET PRACTICE TEST 3

SECTION I

1. Ⓐ Ⓑ Ⓒ Ⓓ Ⓔ
2. Ⓐ Ⓑ Ⓒ Ⓓ Ⓔ
3. Ⓐ Ⓑ Ⓒ Ⓓ Ⓔ
4. Ⓐ Ⓑ Ⓒ Ⓓ Ⓔ
5. Ⓐ Ⓑ Ⓒ Ⓓ Ⓔ
6. Ⓐ Ⓑ Ⓒ Ⓓ Ⓔ
7. Ⓐ Ⓑ Ⓒ Ⓓ Ⓔ
8. Ⓐ Ⓑ Ⓒ Ⓓ Ⓔ
9. Ⓐ Ⓑ Ⓒ Ⓓ Ⓔ
10. Ⓐ Ⓑ Ⓒ Ⓓ Ⓔ
11. Ⓐ Ⓑ Ⓒ Ⓓ Ⓔ
12. Ⓐ Ⓑ Ⓒ Ⓓ Ⓔ
13. Ⓐ Ⓑ Ⓒ Ⓓ Ⓔ
14. Ⓐ Ⓑ Ⓒ Ⓓ Ⓔ
15. Ⓐ Ⓑ Ⓒ Ⓓ Ⓔ
16. Ⓐ Ⓑ Ⓒ Ⓓ Ⓔ
17. Ⓐ Ⓑ Ⓒ Ⓓ Ⓔ
18. Ⓐ Ⓑ Ⓒ Ⓓ Ⓔ
19. Ⓐ Ⓑ Ⓒ Ⓓ Ⓔ
20. Ⓐ Ⓑ Ⓒ Ⓓ Ⓔ
21. Ⓐ Ⓑ Ⓒ Ⓓ Ⓔ
22. Ⓐ Ⓑ Ⓒ Ⓓ Ⓔ
23. Ⓐ Ⓑ Ⓒ Ⓓ Ⓔ
24. Ⓐ Ⓑ Ⓒ Ⓓ Ⓔ
25. Ⓐ Ⓑ Ⓒ Ⓓ Ⓔ
26. Ⓐ Ⓑ Ⓒ Ⓓ Ⓔ
27. Ⓐ Ⓑ Ⓒ Ⓓ Ⓔ
28. Ⓐ Ⓑ Ⓒ Ⓓ Ⓔ
29. Ⓐ Ⓑ Ⓒ Ⓓ Ⓔ
30. Ⓐ Ⓑ Ⓒ Ⓓ Ⓔ
31. Ⓐ Ⓑ Ⓒ Ⓓ Ⓔ
32. Ⓐ Ⓑ Ⓒ Ⓓ Ⓔ
33. Ⓐ Ⓑ Ⓒ Ⓓ Ⓔ
34. Ⓐ Ⓑ Ⓒ Ⓓ Ⓔ
35. Ⓐ Ⓑ Ⓒ Ⓓ Ⓔ
36. Ⓐ Ⓑ Ⓒ Ⓓ Ⓔ
37. Ⓐ Ⓑ Ⓒ Ⓓ Ⓔ
38. Ⓐ Ⓑ Ⓒ Ⓓ Ⓔ
39. Ⓐ Ⓑ Ⓒ Ⓓ Ⓔ
40. Ⓐ Ⓑ Ⓒ Ⓓ Ⓔ
41. Ⓐ Ⓑ Ⓒ Ⓓ Ⓔ
42. Ⓐ Ⓑ Ⓒ Ⓓ Ⓔ
43. Ⓐ Ⓑ Ⓒ Ⓓ Ⓔ
44. Ⓐ Ⓑ Ⓒ Ⓓ Ⓔ
45. Ⓐ Ⓑ Ⓒ Ⓓ Ⓔ
46. Ⓐ Ⓑ Ⓒ Ⓓ Ⓔ
47. Ⓐ Ⓑ Ⓒ Ⓓ Ⓔ
48. Ⓐ Ⓑ Ⓒ Ⓓ Ⓔ
49. Ⓐ Ⓑ Ⓒ Ⓓ Ⓔ
50. Ⓐ Ⓑ Ⓒ Ⓓ Ⓔ
51. Ⓐ Ⓑ Ⓒ Ⓓ Ⓔ
52. Ⓐ Ⓑ Ⓒ Ⓓ Ⓔ
53. Ⓐ Ⓑ Ⓒ Ⓓ Ⓔ
54. Ⓐ Ⓑ Ⓒ Ⓓ Ⓔ
55. Ⓐ Ⓑ Ⓒ Ⓓ Ⓔ
56. Ⓐ Ⓑ Ⓒ Ⓓ Ⓔ
57. Ⓐ Ⓑ Ⓒ Ⓓ Ⓔ
58. Ⓐ Ⓑ Ⓒ Ⓓ Ⓔ
59. Ⓐ Ⓑ Ⓒ Ⓓ Ⓔ
60. Ⓐ Ⓑ Ⓒ Ⓓ Ⓔ
61. Ⓐ Ⓑ Ⓒ Ⓓ Ⓔ
62. Ⓐ Ⓑ Ⓒ Ⓓ Ⓔ
63. Ⓐ Ⓑ Ⓒ Ⓓ Ⓔ
64. Ⓐ Ⓑ Ⓒ Ⓓ Ⓔ
65. Ⓐ Ⓑ Ⓒ Ⓓ Ⓔ
66. Ⓐ Ⓑ Ⓒ Ⓓ Ⓔ
67. Ⓐ Ⓑ Ⓒ Ⓓ Ⓔ
68. Ⓐ Ⓑ Ⓒ Ⓓ Ⓔ
69. Ⓐ Ⓑ Ⓒ Ⓓ Ⓔ
70. Ⓐ Ⓑ Ⓒ Ⓓ Ⓔ
71. Ⓐ Ⓑ Ⓒ Ⓓ Ⓔ
72. Ⓐ Ⓑ Ⓒ Ⓓ Ⓔ
73. Ⓐ Ⓑ Ⓒ Ⓓ Ⓔ
74. Ⓐ Ⓑ Ⓒ Ⓓ Ⓔ
75. Ⓐ Ⓑ Ⓒ Ⓓ Ⓔ
76. Ⓐ Ⓑ Ⓒ Ⓓ Ⓔ
77. Ⓐ Ⓑ Ⓒ Ⓓ Ⓔ
78. Ⓐ Ⓑ Ⓒ Ⓓ Ⓔ
79. Ⓐ Ⓑ Ⓒ Ⓓ Ⓔ
80. Ⓐ Ⓑ Ⓒ Ⓓ Ⓔ
81. Ⓐ Ⓑ Ⓒ Ⓓ Ⓔ
82. Ⓐ Ⓑ Ⓒ Ⓓ Ⓔ
83. Ⓐ Ⓑ Ⓒ Ⓓ Ⓔ
84. Ⓐ Ⓑ Ⓒ Ⓓ Ⓔ
85. Ⓐ Ⓑ Ⓒ Ⓓ Ⓔ
86. Ⓐ Ⓑ Ⓒ Ⓓ Ⓔ
87. Ⓐ Ⓑ Ⓒ Ⓓ Ⓔ
88. Ⓐ Ⓑ Ⓒ Ⓓ Ⓔ
89. Ⓐ Ⓑ Ⓒ Ⓓ Ⓔ
90. Ⓐ Ⓑ Ⓒ Ⓓ Ⓔ
91. Ⓐ Ⓑ Ⓒ Ⓓ Ⓔ
92. Ⓐ Ⓑ Ⓒ Ⓓ Ⓔ
93. Ⓐ Ⓑ Ⓒ Ⓓ Ⓔ
94. Ⓐ Ⓑ Ⓒ Ⓓ Ⓔ
95. Ⓐ Ⓑ Ⓒ Ⓓ Ⓔ
96. Ⓐ Ⓑ Ⓒ Ⓓ Ⓔ
97. Ⓐ Ⓑ Ⓒ Ⓓ Ⓔ
98. Ⓐ Ⓑ Ⓒ Ⓓ Ⓔ
99. Ⓐ Ⓑ Ⓒ Ⓓ Ⓔ
100. Ⓐ Ⓑ Ⓒ Ⓓ Ⓔ

answer sheet

SECTION II

Essay Question 1

answer sheet

Essay Question 2

answer sheet

Essay Question 3

answer sheet

Essay Question 4

answer sheet

Practice Test 3

SECTION I

100 QUESTIONS • 80 MINUTES

Directions: Each of the questions or incomplete statements below is followed by five suggested answers or completions. Select the one that is best in each case and mark the corresponding circle on the answer sheet.

QUESTIONS 1–2 REFER TO THE FOLLOWING EQUATION:

$$C_6H_{12}O_6 + xO_2 = 6H_2O + 6\ CO_2 + \text{energy}$$

1. To balance this equation, how many oxygen molecules (x) are required?

(A) 1
(B) 2
(C) 3
(D) 4
(E) 6

2. The above equation best represents which of the following processes?

(A) Photosynthesis
(B) The Calvin cycle
(C) Glycolysis
(D) The Krebs cycle
(E) Glycolysis and oxidative phosphorylation

3. In the vertebrate eye, how does the eye focus on objects at different distances?

(A) The length of the eye is altered by muscles surrounding the eye in the eye socket.
(B) Optical processing in the brain sharpens the image differently based on perceived distance.
(C) The shape of the lens is altered by contraction of ciliary muscles, changing the focal distance.
(D) Dilation and constriction of the iris changes the pupil diameter and moves the focal point.
(E) The shape of the cornea changes according to stress placed on the vitreous humor.

4. The osmolarity of bony fishes living in fresh water is maintained by which of the following mechanisms?

(A) Swallowing of water and excretion of concentrated urine
(B) Osmotic inflow of water and production of large quantities of dilute urine
(C) Concentration of nitrogenous waste as uric acid
(D) Establishment of iso-osmotic potential with the external environment
(E) Accumulation of urea in tissues to prevent water loss

5. Which of the following processes has a negative change in free energy?

(A) Flagellar movement of a bacterium
(B) Absorption of glucose from an area of 50 mM glucose to a region containing 200 mM glucose
(C) Fermentation of glucose, creating ethanol
(D) Synthesis of a polysaccharide
(E) Rolling a rock up a hill

6. Five liters of a solution of 200 mM pyruvate contains how many moles of pyruvate?

(A) 0.01 moles
(B) 0.2 moles
(C) 0.5 moles
(D) 1 mole
(E) 2.5 moles

7. Which of the following plays the most important role in the formation of secondary protein structure?

(A) Polypeptide bonds
(B) Hydrogen bonds between amino acid side chains
(C) van der Waals interactions
(D) Hydrogen bonds between functional groups in polypeptide backbone
(E) Ionic bonds

QUESTION 8 REFERS TO THE FOLLOWING FIGURE:

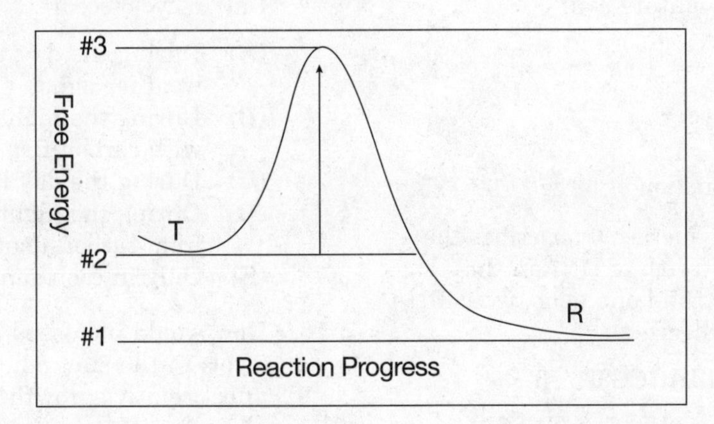

8. This chart describes the progress of a reaction converting T to R in the absence of an enzyme. If an enzyme that catalyzes this reaction is added, how will the reaction be changed?

(A) The energy level of T (#2) will be reduced.
(B) The energy level of R (#1) will be reduced.
(C) The equilibrium of T and R will be changed, increasing the amount of R present at equilibrium.
(D) The difference between #3 and #2 will be reduced.
(E) The backward reaction from R to T will be blocked.

9. Lipids are a more efficient form of energy storage than carbohydrates for long-term energy storage by animals because

(A) Lipids are more highly reduced.
(B) Carbohydrates contain less water.
(C) Lipids are not stable for long-term energy storage.
(D) Lipid storage can lead to heart disease.
(E) Carbohydrates are more readily accessed for short-term energy needs.

10. How do chromatin proteins reach their destination in the cell?

(A) They are transported in membrane vesicles from the endoplasmic reticulum into the nucleus.
(B) They are translated on the nuclear envelope and bud off in vesicles that enter the nucleus.
(C) They are imported into the nucleus through nuclear pores in the nuclear envelope.
(D) They are translated in the nucleus and retained in the nucleus after translation.
(E) They are integrated as transmembrane proteins in the nuclear envelope.

11. The movement of sperm flagella involves which of the following?

(A) Streaming of cytoplasm involving microfilaments
(B) Sliding of microtubules past one another, driven by the motor protein dynein
(C) Rotary movement of basal motor proteins
(D) Actin- and myosin-based filament sliding
(E) Fluid mosaic model of cellular motion

practice test

12. Meiosis of one primary oocyte creates how many mature ova?

(A) 1
(B) 2
(C) 3
(D) 4
(E) 8

13. If RNA polymerase transcribes the following stretch of DNA in the template strand of a gene, what RNA will it synthesize?

DNA: 5'-ATTAGCGTAA-3'

(A) RNA: 5'-TTACGCTAAT-3'
(B) RNA: 5'-TAATCGCATT-3'
(C) RNA: 5'-TTUCGCTUUT-3'
(D) RNA: 5'-UUACGCUAAU-3'
(E) RNA: 5'-UAAUCGCUAA-3'

14. If a flask of sterile nutrient media containing all essential nutrients is inoculated with bacteria from a single colony and then incubated at 37° C, which of the following would best describe the growth of the bacteria?

(A) The bacteria will indefinitely grow exponentially.
(B) The bacteria will compete with other organisms in the flask for limited nutrients.
(C) The bacteria will grow exponentially until density-dependent factors cause the population size to stabilize or fall.
(D) The bacteria will display limited growth until natural selection provides for more efficient bacteria capable of exponential population growth.
(E) The population size will rise initially and then fluctuate around the carrying capacity.

15. In a CAM plant, when does the Calvin cycle occur?

(A) During the day, unless the weather is hot
(B) During the night, in conjunction with carbon fixation
(C) During the day in all conditions
(D) During the night, separately from carbon fixation
(E) Only in cool, sunny weather

16. How would increased carbon dioxide levels in the environment likely affect relative growth of C3 and C4 plants?

(A) The growth of C3 plants will increase because of increased photorespiration, while C4 plants will be unaffected.
(B) The growth of C3 plants will be inhibited because they will use sunlight less effectively.
(C) Growth of C4 plants will be inhibited by the increased carbon dioxide levels.
(D) The growth of C3 plants will be increased by the reduction in photorespiration and improved carbon fixation, while C4 plant growth will show less of an increase.
(E) All plants will increase equally in growth.

17. In fish, which of the following is a consequence of having a two-chambered heart, compared with other vertebrates that have a three- or four-chambered heart?

(A) Fish have a higher pressure in the systemic circulation than the pulmonary circulation.
(B) Fish have a higher pressure in the left atria than in the right atria.
(C) Pressure in systemic fish capillaries is lower than in the gills.
(D) Fish are unable to extract oxygen efficiently using their gills alone and must increase oxygen absorption by swallowing water.
(E) Fish endurance is greater than that observed in amphibians.

18. Smooth muscle does not appear striated under a light microscope because

(A) The sliding filament mechanism of muscle contraction is not used.
(B) The actin and myosin filaments are not aligned in sarcomeres.
(C) Actinomyosin replaces actin in muscle contraction role.
(D) The length of smooth muscle cells does not change during muscle action.
(E) Each cell has a single nucleus.

19. A decrease in oceanic phytoplankton would most likely have which of the following effects on the Earth's chemical cycles?

(A) Evaporation from oceans would be reduced because of reduced transpiration.
(B) More carbon dioxide would accumulate in the atmosphere because of reduced photosynthetic activity.
(C) The rate of denitrification would increase, releasing more nitrate pollutants into the atmosphere.
(D) More carbon would enter deep ocean sinks and sedimentary rocks.
(E) The amount of acid rain would decrease.

20. The surfaces of leaves on a tree can be significantly cooler than the air that is farther from the tree because

(A) Water has a high molar heat of evaporation.
(B) Trees absorb only a limited range of wavelengths for photosynthesis.
(C) Water molecules display a great deal of cohesion and surface tension.
(D) Removing carbon dioxide from the air creates cooling, resulting from a reduced greenhouse effect.
(E) Antenna pigments radiate most of the energy at longer wavelengths.

21. The layer of liquid that lines alveoli in the mammalian lung contains a surfactant. The function of this surfactant is to

(A) create bubbles that improve gas exchange with the atmosphere
(B) drive countercurrent exchange in the lungs
(C) reduce surface tension in the alveoli to prevent them from collapsing
(D) allow oxygen molecules to squeeze between endothelial cells to reach capillaries
(E) keep the surface of alveoli clean

22. Most frogs display which of the following adaptations?

(A) Internal fertilization
(B) Amniotic eggs
(C) Loss of gills in adults
(D) A muscular post-anal tail in adults
(E) A ventral nerve cord

23. Which of the following is most closely related to the ancestor of the plants?

(A) Fungi
(B) Archae
(C) Photosynthetic cyanobacteria
(D) Green algae
(E) Ciliated protists

24. A cancer cell would be most likely to be killed by which of the following elements of the immune system?

(A) Macrophage
(B) Cytotoxic T cell
(C) Helper T cell
(D) Plasma B cell
(E) Cytotoxic B cell

25. Regulation of breathing is *primarily* mediated by which of the following areas of the brain?

(A) Medulla oblongata
(B) Cerebellum
(C) Corpus callosum
(D) Thalamus
(E) Hippocampus

26. Which of the following occurs when a mammal hibernates in winter?

(A) The mammal's temperature set point is reduced, allowing for energy conservation.
(B) The mammal succumbs to extreme cold and is unable to maintain its body temperature because of lack of nutrition.
(C) The lack of activity reduces the mammal's core temperature.
(D) The mammal's core temperature is unaffected during hibernation.
(E) The reduced temperature is a result of other endocrine factors that override the mammal's normal hypothalamic set point.

27. Which of the following is *not* a property of phospholipids?

(A) Spontaneous self-assembly into membranes and vesicles
(B) Storage of energy in adipose tissue
(C) Lateral diffusion in lipid bilayer membranes
(D) Each contain two fatty acid groups
(E) Contain a glycerol and polar group containing phosphate

28. Which of these is an example of the movement of potassium ions through a ligand-gated ion channel?

(A) Passive diffusion
(B) Facilitated diffusion
(C) Secondary active transport
(D) Primary active transport
(E) Osmosis

QUESTIONS 29–30 REFER TO THE FOLLOWING:

A photosynthetic antenna pigment is purified from cells and mixed in buffer; it is then exposed to a range of wavelengths to test absorption. The following absorption spectra are observed:

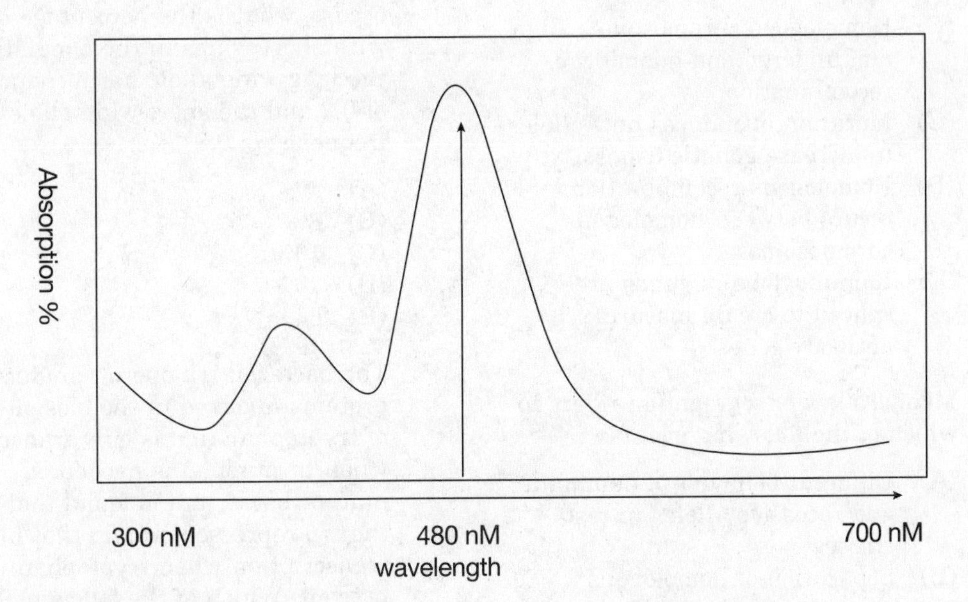

29. A flash of light at 480 nM is also followed by fluorescence of the protein in solution. Which of the following would characterize this fluorescence?

(A) All of the energy that enters the system is re-emitted in the light that is fluoresced.
(B) The fluoresced light will cover equally all measured wavelengths of light.
(C) Light will be fluoresced at the same wavelength at which it is absorbed.
(D) Light will be fluoresced at wavelengths smaller than 480 nM.
(E) Light will be fluoresced mainly at wavelengths greater than 480 nM.

30. The effect of this molecule in photosynthesis is to

(A) absorb light that chlorophyll does not absorb itself and to transfer energy to the photosystem reaction center
(B) reflect sunlight in unproductive wavelengths that are not well-suited for photosynthesis
(C) keep leaves from becoming excessively heated as a result of metabolic reactions
(D) allow photosynthesis to proceed in the absence of chlorophyll
(E) transfer energy directly to the Calvin cycle

31. During meiotic recombination, which of the following occurs?

(A) Sister chromatids align and undergo homologous recombination.
(B) Homologous chromosomes align and undergo non-homologous recombination.
(C) Mutation introduces new alleles to increase genetic fitness.
(D) Homologous recombination occurs between homologous chromosomes.
(E) Immunoglobulin genes are spliced to create mature antibody genes.

32. Mendel's law of segregation refers to which of the following events:

(A) Different behavior of dominant and recessive alleles in test crosses
(B) Separation of homologous chromosomes into different gametes during meiosis
(C) The blending of traits seen in partial dominance
(D) The independent inheritance in dihybrid crosses seen with traits caused by genes on two different chromosomes
(E) The bell-shaped distribution of traits caused by multiple genes

33. Of the following, what do all viruses have in common?

(A) A DNA genome
(B) Envelope membrane proteins
(C) A nucleic acid genome replicated inside the host cell
(D) Lysogenic and lytic stages in their life cycle
(E) Similar shape

34. In a wild population of fruit flies, two alleles for wing length are present—a recessive short-wing allele and a dominant long-wing allele. Assuming that Hardy-Weinberg equilibrium occurs, what is the percentage of flies with short wings in the population if the long-wing allele has a frequency of 0.2 and the short-wing allele has a frequency of 0.8?

(A) 4%
(B) 8%
(C) 16%
(D) 40%
(E) 64%

35. The bacterial trp operon produces proteins involved in the biosynthesis of tryptophan and is only transcribed when tryptophan is needed. A mutant bacterium is found that lacks the trp repressor protein that blocks transcription when tryptophan is present. Which of the following best describes how these bacteria will grow?

(A) The mutant bacteria will grow more rapidly than wild-type bacteria in all conditions.
(B) The mutant bacteria will be unable to grow if tryptophan is not provided.
(C) The mutant bacteria will compete less effectively with wild-type bacteria when tryptophan is provided for the bacteria.
(D) The mutant bacteria will not grow under any condition.
(E) The mutant bacteria will grow more rapidly than wild-type bacteria in the presence of tryptophan.

36. If there are three genes in the trp operon, how would eliminating the trp repressor affect these three genes?
   (A) They will be transcribed at three different rates, all higher than before.
   (B) They will all be transcribed at the same rate and will not respond to the presence of tryptophan.
   (C) They will be transcribed more rapidly but translated more slowly.
   (D) Transcription remains the same for the genes, but translation is repressed.
   (E) They will be repressed in all conditions.

37. Human red blood cells have all of the following traits EXCEPT
   (A) They do not perform transcription.
   (B) They are involved in the transportation of carbon dioxide in blood.
   (C) They are produced from precursor cells in the lymphatic system.
   (D) They do not perform oxidative phosphorylation.
   (E) Their production is stimulated by erythropoietin.

38. The movement of microorganisms toward a light source is called
   (A) Phototropism
   (B) Phototaxis
   (C) Photoactivation
   (D) Chemotaxis
   (E) Chemotropism

39. Bile salts are produced in which organ?
   (A) Pancreas
   (B) Appendix
   (C) Liver
   (D) Gall Bladder
   (E) Small intestine

40. Denitrifying bacteria play which one of the following roles in the nitrogen cycle?
   (A) Convert nitrates to molecular nitrogen, N_2
   (B) Fix nitrogen from N_2 to ammonia and nitrates
   (C) Decompose amino acids in decaying material to release nitrates in soil
   (D) Convert chemical fertilizers into biologically active forms used by plants
   (E) Digest nitrogen-containing plant material in the gastrointestinal tracts of herbivores

41. Adaptations found in ferns but not in mosses include
   (A) Photosystem I
   (B) Xylem
   (C) Gametophytes
   (D) Motile sperm
   (E) Seeds

42. Which of the following have cell walls made of chitin?
   (A) Grasses
   (B) Mushrooms
   (C) Gram-positive bacteria
   (D) Echinoderms
   (E) Insects

43. Hemophilia is a genetic disorder that causes deficient blood clotting. The gene involved in hemophilia has been localized to the X chromosome, and a recessive allele of this gene has been linked to hemophilia. If a man with hemophilia marries a woman who is heterozygous for this allele, what percentage of their female children will have hemophilia?
   (A) 0%
   (B) 25%
   (C) 50%
   (D) 75%
   (E) 100%

44. By what mechanism is carbon dioxide exchanged between cells and the external environment in Platyhelminthes (flatworms)?

(A) Through tracheoles
(B) By oxygen transport protein in hemolymph
(C) By passive diffusion across capillary endothelial cells
(D) By passive diffusion between cells and the external environment
(E) Via exchange across gills

45. The main interaction between fatty acid side chains in lipid bilayer membranes is

(A) van der Waals
(B) hydrogen bond
(C) covalent bond
(D) ionic bond
(E) transition state intermediate

46. The volume of air exchanged by a human during a normal breath at rest is

(A) tidal volume
(B) residual volume
(C) vital capacity
(D) reserve capacity
(E) functional reserve

47. Artificial selection is sometimes used in agriculture, where plants are selectively bred for desired traits such as insect resistance or seed size. The main difference between natural selection and artificial selection is that

(A) Artificial selection requires the introduction of modified genes using recombinant DNA technology.
(B) Artificial selection requires human intervention to create selective pressure.
(C) Only artificial selection will work on plants that lack genetic variation.
(D) Artificial selection works on plants with polyploid genomes.
(E) Natural selection cannot act on the same traits that artificial selection acts on.

48. In cardiac muscle tissue, action potentials are communicated through the tissue by which one of the following mechanisms?

(A) Chemical synapses connect cardiac muscle cells, carrying action potentials between cells after a slight delay.
(B) Electrical synapses interconnect cardiac muscle cells, carrying action potentials directly from cell to cell through gap junctions.
(C) Neurons running throughout cardiac muscle tissue trigger and convey action potentials with each cardiac cycle.
(D) Cardiac muscle is composed of large, multinucleated cells that carry action potentials from one side of the heart to the other.
(E) Hormones carried through the blood act on cardiac muscle cells to trigger action potentials.

QUESTIONS 49–50 REFER TO THE FOLLOWING FIGURE:

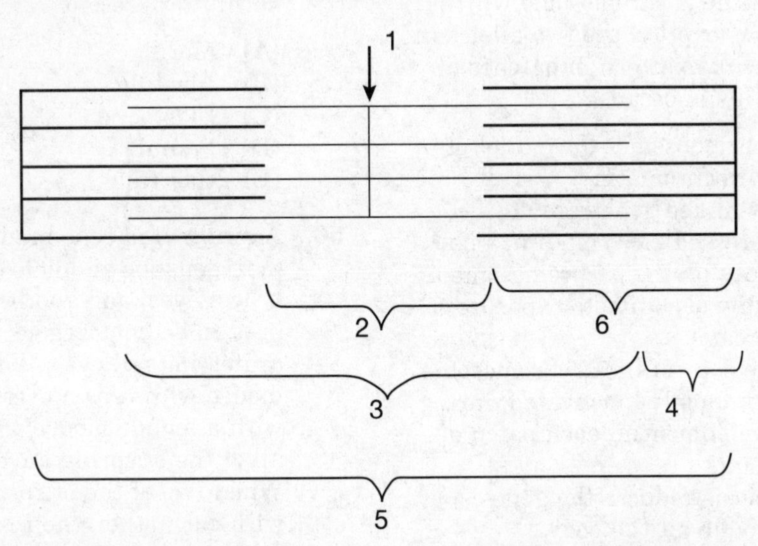

49. The figure depicts a sarcomere in skeletal muscle. Which of the following grows shorter during muscle contraction?

(A) 2
(B) 3
(C) 5
(D) 3 and 5
(E) 2 and 5

50. In this figure, where is tropomyosin localized between muscle contractions?

(A) 1
(B) 2
(C) 3
(D) 4
(E) 6

51. For an ectothermic fish, how will warming water that accompanies climate change affect its oxygen use and availability?

(A) It will regulate its temperature to maintain constant respiratory needs, with increased oxygen in warmer water.
(B) It will regulate its temperature and have reduced oxygen use, with increased oxygen available in warmer water.
(C) Its internal temperature will increase, decreasing its metabolic use of oxygen, with reduced oxygen available in warmer water.
(D) Its internal temperature will increase, increasing its use of oxygen, with less oxygen available in warmer water.
(E) Its internal temperature will increase and it will resort to lactic fermentation to maintain energy levels with less oxygen.

practice test

52. Females who are heterozygous for genes on the X chromosome will sometimes express the two alleles in a mosaic-like pattern in patches of tissue. This is because

(A) X chromosomes have a lot of heterochromatin
(B) Sex-linked genes are always expressed more often in women
(C) Genes on the X chromosome are highly sensitive to exposure to estrogen
(D) One copy of the X chromosome is randomly inactivated during development in each patch of tissue
(E) Meiosis reduces the gene copy throughout the body

53. Down syndrome with trisomy of chromosome 21 results from

(A) Meiotic recombination
(B) Non-disjunction during meiosis 2
(C) Chromosomal translocation
(D) Nonsense mutation
(E) Fertilization of an egg with multiple sperm cells

54. In DNA, which of the following is true?

(A) The percentage of guanine in DNA is always the same in all species.
(B) The total percentage of guanine and cytosine in DNA always equals 50%.
(C) The percentage of adenine always equals the percentage of thymine in DNA.
(D) Some species use uracil instead of thymine in DNA.
(E) The two polynucleotide strands in the DNA double helix run parallel in the same 5' to 3' orientation.

55. In which embryonic stage is mesoderm first present?

(A) Morula
(B) Blastula
(C) Blastocyst
(D) Planula
(E) Gastrula

56. A strain of mice is bred to be homozygous for an allele conferring brown eyes, and another strain is bred to be homozygous for an allele conferring red eyes. When a male mouse with brown eyes is crossed with a female mouse with red eyes, all of the offspring have brown eyes. When two of the offspring are crossed with each other, the results of this cross are:

2 males with red eyes

2 females with red eyes

5 males with brown eyes

7 females with brown eyes

The results of this cross suggest which of the following?

(A) The gene for eye color is on the X chromosome.
(B) Natural selection reduces the prevalence of the allele for red eyes.
(C) The F1 mice are at Hardy-Weinberg equilibrium.
(D) The eye color gene is autosomal and the allele for brown is dominant.
(E) Animals with red eyes do not see as well.

57. Which of the following is involved in the response of cytotoxic T cells to antigens?

(A) Binding of a T cell receptor to a virus in the blood causes T cells to proliferate and kill infected cells.
(B) Helper T cells stimulated by antigen-presenting cells secrete cytokines that activate cytotoxic T cells.
(C) The T cell receptor gene undergoes recombination in response to the presence of an antigen.
(D) The T cell receptor of cytotoxic T cells recognizes antigen in MHC II in antigen-presenting cells.
(E) T cells phagocytose infected cells and destroy them in lysosomes.

58. Different visual receptor cells are sensitive to detecting different color ranges. Which of the following describes the difference in photoreceptor cells that allow the detection of different colors?

(A) Different cells use different second messenger signaling systems in response to light.
(B) Different cells filter light differently, allowing light of different wavelengths to excite photosensitive proteins.
(C) Different cells have visual pigments containing different chemical derivatives of vitamin A.
(D) Cells that respond to different wavelengths are distributed differently around the retina.
(E) Different cells express different genes that provide the protein component of visual pigments.

59. Fish and dolphins have a similar body shape. This is an example of

(A) homologous structures
(B) analogous structure
(C) adaptive radiation
(D) punctuated equilibrium
(E) divergent selection

60. Each immunoglobulin-G protein includes how many polypeptide subunits?

(A) 1
(B) 2
(C) 3
(D) 4
(E) 6

61. Which of the following would respond to volatile organic compounds?

(A) Taste receptors for sweet tastes
(B) Receptors in lateral line of osteicthyes
(C) Baroreceptors in a giraffe
(D) Photoreceptors in a dragonfly
(E) Olfactory receptors in a mouse

62. Many low-abundance recessive alleles persist in the human population for many generations, even if homozygous-recessive individuals do not survive to reproductive age. Why would these alleles persist in the population?

(A) Recombination constantly regenerates the alleles.
(B) Heterozygotes act as carriers for the recessive alleles.
(C) Recessive alleles are not subject to natural selection.
(D) Sexual reproduction provides increased genetic variation for populations.
(E) Independent segregation ensures that the recessive alleles are present in separate gametes.

63. One difference between mitochondria and chloroplasts is that

(A) Mitochondria contain DNA and chloroplasts do not.
(B) Mitochondria divide mitotically and chloroplasts do not.
(C) Mitochondria are surrounded by two membranes but chloroplasts have only one.
(D) Mitochondria use a pH gradient to make ATP and chloroplasts do not.
(E) Mitochondria consume oxygen but chloroplasts produce oxygen.

64. In the following reaction, when the Δ G is 0, which of the following statements is true?

$$A + B \leftrightarrows C + D$$

(A) The reaction will progress until the ΔG is negative.
(B) The forward reaction alone will progress.
(C) Both forward and backward reactions will occur at the same rate.
(D) Neither forward or backward reactions will occur.
(E) The reaction will progress to the right until the ΔG is positive.

65. Information suggesting that RNA may have provided the genome for the earliest forms of life includes

(A) fossilized sections of RNA in remains of early cells
(B) catalytic activity observed in self-splicing introns
(C) use of RNA genomes in some archae
(D) the role of mRNA in membrane formation
(E) the ongoing creation of RNA from abiotic compounds in some clay-like soils

66. How does glucagon affect the liver?

(A) Glycogen synthesis is stimulated.
(B) Uptake of free fatty acids to produce triglycerides is increased.
(C) Glycogen phosphorylase activity is triggered.
(D) The liver releases insulin.
(E) The rate of liver glycolysis will increase.

67. Rhizobium bacteria occupying nodules in legumes perform which of the following?

(A) Photosynthesis
(B) Carbon fixation
(C) Nitrogen fixation
(D) Parasitism of plant tissues
(E) Protection from fungal infection

QUESTIONS 68–69 REFER TO THE FOLLOWING:

A true breeding line of beans with wrinkled blue seeds is crossed with a true breeding line of beans with smooth green seeds. The F1 progeny of this cross all have smooth, green seeds. Two F1 plants are crossed and the following phenotypes are observed in the F2 generation:

Smooth green	200 beans
Wrinkled green	5 beans
Smooth blue	7 beans
Wrinkled blue	40 beans

68. The above data suggest which of the following?

(A) The gene for seed color displays epistasis toward the gene for seed shape.
(B) The two genes are located on the same chromosome.
(C) The genes display independent assortment.
(D) The F1 progeny are homozygous for the green allele.
(E) The beans demonstrate polyploidy.

69. The abundance of the smooth blue phenotype in this cross is related to which of the following?

(A) The number of copies of the seed color gene that are present
(B) The distance between the genes for seed color and seed shape
(C) The exposure of the bean plants to increased sunlight during the experiment
(D) The degree of penetrance between the two genes involved
(E) The rate of mutation in this species of bean

70. Vaccination against a viral disease with a killed virus can protect against a future infection of the virus for many years. The mechanism by which this long-term protection occurs involves

(A) the circulation of immunoglobulins
(B) macrophage activation by antigen in later infections
(C) the proliferation and differentiation of memory B cells during the initial infection
(D) display of the antigen in MHC I
(E) the folding of immunoglobulins around antigen to increase affinity in later infections

71. In males, testosterone is produced primarily by which of the following?

(A) Sertoli cells in the testes
(B) Leydig cells in the seminiferous tubules
(C) Spermatogonia
(D) Prostate gland
(E) Anterior pituitary

QUESTIONS 72–73 REFER TO THE FOLLOWING:

A point mutation in a mouse gene encoding a protein kinase changes a lysine (wild-type) to an arginine codon (Arg22). The version of the gene containing the arginine still produces protein in cells containing this gene. Cells that have one copy of the Arg22 version of the gene have one-half the protein kinase activity of cells with two copies of the wild-type allele, and cells with two copies of the Arg22 gene have no protein kinase activity, although the cells otherwise appear normal. Mice that have one copy of the Arg22 allele and one wild-type allele appear normal, while mice with two copies of the Arg22 allele are not seen in progeny of crosses between heterozygotes.

72. What type of mutation is described?

(A) Missense
(B) Deletion
(C) Frame shift
(D) Nonsense
(E) Translocation

73. Which of the following provides the best explanation for the behavior of the allele in mice?

(A) The mutation is a dominant lethal one.
(B) The mutation disrupts an essential step in protein translation.
(C) The activity of the gene is not required for the basic metabolism of cells, but it plays an essential role in mouse development.
(D) Mutations of lysines to arginines are quite conservative and do not disrupt protein function.
(E) The protein kinase causes apoptosis of cells involved in gamete formation.

74. Blocking the activity of acetylcholinesterase at the neuromuscular junction and not elsewhere in the body does which of the following?

(A) Produces skeletal muscle tetany
(B) Increases the frequency of action potentials
(C) Causes retrograde axonal action potentials
(D) Inhibits muscle contraction
(E) Prevents action potentials from reaching the muscle

75. Which of the following do nematodes possess?

(A) A complete digestive tract but no reproductive or circulatory tissue
(B) A complete digestive tract and an open circulatory system
(C) No circulatory system, with a pseudocoelom
(D) A gastrovascular cavity and gills for respiration
(E) No nerve cells but complete digestive tract

76. A heart attack (myocardial infarction) involves which of the following occurrences?

(A) The disruption of action potential passage from CNS to heart
(B) The disruption of the passage of action potentials from atrium to ventricles
(C) The disruption of blood flow to coronary muscle tissue
(D) Incomplete closure of heart valves, leading to inefficient blood flow
(E) Poor oxygenation of blood in the lungs

77. When does the right atrioventricular valve (the tricuspid valve) close?

(A) When the semilunar (pulmonary) valve closes
(B) When the right atrium contracts
(C) When the right ventricle contracts
(D) During ventricular diastole
(E) When stimulated by the pacemaker region

78. Which of the following is probably true about the evolution of early eukaryotes?

(A) Eukaryotes arose from prokaryotes on multiple occasions.
(B) Eukaryotes and prokaryotes arose independently.
(C) Eukaryotes first acquired the chloroplast, then mitochondria later.
(D) Eukaryotes arose before prokaryotes.
(E) The first eukaryotes acquired organelles from prokaryotes that became endosymbionts.

QUESTION 79 REFERS TO THE FOLLOWING FIGURE:

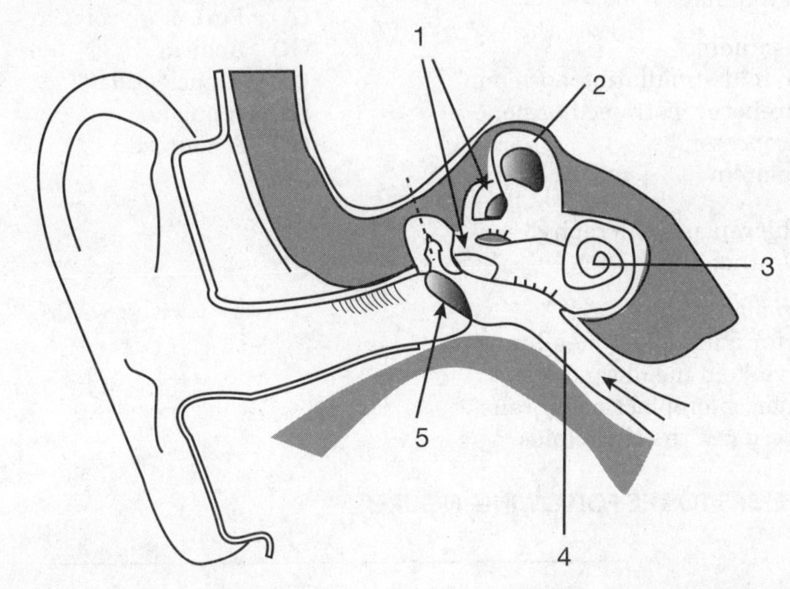

79. The auditory nerve is connected to which numbered structure in the illustration?

(A) 1
(B) 2
(C) 3
(D) 4
(E) 5

80. When a primary consumer eats a primary producer, which of the following does *not* occur with the energy present in the primary producer?

(A) Energy is lost as heat.
(B) Energy is lost in excreted material.
(C) Useful energy is lost in carbon dioxide.
(D) Energy is passed to higher trophic levels through predation.
(E) Some energy is captured in the biological macromolecules that are synthesized by the consumer.

81. Which of the following prevents the stomach epithelium from being damaged by acid secretion?

(A) Mucus
(B) Pepsin
(C) Bicarbonate
(D) Pyloric sphincter
(E) Salivary amylase

82. One of the actions of caffeine is to inhibit the body's enzyme phosphodiesterase, which breaks down cyclic AMP. How might caffeine affect signaling by epinephrine in muscle tissue?

(A) It will prevent cyclic AMP degradation, increasing the effects of epinephrine.
(B) It will increase the release of epinephrine by muscle tissue.
(C) It will bind to the epinephrine receptor, blocking epinephrine responses.
(D) It will reduce cyclic AMP levels in cells, increasing the epinephrine response.
(E) It will increase cyclic AMP levels in cells, blocking the response to epinephrine.

practice test

83. Which of the following is produced in the hypothalamus?

(A) Calcitonin
(B) Thyroid-stimulating hormone
(C) Adrenocorticotropic hormone
(D) Vasopressin
(E) Prolactin

84. In the chloroplast, chlorophyll is located in the:

(A) Stroma
(B) Outer chloroplast membrane
(C) Thylakoid membrane
(D) Inner chloroplast membrane
(E) Interior of the thylakoids

QUESTION 86 REFERS TO THE FOLLOWING FIGURE:

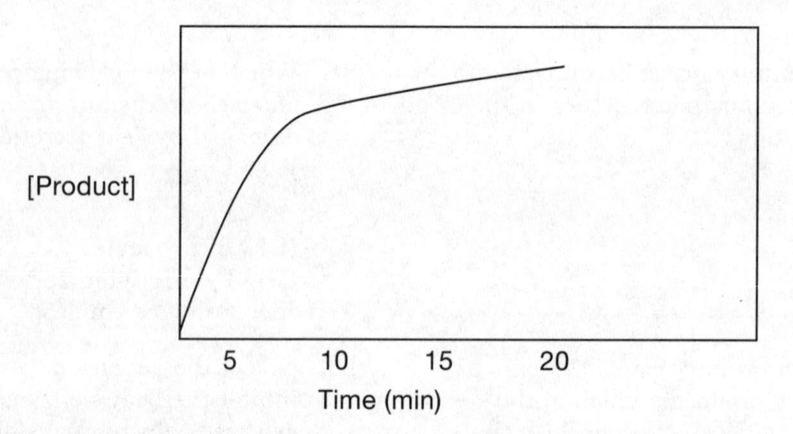

86. Factors affecting the activity of a purified enzyme are examined in a lab, producing the data shown above. Which of the following is the most likely explanation for what is observed in this reaction as more than 10 minutes pass?

(A) The V_{max} is achieved.
(B) The equilibrium for the reaction changes.
(C) The active sites become saturated with substrate.
(D) Product accumulates and acts as a competitive inhibitor for the enzyme.
(E) The enzyme is consumed in the reaction.

85. Ribose is which of the following?

(A) Part of glycolysis
(B) Achiral
(C) A nucleotide
(D) A purine
(E) A pentose

QUESTIONS 87–89 REFER TO THE FOLLOWING FIGURE:

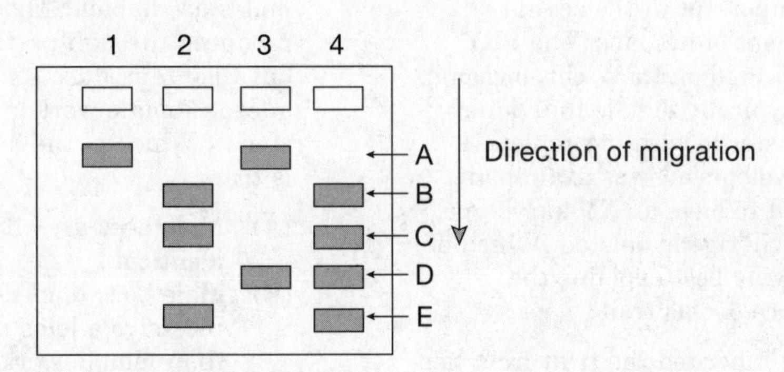

87. A researcher is working to clone a gene and has inserted a fragment of DNA containing this gene into a plasmid, PlasmidX1. The researcher digests four DNA samples in the following manner, and then subjects the samples to agarose gel electrophoresis:

1. PlasmidX1 digested with EcoR1
2. PlasmidX1 digested with BamH1
3. PlasmidX1 with gene inserted, digested with EcoR1
4. PlasmidX1 with gene inserted, digested with EcoR1 and BamH1

The band that contains the gene of interest must be:

(A) Band A
(B) Band B
(C) Band C
(D) Band D
(E) Band E

88. Samples that migrate further have what molecular property?

(A) Higher content of G-C base pairs
(B) Lower molecular weight
(C) Higher charge density
(D) More elongated shape
(E) Overall positive charge

89. How is the DNA in this gel visualized?

(A) Dark field microscopy
(B) Confocal microscopy
(C) UV irradiation of fluorescent dye
(D) Coomassie blue staining
(E) Gram stain

90. Cancer cells found in non-germ-line tissues that are not associated with viral infection are often found to have oncogenes that induce unchecked proliferation in cancer cells. The source of the oncogenes is

(A) mutation of genes present in all cells to regulate the cell cycle
(B) meiotic recombination, which generates new combinations of alleles in cancer cells that alter cell cycle regulation.
(C) cancer cells, which are often polyploid
(D) mutation of mRNA
(E) damaged ribosomes

91. A gene called SRY has been identified as important in the sexual development of humans. The SRY gene is located on the Y chromosome. A phenotypically female individual who was sterile with incomplete ovary development was examined and found to have an XY karyotype, with the SRY gene deleted. Which of the following best explains the inheritance of this trait?

(A) She inherited the trait from her mother.
(B) Her mother displayed parthenogenic reproduction.
(C) The trait is the result of a mutation that occurred in her father's germ line.
(D) The mutation occurred during gastrulation during the person's development.
(E) Non-disjunction occurred during spermatogenesis.

92. The fact that SRY deletion causes a person to develop as a woman suggests which of the following?

(A) The testosterone receptor is not involved in male sexual development.
(B) Women do not produce testosterone.
(C) SRY stimulates female sexual development.
(D) Women have an additional copy of the SRY gene on the X chromosome.
(E) The default sexual development pathway is to be female.

93. In bees, females are diploid and males are haploid. Haploid males are produced through parthenogenesis, but they reproduce sexually with the queen. Female worker bees are sterile. Which of the following results is true?

(A) Male bees are all genetically identical.
(B) Male bees tend to express recessive alleles more often than female worker bees.
(C) Male bees do not contribute genetically to the bee gene pool.
(D) Meiosis is not involved in bee oogenesis.
(E) All bees in a hive are clones.

94. Bacterial restriction enzymes provide which of the following adaptations?

(A) They allow viruses to degrade bacterial genome.
(B) They protect bacteria against bacteriophage infection.
(C) They switch lysogenic to lytic infections.
(D) They stimulate the adaptive immune response.
(E) They stimulate production of the capsule.

95. The radula in mollusks

(A) mates appendages in stationary bivalves
(B) is a beak-like structure in the mollusk's mouth
(C) is involved in chemosensation
(D) scrapes food in gastropods
(E) moves gills

96. Angiosperm fruit develops from which of the following?

(A) Ovule
(B) Ovary
(C) Carpel
(D) Cotyledons
(E) Vacuole

97. The wavelike contraction of smooth muscle in the human gastrointestinal tract is called

(A) homeostasis
(B) peritoneal
(C) esophageal reflux
(D) peristalsis
(E) secretion

98. A man is raised in an environment at sea level and then lives for several months at high altitude in a mountain range. In which one of the following ways might his body adapt to the new environment?

(A) A mutation of his hemoglobin genes provides increased affinity for oxygen.
(B) The body increases production of erythrocytes.
(C) Lymphocytes mature more rapidly.
(D) The walls of the lungs' alveoli between the blood and the atmosphere become thinner.
(E) The rate of carbon dioxide production increases.

99. Severing the corpus callosum in the brain would cause which one of the following?

(A) Lack of emotional responses to stimuli
(B) Loss of memory
(C) Lack of coordination of cerebral hemispheres
(D) Loss of motor responses
(E) Uncoordinated motion

100. The human genome contains about 3 billion base pairs of DNA. In contrast, some salamanders have 100 billion base pairs of DNA in their genome. Salamanders do not appear to have more genes than humans, however. Which of these is the most likely explanation for the larger size of the salamander genome?

(A) The salamander's genome has increased quantities of transposable elements and there is a lack of selection pressure against a large genome.
(B) Selection pressure favors large genomes to improve the efficiency of translation.
(C) Salamander DNA replication machinery is very rapid.
(D) Aquatic environments favor large genomes.
(E) Amphibians are tetraploid.

SECTION II

4 QUESTIONS • 90 MINUTES

Four mandatory questions will be asked. In general, there will be two questions about molecules and cells, one on genetics and evolution, and one on organisms and populations. Some questions may cover more than one of these areas. One or more of the four questions may be designed to test analytical and reasoning skills. Laboratory experiences may be reflected in these questions.

Essay Question 1

The glycolytic enzyme phosphofructokinase catalyzes the following reaction:

Fructose-6-phosphate + ATP ⇆ fructose-1,6-bisphosphate + ADP

This reaction step is a key irreversible step in glycolysis and is subject to regulation. The reaction rate displayed by the enzyme is tested with varying amounts of fructose-6-phosphate, using either a low concentration of ATP present or a high concentration of ATP present. This produces the following data:

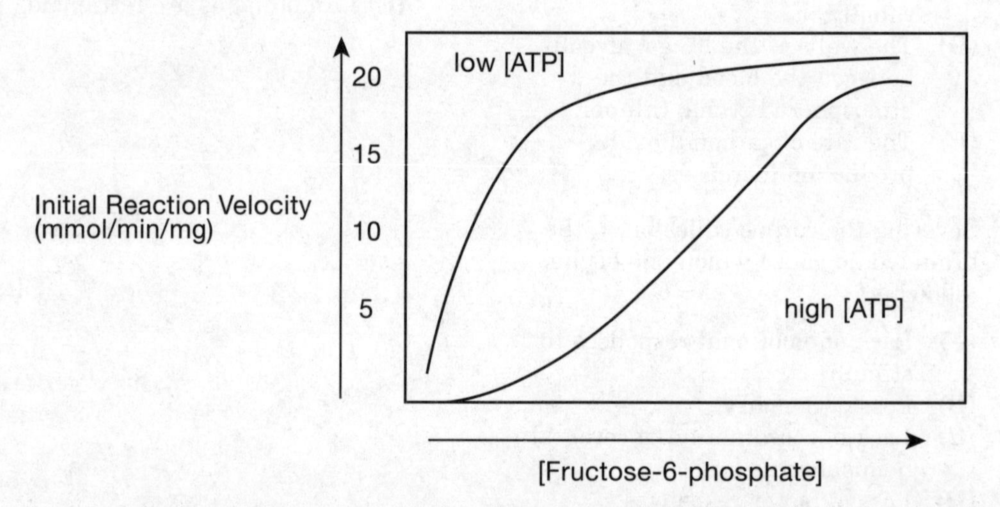

(A) Describe the effect of changing the concentration of fructose-6-phosphate in the presence of a low ATP concentration—including both ends of the concentration range—from low fructose-6-phosphate to high fructose-6-phosphate. Why does the curve level off at high concentrations of fructose-6-phosphate?

(B) Describe a mechanism by which ATP might inhibit enzyme activity when a high concentration of ATP is present.

(C) Describe the effect of inhibition of phosphofructokinase by ATP on metabolism in intact cells. In what way will this affect glucose consumption, oxygen consumption, carbon dioxide production, and ATP production?

Essay Question 2

Describe the impact of three of the following on the stimulation of skeletal muscle contraction by vertebrate motor neurons:

(A) Toxin that blocks voltage-gated sodium channels in neurons
(B) Inhibitor of acetylcholinesterase
(C) Antagonist of acetylcholine receptor on muscle cell
(D) Blockade of gap junctions

Essay Question 3

A researcher examines members of a large family for a disease that affects the nervous system. Individuals affected by the disease begin to display impairments in motor control at age 40, after which time the disease progresses rapidly and is lethal. The researcher observes the following pedigree:

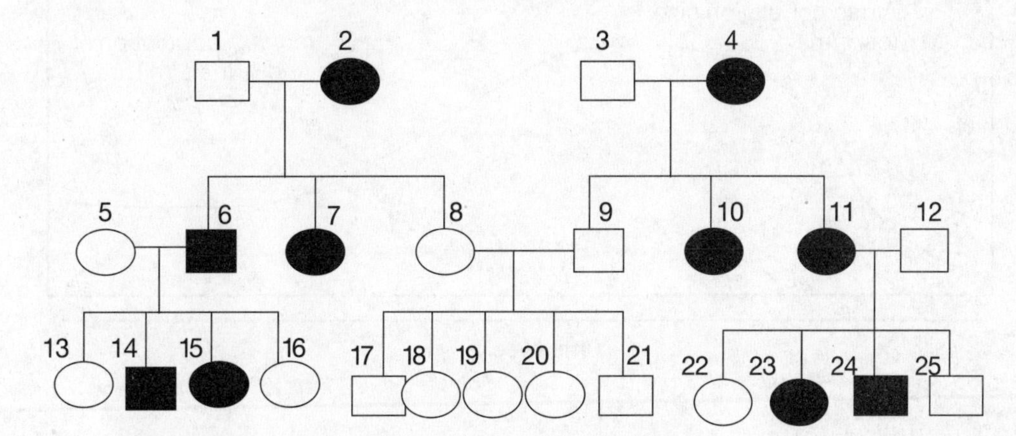

(A) The researcher suspects that an allele of a gene involved in neuronal signaling is involved. It is discovered that Individual 9 is homozygous at this gene, while Individual 4 is heterozygous at this gene. Assuming the disease is genetically inherited and this gene is responsible, what is the mechanism for inheritance of this disease? If Individual 25 is 25 years old, what is the probability that he will contract the disease later in life? How does this lethal disease persist in the population for a great number of generations?

(B) Since the family members in the study all live in the same area and the genetic diversity in this population is limited, the researcher suspects that the environment may play a role in the disease. Explain how two of the following processes would help distinguish between genetic and environmental influences involved in this disease. Describe the limitations of each approach.
1. Studying identical twins.
2. Examining the members of another family in another country whose members also exhibit the disease.
3. Identifying the mouse homolog of this gene, engineering the same mutation in the mouse gene that is present in the human disease allele, and then breeding the mice to see whether they display the same phenotype as the humans who have the disease.

Essay Question 4

A researcher establishes a culture of algae in a lab and supplies light for 12 hours per day. In the first experiment, the size of the algal population and the oxygen content of the media are measured over time. After a period, a daphnia species, *D. ambigua*, is added to the culture and its population size is traced over time; algae and oxygen levels are also monitored.

In a second experiment, another daphnia species, *D. melanica*, is added to the culture at the same time. Both *Daphnia* spp. feed on the algae as their sole source of nutrition in the culture tanks, although in a natural setting their prey are more varied. (*D. melanica* tends to predominate in shallower ponds in the wild, while *D. ambigua* is more abundant in colder ponds.) The data below are observed:

Experiment 1

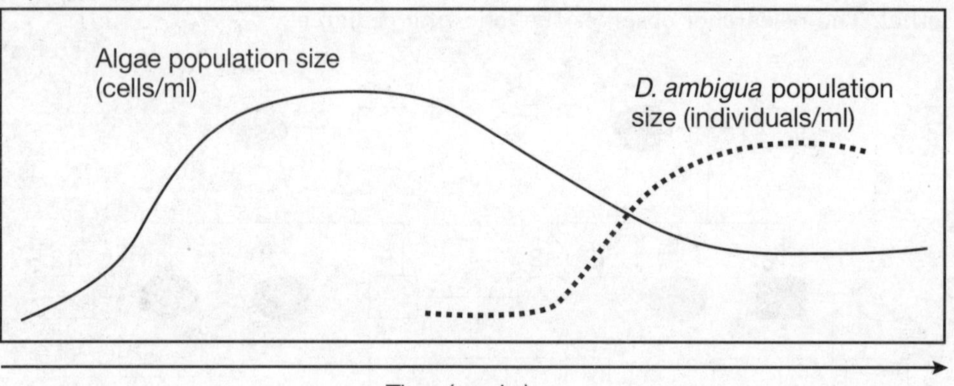

Experiment 2

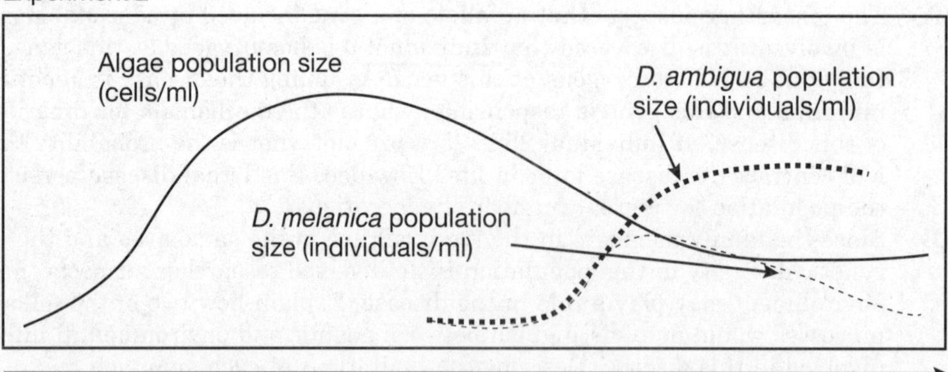

(A) The algal species population size appears to have reached a plateau before the addition of the *Daphnia* spp. What might have caused it to reach a plateau? Describe the different phases of growth of the algal population species in Experiment 1.

(B) What is the relationship between *D. melanica* and *D. ambigua* in this experiment? In a more natural environment, describe how these species may interact.

(C) After performing the initial experiment, the researcher realizes that there is a weakness in the initial experimental design. What might limit any conclusions one can draw about the population growth of *D. melanica* in Experiment 2, and what type of experiment can the researcher perform that will help to draw stronger conclusions?

ANSWER KEY AND EXPLANATIONS

Section I

1. E	21. C	41. B	61. E	81. A
2. E	22. C	42. B	62. B	82. A
3. C	23. D	43. C	63. E	83. D
4. B	24. B	44. D	64. C	84. C
5. C	25. A	45. A	65. B	85. E
6. D	26. A	46. A	66. C	86. D
7. D	27. B	47. B	67. C	87. D
8. D	28. B	48. B	68. B	88. B
9. A	29. E	49. E	69. B	89. C
10. C	30. A	50. E	70. C	90. A
11. B	31. D	51. D	71. B	91. C
12. A	32. B	52. D	72. A	92. E
13. D	33. C	53. B	73. C	93. B
14. C	34. E	54. C	74. A	94. B
15. C	35. C	55. E	75. C	95. D
16. D	36. B	56. D	76. C	96. B
17. C	37. C	57. B	77. C	97. D
18. B	38. B	58. E	78. E	98. B
19. B	39. C	59. B	79. C	99. C
20. A	40. A	60. D	80. C	100. A

1. **The correct answer is (E).** Both sides of the equation must have the same number of each type of atom. If there are 6 oxygen atoms in water and 12 oxygen atoms in carbon dioxide on the right side, there should be 18 oxygen atoms on the left side of the equation as well. To achieve this, there need to be 6 oxygen molecules on the left side, providing a total of 18 oxygen atoms on both sides of the equation:

$$C_6H_{12}O_6 + 6O_2 = 6H_2O + 6\ CO_2 + \text{energy}$$

2. **The correct answer is (E).** This equation shows the complete oxidation of carbohydrates into carbon dioxide, with energy released. This process involves glycolysis in the first step and then oxidative phosphorylation for the complete oxidation to carbon dioxide, with energy captured as ATP.

3. **The correct answer is (C).** The lens is flexible and is attached to ciliary muscles that can contract to "flatten" the lens or relax to allow the lens to return to a more rounded shape. Changing the shape of the lens changes how strongly it refracts light and thereby shifts the focal point. The eyeball and cornea do not change shape to focus vision, although changes in either one may result in reduced visual acuity and the need for corrective lenses. Optical processing does a great deal of the work at many levels to change the way in which visual information is received, processed, and perceived;

however, it does not physically change the focus of the eye. The changes in size of the pupil occur in response to altered brightness, not altered distance of focus.

4. **The correct answer is (B).** Freshwater fish have higher osmotic potential in their tissues than the surrounding water. To maintain osmotic balance, water moves into their tissues by osmosis and is excreted as dilute urine. Swallowing water, excreting salt, and producing concentrated urine are adaptations of bony marine fishes, so answer choice (A) is incorrect. Answer choice (C) is incorrect because fish do not produce uric acid, and this would not help them adapt to freshwater anyway. Bony fish in freshwater cannot achieve the same osmotic potential as freshwater itself; they cannot live with this few solutes in their cells and tissues, so answer choice (D) is incorrect. Sharks can accumulate urea in the marine environment to increase their osmotic potential, and this reduces the loss of water to the environment—but this mechanism would not work as a freshwater adaptation, so answer choice (E) is incorrect.

5. **The correct answer is (C).** Fermentation oxidizes glucose and releases energy. Moving from a high-energy substrate to lower-energy products, the reaction has a negative free energy change. This type of change occurs spontaneously (although not necessarily quickly); in contrast, any process with a positive free energy change will require an energy source. Most activities of living things have a positive free energy change and require energy. Any form of active movement, active transport against a concentration gradient, or biosynthesis all require energy, often involving ATP in some way.

6. **The correct answer is (D).** Molarity is provided in moles per liter. To solve this equation:
   0.2 moles/liter = x moles/5 liters
   0.2 moles/liter $\times$ 5 liters = 1 mole

7. **The correct answer is (D).** Secondary structures such as alpha helices and beta-pleated sheets are formed through the interaction of local groups in the polypeptide backbone and other parts of the protein backbone. The amino acid side chains are not involved.

8. **The correct answer is (D).** Enzymes act by reducing the activation energy of reactions, shown in this chart as the difference between #3 and #2 (indicated by the arrow). They do not change the free energy of reactants (substrates) or products—these are properties of the chemical involved and cannot be changed whether or not an enzyme is present. Similarly, enzymes do not change the equilibrium of a reaction, so answer choice (C) is incorrect. Enzymes accelerate forward and backward reactions, so answer choice (E) is also incorrect.

9. **The correct answer is (A).** Lipids are highly reduced. The more reduced a molecule it is, the greater its energy content when it is oxidized. Lipids also contain less water than carbohydrates, which further concentrates the energy they store in adipose cells. They are very stable, so answer choice (C) is incorrect. Excessive dietary fat can increase the risk of heart disease, but that is not the question here, so answer choice (D) is incorrect. Carbohydrates are usually accessed first for immediate energy needs,

but this does not change the fact that lipids are more efficient for long-term energy storage, so answer choice (E) is incorrect.

10. **The correct answer is (C).** All proteins are transcribed in the nucleus and translated in the cytoplasm or on the rough ER. Proteins localized in the nucleus must return to the nucleus through nuclear pores. Small proteins can move readily through the pores, but larger proteins require a nuclear localization sequence in their polypeptide chain to do so.

11. **The correct answer is (B).** Eukaryotic flagella contain microtubules arranged with motor proteins that drive the microtubules to slide past one another, moving the flagella back and forth in a whiplike motion. Cytoplasmic streaming involves amoeboid motion, as displayed by amoebas or macrophages, so answer choice (A) is incorrect, and answer choice (C) describes how bacterial, not eukaryotic, flagella work. Although actin and myosin do provide some aspects of cytoskeletal action and muscle contraction, they have nothing to do with flagella movement, so answer choice (D) is incorrect. Answer choice (E) is incorrect as well: There is no such thing as the fluid mosaic model of cellular motion.

12. **The correct answer is (A).** Oogenesis involves unequal division of cytoplasm. Each round of meiotic cell division in oogenesis only creates one mature ovum for every primary oocyte that begins the process. The other smaller cells created, called polar bodies, have a full copy of the genome but almost no cytoplasm, and they do not result in viable oocytes.

13. **The correct answer is (D).** RNA is synthesized 5' to 3', running antiparallel to the template strand. Also, RNA contains uracil instead of thymine. The easiest way to solve this one is to write the complimentary RNA sequence above the DNA, and then turn it around to get the 5' to 3' orientation.

14. **The correct answer is (C).** The bacteria will grow exponentially at first, but when nutrients are consumed or toxins accumulate, growth will slow or halt, and eventually the population size may drop. The consumption of limited nutrients or the accumulation of waste are density-dependent; that is, they happen more rapidly with denser populations. No population can grow exponentially indefinitely; some force eventually halts population growth, so answer choice (A) is incorrect. Since the flask was sterile before it was inoculated, answer choice (B) is incorrect. Selection is not the key here because there is no limiting growth at first. The media contain all nutrients required for growth and, at first, there are plenty of nutrients. The population would not fluctuate as a natural population would, so answer choice (E) is incorrect.

15. **The correct answer is (C).** The Calvin cycle always occurs during the day in conjunction with photosynthesis. The difference among C3, C4, and CAM plants is in the timing of carbon fixation. In C4 plants, the Calvin cycle and carbon fixation are in different locations in the plant; in CAM plants, carbon fixation and the Calvin cycle occur at different times of day.

16. **The correct answer is (D).** C3 plants experience increased photorespiration as temperatures increase. Oxygen begins to compete with carbon dioxide for binding to

rubisco. But with higher carbon dioxide concentrations in the atmosphere, photorespiration may be reduced for C3 plants, thereby increasing their rate of growth. C4 plants do not have the added benefit of increased carbon dioxide, since they have already evolved to minimize photorespiration, so their rate of growth would likely be less affected.

17. **The correct answer is (C).** Fish have only two chambers in the heart, so blood is pumped sequentially through two capillary beds: first through the gills to be oxygenated, then through the capillaries in the rest of the body to deliver oxygen and nutrients and remove waste. Since there are two sequential capillary beds, the pressure in the second capillary bed is lower and blood movement can be sluggish, unless aided by muscle movement.

18. **The correct answer is (B).** Smooth muscle cells use actin and myosin filaments as the contractile mechanism, although the way in which that contraction is triggered is distinct from that of skeletal muscle. Also, the fibers of smooth muscle are not aligned in the same regular manner as those in skeletal or cardiac muscle, so it does not have the same striated appearance under a microscope as the other muscle tissue types.

19. **The correct answer is (B).** Phytoplankton make a big contribution to the carbon cycle, taking in carbon dioxide as part of oceanic photosynthesis. They do not have a big effect on the rate of evaporation and certainly not through transpiration (this is found in terrestrial plants). They are not involved in denitrification, and denitrification produces nitrogen gas rather than nitrate pollutants. The longer-term geochemical processes such as the entry of carbon into sinks like sedimentary rocks would not be greatly affected. Acid rain is the result of man's activities, not marine algae.

20. **The correct answer is (A).** Trees lose a great deal of moisture through their leaves as a result of transpiration. The evaporation of water draws heat from the surface of the leaf in the same way that evaporation of sweat cools the skin. The hydrogen bonds between water molecules take a considerable energy to break—energy drawn from the heat of the leaves and the surrounding air as the water evaporates and cools the leaves.

21. **The correct answer is (C).** The alveoli are very small and round and coated with a thin layer of water that is required for gas exchange. The high surface tension of water would draw the walls of the alveoli inward and cause them to collapse into themselves if the surfactant did not reduce the surface tension.

22. **The correct answer is (C).** Frogs are amphibians, which lay moist eggs that lack a thick shell. They lay their eggs in a damp terrestrial or aquatic environment and fertilize the eggs externally. These are not amniotic eggs with shells, such as those of reptiles and birds. A frog begins as a tadpole, with gills and a post-anal tail, but then it undergoes metamorphosis to become terrestrial and loses its tail and gills. Frogs do not have a ventral nerve cord at any stage of development; rather, they have a hollow dorsal nerve cord like all other chordates.

23. **The correct answer is (D).** The structure and function of photosynthesis in green algae is closest to plants. Fungi might superficially resemble plants, but they are not

closely related, and are less so than green algae. Prokaryotes such as archae or cyanobacteria are even less closely related.

24. **The correct answer is (B).** These cells can directly kill cells they recognize as not being "self," including virus-infected cells and cancer cells. Macrophages are part of the innate system and play a major role in phagocytosing bacteria; they are less likely to defeat cancer cells, however. Helper T cells play an essential role in modulating the activity of other cells but will not themselves kill a cancer cell. B cells mediate the humoral response, secreting antibodies and not directly killing other cells. There is no such a thing as cytotoxic B cells.

25. **The correct answer is (A).** The brainstem, including the pons and the medulla, regulate several involuntary homeostatic functions including breathing, heart rate, and digestive reflexes. Where there is some voluntary control over these functions, the cerebrum may also be involved and may temporarily override involuntary control, such as with breathing.

26. **The correct answer is (A).** During winter, animals hibernate to conserve energy, reducing their hypothalamic set point to allow their internal body temperature to fall to a level closer to the temperature of their environment.

27. **The correct answer is (B).** Phospholipids have all of these other traits as part of lipid bilayer membranes, but they do not store energy. That is the function of triglycerides.

28. **The correct answer is (B).** Passive diffusion occurs directly through a membrane without involvement of a protein. Facilitated diffusion includes the diffusion of an ion or other solute down a concentration (or electrochemical) gradient through a protein channel. Active transport involves pumping material against a gradient, and osmosis is the movement of water across a membrane.

29. **The correct answer is (E).** Longer wavelengths have less energy, and the fluoresced light will lose some energy to molecular motion and heating between the time it is absorbed and the time it is fluoresced.

30. **The correct answer is (A).** Antenna pigments absorb light at wavelengths; chlorophyll itself does not. Antenna pigments transfer energy to chlorophyll at the photosystem reaction center.

31. **The correct answer is (D).** During meiotic prophase I, homologous chromosomes line up with one another and homologous chromosomes exchange genetic material with one another, performing homologous recombination.

32. **The correct answer is (B).** The law of segregation refers to the ability of the basic unit of inheritance—the allele—to move into two separate gametes during reproduction, reflecting the separation of homologous chromosomes during meiosis.

33. **The correct answer is (C).** Viruses are varied. Some have an RNA genome; others have a DNA genome. Some have an envelope; others do not. Bacteriophage viruses are well-known for having both lytic and lysogenic life cycles, but not all viruses replicate in this way. One thing they all do have is a nucleic acid genome (RNA or DNA) that

requires replication in the infected host cell. As obligate intracellular parasites, viruses cannot perform this, or any other biosynthetic activity, on their own.

34. **The correct answer is (E).** The short wing phenotype will be seen only in homozygous recessive animals. The abundance of the homozygous recessive genotype will be the allele frequency squared, or $0.8 \times 0.8 = 64\%$.

35. **The correct answer is (C).** The genes of the trp operon are not expressed in wild-type bacteria when they are not needed. If they are, they will produce proteins and tryptophan that is not needed—a waste of metabolic resources that will reduce the ability of the mutant to compete with wild-type bacteria when tryptophan is present.

36. **The correct answer is (B).** The genes in a bacterial operon are all transcribed in the same RNA by RNA polymerase. This coordinates their expression in the correct environmental conditions. Trp repressor blocks transcription when an abundance of tryptophan is present; this avoids the waste of metabolic energy. If the repressor is not present, cells will continuously transcribe and translate all the genes in the operon.

37. **The correct answer is (C).** Like other blood cells, red blood cells are produced from stem cells in bone marrow, not in the lymphatic system. Red blood cells lack a nucleus and mitochondria, so they do not perform transcription, and they do not perform oxidative phosphorylation. They rely on glycolysis and fermentation for energy. They do transport carbon dioxide in blood, expressing carbonic anhydrase that converts carbon dioxide to bicarbonate in blood. Red blood cell production is stimulated by the protein growth factor erythropoietin.

38. **The correct answer is (B).** The roots of each word provide the correct answer. In this case, the word is a combination of *photo-* (light) and *-taxis* (movement). Phototropism is the growth toward light. Photoactivation is activation of a system by light, not movement toward light. Chemotaxis is movement toward a chemical stimulus; chemotropism is growth toward a chemical stimulus.

39. **The correct answer is (C).** They are produced in the liver and stored in the gall bladder for release into the small intestine, where they act as detergents to break up lipid globules and improve digestion.

40. **The correct answer is (A).** Answer choice (B) describes the opposite process, nitrogen fixation. Different organisms would also perform decomposition, so answer choice (C) is incorrect. Chemical fertilizers are available in a form that plants can use without further modification, so choice (D) is incorrect. Animals get nitrogen from plant proteins and perhaps some nucleic acids. Denitrying bacteria are not involved in this process, so answer choice (E) is incorrect.

41. **The correct answer is (B).** Ferns are vascular seedless plants; mosses are nonvascular plants. Mosses also lack xylem, which ferns have. All plants have photosystems, as well as gametophytes, as part of the alternation of generations with sporophytes. Motile sperm are present in both ferns and mosses; they cross the distance to the female gamete. Neither of these groups has seeds, an innovation found in gymnosperms and angiosperms but not less complex plants.

42. The correct answer is (B). Fungi have cell walls made of chitin. Plants have cell walls made of cellulose; bacteria have cell walls of peptidoglycan. Echinoderms don't have cell walls. Insects have chitin in the exoskeleton, but this is not the same as a cell wall.

43. The correct answer is (C). All the female children will receive one X carrying the recessive allele from their father. They have a 50% chance of receiving a second hemophilia recessive allele from their mother. If they have two copies of the recessive allele, they will express the disease.

44. The correct answer is (D). Flatworms such as *Planaria* have small bodies, and all cells are close to the external environment and/or the gastrovascular cavity. Respiratory gases can diffuse to all cells without a specialized respiratory system or oxygen carriers. Movement of respiratory gases always occurs by passive diffusion down a gradient. Flatworms have no tracheoles; these organs are found in insects. Nor do flatworms have oxygen transport proteins. They do not have capillaries since they lack a circulatory system, and they also lack gills.

45. The correct answer is (A). Fatty acid side chains are long alkyl groups. No polar or charged groups occur in this part of the molecule. Fatty acid chains interact through weaker, transitory uneven charge distributions that occur in molecules that are lined up side by side. These weak interactions allow the membrane to be quite fluid and always moving. Transition state intermediates refer to enzyme reactions, not membranes or intermolecular interactions.

46. The correct answer is (A). The residual volume is the amount of air left in the lungs after the strongest possible exhalation, so answer choice (B) is incorrect. The vital capacity refers to the highest volume of air that a person can exhale after the strongest possible inhalation, so answer choice (C) is incorrect. The terms "reserve capacity" and "functional reserve" are not commonly used in reference to lung function.

47. The correct answer is (B). Artificial selection is similar to natural selection, but the former mechanism involves selective pressure by humans rather than by other elements of an ecosystem. Artificial selection does not involve recombinant DNA technology and is distinct from genetic engineering. Genetic engineering does not use selective breeding; instead, it inserts genes or directly modifies the genome to engineer the desired traits. If a plant only reproduces asexually and lacks genetic variation, then there is no genetic variation for selection of any sort to act upon. Any form of selection can work on polyploid plants, just as with any other plant type. Nothing about natural selection prevents artificial selection from acting on the same traits. It is unlikely that the selective force exerted by humans will exactly mimic natural selection or that the endpoint would be the same. However, in principle they can act on the same traits.

48. The correct answer is (B). Cardiac muscle cells are connected by electrical synapses, with gap junctions linking cells and carrying action potentials from the cytoplasm of one cell directly to the cytoplasm of another. They do not have chemical synapses with release of neurotransmitter across a synaptic cleft between cardiac muscle cells, so answer choice (A) is incorrect. Neurons innervate the pacemaker region of the heart to affect the heart rate, but they do not convey action potentials throughout the heart with each cardiac cycle, so answer choice (C) is incorrect. A cardiac muscle cell has a single nucleus; a skeletal muscle cell is multinucleated, so answer choice (D) is incorrect. Hormones can modify the heart rate and alter the rate at which the pacemaker region initiates action potentials, but hormones do not transmit information-carrying action potentials.

49. **The correct answer is (E).** In the sliding filament mechanism of muscle contraction, actin and myosin fibers slide past one another during each muscle contraction. This makes the overall length of the complete sarcomere decrease, drawing the two ends of #5 together to make it shorter. It also decreases the length of the middle region (#2), where only myosin is present.

50. **The correct answer is (E).** Tropomyosin binds to actin filaments, blocking myosin heads from binding to the actin. When calcium is present, troponin and tropomyosin move, revealing binding sites for myosin and allowing muscle contraction to occur.

51. **The correct answer is (D).** Fish are generally ectotherms, which do not regulate their internal temperature. Rather, their internal temperature is mostly determined by their environment. With warmer water, the internal temperature of a fish rises and the rate of metabolic reaction increases, thus increasing oxygen use. However, warmer water holds less dissolved oxygen than cold water so, although fish require more oxygen, they have less present.

52. **The correct answer is (D).** During the course of development, X inactivation randomly turns off one copy of the X in each cell, so patches of tissue have one X inactivated while other patches have the other copy inactivated. Two X chromosomes carrying different alleles of genes may result in patchy expression of the two different alleles.

53. **The correct answer is (B).** During meiosis, chromosomes first undergo DNA replication, then separation of homologous chromosomes in meiosis I and sister chromatids in meiosis II. If chromosomes do not segregate normally during meiosis, a gamete may end up with more than one chromosome. If this gamete is involved in fertilization, the zygote will have three copies of one of the chromosomes, rather than two copies. In the case of Down syndrome, three copies of chromosome 21 are present. Meiotic recombination will exchange sections of chromosomes between homologous chromosomes, but it cannot create an extra copy of a whole chromosome. Translocation moves a piece of a chromosome, but does not result in an extra copy of a whole chromosome. A nonsense mutation only changes a single base pair, and multiple fertilizations would produce too many copies of all of the chromosomes rather than just one.

54. **The correct answer is (C).** Adenine and thymine base-pair with one another, so equal quantities of A and T must always be present in double-stranded DNA. The amount of guanine in DNA can vary greatly among species and does not need to remain constant. Similarly, G and C together are not fixed as a specific percent of DNA. They will change in percentage from gene to gene and genome to genome. Uracil is found in RNA, and thymine is found in DNA. The two strands in double-stranded DNA are antiparallel, meaning that they run in opposite directions.

55. The correct answer is (E). During gastrulation, the embryo is transformed from two germ layers to three germ layers with mesoderm formation. The morula is a simple solid ball of cells at early cleavage stage, and the blastula is a hollow ball of cells. The blastocyst is the mammalian embryo at an early stage and still has just two germ layers. Planulas are the larvae of some organisms such as cnidarians.

56. The correct answer is (D). Members of the F1 generation would all be heterozygotes. Since all of the F1 had brown eyes, the allele for red must be recessive. If two heterozygotes are crossed, about 25% of the offspring of this cross ($\frac{1}{2} \times \frac{1}{2}$) should be homozygous recessive. No sex linkage is apparent, since roughly equal ratios are observed in males and females. Natural selection is not relevant for this question because this is a controlled mating situation, and Hardy-Weinberg does not apply because this is not an unperturbed population.

57. The correct answer is (B). T cells do not recognize or respond to antigen in blood. T cell receptors go through recombination before they are ever in contact with antigen, creating a population of T cells with fairly random antigen affinities. From these, specific clones are selected and amplified by antigen exposure. Helper T cells recognize MHC II in antigen-presenting cells. T cells do not kill cells by engulfing them, as macrophages do. Cytotoxic T cells can kill virus-infected cells by expressing proteins that induce apoptosis and by inserting pores in the infected cell's membrane to cause cell lysis.

58. The correct answer is (E). Different cone cells (the photoreceptors that distinguish color) express different photopsin genes, each of which is sensitive to different wavelengths of light when joined to the retina. All the cells otherwise function in the same manner.

59. The correct answer is (B). Natural selection drove this shape as the optimal shape for movement through water. But fish and dolphins evolved independently to have this shape, so it is analogous and not homologous.

60. The correct answer is (D). Four subunits, two light chains, and two heavy chains are all joined by disulfide bonds.

61. The correct answer is (E). Volatile organic compounds travel through the air, triggering a response in olfactory receptors and are perceived as smells. Sweet tastes are produced by extremely hydrophilic sugars—compounds with low vapor pressure that are not volatile and are not generally perceived by taste receptors. Receptors in the lateral line are involved in the sensing movement and vibration in water. Baroreceptors detect blood pressure (root *baro-* = pressure, like barometer). Photoreceptors sense light in the eye.

62. The correct answer is (B). If the alleles are not expressed in heterozygotes, then there is selection against the allele only in homozygotes, not in heterozygotes. There are far more heterozygotes than homozygotes for a rare recessive allele, so selective pressure against the allele works very slowly.

63. The correct answer is (E). Both organelles have their own genome, they both divide by binary fission, they both have two membranes, and they both create a pH gradient that drives ATP production. Only mitochondria consume molecular oxygen as part of oxidative phosphorylation, however, while chloroplasts produce molecular oxygen.

64. **The correct answer is (C).** If $\Delta = 0$, then the reaction is at equilibrium. Being at equilibrium does not mean that everything is static, however. The net amounts of reactants and products in the reaction will stay constant, but this is a result of the rate of the forward reaction being equal to the rate of the backward reaction at equilibrium.

65. **The correct answer is (B).** Some introns have catalytic activity, including the ability to splice themselves out. To survive as a genome, the first RNA genomes probably replicated themselves, acting as both enzyme and genome. RNA molecules cannot survive intact in fossils. Archae have DNA genomes, as do bacteria and eukaryotes. mRNA does not form membranes. There is no abiotic creation of RNA occurring now, because the conditions of the early Earth are no longer present.

66. **The correct answer is (C).** Glucagon is released by the pancreas when blood glucose levels are low. Glucagon activates glycogen phosphorylase in the liver, which breaks down liver glycogen to release glucose into the blood. The liver does not store triglycerides in this situation, and it does not produce insulin. Also, the liver would have to increase gluconeogenesis but not increase glycolysis to produce glucose. Increasing glycolysis would waste energy if it were happening simultaneously with gluconeogenesis.

67. **The correct answer is (C).** The rhizobium bacteria in root nodules of legumes (peas, beans, clover) fix nitrogen from the atmosphere. The relationship is mutually rewarding: the plant receives biologically available nitrogen and the bacteria receive nutrients from the plant.

68. **The correct answer is (B).** The two genes display linkage in this cross. The F1 plants are all heterozygous for both alleles, as would be the case with or without linkage, so these data are not informative. In the F2 cross, though, with two genes displaying independent assortment and recessive and dominant alleles at both genes involved as described, the phenotypes would be present at a ratio of 9:3:3:1 if the two genes displayed independent assortment. The observed ratios are much closer to 3:1, which would occur if the genes were so closely linked on a chromosome that no linkage occurred between them. The wrinkled-green and smooth-blue phenotypes are recombinant.

69. **The correct answer is (B).** The greater the distance between the two genes, the larger the number of recombination events that occur between them. This increases the abundance of the recombinant phenotypes in the F2 generation.

70. **The correct answer is (C).** Vaccines stimulate an immune response against antigen, causing the clonal selection and amplification of B cells that recognize the antigen. Some of these B cells become plasma cells that circulate for a relatively brief time, actively producing antibody. Others become longer-lived memory cells that persist for years; they may be reactivated if the same antigen reappears. Macrophages are part of innate immunity; their response to later infections is not altered by the vaccine. The antibodies produced in response to the vaccine are cleared before long—they do not last

years like the vaccine. MHC I, the mechanism by which viral-infected cells throughout the body display antigen to cytotoxic T cells, will not be involved. Antibodies do not fold around antigen.

71. **The correct answer is (B).**

72. **The correct answer is (A).** This is a change in a single codon, which would change a single base pair, making this a missense mutation.

73. **The correct answer is (C).** Cells appear unaffected by the mutation even when both copies of the gene are disrupted, but animals cannot develop normally to birth if both copies are disrupted. The mutation is not dominant lethal, because animals survive and appear normal if they have one normal copy. Protein translation cannot be disrupted; if it is, cells cannot survive without protein kinase. While it is true that a lysine to arginine mutation is conservative—these two amino acids are chemically similar and such a mutation often has minimal or no impact on a protein's function—in this particular case, the result is otherwise. No data about gamete formation exist, and the reason homozygous Arg22 mice are not born must have to do with a more fundamental developmental defect.

74. **The correct answer is (A).** If acetylcholinesterase is blocked, acetylcholine is not cleared from the synapse after each action potential at the neuromuscular junction. The muscle will be constantly stimulated and will continue contracting uncontrollably. Blocking acetylcholine at the neuromuscular junction will not affect how action potentials reach the neuromuscular junction.

75. **The correct answer is (C).** Nematodes are quite small and have no circulatory system, exchanging material directly with the exterior and with the complete digestive tract. They have nerve and reproductive cells but no gills or respiratory tissues.

76. **The correct answer is (C).** A heart attack occurs when deposits or clots block flow through the arteries that supply the heart. The lack of blood damages heart tissue and prevents the heart from pumping normally. The cardiac cycle does not require stimulation by the nervous system. It is initiated from within the heart in the pacemaker region. Action potentials are relayed from atria to ventricles using special fibers in the wall between them—a system that is not in itself the cause of heart attacks. A leaky heart valve causes a murmur, not a heart attack. The blood is oxygenated when a heart attack occurs, but the oxygenated blood does not reach cardiac muscle because of a blocked coronary artery.

77. **The correct answer is (C).** When the ventricles contract, the semilunar valves open to allow blood to flow out of the ventricles and into the major arteries. At the same time, the atrioventricular valves close to prevent blood from flowing backward into the atria. When the ventricles contract, the atrioventricular valve must be closed to allow blood to flow forward. During ventricular diastole the ventricles are relaxed, and the atrioventricular valve can open. Valve openings are not caused by the pacemaker region.

78. **The correct answer is (E).** Mitochondria and chloroplasts were once free-living prokaryotes that became endosymbionts inside the first eukaryotic cells. All eukaryotes

are probably derived from the same common ancestor. All living things, including prokaryotes, archae, and eukaryotes, arose from a common ancestor even further back in time. The mitochondria likely evolved first, since all eukaryotes have mitochondria, but only some of them (plants and algae) have the chloroplast.

79. **The correct answer is (C).** The cochlea contains the sensory receptors that detect sound and respond to it by creating action potentials that are conveyed by the auditory nerve to the brain.

80. **The correct answer is (C).** Carbon dioxide does not contain any useful energy. Carbon dioxide is the most oxidized form of carbon. When an animal consumes a plant, not all of the energy in the plant is captured in the animal. Some is lost to the environment as heat energy; some is lost in the indigestible material in feces or urine. Some of the energy that is not lost is captured by the consumer. This helps to build macromolecules, moving the energy from one trophic level to another.

81. **The correct answer is (A).** The mucus secreted in the stomach coats the epithelium and helps to protect the cells lining the stomach. Pepsin is a protease in the stomach that plays no role in protection against acid. Bicarbonate helps neutralize stomach acidity once the stomach's contents move to the small intestine, but it does not protect against acidity in the stomach. The pyloric sphincter controls the movement of food from the stomach into the small intestine. Salivary amylase digests starch in the mouth.

82. **The correct answer is (A).** Epinephrine provokes what is sometimes called the "fight or flight response" including mobilizing tissues such as muscle for rapid action. It binds to a cell surface receptor in target cells, activating adenylate cyclase and increasing production of the second messenger cyclic AMP. Phosphodiesterase helps turn off this signal by degrading cyclic AMP. If phosphodiesterase is inhibited by caffeine, then cyclic AMP will stay higher than it would otherwise, increasing and prolonging the response to epinephrine.

83. **The correct answer is (D).** Vasopressin is synthesized by cells in the hypothalamus, then transported in neuronal extensions to the posterior pituitary, where it is released. Calcitonin is present in the thyroid gland, and TSH, prolactin, and ACTH are present in the anterior pituitary.

84. **The correct answer is (C).** Chlorophyll and the antenna molecules in the photosystems are in the thylakoid membranes. The Calvin cycle takes place in the stroma, the fluid surrounding the thylakoids inside chloroplasts. The chloroplast membranes are not the site of photosynthesis.

85. **The correct answer is (E).** Ribose is a five-carbon sugar, called a pentose. It is part of RNA and ribonucleotides, but it is not itself a complete nucleotide and it is not a purine base. It has multiple chiral carbons, so it is not achiral.

86. **The correct answer is (D).** With time, the rate of product accumulation slows. This suggests that something may be inhibiting the enzyme activity or that the substrate amount is low. Product inhibition is the most reasonable explanation here. It occurs frequently in enzyme reactions, in which the product either acts as an inhibitor or the

rate of the backward reaction increases as the concentration of product increases. This graph resembles that of the one that traces the effect of changing substrate concentration on the initial enzyme reaction rate. However, this one depicts the amount of product accumulating in a reaction over time. This is not the initial reaction velocity, and substrate concentration is not being varied. By definition, enzymes are catalysts and are not consumed in reactions. Equilibrium does not change as the reaction progresses, but the reaction will move toward equilibrium.

87. **The correct answer is (D).** Lanes 1 and 3 are digested with EcoR1. Lane 1 has plasmid only; lane 3 has plasmid with gene inserted. Band A is present with plasmid only; band D appears in lane 3 when the gene is inserted. The same comparison of lanes 2 and 4 supports the same conclusion, with the appearance of band D when the gene is present while the other bands stay the same. The gene must be contained in an EcoR1 fragment, with restriction sites on both sides of the fragment. It must not contain any BamH1 sites, however, since its appearance is the same in the lane digested with EcoR1 alone (lane 2) as it is in the sample digested with both enzymes (lane 4).

88. **The correct answer is (B).** DNA is negatively charged and moves through agarose gels toward the positive electrode. When DNA is linear, one DNA has pretty much the same shape as other DNA molecules, moving like a snake through the gel. The higher the molecular weight of a DNA molecule, the more it gets trapped on the gel as it tries to move through it, moving more slowly than shorter DNAs that can proceed more easily and quickly. The amount of G-C base pairs and the sequence of the DNA do not matter in an agarose gel. Movement is determined by the negative charges in the phosphate backbone and the length of the molecule. All linear DNA molecules have roughly the same charge density—the number of charges per base pair. All DNA molecules have similar shapes. DNA is never positively charged.

89. **The correct answer is (C).** DNA in agarose gels is visualized by staining it with a fluorescent dye that sticks to DNA, usually binding between the stacked up bases in the middle. The dye can then be made to glow in the dark (fluoresce) under a UV light. DNA is a molecule; it is unseen by visual microscopy. Coomassie blue is used to stain proteins in a protein gel, and Gram stain is used to stain bacteria.

90. **The correct answer is (A).** Generally, oncogenes are mutated versions of cellular genes that help to regulate the cell cycle. When the cell cycle is not properly regulated, the normal controls that prevent excessive proliferation are lost and cancer can result. Meiotic recombination cannot be involved here because it occurs only in germ-line cells. Cancer cells are often polyploidy, but this does not explain the origin of oncogenes. mRNA cannot be mutated; only DNA can. Ribosome damage would not produce heritable mutations.

91. **The correct answer is (C).** The mutation in SRY makes a person sterile, so the options for its origin are limited. It will not be passed to the next generation and it could not have come from the mother, so it must have come from the father. Because the father did not display this trait, the deletion must have occurred in the father's germ line.

92. The correct answer is (E). The SRY gene causes male development. When the gene is deleted and its signal is no longer present, the embryo develops as a female. There is no information linking the actions of SRY directly to testosterone production or signaling. The information given runs counter to answer choices (C) and (D).

93. The correct answer is (B). Bees in a hive have complex genetics, but it is not necessary to sort through these genetics to find an answer to the question. If males are haploid, they express recessive alleles that often remain masked in diploid females who are usually heterozygotes. Males are not all identical and they do contribute to the gene pool. Bees use sexual reproduction to produce genetic variation, including meiosis in producing eggs.

94. The correct answer is (B). Restriction enzymes are the "molecular scissors" used by molecular biologists to cut DNA at specific sequences. They were discovered in DNA, where they do not cut the bacterial genome (it is modified to prevent this). However, they do cut the DNA of a virus that threatens to infect the bacteria.

95. The correct answer is (D). The radula is a rasping, tongue-like structure near the mouth that cuts and scrapes food to move it into the mouth.

96. The correct answer is (B). Individual seeds develop from ovules inside the ovary, but the fruit develops from the whole ovary in the flower.

97. The correct answer is (D). Peristalsis is the contraction of muscle rings around portions of the GI tract. This action moves food through the esophagus, stomach, and intestine.

98. The correct answer is (B). When at a higher altitude for a long period, increased erythropoietin is released. This triggers an increase in red blood cell (erythrocyte) production. Genes do not mutate in response to environmental stimulus. Lymphocytes do not carry oxygen and are not part of the response to high altitudes. The walls in alveoli cannot become thinner, because they are already at minimal thickness to allow easy diffusion between blood and air. Living at a high altitude does not increase the rate of metabolic carbon dioxide production.

99. The correct answer is (C). The corpus callosum connects the left and right hemispheres of the cerebral cortex. Both hemispheres are somewhat specialized and coordinate functions such as the perception of images and the association of words with images. Severing the corpus callosum blocks the integration of information between the left and right sides of the cortex.

100. The correct answer is (A). Such large genomes are not the result of a varying gene number; humans have at least as many genes as other vertebrates, if not more. The increased amount of DNA in species such as salamanders seems to be related to a proliferation in the quantity of repetitive DNA, including transposons and retrotransposons. If the selective pressure against having so much DNA in the genome is not strong enough, then the genome may grow larger, as in this case. Even if amphibians were tetraploid, this fact would only explain why they have twice as much DNA per cell than humans, not why their genome is so much larger.

Section II

SUGGESTIONS FOR ESSAY QUESTION 1

(A) Fructose-6-phosphate is a substrate of phosphofructokinase. The curve plots the effect of changing substrate concentration on the reaction velocity of the enzyme-catalyzed reaction. When a low concentration of ATP is present, this relationship resembles the typical effect of a simple substrate-enzyme interaction. At low substrate concentrations, most of the enzyme's active sites are empty and a linear relationship exists between the amount of substrate added and the rate of the initial reaction. As enzyme active sites start to fill with increasing substrate concentration, the rate of the reaction increases more slowly. At very high substrate concentrations, all of the active sites are filled with substrate, the enzyme is saturated, and the rate cannot increase any further. One of the hallmarks of enzyme-catalyzed reactions is that they are saturable with increasing substrate concentration. The reaction velocity at this point, where all enzyme active sites are occupied, is called the V_{max}.

(B) In addition to fructose-6-phosphate, ATP is also a substrate for the enzyme. If a simple enzyme-substrate relationship existed between ATP and phosphofructokinase, then adding more substrate would simply increase the reaction rate, as occurs with fructose-6-phosphate. The opposite occurs in this case, however. At a high concentration of ATP, the enzyme reaction is inhibited, compared with that of a low ATP concentration. ATP is acting as an inhibitor. Since ATP is a substrate, its action as an inhibitor must occur through a different binding site apart from the active site. Binding to a different site, it must propagate a change in conformation through the enzyme to change the activity of the active site. Such a mechanism of enzyme inhibition is called allosteric, occurring through a site other than the active site

(C) As the ATP concentration increases, the activity of phosphofructokinase is inhibited. This enzyme catalyzes a key regulatory step in the glycolytic step. As phosphofructokinase is inhibited, glycolysis slows and less glucose is consumed. If glycolysis slows, then less pyruvate is produced and the Krebs cycle slows. Slowing of the Krebs cycle also slows carbon dioxide production and slows the reduction of the high-energy electron carriers NADH and $FADH_2$, which in turn reduces the rate at which they transfer electrons to the electron transport chain. When fewer electrons are transferred to the electron transport chain, fewer protons are pumped across the inner mitochondrial membrane and less oxygen is consumed as the final electron acceptor. When production of the pH gradient slows, ATP production slows, reducing the concentration of ATP inside the cell.

SUGGESTIONS FOR ESSAY QUESTION 2

(A) Voltage-gated sodium channels are responsible for the propagation of action potentials along the neuronal axon. If voltage-gated sodium channels are blocked, then action potentials cannot propagate along motor neuron axons, action potentials will not reach muscle, and motor neurons cannot stimulate contraction of skeletal muscle.

(B) Acetycholine is the neurotransmitter released by motor neurons at the synapse formed with skeletal muscle at the neuromuscular junction. Acetylcholine binds to receptors on the muscle cell, triggering the action potential in the muscle cell that causes contraction of the muscle cell. Acetylcholinesterase is an enzyme in the synaptic cleft that cleaves acetylcholine and inactivates it. If acetylcholinesterase is inhibited and cannot inactivate acetylcholine, then the acetylcholine will remain in the synaptic cleft, continue binding to receptors, and cause an excessive response to nervous stimulation, thereby preventing the contraction from ending when it normally would in the absence of the inhibitor.

(C) Acetylcholine receptors are found on the skeletal muscle cell membrane at the synapse with the motor neuron. The receptors open ion channels in the muscle cell, triggering a muscle cell action potential that causes muscle contraction. If the acetylcholine receptors are blocked, the muscle cannot contract. The antagonist of the receptor will dampen or block the ability of the nervous system to cause muscle contraction.

(D) Gap junctions do not play a role in the contraction of skeletal muscle. Skeletal muscles do not use electrical synapses between cells. Each cell is stimulated with a chemical synapse with motor neurons. Blockage of gap junctions will not affect how motor neurons carry action potentials or how skeletal muscle responds to stimulation. Cardiac muscle does use electric synapses to carry action potentials directly from cell to cell, but that's another question.

SUGGESTIONS FOR ESSAY QUESTION 3

(A) The pattern of inheritance indicates that the disease is expressed in every generation. In addition, we are given the genotypes of two individuals, one who expressed the trait and another who does not. If a heterozygous individual expressed the trait, then the allele of this gene causing disease must be a dominant allele. Everything in the pedigree is consistent with a dominant allele causing the disease. If the allele is dominant, then the mother of #25 is heterozygous for the dominant disease allele and the father is homozygous for the recessive wild-type allele. The probability of having the disease allele comes from the mother. She has two different alleles, so the probability that #25 will get the disease allele from his mother is 50%.

A dominant lethal allele will not normally persist in the population if it is expressed before reproductive maturity. In this case, individuals do not express the disease until they are older than 40 years and have had time to reproduce. There is not much selection, therefore, against the disease allele.

(B) **Studying identical twins.** Since it is not possible to manipulate humans to determine how we are influenced by genes and environment, biologists observe humans to try to address these questions. Identical twins are often studied as a unique opportunity to separate the role of genes and environment because they possess exactly the same genotype at all genes. Any differences between identical twins must be the result of environment. Twins raised together in the same family are exposed to the same environment. But in some cases twins are separated and placed in different environments. Geneticists often seek out the latter to try to determine if their expressions of complex traits are similar. If twins raised together have the same phenotype but those who are raised apart have a different phenotype, then it becomes clear that the environment must play an important role in that trait. If twins always have the same phenotype whether they are raised together or separately, then it's clear that genes must play a dominant role in the expression of that trait. A limitation to these conclusions is that a limited number of twins have been raised apart and studied by geneticists. In addition, for these sets of twins, geneticists obviously have no control over the type of environment in which they were raised, so investigators must rely on interviewing relatives, friends, and acquaintances in an attempt to piece together the influences of culture and environment.

Examination of another family in another country whose members also display the disease. Genotyping large numbers of people and examining their phenotype and pedigree may allow geneticists to determine whether the relationship between genes and a disease transcends differences in cultures and local environments in a family. Also, studying people who live in the same general environment helps investigators determine how traits relate to genes and that environment. If the environment plays a major role in a trait, that trait should not vary greatly from family to family. A limitation is that culture and environment are not simple or controllable elements. Also, people often move to other environments or locations. Because it is not a controlled experiment, it is difficult to determine which variable is responsible for which trait, particularly for traits in which multiple genes are involved or in which genes and the environment both play a role.

Identifying the mouse homolog of this gene, engineering the same mutation in the mouse gene that is present in the human disease allele and then breeding the mice to see whether they display the same phenotype as humans with the disease. Mice reproduce rapidly, and their genetics are well understood, making them a preferred model for the study of mammalian genetics. The genetics and environment of lab animals can be carefully controlled from birth, allowing a much cleaner background on which to determine the role of genes in a trait. This is particularly true compared to examining a human population at large, where a researcher has no control over the complex environmental factors to which humans and their genes are exposed. However, while mice are mammals and have genes and traits similar to humans, they are obviously not humans. The function of genes in mice is not always the same as that of humans, so the response of mice to the mutation that occurs in humans may not be the same. If the mice with the introduced mutation develop a disease that appears similar to the human disease, this may be highly suggestive that genetics are involved in the disease. If no disease is observed in the mice, however, this does not necessarily mean that the human gene is not involved in the human disease. Whatever results are observed in mice must be considered in light of the differences in genes and physiology between mice and humans.

SUGGESTIONS FOR ESSAY QUESTION 4

(A) When the population is first introduced into the environment, resources are abundant relative to the number of algae present, so the algal population grows rapidly, perhaps exponentially at first. As the population grows, density-dependent factors begin to limit further population growth. For the algae, this might be biologically available nitrogen, other nutrients, or perhaps the amount of sunlight cells receive. The denser the culture, the more some cells block light from other cells in the culture. Eventually, the culture reaches a plateau and grows no further. This seems to be the carrying capacity of the culture—the point at which the rate of births and the rate of deaths are equal. With the introduction of a predator species, the algal population drops until it seems to reach a new, lower carrying capacity, in balance with the *Daphnia* predator population.

(B) In this experiment, both *Daphnia* species are forced to live in the same habitat, and eat the same food source. The *Daphnia* are in competition with each other. In a more natural ecosystem, these two species of *Daphnia* would occupy somewhat different niches. Species that occupy the same niche either compete with each other (in which case one species competes less effectively and declines), or the two species evolve to occupy different niches. Two ways in which a species could evolve to occupy a different niche are changing its food source to prevent niche overlap and changing its range. The information provided here suggests that these species already occupy different niches in the wild but are forced artificially into the same niche in this experiment. Under different laboratory conditions, *D. melanica* may have competed more effectively.

(C) A difficulty with the first two experiments was an omission in the initial experimental design: The growth of *D. melanica* alone with algae was not included as a comparison to the growth of the two *Daphnia* species together. Although it seems likely that the decline in the *D. melanica* population when the two species were mixed together resulted from competition, it is possible that a different factor was involved and was not appreciated in the initial experiments because this experimental control was not included.

APPENDIX

COLLEGE-BY-COLLEGE GUIDE TO AP CREDIT AND PLACEMENT

College-by-College Guide to AP Credit and Placement

appendix

For the past two decades, national and international participation in the AP Program has grown steadily. Colleges and universities routinely award credit for AP test scores of 3, 4, or 5, depending on the test taken. The following chart indicates the score required for AP credit, how many credits are granted, what courses are waived based on those credits, and other policy stipulations at more than 400 selective colleges and universities.

Use this chart to discover just how valuable a good score on the AP Biology Exam can be!

School Name	Required Score	Credits Granted	Course Waived	Stipulations
Agnes Scott College (GA)	4–5	4	None	
Albion College (MI)	4–5			A score of 4 or 5 on an AP exam automatically earns college credit and may be counted toward advanced placement in that area. Depending on the department, a grade of 3 may result in advanced placement and, at the discretion of the department, may also be granted as college credit.
Allegheny College (PA)				Allegheny can accept up to a total of 20 credits maximum for Advanced Placement, International Baccalaureate, CLEP and "college in high school" courses combined.
Alma College (MI)	4	4	BIO 180	
American University (DC)	4–5	8	BIO 110G & BIO 210G	
Asbury College (KY)	3	3	BIO 100 & BIO 101	
	4	8	BIO 161 & BIO 162	
Augustana College (IL)	3	3	BI 101	
Augustana College (SD)	4–5	4	BIO 110 or BIO 120	
Austin College (TX)	4–5		Biology Elective	
Azusa Pacific University (CA)	3	4	Nature core, BIOL 151 (General Biology I), and non-GS (General Studies) elective units	
Baldwin-Wallace College (OH)	4–5	4	BIO ELE	

School Name	Required Score	Credits Granted	Course Waived	Stipulations
Bates College (ME)	3		1 Unspecified Course	
	4–5		2 Unspecified Course	
Baylor University (TX)	4		Biology 1401	
	5		Biology 1305–1105 & 1306–1106	
Belmont University (TN)	3–4		BIO 1010	
	5		BIO 1020	
Benedictine University (IL)	4	3	BIOL 108	
Bethel University (MN)	3			
Birmingham-Southern College (AL)	4	4	BI 105 & BI 115	
Boston College (MA)	4	6	2 Natural Science Core.	
Boston University (MA)	4	4	BI 105, BI 107 or BI 108	
	5	8	BI 107 & BI 108 or BI 105 & BI 107	
Bowdoin College (ME)	5	1		Students who receive AP scores of 5 may enroll in Biology 109. Upon successful completion of Biology 109, with a grade of B- or better, one AP credit will be awarded.
Bradley University (IL)	4–5	6	Biology 121 & Biology 122	
Brandeis University (MA)	5	1	None	
Brigham Young University (UT)	3–5	6	Biol 100	

School Name	Required Score	Credits Granted	Course Waived	Stipulations
Bryn Mawr College (PA)	4	2	One Div. 2 Lab.	The AP Biology exam is not the equivalent of BIOL 101 and 102. It may not be used toward the major in Biology. A student wishing to enter 200-level biology courses without having taken 101–102 must take placement exams administered by the Biology Dept.
Bucknell University (PA)	4–5	1	BIOL 121 or BIOL 111	
California Polytechnic State University, San Luis Obispo (CA)	3		BIO 111 or BIO 115 plus electives	
	4–5		BIO 111 or 115 or 151 or 161 or BIO 213/ENGR 213 plus electives	
Calvin College (MI)	4–5	4	Biology 111	
Canisius College (NY)	3–4	3	Free Elective	
	5	6	BIO 101 & BIO 102 labs	
Carleton College (MN)	3–4	6		Placement into either a) Biology 125 fall term (Dyad, Triad), b) the winter term offering of Biology 125, c) Biology 126.
	5	6		6 credits granted that count toward the biology major and placement is awarded into Biology 126.
Carroll College (WI)	3	4	Biology 100 satisfies LSP 2	
Case Western Reserve University (OH)	4	3	BIOL 114	

School Name	Required Score	Credits Granted	Course Waived	Stipulations
Case Western Reserve University *—continued*	5	4	BIOL 214	Students whose AP Biology course had limited laboratory instruction may accept credit in BIOL 114 and enroll in BIOL 214.
Cedarville University (OH)	4	3.5	BIO 1000	
	5	4	BIO 1110	
Central College (IA)	3–5	8	BIOL 130q	
Centre College (KY)				We participate in the Advanced Placement program of the College Board. Entering freshmen presenting score reports of 4 or 5 on the Advanced Placement examinations will be notified by the College regarding placement and/or course credit on positive recommendation from the appropriate Centre academic program committee.
Chapman University (CA)	4	4	SCI 100, BIOL & Waiver SCI 100	
Christendom College (VA)				Acceptance of advanced placement credit is not automatic; petition for acceptance of AP credit must be made to the Academic Dean. A score of 4 or 5 in any of several fields will be granted liberal arts elective credit, to the limit of 9 semester hours toward fulfillment of degree requirements. On the recommendation of the relevant department chairman, the Academic Dean may grant credit toward fulfillment of certain departmental major requirements.
Claremont McKenna College (CA)	4–5		AP Biology Elective	

School Name	Required Score	Credits Granted	Course Waived	Stipulations
Clarkson University (NY)	4		BY152/ BY154 Life's Diversity	
	5		BY152/ BY154 Life's Diversity and BY153/ BY155 Cellular and Molecular Biology	
Clemson University (SC)	3	8	BIOL 103 & BIOL 104	
	4–5	10	BIOL 110 & BIOL 111	
Colby College (ME)	4–5	6	BI 163 & BI 164	
College of Charleston (SC)	3–5	4 or 8	BIOL 111 & 111L or BIOL 112 & 112L	
The College of New Jersey (NJ)	4	1	Liberal Learning Science (Biology Majors Elective only)	
College of Saint Benedict (MN)	3	None	BIOL 121	
	4	4	BIOL 121	
	5	8	BIOL 121 & BIOL 221	
The College of St. Scholastica (MN)	3–5	4	BIO 1120	

School Name	Required Score	Credits Granted	Course Waived	Stipulations
The College of William and Mary (VA)	5	4	BIO 100 & BIO 200, 203, or 204 or BIO 102 & BIO 200, 203, or 204	
Colorado Christian University (CO)	3–5	8	BIO101, 102, 111, 112 except pre-professional	
The Colorado College (CO)	4–5	1	BY 105 or BY 109	
Colorado School of Mines (CO)	5	8	Biology 1 & Biology 2 Labs	
Colorado State University (CO)	4	4	LSCC 102	
	5	8	LSCC 102 & LS 103	
Columbia College (NY)	4–5	3	No Exemption	
Columbia University, The Fu Foundation School of Engineering and Applied Science (NY)	4–5	3	No Exemption	
Concordia College (MN)	3	1	BIO 101N	
Converse College (SC)	3	4		
Cornell University (NY)	4	4		Contact the Biology Department for placement information.
	5	8	Placement out of all introductory courses	
Cornerstone University (MI)	3	None	BIO 111	A test score of 3 permits a waiver of the course, a test score of 4 or 5 permits a waiver of the course and awards credit. Students may earn a maximum of 30 credits by exam.
	4–5	8	BIO 111	

School Name	Required Score	Credits Granted	Course Waived	Stipulations
Covenant College (GA)	4–5	8	BIO 111 & BIO 112	
Creighton University (NE)	3–4	3	BIO 149	
	5	8	BIO 211 & BIO 212	
Dartmouth College (NH)	5		One unspecified Biology credit	
Denison University (OH)	4–5		BIOL 100	
DePauw University (IN)	4–5	4	BIO 135	
Dickinson College (PA)	4–5		1 BIOL 100-level courses w/lab	Although students receive credit for 1 BIOL 100-level course w/lab, this credit does NOT fulfill the QR requirement.
Drake University (IA)	4–5	8	BIO 1 & BIO 18	
Drury University (MO)	3–5	4	BIOL 110	
Duke University (NC)	4–5		Biology 19	
Duquesne University (PA)	3–5	8	BIO 111/113 or BIO 112/114, Biology 1, 2	
Elizabethtown College (PA)	3–4	4	BIO 101	
	5	8	BIO 101 & BIO 102	
Elmira College (NY)	3	4	BIO 1020	
	4–5	8	Registrar and Biology faculty to determine placement	
Elon University (NC)	4	4	BIO 111 & BIO 113	
	5	7	BIO 111, 112 & 113	

School Name	Required Score	Credits Granted	Course Waived	Stipulations
Embry-Riddle Aeronautical University (AZ)	3–5	3	Physical Science Elective	
Emory University (GA)	4–5	4	Biology 120	Emory College grants 4 semester hours of college credit for each score of 4 or 5 on examinations of the Advanced Placement (AP) Program of the College Entrance Examination Board.
Fairfield University (CT)	4–5	4	BI 170	
Florida Institute of Technology (FL)	4	4	General Biology 1	
	5	8	General Biology 1 & General Biology 2	
Florida International University (FL)	3	4	BSC UCC1 / UCC1L	
	4	4	BSC 1010 / 1010L	
	5	8	BSC 1010 / 1010LBSC 1011 / 1011L	
Florida State University (FL)	3	4	BSC 1005 & BSC 1005L	
	4	4	BSC 2010 & BSC 2010L	
	5	9	BSC 2010 & BSC 2010L & BSC 2011 & BSC 2011L	
Franklin & Marshall College (PA)	4–5		General Elective or BIO 110	
Furman University (SC)	4–5	4	BIO 11	
George Fox University (OR)	3	3	BIOL 100	

School Name	Required Score	Credits Granted	Course Waived	Stipulations
George Fox University—*continued*	4–5	8	BIOL 101 & BIOL 102	
Georgetown University (DC)	4	3	Principles of Biology	These credits may count as a free elective credit. Most NHS majors are required to take Human Biology I & II.
	5	4	Principles of Biology	These credits may count as free elective credit. Most NHS majors are required to take Human Biology I & II.
The George Washington University (DC)	4–5	8	BISC 13–14	
Georgia Institute of Technology (GA)	5	4	BIOL 1510	
Gonzaga University (WA)	4–5	4	Biology 100/100L	
Gordon College (MA)	4–5	4	Core Non-Lab Life Science	Not more than 32 credits may be credited toward a Gordon degree.
Goshen College (IN)	3–5	3	Gen Ed Biological Science	
Grinnell College (IA)	4–5			
Grove City College (PA)	4–5	4	BIOL 102	Students with a score of 5 may receive 4 additional credits for BIOL 101 with Biology departmental permission.
Gustavus Adolphus College (MN)	4–5		BIO 101	Students who score a 4 or 5 on the College Board's Advanced Placement exams will receive credit toward graduation from Gustavus.

School Name	Required Score	Credits Granted	Course Waived	Stipulations
Hamilton College (NY)	4–5	1	BIO 110	Students having obtained a score of 4 or 5 on the Biology AP exam will receive 1 credit after placement in and completion of a course beyond Bio 110 with a minimum grade of C- or better in that course. The credit is for exemption from an introductory semester of college-level biology.
Hanover College (IN)	4–5		Bio 121	
Harding University (AR)	4–5	4	BIOL 121	
Haverford College (PA)				The registrar will award one course credit for an AP score of 5 and one-half course credit for a score of 4. No credit is awarded for scores under 4. The maximum AP credit awarded to any student may not exceed 4 course credits.
Hendrix College (AR)	4–5		BIOL 101	
Hope College (MI)	4–5	4	Biol 240	
Illinois College (IL)	3–5	4	BIOL 105	Illinois College grants appropriate academic credit for scores of 4 and 5 on Advanced Placement (AP) Examinations administered by the College Entrance Examination Board (CEEB). For most AP examinations scores of 3 also qualify for credit. Applicants should request that their scores be reported to the IC Registrar's Office.
Illinois Institute of Technology (IL)	4–5	6	BIOL 107 & BIOL 109	
Iowa State University of Science and Technology (IA)	4–5		Biology 100	

School Name	Required Score	Credits Granted	Course Waived	Stipulations
Ithaca College (NY)	3	6	BIOL-11500 Essentials of Biology BIOL-1xxxx	
	4	6	BIOL-11500 Essentials of Biology BIOL-1xxxx biology elective	
	5	8	BIOL-12100, BIOL-12200 Principles of Biology: may receive course-specific credit provided there is evidence of laboratory component and a passing score on a dept. exam	
James Madison University (VA)	3	4	GISAT 113 & BIO 000	
	4	8	GSCI 103 & GSCI 104 & BIO 000 or GISAT 113 & BIO 000	BIO 000 does not count toward major or minor requirements in Biology or toward general education requirements but is elective credit toward a degree.
	5	8	GSCI 103, GSCI 104 & BIO 000 or GISAT 113 & BIO 000	For Non-Majors

School Name	Required Score	Credits Granted	Course Waived	Stipulations
James Madison University—*continued*	5	8	BIO 114 & BIO 124	For Majors
John Brown University (AR)	3		BIO 1003	
	4–5		BIO 1124 or BIO 1134 or BIO 2134	
John Carroll University (OH)	3–4	4	BL 112, 112L	
	5	8	BL 155–158	
The Johns Hopkins University (MD)	4–5	8		
Juniata College (PA)	4	3	Natural Science Elective	
	5		BI 111 and/or BI 113	
Kenyon College (OH)	3			See Department for Placement.
	4	.5		See Department for Placement.
	5	.5	BIOL 113	
Kettering University (MI)	4–5	4	BIOL 141 & BIOL 142	
Knox College (IL)	4–5		BIOL 120	
Lafayette College (PA)	4–5	2	Biology 101 & Biology 102	
Lake Forest College (IL)	3–5		BIOL 105	
Lebanon Valley College (PA)	4–5	8	BIO 101/102 (Non-science majors) or BIO Elective (Science majors)	

School Name	Required Score	Credits Granted	Course Waived	Stipulations
Lehigh University (PA)	4–5	4	EES 31 & 22	
LeTourneau University (TX)	3	3	BIOL 1113	Credit for laboratories in the natural sciences could be awarded on demonstrated mastery of equivalent college level laboratory experiences.
	4–5	6	BIOL 1113 & BIOL 1123	
Lewis & Clark College (OR)	4		Credits granted toward graduation are not used as elective credits toward the major.	
	5	4	Biology 141 or Biology 151	
Lipscomb University (TN)	3–5	3	BY 1003	
Louisiana State University and Agricultural and Mechanical College (LA)	3	3	BIOL 1001 or BIOL 1201	
	4–5	4	BIOL 1201 & BIOL 1208	
Loyola Marymount University (CA)	4–5	6	No specific course	
Loyola University Chicago (IL)	4–5	8	BIOL 101 & BIOL 102 & labs	
Loyola University New Orleans (LA)	4–5		BIOL A106	
Lycoming College (PA)	3	4	BIO 106	
	4	4	BIO 110	
	5	8	BIO 110 & BIO 111	
Macalester College (MN)	5		BIOL 194	

School Name	Required Score	Credits Granted	Course Waived	Stipulations
Marietta College (OH)	4–5	4	BIOL 101	
Marist College (NY)	3	4	BIOL 130L	Non-Science Majors
	4–5	8	BIOL 130L & BIOL 131L	Non-Science Majors
	4–5		Science Majors must contact the Biology Dept.	For Science Majors
Marquette University (WI)	4	3	BIOL 5	
	5	6	BIOL 1 & BIOL 2	
Massachusetts Institute of Technology (MA)	5	12	subject 7.012	If you have received a 5 on the Biology AP exam yet still enroll in one of the introductory biology subjects (7.012, 7.013, or 7.014), you will receive nine units of general elective credit, along with twelve more units of credit toward a passing grade in 7.012, 7.013, or 7.014. For scores lower than 5, no credit is given.
McGill University (QC)	4–5	6	BIOL 111 & BIOL 112	
Messiah College (PA)	3–5	3	BIOL 102	For non-biology dept. majors
	4–5	3	BIOL 200	For biology dept. majors
Michigan State University (MI)	3	No credit	BS 110, 111 & 111L	
	4–5	8	BS 110, 111 & 111L	
Michigan Technological University (MI)	3	4	unassigned biology credit.	For Biology and clinical science laboratory majors
	3	4	BL 1040	For all non-biology majors
	4–5	8	BL 1040 & BLU	For all non-biology majors

School Name	Required Score	Credits Granted	Course Waived	Stipulations
Michigan Technological University—*continued*	4–5	8	unassigned biology credit.	For Biology and clinical science laboratory majors
Middlebury College (VT)	5	1	N/A	
Mills College (CA)	4	1		
	5	1.25		
Mississippi College (MS)	3–5	8	BIO 111 & BIO 112	
Missouri State University (MO)	3	4	BIO 102	
	4–5	4	BIO 121	
Moravian College (PA)	4			
Mount Saint Vincent University (NS)	4–5		BIOL 1152 & BIOL 1153	
Murray State University (KY)	3–4	7	BIO 101 & BIO 115	
	5	11	BIO 101 & BIO 115 & BIO 116	
Nebraska Wesleyan University (NE)	4–5	4	Biology 001	
New College of Florida (FL)	4–5			Advanced Placement (AP) exams (with math score of 3; other subject areas with scores of 4 or 5) may be applied toward completing LAC expectations.
New Jersey Institute of Technology (NJ)	4	4	BIO 101	
	5	8	BIO 101 & BIO 102	
New Mexico Institute of Mining and Technology (NM)	3–5	6	Biology 100 & Biology 111	
New York University (NY)	4–5	8		

School Name	Required Score	Credits Granted	Course Waived	Stipulations
North Carolina State University (NC)	3			Students must enroll in BIO 181/183 or ZO 150/160 depending on curriculum. At the beginning of the semester, students must request a departmental placement exam in Biology. If they pass that exam, they will be granted 8 hours credit for BIO 181/181L and BIO 183/183L.
	4–5	8	BIO 181/181L & BIO 183/183L	
North Central College (IL)	4–5	3.5	BIO 100	
Northwestern College (IA)	4–5	4	BIO 102	
Northwestern College (MN)	3	4	BIO 1011	
	4–5	8	BIO 1011 & BIO 1012	
Northwestern University (IL)	5	1	Biology 1X	
Oberlin College (OH)	4	4	Natural Science Credit	
	5	4	Biology 605	
Oglethorpe University (GA)	3	4	Gen 102 Natural Science	
	4–5	4	Gen 102 Natural Science & BIO 102	
Ohio Northern University (OH)	3	4	BIOL 121	
	4–5	8	BIOL 121 & BIOL 122	
The Ohio State University (OH)	3	5	Biology 101	
	4	5	Biology 113	Req'd Prerequisite for Biomolecular Option.

School Name	Required Score	Credits Granted	Course Waived	Stipulations
The Ohio State University—*continued*	5	10	Biology 113 & Biology 114	Both required for Pre Med.
Ohio Wesleyan University (OH)	4–5	1		A student who achieves a 4 or better on the Advanced Placement Examination in Biology will be awarded 1 credit toward graduation for a general biology course. In order to get an exemption from an introductory course, a student must take and pass a placement exam during freshman fall orientation.
Oklahoma Baptist University (OK)	4	3	GNSC 1124	
Pacific Lutheran University (WA)	4–5	4	BIOL 111	
Penn State University Park (PA)	4	4	Biology 011 & Biology 012	
	5	4	Biology 110	No credits are awarded for Biology 220, Biology 230, and Biology 240
Pepperdine University (CA)	3–5	4	Laboratory Science GE Requirement	
Point Loma Nazarene University (CA)	3	3	Elective	
	4	4	BIO 101	
	5	4	BIO 101	Biology majors who receive a score of “5” must meet with their adviser before credit can be determined. The adviser will then choose whether to award credit for BIO210 or BIO120 depending on the students’ strengths.
Polytechnic University, Brooklyn Campus (NY)	4	4	BŞ in Liberal Studies Degree Program	

School Name	Required Score	Credits Granted	Course Waived	Stipulations
Polytechnic University, Brooklyn Campus—*continued*	5	4	Life Sciences 1004	
Pomona College (CA)	4–5	1		
Presbyterian College (SC)	4	8	BIOL 101/101L & BIOL 102/102L	
Princeton University (NJ)	5	2	MOL 101 & EEB/ MOL 210 or 211	
Providence College (RI)	4–5		Natural Science Core Group II or free elective	
Purdue University (IN)	4	4	BIOL 110	
	5	8	BIOL 110 & BIOL 111	
Queen's University at Kingston (ON)	4–5	3	BIOL 102 & BIOL 103	
Reed College (OR)	4–5	1		There is no advanced placement because the department believes that all biology students should be exposed to the instructors, concepts, and methods in Reed's first-year, team-taught course.
Rensselaer Polytechnic Institute (NY)	4	4	BIOL 1000	
	5	4	BIOL 1010	
Rhodes College (TN)	4–5	9	Biology 130, 131, 140, 141	
Rochester Institute of Technology (NY)	3–5		1001–201,202,203, 205,206,207	For Non-biology majors

School Name	Required Score	Credits Granted	Course Waived	Stipulations
Rochester Institute of Technology—*continued*	4–5		1001–201,202,203, 205,206,207	For Biology majors
Rose-Hulman Institute of Technology (IN)	4	4	AB 101	
	5	4	AB 110	
Rutgers, The State University of New Jersey, New Brunswick (NJ)	4–5	8	119: 101,102	
Saint Francis University (PA)	3		BIO 111 or BIO 112	Science majors who receive a grade of "3" on the Advanced Placement Biology test will have the choice of waiving either Biology 111 (General Zoology) or Biology 112 (General Botany).
	3		BIO 101 & BIO 103	Non-science majors receiving a grade of "3" on the Advanced Placement Biology test will be waived from taking Biology 101 (General Biology) and Biology 103 (Environmental Studies).
	4–5		BIO 101 & BIO 103	Non-science majors receiving a grade of "4" or "5" on the Advanced Placement Biology test will receive credit for taking Biology 101 and Biology 103.
	4–5		BIO 111 or BIO 112	Science majors who receive a grade of "4" or "5" on the Advanced Placement Biology test will have the choice of receiving credit for either Biology 111 or Biology 112.
Saint John's University (MN)	3	None	BIOL 121	
	4	4	BIOL 121	
	5	8	BIOL 121 & BIOL 221	
Saint Louis University (MO)	5	8	BL A 104, 106	
Saint Mary's College of California (CA)	3		BIOL 50 & BIOL 51	

School Name	Required Score	Credits Granted	Course Waived	Stipulations
Saint Mary's College of California—*continued*	4–5		BIOL 90 & BIOL 91	
St. Norbert College (WI)	3–5	4	BIOL 100	
St. Olaf College (MN)	4–5	1	BIO 126	
Samford University (AL)	3–5	4	Biology 105	
San Diego State University (CA)	3–5	6	Biology 100/100L & 2 units of Biology 299	
Santa Clara University (CA)	4–5	4	University Core natural science requirement without lab.	
Seattle Pacific University (WA)	4–5	5	BIO 2101	
Seattle University (WA)	4	5	BIOL 161 & BIOL 171	
	5	15	BIOL 161, 171, 161, 172, 163 & 173	
Sewanee: The University of the South (TN)	4–5	4		AP test scores of 4 or 5 and IB test scores of 5 or higher on higher level exams, which do not represent the same academic area, will earn semester hours of credit for entering students. Credit is not given for organization and management studies or for studio art courses. Credit for one elective course (four semester hours) may be earned. Students may earn up to 8 full-course/32-semester-hour credits through AP.
Simpson College (IA)	3–5	8	Biology 110 & Biology 111	

School Name	Required Score	Credits Granted	Course Waived	Stipulations
Southern Methodist University (TX)	4–5	8	BIOL 1401 & BIOL 1402	
Southwest Baptist University (MO)	3–5	4	BIO 1004	
Southwestern University (TX)	4–5	4	BIO 50–102, 112, 122, 162	
State University of New York at Binghamton (NY)	3	4	Elective	
	4	4	BIOL 118	
	5	8	BIOL 118 & BIOL 117 or BIOL 118 & Elective	
State University of New York College at Geneseo (NY)	3–4		BIOL 1TR	For Non-biology Majors.
	5		BIOL 117 & BIOL 119	For Biology Majors. Labs (BIOL 118, 120) must be taken at Geneseo. Biol & Biochem majors advised to refuse credit for scores lower than 5.
	5		BIOL 117 & BIOL 119	For Non-biology Majors.
Stetson University (FL)	4	8	BY 101 & BY 102	
Stevens Institute of Technology (NJ)	4–5	4	CH 281 & CH 282	
Stonehill College (MA)	4–5	8	Biological Principles 1 & 2	
Stony Brook University, State University of New York (NY)	3	3	BIO 150	
	4–5	4	BIO 150	
Susquehanna University (PA)	4	4	BIOL 010	

School Name	Required Score	Credits Granted	Course Waived	Stipulations
Swarthmore College (PA)	5	1	BIOL 001 & BIOL 002	
Syracuse University (NY)	4–5	6	BIO 200	Counts as a sequence in natural sciences and mathematics.
Tabor College (KS)	3	8	BI 107 & BI 108	
Taylor University (IN)	3	5	BIO 100 (General Biology w/Lab)	
	4	4	BIO 101 (Principles of Cell Biology)	
Tennessee Technological University (TN)	4	4	BIOL 1010 & BIOL 1110	
	5	8	BIOL 1010 and 1020 or BIOL 1110 and 1120	
Texas A&M University (TX)	4	8	BIOL 111 & BIOL 112	
Texas Christian University (TX)	3	3	10003	
	4	4	10504	
	5	8	10504 & 10514	
Texas Tech University (TX)	3	8	BIOL 1401 & BIOL 1402	
	5	8	BIOL 1403 & BIOL 1404	
Towson University (MD)	3	4	BIOL 115	
	4–5	8	BIOL 201 & BIOL 202	
Trinity University (TX)	4	3	BIOL 2301	

School Name	Required Score	Credits Granted	Course Waived	Stipulations
Truman State University (MO)	3	4	BIOL 100	
	4–5	4	BIOL 107	
Tufts University (MA)	5			Consult department regarding placement; normally, placement will be in Biology 13 or 14; one course credit can count toward the major. Students who take both Biology 13 and 14 will not also be awarded acceleration credit.
Tulane University (LA)	4	4	CELL 103/106	
	5	7	CELL 101 or 103/106, EBIO 101/111	
Union University (TN)	3–5	4	BIO 100	
University at Buffalo, the State University of New York (NY)	3	7	Elective credit for graduation	
	4–5	7	BIO 129 & BIO 130	
The University of Alabama in Huntsville (AL)	3	4	BYS 119	
	4–5	8	BYS 119 & BYS 120	
The University of Arizona (AZ)	3	4	Lower Division Dept. elective credit	
	4–5	8	ECOL 181 R/L & ECOL 182	
University of Arkansas (AR)	3		BIOL 1543H/ 1541M	
	4		BIOL 1543/1541L	
	5		BIOL 1543H/ 1541M	

School Name	Required Score	Credits Granted	Course Waived	Stipulations
University of California, Davis (CA)	3–5	8	Biological Sciences 10	Biological Sciences 1A is the first course taken by most students contemplating majors in the life sciences.
University of California, Irvine (CA)	3–5	8	Elective credit only.	For Bio Sci majors
	3–5	8		One Biological Sciences course toward category II of the UCI breadth requirement (for non-bio sci majors)
University of California, Los Angeles (CA)	3–5	8	LIFE SCI unassigned	
University of California, Riverside (CA)	3–5	8		
University of California, San Diego (CA)	3	8	Biol 10, may take Biol 1, 2, 3 for cred.	
	4–5	8	Bild 1 & Bild 2	
University of California, Santa Barbara (CA)	3–5	8	EEMB 20, MCDB 20, & Natural Science 1C	
University of California, Santa Cruz (CA)	3–5	8	Satisfies one "IN"	
University of Central Arkansas (AR)	4		BIOL 1441	
University of Central Florida (FL)	3		BSC 1005/1005L	
	4		BSC 2010C	
	5		BSC 2010C and BSC 2011C	

School Name	Required Score	Credits Granted	Course Waived	Stipulations
University of Chicago (IL)	5		1 year Fundamental BIO SCI	Students with AP 4 who plan to concentrate in the biological sciences or prepare for the health professions must register for a Fundamental Sequence (BioSci 170s, 180s, or 190s). These students may use their AP credit in electives. Students with AP 5 place out of the general education requirement and into the two-quarter 240s sequence in the concentration.
University of Colorado at Boulder (CO)	4–5	8	EPOB 1210, 1220, 1230, and 1240	
University of Connecticut (CT)	4–5	8	BIOL 107 & BIOL 108	
University of Dallas (TX)	3–5	8	BIO 1311, 1111, 1312, 1112	Science majors should consult with departments and pre-med adviser before taking credit or placement.
University of Dayton (OH)	4	4	BIO 151/151L	
	5	8	BIO 151/151L & BIO 152/152L	
University of Delaware (DE)	3	4	BISCI 104	
	4–5	4	BISCI 208	
University of Denver (CO)	4	8		
	5	12	4 NATS	
University of Evansville (IN)	4–5	4	Biology 107	
University of Florida (FL)	3–4	4	BSC 2007/2009L	
	5	8	BSC 2010/2010L & BSC 2011/2011L	

School Name	Required Score	Credits Granted	Course Waived	Stipulations
University of Georgia (GA)	3	4	BIOL 1103/1103L	
	4–5	8	BIOL 1107 & BIOL 1108	
University of Illinois at Chicago (IL)	3–5	10	BIOS 100 & BIOS 101	
University of Illinois at Urbana–Champaign (IL)	3	3	Integrative BIO 100	
	4–5	8	Integrative Biology 150 and Molecular & Cellular Biology 150	
The University of Iowa (IA)	3–5	4		Duplicates 002:021 Human Biology. May be applied to General Education Program: natural sciences (nonlab) hours; or used as elective hours. NOTE: If you score 5, consult the department for possible equivalency credit for 002:002 Introductory Animal Biology.
University of Kansas (KS)	4	3	BIOL 100	For non-biology majors
	5	4	BIOL 100 & BIOL 102	For non-biology majors
	5	4	BIOL 150	
University of Kentucky (KY)	3	3	BIO 102 & BIO 103	
	4–5	5	BIO 150, 151, 152, & 153	3 credit hours each for BIO 150, 152 with a grade of CR. 2 credit hours each for BIO 151, 153 with a grade of CR.
University of Mary Washington (VA)	3	4	BIOL ELEC	
	4–5	8	BIOL 0121 & BIOL 0122	

School Name	Required Score	Credits Granted	Course Waived	Stipulations
University of Maryland, College Park (MD)	4	8	BSCI 105 & LL elective	BSCI fulfills a major requirement in all Life Sciences; it also fulfills CORE-Lab (Life) Science requirement. Contact the College of Life Sciences for placement.
	5	8	BSCI 105 & BSCI 106	
University of Miami (FL)	3–4	3	BIL 101	
	5	10	BIL 150, 151, 160, 161.	
University of Michigan (MI)	3	4	Biology 100	
	4–5	5	Biology 162	
University of Minnesota, Morris (MN)	3–5	8		Intro Biology with Lab; may not count in major
University of Minnesota, Twin Cities Campus (MN)	3–5	4	Biology 1009	
University of Missouri–Columbia (MO)	4–5	5	Bio Sci 1500	
University of Missouri–Kansas City (MO)	3	3	Biology 102	
University of Missouri–Rolla (MO)	4	3	Biological Sciences 110	
	5	5	Biological Sciences 110 & 112	
University of Nebraska–Lincoln (NE)	4–5	3	Biology 101	Currently, laboratory credit for Biology 101L is not available through advanced placement.
The University of North Carolina at Asheville (NC)	3	3	BIOL 109	
	4–5	5	BIOL 105	
The University of North Carolina at Chapel Hill (NC)	3–4	4	Biology 101/101L	

School Name	Required Score	Credits Granted	Course Waived	Stipulations
The University of North Carolina at Chapel Hill —*continued*	5	8	Biology 101/101L & Biology 279/279L	
The University of North Carolina Wilmington (NC)	3	4	BIO 105	
University of North Florida (FL)	3–4	4	BSC 1005C	
	5	8	BSC 1010C & BSC 1011C	
University of Oklahoma (OK)	3		ZOO 1114	
	4–5		ZOO 1114 & ZOO 1121	
University of Pennsylvania (PA)	5	1	Biology 091	No AP credit is given for laboratory work
University of Pittsburgh (PA)	3	3	BIOL 0110	
	4	4	BIOL 0110 & BIOL 0111	
	5	8	BIOL 0110, 0120, 0111, 0121	
University of Rhode Island (RI)	3–5	8	Biology 101 & Biology 102	
University of Richmond (VA)	4–5	4	BIOL 102	
University of Rochester (NY)	4–5	4		Cannot use credits to meet one of the requirements of a BIO major
University of San Diego (CA)	3–5	3	Bio 101 w/lab	
The University of Scranton (PA)	4–5		BIOL 141,141L BIOL 142,142L	No credit can be given to Biology majors, Biomathematics majors, or students who will eventually apply for admission into the Physical Therapy graduate programs. Students interested in Pre-Med should consult with their Advisor.

School Name	Required Score	Credits Granted	Course Waived	Stipulations
University of South Carolina (SC)	3		BIOL 101 & BIOL 101L	
	4–5		BIOL 101, 101L, 102, and 102L	
University of St. Thomas (MN)	4–5		BIOL 101	For non-biology or biochemistry majors.
	4–5		BIOL 201	For Biology majors. Students will be required to take BIOL 201 Lab at St. Thomas
The University of Tennessee at Chattanooga (TN)	3–5	8	BIOL 121 & BIOL 122	
The University of Texas at Austin (TX)	5		BIO 311C & BIO 311D	
The University of Texas at Dallas (TX)	3		3 SCH free electives	
	4		BIOL 2312 & BIOL 2112	
	5		BIOL 2311 + 2111, BIOL 2312 + 2112, BIOL 2281	
University of the Pacific (CA)	4–5	4	BIOL 041	
University of the Sciences in Philadelphia (PA)	4–5			
University of Tulsa (OK)	4	8	BIOL 1603 & 1601, BIOL 1703 & 1701	
University of Utah (UT)	3	6	1 Science IE	IE=Intellectual Exploration
University of Virginia (VA)	4–5	6	BIOL 201 & BIOL 202	No laboratory credit given
University of Washington (WA)	4–5	10	BIOL 161 & BIOL 162	First two quarters of general biology. Counts toward Natural World general education requirement for graduation

School Name	Required Score	Credits Granted	Course Waived	Stipulations
University of Wisconsin–La Crosse (WI)	3	4	Elective Credit	
	4–5	4	Biology 103	
University of Wisconsin–Madison (WI)	3	3	Biological Science Electives	
	4–5	5	Biological Science Electives	
University of Wisconsin–River Falls (WI)	3	3	Biology Elective (or BIOL 100 if not Biology Major).	
	4–5	3	BIOL 150	
Ursinus College (PA)	4–5			
Valparaiso University (IN)	3	4	BIO 171	Biology majors are encouraged to repeat BIO 171 if they receive a score of 3 on the AP Biology exam.
Vanderbilt University (TN)	4–5	4	Biological Sciences 100	
Vassar College (NY)	4–5			You will receive one unit of 100-level Biology credit toward graduation, you may elect 200-level Biology courses once you have completed BIOL 106, and BIOL 105 will not be a requirement for your Biology major.
Villanova University (PA)	4–5		BIO 20	
Virginia Military Institute (VA)	3–5	8	BI 101 & BI 102	No credit for score of 3 for Biology Majors.
Virginia Polytechnic Institute and State University (VA)	4	4	BIOL 1006 & 1016	For non-biology Majors
	4	4	BIOL 1106 & BIOL 1116	For Biology Majors

School Name	Required Score	Credits Granted	Course Waived	Stipulations
Virginia Polytechnic Institute and State University—*continued*	5	8	BIOL 1005, 1006, 1015, 1016	For non-biology Majors
	5	8	BIOL 1105, 1106, 1115, 1116	For Biology Majors
Wake Forest University (NC)	4–5	4	BIO 111	
Wartburg College (IA)	3–5		BI 101	
Washington and Lee University (VA)	5		BIOL 15AN (4-credit 100-level GE5a elective)	
Washington College (MD)	4–5		BIO 111 & BIO 112	
Wellesley College (MA)	5	1	No Exact Equivalent	
Wesleyan University (CT)	4–5	1	BIOL 181 or BIOL 182	
Western Washington University (WA)	3–5	8	Biology 101 & Biology 102	
Westminster College (UT)	4–5	8	LETRN 001 & BIOL 100T	
Westmont College (CA)	4–5	4	BIO 005	
Whitman College (WA)	4	3	3 credits General Electives.	Cannot be applied toward biology credit.
	5	4	BIO 111	
Whitworth University (WA)	4	3	BI 150	
	5	4	BI 150 & BI 154	
Willamette University (OR)	4–5	4	BIOL 110	These course credits do not satisfy Mode of Inquiry general requirements.
William Jewell College (MO)	4	4	BIO Elective	

School Name	Required Score	Credits Granted	Course Waived	Stipulations
Williams College (MA)	5		Dept. assigns	Placement depends on department exam results. Students must obtain an AP score of 5 to be eligible to take the department exam.
Wofford College (SC)	4	8	Biology 101 & Biology 102	
Worcester Polytechnic Institute (MA)	4–5		BB 1000 (elective)	1/3 credit is earned and posted to WPI transcript by the Registrar's office for examination with the appropriate grade. This AP credit is listed as L on the transcript.
Xavier University (OH)	4–5	6	BIOL 160 & BIOL 162	Those taking BIOL 161 (lab) must audit BIOL 160; those taking BIOL 163 (lab) must audit BIOL 162.
Yale University (CT)	5	1		

NOTES

NOTES

NOTES

NOTES

NOTES

NOTES

NOTES

NOTES

Peterson's Book Satisfaction Survey

Give Us Your Feedback

Thank you for choosing Peterson's as your source for personalized solutions for your education and career achievement. Please take a few minutes to answer the following questions. Your answers will go a long way in helping us to produce the most user-friendly and comprehensive resources to meet your individual needs.

When completed, please tear out this page and mail it to us at:

Publishing Department
Peterson's, a Nelnet company
2000 Lenox Drive
Lawrenceville, NJ 08648

You can also complete this survey online at **www.petersons.com/booksurvey.**

1. What is the ISBN of the book you have purchased? (The ISBN can be found on the book's back cover in the lower right-hand corner.) ____________________

2. Where did you purchase this book?

- ❑ Retailer, such as Barnes & Noble
- ❑ Online reseller, such as Amazon.com
- ❑ Petersons.com
- ❑ Other (please specify) ____________________

3. If you purchased this book on Petersons.com, please rate the following aspects of your online purchasing experience on a scale of 4 to 1 (4 = Excellent and 1 = Poor).

	4	3	2	1
Comprehensiveness of Peterson's Online Bookstore page	❑	❑	❑	❑
Overall online customer experience	❑	❑	❑	❑

4. Which category best describes you?

- ❑ High school student
- ❑ Parent of high school student
- ❑ College student
- ❑ Graduate/professional student
- ❑ Returning adult student
- ❑ Teacher
- ❑ Counselor
- ❑ Working professional/military
- ❑ Other (please specify) ____________________

5. Rate your overall satisfaction with this book.

Extremely Satisfied	Satisfied	Not Satisfied
❑	❑	❑

6. Rate each of the following aspects of this book on a scale of 4 to 1 (4 = Excellent and 1 = Poor).

	4	3	2	1
Comprehensiveness of the information	❑	❑	❑	❑
Accuracy of the information	❑	❑	❑	❑
Usability	❑	❑	❑	❑
Cover design	❑	❑	❑	❑
Book layout	❑	❑	❑	❑
Special features *(e.g., CD, flashcards, charts, etc.)*	❑	❑	❑	❑
Value for the money	❑	❑	❑	❑

7. This book was recommended by:

❑ Guidance counselor
❑ Parent/guardian
❑ Family member/relative
❑ Friend
❑ Teacher
❑ Not recommended by anyone—I found the book on my own
❑ Other (please specify) ______________________

8. Would you recommend this book to others?

Yes	Not Sure	No
❑	❑	❑

9. Please provide any additional comments.

____

____

____

____

____

Remember, you can tear out this page and mail it to us at:

Publishing Department
Peterson's, a Nelnet company
2000 Lenox Drive
Lawrenceville, NJ 08648

or you can complete the survey online at **www.petersons.com/booksurvey.**

Your feedback is important to us at Peterson's, and we thank you for your time!

If you would like us to keep in touch with you about new products and services, please include your e-mail address here: ______________________